AF509978

Potassium Use in Agriculture

Potassium Use in Agriculture

Julia Lindsey

Callisto Reference
www.callistoreference.com

Callisto Reference,
118-35 Queens Blvd., Suite 400,
Forest Hills, NY 11375, USA

Visit us on the World Wide Web at:
www.callistoreference.com

© Callisto Reference, 2023

This book contains information obtained from authentic and highly regarded sources. All chapters are published with permission under the Creative Commons Attribution Share Alike License or equivalent. A wide variety of references are listed. Permissions and sources are indicated; for detailed attributions, please refer to the permissions page. Reasonable efforts have been made to publish reliable data and information, but the authors, editors and publisher cannot assume any responsibility for the validity of all materials or the consequences of their use.

ISBN: 978-1-64116-750-5 (Hardback)

Trademark Notice: Registered trademark of products or corporate names are used only for explanation and identification without intent to infringe.

Cataloging-in-Publication Data

Potassium use in agriculture / Julia Lindsey.
p. cm.
Includes bibliographical references and index.
ISBN 978-1-64116-750-5
1. Potassium in agriculture. 2. Agriculture. 3. Fertilizers. 4. Soil fertility. I. Lindsey, Julia.
S587.5.P6 P68 2023
631.83--dc23

Table of Contents

Preface

I am honored to present to you this unique book which encompasses the most up-to-date data in the field. I was extremely pleased to get this opportunity of editing the work of experts from across the globe. I have also written papers in this field and researched the various aspects revolving around the progress of the discipline. I have tried to unify my knowledge along with that of stalwarts from every corner of the world, to produce a text which not only benefits the readers but also facilitates the growth of the field.

Potassium is an essential macronutrient required for the growth and metabolism in plants. Its deficiency causes poor root development, slow growth, low disease resistance, delayed maturity, insufficient seed production, smaller seed production and lower yields. The potassium concentration in soil is little as a major part of potassium exists in insoluble form. It has been further depleted due to the rapid development of agriculture along with an extensive use of fertilizers. In order to mitigate the effects of potassium deficiency, several strains of bacteria and fungi have been identified for their potential of high potassium solubilization. Pisolithus, Cenococcum, Piloderma, Bacillus, Paenibacillus, Acidithiobacillus, Pseudomonas, Burkholderia, Aspergillus and Clostridium are some of the species that release large amounts of potassium from different minerals. These species can potentially enhance the productivity of many crops. This book includes some of the vital pieces of work being conducted across the world on various topics related to the role of potassium in agriculture. Agronomists, educators and students in this field will be assisted by this book.

Finally, I would like to thank all the contributing authors for their valuable time and contributions. This book would not have been possible without their efforts. I would also like to thank my friends and family for their constant support.

Julia Lindsey

1

Potassium (K) Cycle: A Recommendation Framework

Sylvie M. Brouder, Jeffrey J. Volenec, and T. Scott Murrell

Abstract Nutrient recommendation frameworks are underpinned by scientific understanding of how nutrients cycle within timespans relevant to management decision-making. A trusted potassium (K) recommendation is comprehensive enough in its components to represent important differences in biophysical and socioeconomic contexts but simple and transparent enough for logical, practical use. Here we examine a novel six soil-pool representation of the K cycle and explore the extent to which existing recommendation frameworks represent key plant, soil, input, and loss pools and the flux processes among these pools. Past limitations identified include inconsistent use of terminology, misperceptions of the universal importance and broad application of a single soil testing diagnostic, and insufficient correlation/calibration research to robustly characterize the probability and magnitude of crop response to fertilizer additions across agroecozones. Important opportunities to advance K fertility science range from developing a better understanding of the mode of action of diagnostics through use in multivariate field trials to the use of mechanistic models and systematic reviews to rigorously synthesize disparate field studies and identify knowledge gaps and/or novel targets for diagnostic development. Finally, advancing evidence-based K management requires better use of legacy and newly collected data and harnessing emerging data science tools and e-infrastructure to expand global collaborations and accelerate innovation.

S. M. Brouder (✉) · J. J. Volenec
Department of Agronomy, Purdue University, West Lafayette, IN, USA
e-mail: sbrouder@purdue.edu; jvolenec@purdue.edu

T. S. Murrell
Department of Agronomy, Purdue University, West Lafayette, IN, USA

African Plant Nutrition Institute and Mohammed VI Polytechnic University, Ben Guerir, Morocco
e-mail: s.murrell@apni.net

1.1 Overview of the Potassium Cycle

Nutrient recommendation frameworks are underpinned by scientific understanding of how nutrients cycle within timespans relevant to management decision-making. The cyclic nature of K transfers and transformations in crop production can be shown by a diagram depicting pools of K in the soil-plant system and the fluxes of K between those pools within a given volume of soil for a specified period of time (Fig. 1.1). Time scales typically reference a crop within a season or a sequence of crops within a relatively short period of time (2–4 years) for which a single or small suite of interrelated management decisions will be made. The horizontal spatial extent may range from an individual plant to an entire farm enterprise but traditionally has emphasized the "field scale," reflecting a farmer's predetermined management unit. The vertical spatial boundaries typically range from the top of the crop canopy down into the soil to the depth of crop rooting. Therefore, the spatial and temporal extents of interest include all system components that are intrinsic to the soil and the site as well as those that are influenced by management and crop development. Together, these components directly influence crop productivity.

Pools in the K cycle (Fig. 1.1) are categorized as inputs (pool 1), outputs (pools 2–5), plant pools within the cycle boundaries (pools 6–7), and those within the soil

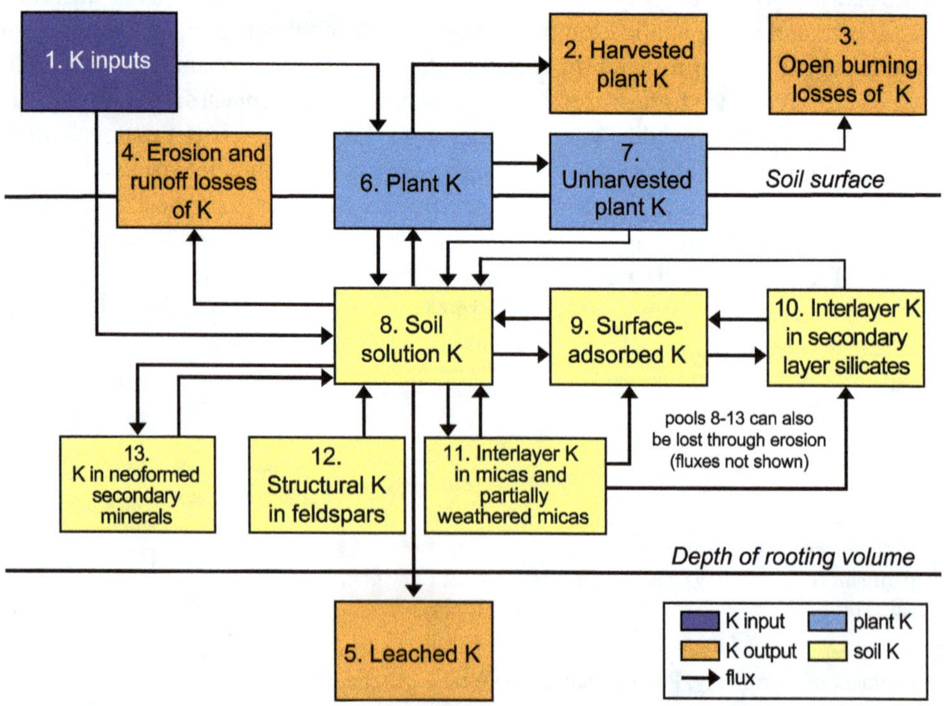

Fig. 1.1 The K cycle. Pools are denoted by rectangles and are quantities of K in one or more types of locations. Fluxes are denoted by arrows and are movements of K from one pool to another. This cycle depicts six pools of soil K (referenced herein as the six soil-pool model)

(pools 8–13). These pools are examined in depth throughout this book; their abbreviations relevant to this chapter are provided in Table 1.1. The pools are as follows:

Pool 1. K inputs is *the total quantity of K originating outside a given volume of soil that moves into that volume.* Inputs include organic and inorganic fertilizer additions (K_{Fert}, Table 1.1); K in crop residues brought onto the field from other areas; K in precipitation; K in irrigation water; K transported to the soil volume via runoff and erosion; K brought in as seeds, cuttings, or transplants; and atmospheric deposition (Chap. 2). This pool is the sum of all of these inputs. Inputs may occur directly to the soil or, as is the case with foliar fertilizer applications, directly to the plant.

Pool 2. Harvested plant K is *the quantity of K in plant material removed from a given area* (K_{Harv}). Such losses occur as the K in the desired plant products is removed from the field—products such as grains, forages, fruits, vegetables, nuts, ornamentals, and fibers (Chap. 3).

Pool 3. Open burning losses of K are *the total quantity of K lost from the unenclosed combustion of materials.* When crop residues left in the field (pool 7) are burned, K in soot and ash is lost if they move offsite (Chap. 3).

Pool 4. Erosion and runoff losses of K group three losses of K: surface runoff, subsurface runoff, and erosion (Chap. 3). **Surface runoff K** is *K^+ in water moving laterally over the soil surface in the direction of the slope.* **Subsurface runoff K** is *K^+ in water that infiltrates the soil surface to shallow depths and then moves laterally in the direction of the slope.* **Erosion loss of K** (K_{Erode}) is *K lost from the lateral or upward movement of soil particles out of a given volume of soil.* Although not depicted by arrows in the diagram, erosion and runoff losses include both soil solution K (pool 8) and K in the soil solids (pools 9–13).

Pool 5. Leached K (in soil) is *the quantity of K displaced below the rooting depth by water percolating down the soil profile* (K_{Leach}) (Chap. 3). Potassium dissolved in the soil solution (K_{Soln}) is subject to leaching, shown by the single flux arrow connecting pool 8 to pool 5. Although not depicted in the diagram, K can also be lost with clay colloids translocating to subsurface horizons.

Pool 6. Plant K is *the total quantity of K accumulated in the plant* (K_{Plant} or $K_{TotPlant}$ as discussed below). Total accumulation considers both aboveground organs, such as stems, leaves, flowers, and fruit, and belowground organs, such as roots, rhizomes, and corms. Nearly all of plant K is taken up by roots from the soil solution, shown by the arrow from pool 8 to pool 6 (Chap. 4). **Influx** is *the movement of K from outside to within a tissue*, in this case from the soil solution into the roots (Barber 1995). Small quantities of K may also move out of plant roots back into the soil solution. **Efflux** is *the movement of K from within to outside a tissue* (Barber 1995; Cakmak and Horst 1991). Efflux is denoted by the arrow drawn from pool 6 to pool 8. Plants differ in their efficiencies of K uptake and utilization (Chap. 5). Potassium can also enter the plant via leaf penetration, shown by the arrow from pool 1 to pool 6 (Marschner 2012). Potassium deposited on leaf surfaces by foliar fertilization, throughfall (Eaton et al. 1973), and

Table 1.1 Definitions of abbreviations as used in the text. Comments are included to provide additional clarity and context

Term	Abbreviation	Definition
Plant K	K_{Plant}	The K content in all plant biomass, usually reported on a dry matter basis, that is newly taken up within the growth cycle/season (the new plant nutritional need). It is measured at the point of maximum accumulation, ideally representing optimal growth and realizing maximum economic yields but without luxury consumption. Note, in annuals, K_{Plant} is often only assessed in aboveground tissues only
Total plant K	$K_{TotPlant}$	The total quantity of K that a plant/crop requires within the growth cycle/season, measured at the point of maximum accumulation representing optimal growth and realizing maximum economic yields. For annuals, $K_{TotPlant} = K_{Plant}$
Total organ K	$K_{OrganTot}$	The total quantity required by a specific organ pool of a plant/crop (leaves, stem/stalk, ear/fruiting body, root system, etc.)
Internally translocated K	K_{Trans}	Portion of $K_{PlantTot}$ or $K_{OrganTot}$ that is met by internal K cycling and translocation within a crop/plant
Plant-available indigenous K	K_{Soil}	The quantity of native K supplied by the soil without new K additions that is newly taken up by a plant/crop within a growth cycle/season
Harvested K	K_{Harv}	The quantity of K in plant material removed from a given area by crop harvest
Unharvested K	K_{UnHarv}	The amount of K returned to the soil in unharvested residues after the growing season/cycle
K concentration in harvested tissue(s)	$K_{AveConc}$	Average value of the K concentration in all harvested tissues that can be used to estimate K_{Harv} from yield mass
Solution K	K_{Soln}	The quantity of K in the soil solution, i.e., the soil pool from which plant/crop roots take up K
Solution soil test K	STK_{Soln}	A quick assay of K_{soln} that may be used in conjunction with quick assays of other soil pools to measure flux
Surface-adsorbed K	K_{Surf}	The quantity of K associated with negatively charged sites on soil organic matter, planar surfaces of phyllosilicate minerals, and surfaces of iron and aluminum oxides
Exchangeable K	K_{Exch}	The K extracted from a sample of soil via cation exchange using a solution of a specified composition (e.g., NH_4^+ or Na^+ salts) under a specific set of conditions. It is typically assumed that release occurs primarily from K_{Surf}, although release can also come from K in readily accessible interlayer positions of soil minerals. To provide an agronomic interpretation, K_{Exch} must be correlated to crop uptake and/or yield (see STK_{Exch})
Useable fraction of exchangeable K	E_{Exch}	Efficiency or fraction of K_{Exch} that is taken up by the root system of the crop/plant during the growing season/cycle
Exchangeable soil test K	STK_{Exch}	The total quantity of K in a soil sample that is released by a routine exchanging test protocol that has been

(continued)

Table 1.1 (continued)

Term	Abbreviation	Definition
		correlated with crop K uptake and/or yield (e.g., $STK_{Exch} = K_{Exch} \times E_{Exch}$), typically also inclusive of K_{Soln}
Quantity of K in solid-phase pool *j*	K_j	The total quantity of K in pool "*j*," where *j* corresponds to K in secondary layer silicates, micas and partially weathered micas, structural K in feldspars, or K in neoformed secondary minerals (pools 10–13, Fig. 1.1)
Useable K fraction in pool *j*	E_j	The efficiency or fraction of pool K_j that is taken up by the plant root
Plant-available nonexchangeable K	$K_{NonExch}$	Quantity of K in secondary layer silicates, micas, feldspars, and neoformed secondary minerals (K_j) accessed by plant/crop roots during a growth season or cycle (e.g., $K_{NonExch} = \sum_{j=\text{pool } 10}^{j=\text{pool } 13} K_j \times E_j$, Fig. 1.1)
Nonexchangeable soil test K	$STK_{NonExch}$	A chemical assay of potential seasonal contributions to K_{Soln} from soil pools 9–13 in Fig. 1.1. (Note: In any routine laboratory test, the protocol will also assess K_{Exch} contributions)
Erosion K losses	K_{Erode}	The quantity of K lost from the movement of soil particles out of a given volume of soil
K lost to leaching	K_{Leach}	The quantity of K displaced below the rooting depth by water percolating within the soil profile
Fixed K	K_{Fixed}	The quantity of K that has moved to interlayer positions in phyllosilicate minerals and has subsequently become unavailable to plants during the growing season/cycle
Precipitated K	K_{Precip}	Soil K that becomes unavailable to the crop/plant during the growing season/cycle following precipitation as neoformed secondary minerals
Lost K	K_{Lost}	The sum of K losses via all loss pathways (K_{Erode}, K_{Leach}, K_{Fixed}, and K_{Precip}) of soil K or K added either as inorganic/organic fertilizer or as unharvested residues (K_{UnHarv})
Fertilizer K	K_{Fert} or K_{FertX}	The input requirement for K fertilizer to meet the portion of plant K demand (K_{Plant}) not supplied by the soil while accounting for expected loss processes (K_{Fert}) or a specific input of fertilizer at rate "X" (K_{FertX})
Efficiency of fertilizer K uptake	E_{Fert} or E_{KAdded}	The efficiency of any single fertilizer application (E_{Fert}) or net K addition, including the sum of K_{Fert} and K_{UnHarv} (E_{KAdded}), that reflects the non-indigenous (soil) portion of system K that is taken up by the plant within the coming growth season/cycle. The efficiency factor adjusts the new K requirement upward from the net plant nutrition need ($K_{Plant} - K_{soil}$) and accounts for losses in plant-available K (K_{Lost}) following K addition
K uptake with rate "X"	$K_{Soil + X}$	Total plant K uptake at a specific fertilizer rate "X"
K recovery efficiency	RE_K	The apparent recovery efficiency for a specific fertilizer rate "X" (e.g., $RE_K = (K_{Soil + X} - K_{Soil})/K_{FertX}$), which is an approximation of E_{KAdded}

atmospheric deposition (Chap. 2) contribute secondarily to plant K via this pathway.

Pool 7. Unharvested plant K (Chap. 6) is *the quantity of plant K returned to the soil volume* (K_{UnHarv}). This return is normally associated with K leached from dead plant material such as pruned branches and leaves in orchards, chaff from machine harvest of grain crops, terminated cover crops, and plant residues left from previous crops. **Leached K** is *the quantity of K removed from plant tissues by water including rain and dew.* Leaching of K can be more rapid from senescing tissues where membranes and cell walls are damaged prior to moisture exposure (Burke et al. 2017). Although also considered leaching, guttation is another process that may lead to loss of K. **Guttation** is *an exudation of xylem sap from leaves due to root pressure* (Taiz et al. 2018). Guttation loss occurs from living tissue.

Pool 8. Soil solution K (Chap. 7) is *the quantity of K dissolved in the aqueous liquid phase of the soil* (K_{Soln}) (Soil Science Glossary Terms Committee 2008). It is present as the cation K^+. Plants take up K^+ only from this pool, denoted by the flux arrow from pool 8 to pool 6. Many soil pools (pools 9–13) contribute K to K_{Soln}.

Pool 9. Surface-adsorbed K is *the quantity of K associated with negatively charged sites on soil organic matter, planar surfaces of phyllosilicate minerals, and surfaces of iron and aluminum oxides* (K_{Surf}) (Chap. 7). Surface-adsorbed K enters the soil solution the most readily and is therefore considered the most plant-available of the soil K pools. As shown by the bidirectional fluxes between pools 8 and 9, K_{Soln} can become K_{Surf} and vice versa.

Pool 10. Interlayer K in secondary layer silicates is *the quantity of K bound between layers of phyllosilicate minerals that are weathering products of primary minerals* (Chap. 7). Secondary layer silicates are formed primarily by transformations of micas and feldspars. The strength of the bonds that K^+ forms in interlayers varies by mineral; therefore, not all interlayer K has the same degree of plant availability. Potassium may move from interlayers to surface sites (arrow from pool 10 to pool 9) or directly to K_{Soln} (arrow from pool 10 to pool 8). Additionally, K_{Surf} may move to the interlayers (arrow from pool 9 to pool 10).

Pool 11. Interlayer K in micas and partially weathered micas (Chap. 7) is *the quantity of K bound between layers of primary mica minerals that are in various stages of chemical and physical breakdown.* Important K-containing micas are biotite and muscovite. Micas do not release plant-available K until chemical or physical forces act upon them. As phyllosilicate micas weather, the edges of their sheets open as hydrated cations replace the dehydrated K^+ originally in the structures. Potassium from these edges may go into the K_{Soln} (arrow from pool 11 to pool 8) or to K_{Surf} (arrow from pool 11 to pool 9). The loss of K near the edges of mica crystals is concomitant with the loss of internal negative charge in the crystal and leads to the formation of secondary minerals. Interlayer K in micas can become interlayer K in secondary layer silicates (arrow from pool 11 to pool 10).

Pool 12. Structural K in feldspars (Chap. 7) is *the quantity of K in structures of tectosilicate minerals, mainly feldspars, and feldspathoids*. The K in these minerals is not bound as strongly as the other elements in the structures. At exposed surfaces, dissolution of the structures can allow other cations in the soil solution to exchange for K^+, moving K^+ into the solution (arrow from pool 12 to pool 8).

Pool 13. Neoformed K minerals are *newly formed minerals created from the reaction of soil solution K with other soil solution ions* (Chap. 7). An example is taranakite, a mineral formed by the reaction of K with phosphorus fertilizer compounds under acidic, saturated solution conditions (Lindsay et al. 1962). Potassium in neoformed secondary minerals can be both a source of K to K_{Soln} as plants deplete K_{Soln} (arrow from pool 13 to pool 8) and a sink for K as newly added K is precipitated out of K_{Soln} (arrow from pool 8 to pool 13).

1.2 Philosophy of a Potassium Recommendation

Cash et al. (2003) stated that to be effective, scientific information has to be credible, salient, and legitimate. Applied to a K recommendation, it is credible when it is scientifically adequate and based on sufficient evidence. It is salient when it addresses the needs of the decision-makers. It is legitimate when it is unbiased and respects stakeholders' values and beliefs. Both scientists and practitioners alike want a recommendation to be "accurate," in that it provides realistic estimates of costs and benefits, with associated levels of confidence, for a given K management option and a given set of conditions that are specific to the user. From the scientific perspective, accuracy is achieved when (1) the individual components that make up the recommendation (i.e., modification for crop, soil type, agroecozone, etc.) are consistent with relevant scientific theory and (2) research has been conducted under a sufficient number of representative conditions and environments that the statistical precision and accuracy of the recommendation can be explicitly given for its inference space. For practitioners, accuracy requires that the recommendation be credible in that it makes sense out of what is observed and that the components themselves can be observed, explained, and understood. Additionally, practitioners desire a recommendation that is customizable to individual contexts (management, environment, whole-farm profitability, etc.) and is not only focused on the cost and benefits associated with a single crop's response to a single nutrient. Finally, practitioners expect recommendations to be reasonably successful in predicting crop production outcomes.

Philosophically, this suggests a three-legged stool model for building a recommendation, where simultaneous consideration is given to (1) the crop-soil K cycle, (2) the ancillary or secondary biophysical factors that can influence the crop-soil K cycle but are often the subject of separate recommendations, and (3) socioeconomic factors that encompass farmer short- and long-term objectives, goals, and preferences (Fig. 1.2). Thus, there is more to a recommendation than understanding the crop- and soil-specific attributes of the K cycle, and all three legs must be subjected

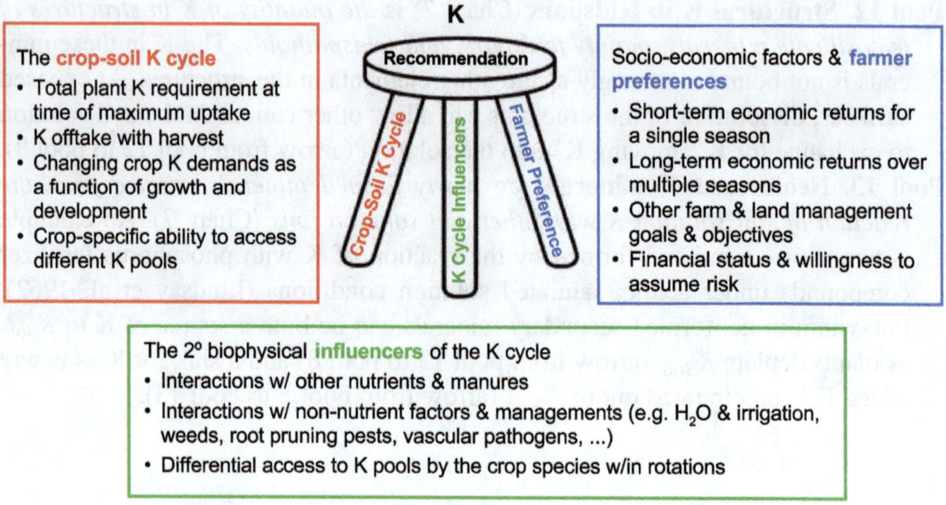

Fig 1.2 A "three-legged stool" conceptualization of the essential considerations of a credible, salient, and legitimate K recommendation

to rigorous analysis. The remainder of this chapter will focus primarily on the crop-soil K cycle and its biophysical regulators. These are the legs of the stool that have been the subject of most agronomic research conducted to develop K recommendations.

1.3 Challenges with Common Potassium Recommendation Terminology

In most soil fertility references and management guides (e.g., Havlin et al. 2014), soil K has been represented as residing in four distinct pools (Fig 1.3a). The K_{Soln} and exchangeable K (K_{Exch}, Table 1.1) pools have long been considered the major in-season source of nutrients to plants and crops and the major foci of research to develop soil testing protocols. The remaining two soil pools in traditional K cycle diagrams were the structural K in primary minerals and the interlayer K in secondary clay minerals. However, as discussed in the remainder of this chapter and in Chap. 7, this traditional four soil-pool model and the accompanying terminology have created confusion in understanding the plant-soil K cycle and its use as a foundation to recommendations. The four soil-pool model uses terminology that confounds the mechanisms of extraction protocols with the actual source pools (e.g., K_{Exch} for the quantity of K extracted via cation exchange versus nonexchangeable K (K_{NonExch}, Table 1.1 and pool B, Fig. 1.3a) for any additional K that may be assessed by using other extractants). Additionally, it lumps together micas and feldspars, does not consider neoformed minerals, and suggests that primary minerals are not important

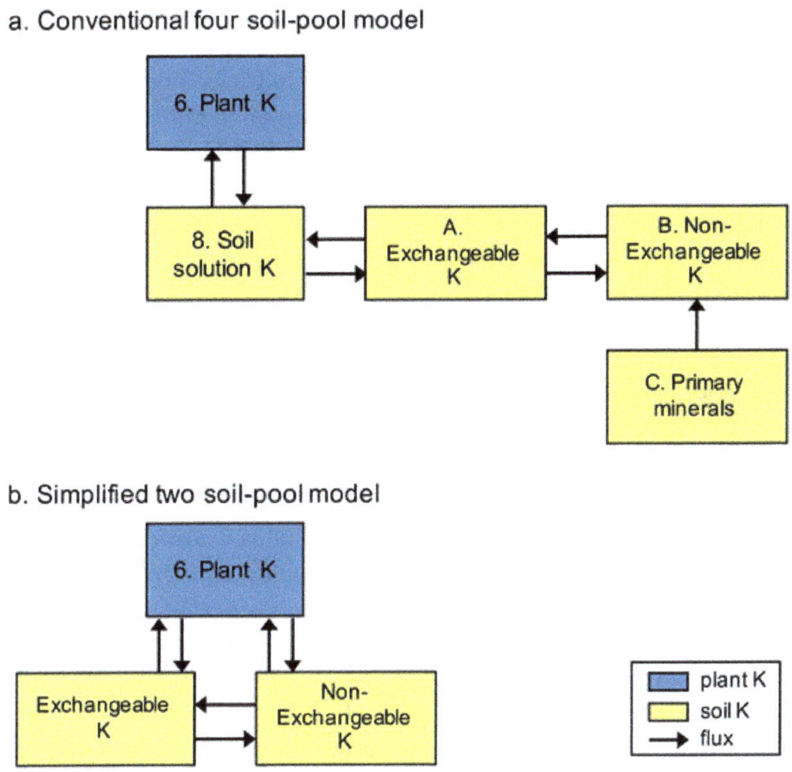

Fig. 1.3 Two simplified K cycles depicting only the relationships among soil K and plant K pools. Other pools have been omitted for simplicity. Pools are denoted by rectangles and are quantities of K in one or more types of locations. Fluxes are denoted by arrows and are movements of K from one pool to another. Pools 6 and 8 retain the numbering from Fig. 1.1. The upper cycle (**a**) is the conventional four soil-pool model. The lower two soil-pool cycle (**b**) is a simplified model that conceptualizes plant-available K as coming from pools measured by a soil test that extracts K with cation exchange (exchangeable and soil solution K) and not measured by such a soil test (nonexchangeable K)

contributors to plant nutrition within typical management timelines (Fig. 1.1, pools 11, 12, and 13).

The consensus among contemporary K researchers is that inconsistency in terminology creates challenges in communicating among scientists and in achieving a recommendation that is understandable and credible to the practitioner. The six soil-pool model (Fig. 1.1) is intended to be more explicit in clarifying potential contributors to K_{Soln} during a growing season and potential sinks for K added as fertilizers, manures, and returned residues and to result in more defensible recommendations. Although more complex, it is expected to alleviate the array of confusions generated by the four soil-pool model and thereby facilitate a more informed understanding of soil assays by mode of action and use of the K cycle to improve research that supports K recommendations.

Terms associated with the four soil-pool model known to be differentially used and prone to creating confusion in trying to communicate the science underpinning a recommendation, along with their clarifying definitions used in this book, include:

- The use of "exchangeable" or "replaceable" to characterize K that is removed with a cation exchange assay that is often assumed to represent just surface-adsorbed K (Fig. 1.1, pool 9). To clarify this term, Chap. 7 defines **exchangeable K** as *the K extracted from a sample of soil via cation exchange using a solution of a specified composition under a specific set of conditions*. As Chap. 8 explains, exchangeable K, when measured by ammonium cations in the soil test extractant, is surface-adsorbed K as well as K in readily accessible interlayer positions of soil minerals. Other exchanging cations (e.g., Na^+, Ba^{2+}, or Ca^{2+}) or extraction conditions may not extract the same amount of K from a soil sample as ammonium does.
- Various uses of the term "fix" (i.e., fixed K, K fixation, a K-fixing soil) to characterize movement of K into interlayer sites of clay minerals, rendering the ion less accessible to the soil solution and therefore less plant-available (Fig. 1.1, movement of K from pools 8 or 9 into pools 10 or 11). The term has also been used to characterize a soil factor expected to reduce the efficiency of a fertilizer K application. Chapter 7 defines **potassium fixation** as *hydrated K^+ ions moving to interlayer positions in phyllosilicate minerals and then dehydrating as the mineral layers contract. In this position, the K^+ is unavailable to plants.*
- "Nonexchangeable" and "exchangeable" terms are sometimes used to classify the relative ease with which K^+ on the cation exchange complex of a soil can be replaced by other cations (ammonium (NH_4^+), magnesium (Mg^{2+}), and calcium (Ca^{2+})) in the soil solution. But the terms are often used synonymously with either fixation terms or analytical protocols such as nitric acid-extractable K versus ammonium acetate K, respectively (McLean and Watson 1985). Chapter 7 defines nonexchangeable K as soil K that is not measured by soil tests that rely on exchange or displacement of K by another cation. In this chapter we use the abbreviation K_{NonExch} to represent the plant-available portion of nonexchangeable K that is accessed by plant/crop roots within seasons or over a few years and might, thus, be a target of a measurement to support a K recommendation.
- Various terms characterizing "available" K (i.e., bioavailable, plant-available, etc.). These terms have been used to describe both the concentration and quantity of K extracted by a soil test protocol (that has been found to be correlated with plant uptake) and the proportion of a crop's K requirement that can be seasonally accessed from the indigenous soil K supply by crop roots. In this and the following chapters, plant-available and bioavailable K are used interchangeably.
- Other terms characterizing outcomes or states of processes such as K "holding capacity" and aspects of efficiency (i.e., uptake, recovery, physiological, agronomic, etc.). These terms are generally assumed to be quantitative, but the mathematical representation can vary in important ways among the users of the term(s). Chapter 3 defines **K holding capacity** as *the maximum quantity of K that can be retained by a given volume of soil*. Chapters 5 and 11 define several commonly used metrics of K efficiency.

The examples above point to efforts by other authors in this book to clarify terms.

1.4 Considerations for Recommendations Derived from the Mass Balance Approach to the Potassium Cycle

In principle, a fertilizer recommendation derived from a plant-soil nutrient cycle such as shown in Fig. 1.1 could focus on all or a subset of the pools and fluxes identified, with pool and/or flux choice reflecting the desired degree of fidelity of the recommendation in space and time. In practice, many nutrient recommendation frameworks use a mass balance approach that considers the predominant pools of nutrient supply and demand and represents complex flux processes as fractions of pools that can reasonably be considered as interacting within the context of a crop season or short sequence of crops for which a management intervention is planned. Thus, the dimension of real time is effectively removed as a direct variable. In the literature, this approach has been most explicitly described for N and directly applied to N management (Stanford 1973; Morris et al. 2018), but the approach is generic and can be applied to any nutrient. In comparing the basic information required for optimizing both yields and fertilizer recovery as identified for N by Stanford (1973), commonalities for K include:

1. For plants or crops that attain the yield expected for a given environment, the internal requirement for K newly taken up from the soil (plant K or K_{Plant}: Fig. 1.1, pool 6) assessed at the point of maximum accumulation and inclusive of all plant tissues including roots (Fig. 1.4).
2. The amount of K that a plant or crop can obtain from the plant-available indigenous soil K supply (K_{Soil}).
3. The understanding that the quantity of fertilizer applied (K_{Fert}) must be higher than the difference between K_{Plant} and K_{Soil} to reflect the reality that recovery efficiency of fertilizer will most likely be reduced from 100% by a variety of practical management considerations and common soil and other environmental conditions.

For K, the basic Stanford equation is

$$K_{\text{Fert}} = (K_{\text{Plant}} - K_{\text{Soil}})/E_{\text{Fert}} \tag{1.1}$$

Or by rearranging

$$K_{\text{Fert}} \times E_{\text{Fert}} = K_{\text{Plant}} - K_{\text{Soil}} \tag{1.2}$$

where E_{Fert} is the efficiency with which the K fertilizer is maintained as available for uptake by the plant or crop (i.e., if 75% is taken up within the growing season, then $E_{\text{Fert}} = 0.75$).

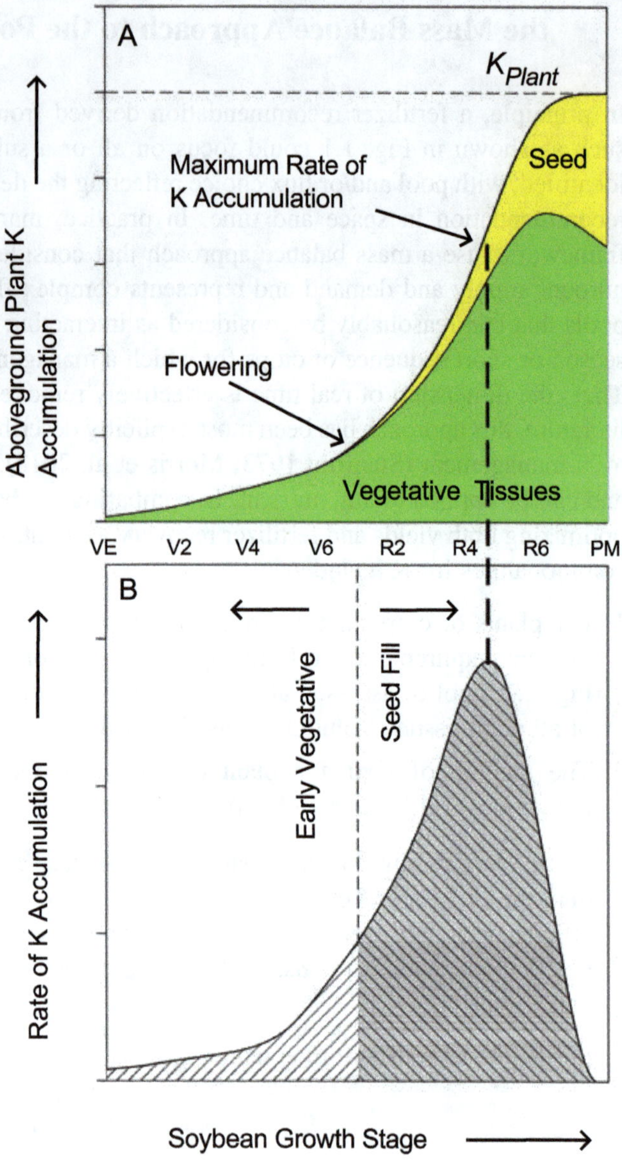

Fig. 1.4 (**a**) Total quantity of aboveground K accumulation and (**b**) the corresponding rate of accumulation by soybean (*Glycine max* (L.) Merr.) shown as a function of crop growth stage from vegetation emergence (VE) through physiological maturity (PM). Generalized curves are derived from Fernández et al. (2009)

1.4.1 Exploring and Characterizing $\mathbf{K}_{\mathit{Plant}}$*: Understood and Easily Assessed?*

Potassium in plants performs an important array of functions, ranging from enzyme activation to its outstanding role in plant-water relations that drives everything from cell extension to the functioning of stomates and control of leaf gas exchange (Marschner 2012). Indeed, it is the wide diversity of functions as well as the high quantities of K required by plants along with an apparent lack of toxicity effect at high tissue K concentrations that make plant K requirement particularly difficult to determine. When grown in soil with abundant K supply, luxury consumption of K in vegetative tissues and fleshy fruits can occur. Thus, identifying robust, species-specific values for K_{Plant} requires rate studies conducted in representative environments where results clearly delineate K_{Plant} at K sufficiency versus K excess conditions.

At sufficient but not excessive soil K supply, plant uptake in annual crops is roughly sigmoidal, with peak accumulation in aboveground tissue occurring at or before physiological maturity (Fig. 1.4a) and often well before actual harvest when some senescence of vegetative tissues has typically already occurred. For example, both Fernández et al. (2009) and Gaspar et al. (2017) document the relatively low level of K accumulation occurring during vegetative growth of soybean (*Glycine max* (L.) Merr.), followed by large increases in aboveground K during reproductive growth. In determinate annuals, this pattern may be shifted such that much of the uptake occurs prior to reproduction (e.g., maize; Ciampitti et al. 2013; Wu et al. 2014). The Fernandez et al. work (2008, 2009) also documents both the influence of soil moisture on temporal patterns of K accumulation and the importance of K rate studies to accurately determine K_{Plant} (Fig. 1.5a). In this example, rainfed soybean experienced a pause in K accumulation during seed development that corresponded with a 10-day period with <5 mm rainfall. During this time, gravimetric soil moisture content fell to <0.15 g g^{-1} in the surface (0–20 cm) soil, well below field capacity (25 g g^{-1}). Further, soybean grown on high-testing soils, with 290 mg K kg^{-1} in the top 10 cm of the profile, accumulated >50% more K_{Plant} by R6 than did soybean grown on medium-testing soils (135 mg K kg^{-1}), although yields and K removal in seed were the same on the high- and medium-testing soils.

In annuals, K_{Plant}, the plant's K requirement for new uptake from the soil for optimal growth throughout the growing season, is essentially the same as the total quantity of K in plant tissue (roots plus shoots, $K_{TotPlant}$). It is important to note that, in practice, most of the characterizations of K_{Plant} for annual grain, seed, and forage crops reflect measurements made on aboveground tissues (e.g., Fernández et al. 2009; Gaspar et al. 2017) with an assumption that belowground K in fibrous and taprooted root systems is relatively minor in comparison to quantities required by aboveground tissues. Although reports of root K contents are sparse and values are prone to variation due to experimental artifacts, Barber (1995) suggested that, in general, concentrations of K in roots are similar to that in shoots. If true, shoot K content and biomass shoot-to-root (S:R) ratios can be used to estimate root K. Studies of dry matter partitioning are also sparse but report that K in annual root systems may range from much less than 10 to over 20% of K_{Plant}. Amos and Walters (2006) integrated results from 45 studies on maize root biomass and identified anthesis as the point of maximum root biomass (31 g plant^{-1}) and a mean S:R dry weight ratio of 6:3 at physiological maturity, but variation among studies was pronounced. These authors also reported that several field studies had S:

Fig. 1.5 Potassium accumulation in (**a**) soybean (*Glycine max* (L.) Merr.) and (**b**) Miscanthus × giganteus (*Miscanthus* × *giganteus* J. M. Greef, Deuter ex Hodk., Renvoize). Error bars indicate the standard deviation (**a**) and error (**b**) of the mean. All aboveground tissues of soybean were pooled from plots with above optimal (high), optimal (medium), and deficient (low) K fertility status (modified from Fernández et al. 2009). Miscanthus data were collected over two sequential growing seasons from K-sufficient plots. (modified from Burks 2013)

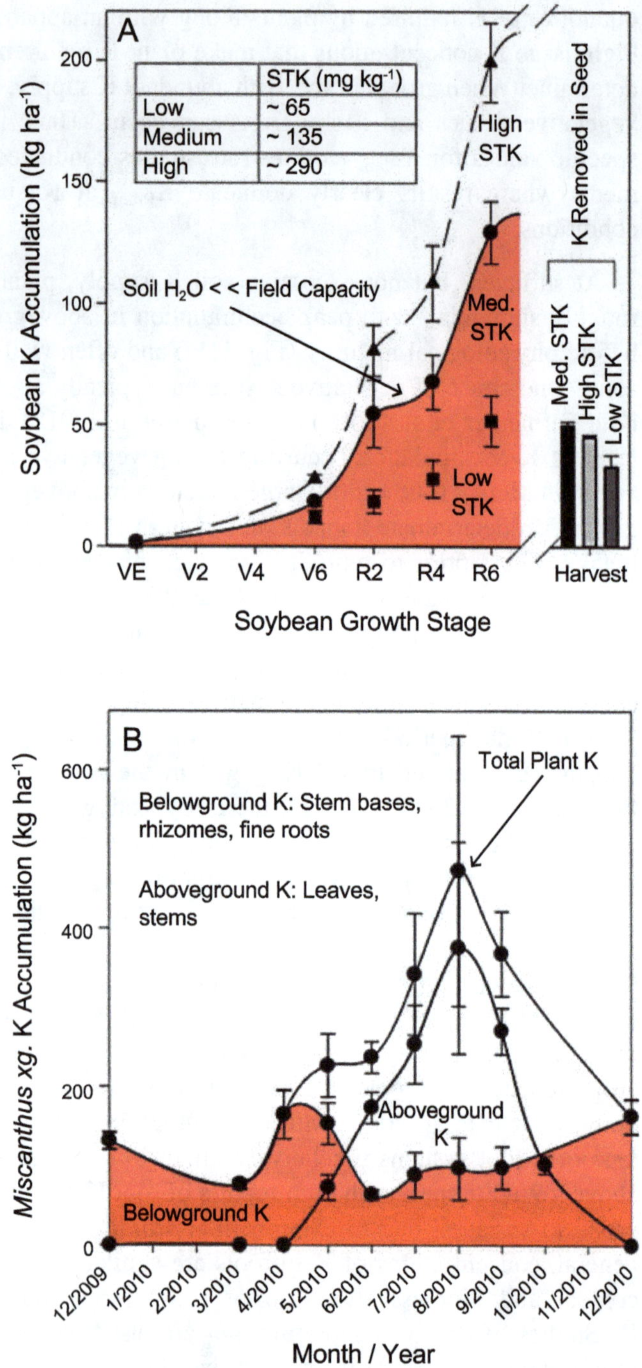

Rs of <5 at maturity. In contrast, a recent field study of maize hybrids from the 1950s to the present day reports S:Rs of approximately 10 and 25 at tasseling and maturity, respectively (Ning et al. 2014). Amos and Walters (2006) noted an array of factors from genetic and environmental to sampling method artifacts can influence S:

Rs. Studies on soybean and wheat S:Rs at or near maturity are also variable (e.g., soybean 5.3 (Mayaki et al. 1976) and 11 (Brown and Scott 1984); wheat 5+ (Hocking and Meyer 1991) and 13 (van Vuuren et al. 1997)).

For crops with large underground storage organs that are the harvestable yield, the storage organs are generally included in K_{Plant} determination although fine roots are not (e.g., potato (*Solanum tuberosum* L.), sugar beet (*Beta vulgaris* L.), and other vegetables (Greenwood et al. 1980)). For perennial crops and trees, the K that a plant must newly acquire from soil for a given growing season (K_{Plant}) is different from total plant K accumulation (K_{PlantTot}). In perennial crops, total plant K accumulation includes both new K uptake (K_{Plant}) and K that is internally recycled from storage organs within a production cycle (K_{Trans}). To find K_{Plant}, the internally recycled K must be subtracted from total K accumulation, according to Eq. (1.3):

$$K_{\mathrm{Plant}} = (K_{\mathrm{PlantTot}} - K_{\mathrm{Trans}}) \tag{1.3}$$

For *Miscanthus* × *giganteus* (*Miscanthus* × *giganteus* J. M. Greef, Deuter ex Hodk., Renvoize), a high-biomass grass that can be used for cellulosic bioenergy production, Burks (2013) found that root reserves (primarily in rhizomes) could account for 58% of the 175 kg K ha^{-1} required for the first 2 months of shoot growth (Fig 1.5b, April–May). However, despite continued large accumulations of K in leaves and stems, K root reserves were partially replenished during the remaining summer months suggesting new uptake of soil K from June to August contributed to both above- and belowground K status. Additional important points about K_{Plant} illustrated by this study include (1) the high degree of variation (large standard error values) in K requirement of vegetative tissues and (2) the challenges of using time series snapshots to fully characterize K_{Plant} from the perspective of organ- or tissue-level mass balances. In theory, a more nuanced expression of K_{Plant} could capture the sum of the specific demands of n different organs or tissues where

$$K_{\mathrm{Plant}} = \sum_{i=1}^{n} \left(K_{\mathrm{OrganTot}_i} - K_{\mathrm{Trans}_i}\right) \tag{1.4}$$

and K_{OrganTot_i} is the specific K requirement of every "*i*th" organ and K_{Trans_i} represents the portion of K received by the "*i*th" organ from any plant part that can temporarily store K. In practice, sequential mass balance measurements such as those made by Burks (2013) cannot easily distinguish on an organ basis whether accumulating K comes from new uptake or internal translocation. Thus, expanding Eq. (1.3) to Eq. (1.4) to represent internal K allocation at an organ level would likely do little to improve estimates of K_{Plant} when the purpose is to identify the general requirement of a plant or crop for new K uptake from soil for a specific season or cropping cycle.

Two additional plant K fractions are relevant to a mass balance approach to developing a fertilizer rate recommendation: the K removed with the harvested grain or organ (K_{Harv}) and the K in any plant residues that are returned to the soil (K_{UnHarv}, discussed below with K_{Soil}). In recommendation systems that seek to raise K_{Soil} to sufficiency (e.g., $K_{Soil} = K_{Plant}$) and maintain it at this level, K_{Harv} is used to directly estimate K_{Fert} for applications to soils at K sufficiency or above (e.g., Vitosh et al. 1995). Historically, it has often been assumed that the fraction of K_{Plant} that is in the harvestable yield is constant and therefore the yield mass can be used as a proxy measure for K_{Plant}. For grain crops, seed K contents are highly conserved, and, thus, grain yields once adjusted to a standardized moisture contents can be used as a robust proxy measure of K removal in grain crops, provided crops are K sufficient (e.g., Brouder and Volenec 2008). In contrast, when the economic yields are vegetative tissues or fleshy fruits, harvested dry weights provide much less precise estimates of K_{Harv} (e.g., *Miscanthus* (Fig. 1.4b), Burks 2013; switchgrass (*Panicum virgatum* L.), Woodson et al. 2013)). For annual grains, low and highly variable harvest indices (HI) for K (K content of grain divided by K_{Plant}) limit the use of simple back calculations of K_{Plant} where

$$K_{Plant} = f(\text{Grain Yield}) \tag{1.5}$$

For N, the presumption of this relationship coupled with additional, simplistic assumptions about proportionality between the optimum fertilizer rate and plant N requirement led the Stanford (1973) model to be dubbed a yield-based approach to fertilizer recommendations. However, recent, rigorous analyses of the Stanford model have highlighted the pitfalls of reducing a mass balance understanding of a nutrient cycle to a recommendation largely derived from estimation of attainable yields or yield goals. Morris et al. (2018) note that, while logical to farmers, yield goal recommendations for N are limited by uncertainties in predicting realistic yields, relationships between uptake and yield, soil N supply, and important interactions among genetics, management, and environment. For K, the early emergence of soil testing as a tool for K management shifted the focus of historic recommendations away from yield goals as a major driver of K_{Fert} calculation. However, as K recommendations are revisited, agronomists should be mindful that similar uncertainties can be expected to plague such an overly reductionist approach to deriving a recommendation for K_{Fert} from the K cycle (discussed further below in Sect. 1.4.4.1).

1.4.2 *Exploring and Characterizing* K_{Soil}*: Was Bray Right?*

In his original report on a sodium acetate-nitric acid procedure for K in soils, Bray (1932) commented that the amount of replaceable K (easily exchangeable surface-adsorbed K, K_{Surf} (pool 9, Fig. 1.1)) in soils was generally considered to be a source of K to the soil solution and therefore available for plant growth. Although he noted other factors potentially influencing soil K that is “given up to the soil solution” (e.g., base cation exchange capacity), Bray identified replaceable K as the most important. This assumption remains the foundation for most K recommendations that involve soil testing. Indeed, much of the subsequent research conducted in the intervening decades has focused on relating simple chemical tests characterizing the quantities of K released from solid phases by an exchanging or displacing cation (K_{Exch}; common exchanging cations are sodium (Na^+) and NH_4^+) to fertilizer requirements (Bray 1944; Chap. 8). This approach results conceptually in a two soil-pool model (Fig 1.3b) and the associated common misperception that

$$K_{Soil} \cong K_{Surf} \cong K_{Exch} \tag{1.6}$$

where K_{Exch} is converted from a concentration measured by the chemical test to a mass of nutrient per unit land area by multiplying it by the approximate mass of a furrow slice of soil. In the USA, historic tabular recommendations often used the furrow slice conversion (mass of soil to depth of 20 cm (8 in)) interchangeably with a measured concentration (e.g., Vitosh et al. 1995). However, in reality, plant roots may not access all K in the furrow slice, and they may also access K from deeper in the soil, reflecting root distribution patterns (Chap. 8). Traditionally, recommended assays for K_{Exch} have followed extensive field work to identify good correlations with crop response to K fertilizer. Normally, the recommended procedures do not separate out the quantity of soluble or solution-phase K (K_{Soln}; Fig. 1.1, pool 8) because the amount of K_{Soln} is generally minor when compared to the amounts of K_{Surf} (Doll and Lucas 1973; Knudsen et al. 1982). For example, an analysis of selected US soils found quantities of K_{Soln} varied between 1 and 30+ mg K kg^{-1} soil under well-moistened conditions, while corresponding K_{Exch} levels ranged from approximately 20 to 850+ mg K kg^{-1} soil (Brouder 2011). Regardless, K measured in a soil test of K_{Exch} (STK_{Exch}) is not the total K_{Exch} or an estimate of the quantitative sum of K_{Surf} and K_{Soln} but rather an index of the fraction or efficiency of this pool (E_{Exch}) that can be accessed by crop roots during a given growing season of crop cycle where

$$STK_{Exch} \cong K_{Exch} \times E_{Exch} \tag{1.7}$$

A large body of subsequent research generally supported Bray's assertion of K_{Exch} as the primary tool to assess K_{Soil} for many soils, but contradictory studies showing poor correlation between crop uptake and yield and various measures of K_{Exch} have also frequently occurred. For example, K studies of cotton grown on vermiculitic soils in CA, USA, found relatively poor relationships between yield and K_{Exch} (NH_4Cl-extractable K; Cassman et al. 1989). Related work demonstrated that K_{soil} could be better predicted from assays targeting the soil solution's relationship to the nonexchangeable soil K pools ($K_{NonExch}$; Cassman et al. 1990; Brouder and Cassman 1994). Similarly, in a recent examination of soil K supply in US Midwest maize production, Navarrete-Ganchozo (2014) found K_{Exch} to be an insensitive predictor of soil K balance. This long-term K rate study demonstrated that, for some soils, the regionally accepted protocol for assessing K_{Exch} (NH_4OAc-Ext.; Brown 1998) failed to find differences in surface (0–20 cm) soil K dynamics, although cumulative crop K balances ranged from more than −400 to 400 kg K ha^{-1}

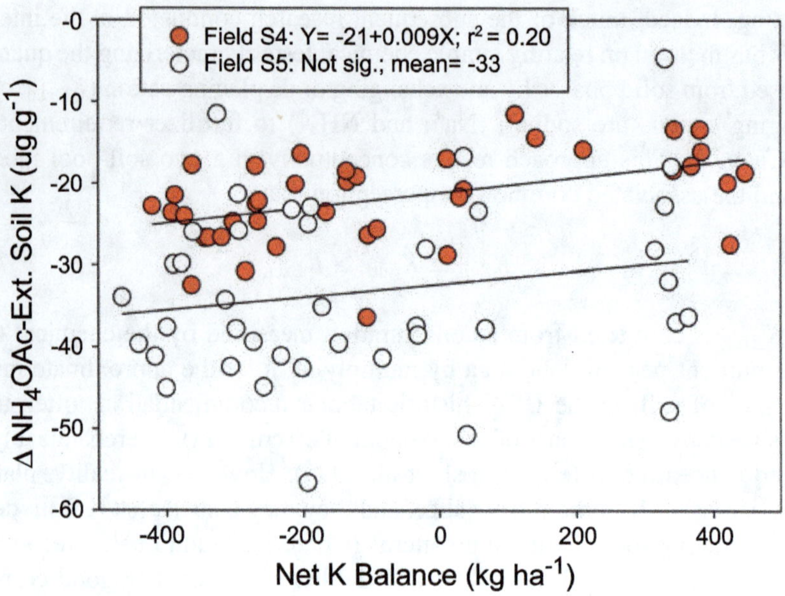

Fig. 1.6 Change in ammonium acetate extractable K (NH$_4$OAc-Ext. K) in surface soil (0–20 cm) between 1997 and 2010 plotted for two fields (S4 and S5) as a function of the net K balance (cumulative fertilizer K applied minus K removed in grain yields). Fertilizer K was added annually and biennially for cumulative applications ranging from 0 to 900 kg K ha^{-1} with all applications ending in 2002. (modified from Navarrete-Ganchozo 2014)

(Fig. 1.6). Such soils require expanding the characterization of K_{Soil} to be inclusive of any net contributions (flux from > flux into) from interlayer K, structural K, and/or K in neoformed secondary minerals (Fig. 1.1, pools 10–13).

The theoretical function for the sum of soil pools "*j*," where "*j*" represents pools 10 to 13 in Fig. 1.1, follows that for K_{Exch} (Eq. 1.7), where for each pool *j*, there is a quantity available (K_j) and an efficiency factor (E_j) to characterize the fraction of the use of pool "*j*" by the plant, crop or crop sequence in the growing season/cycle. The sum of the products of quantities and fractions for all for pools should represent the fraction of K_{NonExch} that is plant-available and accessed by plant roots:

$$K_{\text{NonExch}} = \sum_{j=\text{pool } 10}^{j=\text{pool } 13} K_j \times E_j \tag{1.8}$$

and

$$K_{\text{Soil}} = \text{STK}_{\text{Exch}} + K_{\text{NonExch}} \tag{1.9}$$

It should be noted that although K_{NonExch} has frequently been identified as an important contributor to K_{Soil}, the contributions of specific pools have been inferred from the soil test assays that have correlated well with plant or crop performance and from general knowledge of soil mineralogy. For example, chemical boiling in strong acid (1 M HNO_3) has been used to measure $K_{NonExch}$ (Knudsen et al. 1982; McLean and Watson 1985) and has frequently been inferred as releasing fixed K from interlayer locations, but it may also cause dissolution of K-bearing minerals (Barber and Matthews 1962; Martin and Sparks 1983). From a routine soil testing perspective, considering a framework that uses a two soil-pool model based on STK_{Exch} and an additional, routine or periodic test for potential contributions of $K_{NonExch}$ to K_{Soln} ($STK_{NonExch}$) may be feasible where

$$K_{Soil} \cong f(STK_{Exch}, STK_{NonExch}) \tag{1.10}$$

where $STK_{NonExch}$ is evaluated from a quick chemical assay of potential seasonal contributions of soil pools 9–13 (Fig. 1.1) to K_{Soln} and therefore seasonal crop uptake. Although selection of $STK_{NonExch}$ based on mode of action could render this approach tailorable to soils with differing mineralogies, further partitioning of $K_{NonExch}$ for attribution to a specific K pool may be of little practical value while adding significantly to analytical costs.

Alternatively, as proposed by Cassman et al. (1990), a two-pool model for K-supplying power could focus on the relationship between K_{Soln} concentrations and buffering power as assessed by a $STK_{NonExch}$ single-point assay or fixation isotherm such that

$$K_{Soil} = f(STK_{Soln}, STK_{NonExch}) \tag{1.11}$$

where STK_{Soln} could be assessed by an assay of K concentration in saturated paste or diluted soil solution extractions (Rhoades 1982) such as 0.01 M $CaCl_2$ as proposed by Cassman et al. (1990). In theory, this last approach could approximate soil buffer capacity or the "quantity-intensity" relationship—the ability of the solid phase to replenish solution concentrations depleted by root uptake. In practice, single point measurements of STK_{Soln} and $STK_{NonExch}$ provide only static insights into relative concentrations at the time of measurement and do not necessarily give insights into dynamic interactions among pools. At a minimum, the latter would require multiple assessments over time, and cost would scale accordingly. As Chap. 8 discusses, the mixed bed cation-anion exchange resin method provides a strong sink for K and can be used to estimate the rate of solution K replenishment in response to depletion by plant uptake. In most regions where soil testing has been developed for the purposes of making a K recommendation, STK_{Soln} and $STK_{NonExch}$ tests have been used to explain the failure of STK_{Exch} to predict yield and fertilizer sufficiency but have not been subject to the extensive field correlation-calibration efforts necessary to develop them as the foundation for a recommendation. Regardless, current inconsistencies in the ability of potential routine measures of K_{Soln} and $K_{NonExch}$ to improve recommendations with or without K_{Exch} as a covariate suggest that other covariate measures are needed for accurate prediction of K_{Soil} from soil testing information (e.g., Eqs. 1.10 and 1.11).

1.4.3 *Exploring and Characterizing K_{Fert} and E_{Fert}: Important but Generally Overlooked?*

In general, understanding E_{Fert} has been much less of a research concern for K as compared to N or P, not because K_{Fert} cannot be lost from or retained by soil, but because unrecovered K fertilizer has not been identified as an environmental pollutant. Most commonly available inorganic K fertilizers (soluble K salts of chloride, nitrate, or sulfate) dissolve rapidly in water, and insolubility, per se, is not considered an availability issue outside of multi-nutrient mixtures for fertigation. Thus, the K content of an inorganic fertilizer can be expected to appear in the soil solution (Fig. 1.1, pool 8) very rapidly after application, provided soil conditions are not too dry. Likewise, K introduced to soils in organic materials is also in the K^+ form and should appear rapidly in the soil solution as materials leach. The microbial mineralization-immobilization activity that confounds N recommendations is not a consideration. Thus, the major processes with potential to routinely reduce efficiency of fertilizer application are erosion and runoff (K_{Erode}, pool 4), leaching below the root zone (K_{Leach}, pool 5), and internal sink processes of fixation (K_{Fixed}) into interlayer sites (pools 10 and 11) or precipitation (K_{Precip}) as neoformed secondary minerals (pool 13). Thus:

$$E_{\text{Fert}} = \left(K_{\text{Fert}} - K_{\text{Leach}} - K_{\text{Erode}} - K_{\text{Fixed}} - K_{\text{Precip}}\right)/K_{\text{Fert}} \tag{1.12}$$

An early review of the literature suggested K_{Fert} leaching losses to be negligible on silt loam and heavier textured soils in US states of the Midwest and West (Munson and Nelson 1963). Even when such soils require subsurface agricultural tiles to improve drainage, losses of applied K appear low. Bolton et al. (1970) reported mean annual K concentrations of 0.95 and 1.23 mg L^{-1} in drainflows of unfertilized and fertilized plots, respectively, with corresponding load losses of 0.6 and 1.1 kg K ha^{-1} $year^{-1}$. More recent work found similar drainflow concentrations with no apparent impact of inorganic fertilizer rate (0 vs. 260 kg K ha^{-1}), source (manure vs. KCl), or cropping system (Fig. 1.7). However, applications to sandier soils are much more susceptible to leaching with the magnitude of loss expected to be proportionate to percolate volume with fluctuations reflecting timing, quantity and intensity of rainfall events, and quantities of K added (Bertsch and Thomas 1985; Chap. 3). Likewise, highly weathered tropical and subtropical soils are subject to higher leaching losses (Malavolta 1985), and quantities of K_{Fert} lost to leaching are likely widely variable. For such situations, soil fertility textbooks suggest that losses of 30% or more may occur without management to reduce loss, such as lower annual rates or split applications (e.g., Havlin et al. 2014), but peer-reviewed literature reports remain sparse.

Similarly, K_{Fert} lost to erosion and surface runoff can be impacted by the nature of precipitation events as well as by field slopes, fertilizer placement, and degree of incorporation. In 1985, Bertsch and Thomas (1985) characterized K losses to erosion of temperate soils as understudied and perhaps of greater magnitude than expected.

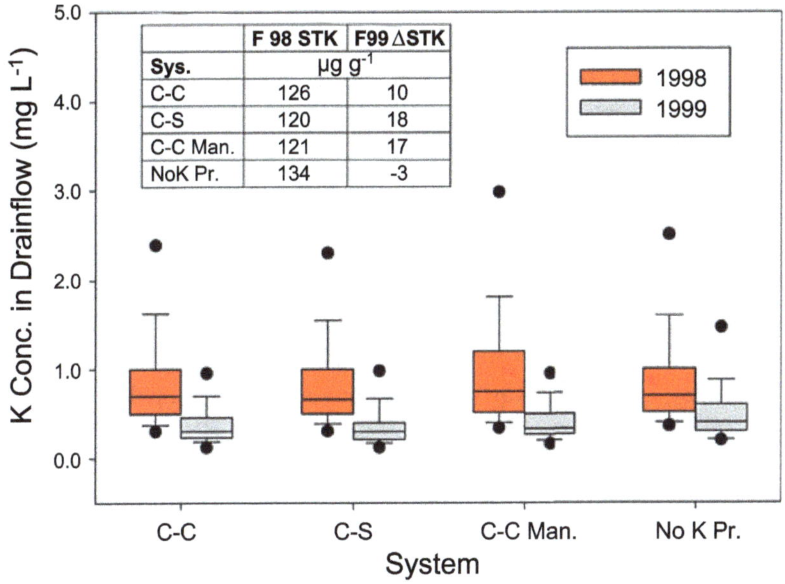

Fig. 1.7 Daily concentration in tile drainflows at the Purdue University Water Quality Field Station in 1998 and 1999 calendar years. Inorganically fertilized continuous corn (C–C; 20 treatment plots) and corn-soybean (C–S; 24 plots) rotations received 260 kg K ha^{-1} as KCl in May 1999. Data for C–C plots receiving regular fall or spring applications of swine manure (C–C man; 8 plots) and an unfertilized restored prairie (no K Pr., 4 plots) are also shown. Sample number varies from 159 (no K Pr., 1998) to 987 (C–S, 1999) daily observations; box plots show 25th, median, and 75th percentile values, error bars indicate 10th and 90th percentiles, and solid circles are 5th and 95th percentiles. Inset table shows the mean soil test levels in the fall of 1998 and the change measured in the subsequent fall 1999 sample. (Brouder, unpublished data)

While not suggesting any specific proportion of K_{Fert} lost by this mechanism, these authors noted that conservation tillage should be helpful in reducing losses. In their chapter on K, Havlin et al. (2014) neither show K_{Erode} as a loss mechanism in their rendition of the K cycle nor discuss it in their text. In general, it appears little effort has been put into understanding K_{Erode} as a significant factor reducing the efficiency of K_{Fert}, although it may be an important consideration under an array of environmental and management conditions. For further discussion of K_{Leach} and K_{Erode}, see Chap. 3.

Unlike K_{Erode}, much attention has been paid in the literature to $\mathrm{K_{Fixed}}$, especially in $\mathrm{STK_{Exch}}$ correlation-calibration studies when fertilizer rates well in excess of K_{Harv} do not appear to build $\mathrm{STK_{Exch}}$ as expected. For example, in the classic Cassman work discussed above (Cassman et al. 1989), cotton (*Gossypium hirsutum* L.) grown on soils with a history of intensive cropping without K fertilization was deficient in K but failed to respond to moderate rates of K_{Fert}. With repeated K additions, these authors observed an increase in apparent K_{Fert} uptake efficiency, which they attributed to partial saturation of K fixation by earlier fertilizer applications. Likewise, Navarrete-Ganchozo (2014) demonstrated that on K-fixing soils, the residual value of K_{Fert} was unobservable several years after halting aggressive fertilizer additions even though large positive input balances remained (Fig. 1.6). As reviewed by Brouder (2011), when K is added to soils with 2:1 layer silicate clays (e.g., weathered micas, smectite, and vermiculite), nonhydrated K can fit into spaces between interlayer surfaces of clay mineral silicate sheets. In minerals with high charge density, K fixation between adjacent sheets can stabilize the overall structure and depress subsequent release of K_{Fixed} (Barber 1995). Factors decreasing the extent of K fixation include the presence of oxide precipitates on clay surfaces or in interlayer positions (e.g., Rich and Obenshain 1955; Horton 1959; Rich and Black 1964; Page and Ganje 1964), increased soil organic matter, especially mobile humic acid fractions (Cassman et al. 1992; Olk and Cassman 1995), and NH_4^+ addition (Bolt et al. 1963; Lumbanraja and Evangelou 1992; Brouder and Cassman 1994). In contrast, wet-dry cycling can drive net K movement into fixed positions, thereby reducing fertilizer efficiency (Olk et al. 1995; Zeng and Brown 2000).

Not surprisingly, in regions with long histories of soil test calibration work, suggestions to assess K_{Fixed} have focused on the same assays proposed for understanding of contributions of K_{NonExch} to K_{Soil} (discussion of Eqs. 1.10 and 1.11). For example, Murashkina et al. (2007) examined the standard NH_4OAc-Ext assay for STK_{Exch}, a modification of the K isotherm method (Cassman et al. 1990), the 5-min sodium tetraphenylboron (TPB) assay, and soil texture with the goal of developing a quick test for routine determination of K_{Fixed}. Because it involves a precipitation reaction that removes K from the soil solution as a sink (like a root would be), the TPB extraction has been advanced as a more mechanistic protocol when compared to strong acid extractions (1 M HNO_3) (Cox et al. 1999). Similarly, Murashkina et al. (2007) found the modified Cassman method was a rapid and reliable method for predicting fixation potential, while TPB was identified as useful with a measure of STK_{Exch} to predict K already fixed or a reduction in fixation potential. It is important to note that isotherm assays cannot distinguish between the mechanisms of K loss from the soil solution to K_{Fixed} and K_{Precip} much as TPB or strong HNO_3 extraction cannot distinguish among pools contributing to K_{NonExch}. Regardless, despite research identifying protocols to assess K_{Fixed} and/or K_{Precip}, recommendation frameworks have yet to be modified to include an explicit consideration of these measures in an estimate of E_{Fert}.

Finally, K returned in unharvested residues (K_{UnHarv}) is significant and should reduce K_{Fert} when compared to a system where residues are removed. But the efficiency of this return is understudied, and the K_{UnHarv} fraction has been largely ignored as an explicit factor in recommendation frameworks. It is interesting to note that Stanford (1973) did not explicitly consider N returned in residues even though N management research focused on residue N contributions to subsequent crops was already being actively pursued. For example, Shrader et al. (1966) had demonstrated that maize (*Zea mays* L.) following oats (*Avena sativa* L.), meadow, and soybean acquired >80 kg N ha^{-1} in fertilizer equivalent from the residues when compared to maize grown without rotation. In their comprehensive expansion of Stanford's equation, Morris et al. (2018) explicitly include a fertilizer equivalent factor for the soil N supply attributable to the legume and an efficiency factor for the fraction that may be taken up by the plant.

In living plants, K is not metabolized and forms only weak complexes in which it is readily exchangeable (Wyn Jones et al. 1979). Thus, K_{UnHarv} left on or in the soils in residues will be returned to the soil solution as the tissues decompose. K_{UnHarv} can represent a significant input for the next crop in a rotation. For example, modern maize and soybean varieties can return quantities of K ranging more than 40–200 and 80–150 kg K ha^{-1}, respectively, depending on growing conditions and degree of luxury consumption (Fernández et al. 2009; Wu et al. 2014; Gaspar et al. 2017; Zhang et al. 2012). As discussed above (Sect. 1.4.1), K HI in grain crops can be expected to be highly variable. Wu et al. (2014) report K HI values ranging more than 0.1 to 0.45 for maize grown under optimum and super-optimum soil K (CV = >25%). Thus, the limitations of using grain dry matter yields to estimate K_{Plant} (Eq. 1.5; Sect. 1.4.1) extend to estimating K_{UnHarv}. Further, research to date is insufficient to determine whether E_{Fert} for inorganic fertilizers and manures could also apply to residues. In temperate systems with conservation or no-tillage, sparse data suggest as much as 80 and 90% of K in maize and soybean residues at the soil surface, respectively, may be leached from the residue by planting time in the following spring (Oltmans and Mallarino 2015); presumably the remaining residue K will be leached from the residue within the growing season. Once residue K enters the soil solution, factors that reduce the efficiency of added K (K_{Leach}, K_{Erode}, K_{Fixed}, and K_{Precip}) can be expected to be similar to those affecting K_{Fert}. However, how residue is handled will likely impact losses in surface runoff. Simulated rainfall studies suggest K concentrations in runoff could initially be substantially higher from no-till soybean fields compared to conventionally tilled soybean fields (Bertol et al. 2007), but typical erosion-related losses of K_{UnHarv} for common residues and their managements are largely unknown.

In sum, with the caveat that almost all the K_{UnHarv} enters the soil solution, the basic Stanford equation (Eq. 1.2) can be expanded for K to

$$E_{KAdded} \times (K_{Fert} + K_{UnHarv}) = K_{Plant} - K_{Soil} \quad (1.13)$$

where a generic efficiency factor (E_{KAdded}) applies to both K_{Fert} and K_{UnHarv}. Then K_{Fert} is estimated as

$$K_{Fert} = [(K_{Plant} - K_{Soil}) \div E_{KAdded}] - K_{UnHarv} \quad (1.14)$$

As indicated by the discussion above, E_{KAdded} is challenging to measure directly and has been approximated as apparent crop recovery efficiency (RE_K; Cassman et al. 2002). At a minimum, determination of this factor requires omission plots to compare K_{Soil} to total plant uptake ($K_{Soil\,+\,X}$) at a specific fertilizer rate "*X*" (K_{FertX}), using Eq. (1.2) rearranged as

$$E_{KAdded} \cong RE_K = (K_{Soil+X} - K_{Soil})/K_{FertX} \quad (1.15)$$

This approach aggregates the disparate impacts of crop-specific K_{UnHarv}, soil mineralogy, and agroecozone environmental parameters that can influence E_{KAdded}

(Chaps. 4 and 5). With enough data, this differential approach could permit identification of categorical classes for E_{KAdded} for routine use in recommendation frameworks. To date, however, existing data resources are too fragmented and incomplete for a robust implementation of this approach. Many K rate studies have not assessed K_{Plant} (as discussed in Sect. 1.4.1). Still, analysis of existing data might be sufficient to determine if this approach to understanding efficiency could be useful in an expanded mass balance recommendation framework.

1.4.4 Potassium Recommendations Without Soil Tests

Although soil tests are widely employed as the basis for K recommendations, there are many places and circumstances where they are not used. Soil testing services are not ubiquitous, and many areas do not have the needed facilities, logistics, or quality control mechanisms in place. Additionally, even when soil testing is available, a particular farmer may have no recent soil test information in hand at the time when a recommendation needs to be made. There are also situations where soil tests, even if available, do not provide reliable diagnostic information. We discuss two K recommendation approaches that can be used either with or without soil test information.

1.4.4.1 Recommendations Based on Nutrient Removal

When plant biomass is removed from a field, the K contained in that biomass is also removed. To maintain K levels in soils, K needs to be added to replace the K removed (K_{Harv}). A **maintenance fertilizer rate** is *the amount of K that replaces K removed by crop harvest.* This rate can be expressed as $K_{Fert} = K_{Harv}$. Ideally, removal is measured by analyzing samples of the harvested biomass for dry matter content and nutrient concentration. Most common, however, is to estimate the quantities of K removed, using average values of K concentrations ($K_{AveConc}$) published in recommendation guidance documents, such as Vitosh et al. (1995) and Mallarino et al. (2013). In production settings, these averages are typically treated as constants. The maintenance rate is estimated akin to Eq. (1.5) as

$$K_{Fert} = K_{Harv} = (\text{Yield}) \times K_{AveConc} \tag{1.16}$$

As discussed in Chap. 3, treating $K_{AveConc}$ as a constant does not acknowledge any variability, leading to maintenance rates that may not accurately replace the K that was removed by harvest.

Maintenance rates are necessary for sustaining soil K levels; however, as Olson et al. (1987) observed, they do not consider the economically optimum rate of K for a given cropping season. To examine the implications, we look at two scenarios at opposite ends of the spectrum.

First, we consider the case when levels of plant-available K in soil are already adequate or nearly adequate for expected levels of crop production. In this case, the cost of K applied at maintenance rates will not be profitable for that season, since the yield and revenue increases needed to recover the costs will not be realized. Farmers who own the land that is fertilized may have sufficient capital and long-term soil management objectives to absorb this cost in the short term to realize the longer-term gains of sustained soil fertility and crop productivity. However, if the farmer who is paying for the fertilizer is renting the land, and the rental agreement does not have provisions for the farmer to recover this cost, then the landowner, not the farmer, will be the one to gain from the maintenance application.

Second, we consider the case when levels of plant-available K are very low. In this case, a maintenance rate may be too low to realize the fully attainable crop responses and revenue increases. In recommendation approaches using algorithms to optimize net returns in one cropping season, recommended K rates can be above maintenance rates (calculated from Kaiser et al. 2018). Using maintenance rates in this scenario leads to two missed opportunities: (1) realizing the full yield and revenue increases possible and (2) increasing soil fertility for the subsequent season or seasons. When application rates exceed maintenance rates, the K supply in the soil increases, assuming that there are no losses to pools 3–5 or 10 in Fig. 1.1.

Maintenance rates have been combined with soil test information in some recommendation systems (Vitosh et al. 1995). In those algorithms, they are recommended when levels of soil fertility have reached levels that are considered optimum for crop production.

1.4.4.2 Recommendations Based on Plant Nutrient Uptake and Yield

Perhaps the most well-developed approach that can be used with or without soil test information is the Quantitative Evaluation of the Fertility of Tropical Soils (QUEFTS) model (Janssen et al. 1990). It was designed as a land productivity evaluation tool, with predicted maize yield as the primary model output. Yield was predicted from both quantities of potentially available soil nutrients and plant nutrient uptake (originally developed for N, P, and K). QUEFTS was developed from data from Kenya but has since been widely used as a framework for K recommendations in other countries and for a variety of crops, including banana (*Musa* spp.; Nyombi et al. 2010), cassava (*Manihot esculenta* Crantz; Byju et al. 2012), maize (*Zea mays* L.; Janssen et al. 1990), oilseed rape (*Brassica napus* L.; Cong et al. 2016), peanut (*Arachis hypogaea* L.; Xie et al. 2020), potato (Kumar et al. 2018), radish (*Raphanus raphanistrum* subsp. *sativus* (L.) Domin; Zhang et al. 2019), rice (*Oryza sativa* L.; Witt et al. 1999), soybean (Jiang et al. 2019), sweet potato (*Ipomoea batatas* (L.) Lam.; Kumar et al. 2016), taro (*Colocasia esculenta* (L.) Schott; Raju and Byju 2019), and wheat (*Triticum aestivum* L.; Chuan et al. 2013). These crops rely on new uptake of nutrients from the soil during the season to meet most, if not all, of their total uptake requirements. To our knowledge, tree crops have not been evaluated with QUEFTS, likely because of the logistical challenges of measuring uptake. We could also find no examples where perennial forages were evaluated with QUEFTS.

QUEFTS models how the potentially available soil supplies of three nutrients interact to affect total uptake and yield. Using N, P, and K as an example, the uptake of K is predicted from potential soil supplies of K and N; then it is predicted a second time from potential soil supplies of K and P. The lower of the two K uptake estimates is used, since it is considered to be more efficient (Witt et al. 1999) and limits the bias introduced by luxury consumption. From this lower uptake, two yield estimates are made, based on lower and upper boundary lines encompassing observations (usually numbering in the hundreds) of K uptake and yield. The slope of each boundary line is yield divided by total nutrient uptake. This slope is termed "internal efficiency." The upper boundary line represents yields associated with maximum K dilution, and the lower boundary line represents yields associated with maximum K accumulation. The creation of paired boundary lines is repeated for the other two nutrients. A systematic comparison of all yield estimates, considering two nutrients at a time, results in a final, average yield estimate.

Liebig's law of the minimum is a fundamental concept in the way QUEFTS evaluates nutrient interactions. Yields are limited by the most limiting of the three nutrients. For a given yield, if total K uptake is lowest (maximum dilution), and uptake of N and P are higher, then K is considered yield-limiting. If total K uptake is highest (maximum accumulation), then one or both of the other nutrients may be limiting, or luxury consumption may be occurring. The model makes it possible to determine optimum uptake levels of nutrients that keep any one nutrient from being limiting (Witt et al. 1999).

When QUEFTS was developed, it relied on soil tests as measures of potentially available soil nutrients. Interestingly, the predictor of potentially available K supply was not exchangeable K alone, but exchangeable K combined with pH, organic carbon, and cation exchange capacity (Janssen et al. 1990). In a later application of the QUEFTS model to irrigated rice systems, Dobermann et al. (1996) developed the concept of effective soil K-supplying capacity. They defined it as "...the amount of K a crop takes up from indigenous resources under optimum conditions—i.e., when all other nutrients are amply supplied and only K is limiting...." Operationally, they measured it from total K uptake in rice grown where N and P had been applied, but not K. Analogous calculations were done for N and P. The emphasis was on the quantities of nutrients the plant actually took up from the soil, rather than what the soil could potentially provide. The collection of large quantities of nutrient uptake data enabled the application of QUEFTS to situations where soil tests were not available.

The primary measurements that are needed by QUEFTS are yield, uptake, and rates of fertilizer applied. In experimental trials, to calculate the effective soil nutrient capacities of each nutrient, three treatments are required: one that omits N (PK plot), another that omits P (NK plot), and a third that omits K (NP plot). To calculate the recovery efficiency of a given nutrient (Eq. 1.15), the total uptake in the omitted plot (NP, for instance, or K_{Soil} in Eq. 1.15) is subtracted from that of the plot where all three nutrients were applied (NPK or $K_{Soil + X}$ in Eq. 1.15) and then divided by the rate of the given nutrient (K in this example or K_{FertX}) applied in the NPK treatment. This data requirement of QUEFTS led to a basic four-treatment "omission plot" design (PK, NK, NP, NPK) that has been deployed widely, both on research stations and farmers' fields. Large databases have been developed from these trials (like the IRRI Mega Project referenced in Witt et al. (1999)), providing proxy data that can be used when local data do not exist.

Although QUEFTS was developed as a land evaluation tool, its framework has been expanded upon to develop recommended rates of N, P, and K. To do this, Guiking et al. (1995) considered the change in uptake and yield that resulted from applying fertilizer. Unfertilized yield was predicted from the effective soil K-supplying capacity—the uptake where no K was applied. Guiking et al. (1995) then considered what happened under fertilization. Fertilization was added to the soil K supply; however, only a fraction of the fertilizer rate (E_{Fert} or recovery efficiency) was taken up by the plant. By multiplying the fertilizer rate by E_{Fert}, the change in plant uptake was estimated. This higher uptake was then used by QUEFTS to estimate fertilized yield. Based on the yield increase and the quantity of fertilizer applied to generate that increase, economic returns could be calculated and economically optimum rates determined through iteration.

More recently, Pampolino et al. (2012) developed Nutrient Expert®, a software tool that builds upon QUEFTS to generate nutrient recommendations for cereal farmers. Nutrient Expert is built on a large database of data from omission trials. In these trials, data collected include yield, nutrient uptake, soil test information (where it exists), crop sequence, crop residue management practices, soil fertility status, and water management information. Information on farmers' existing yields are used as background for conducting omission plot trials. Nutrient Expert® integrates all of these factors to create nutrient recommendations that consider attainable yield levels, expected crop responses, nutrient input and output balance, production risks, and economic returns.

1.5 Diagnostics Development: The Undelivered Promise of "Big Data"

Before discussing pathways to improve the use of the K cycle in recommendations, it is informative to move beyond the existing tools (e.g., a calibrated K exchange test) to reflect on the process of diagnostic development and the extent to which any single measurement can embody all the knowledge necessary to make a decision. The use of soil testing as an essential tool in managing fertilizer has its foundations in the seminal work conducted in the 1920s and 1930s (Melsted and Peck 1973). In the case of soil testing-based approaches to K fertilizer recommendations, the majority of research focused on developing chemical extraction procedures for assessing quantities of plant-available K in crop rooting zones. As remarked by Colwell (1967), "A measurement qualifies to be termed a soil test for a particular nutrient if, and only if, it provides information on the fertilizer requirement of a crop for that nutrient." Generally, the term correlation has been used to characterize the process of relating nutrient uptake and/or yields to the quantity of a nutrient extracted by a particular soil test, while calibration refers to the experimentation needed to characterize the meaning of the soil test result for a crop response (Dahnke and Olson 1990). Much of the initial research on calibration of soil K tests focused on delineating responsive from non-responsive soils and categories for major differences in degrees of crop response (Bray 1944; Cate and Nelson 1971; Olson et al. 1958; Rouse 1967). Research in the latter decades of the twentieth century has tended to focus on the actual quantities of fertilizer needed to obtain maximum or profitable yields and experiments relating yields to fertilizer increments (Dahnke and Olson 1990; Welch and Wiese 1973). Regardless, any individual measurement must be underpinned by sufficient correlation and calibration research if it is to meet Colwell's criterion of providing generalizable information across space and time on a crop's fertilizer requirement. Certainly it would be surprising to identify a result that is universally useful and applicable.

1.5.1 Data Limitations: Historic and Current

The development of a robust diagnostic is inherently a "big data" enterprise; indeed most K measurement(s), whether soil or tissue assays, were never intended to embody all knowledge necessary to guide management. Common approaches and best practices for soil test correlation and calibration are well documented in the literature (Cope and Rouse 1973; Hanway 1973; Melsted and Peck 1973; Dahnke and Olson 1990), but recurring themes dating from the earliest of these reports are (1) the need for large numbers of field studies and sufficient data to quantify the effects of important controlled (e.g., crop or tillage) and uncontrolled (e.g., weather) system variables and their interactions and (2) the inadequacy of resources to acquire those data. In their overview of soil testing principles Melsted and Peck (1973) remarked on the tenuous nature of the relationship between yield and the corresponding level of an available nutrient in the soil. In 1967, Tisdale stated that research concerning soil test development was under-supported because science administrators viewed as low priority the "unravelling of a highly complex functional relationship existing among plant growth, plant nutrient supply, and numerous environmental factors." Hanway (1973) echoed this sentiment, implying that the making of agronomy into a quantitative science was impeded by the ability to conduct "adequate field experiments to provide the data required. Such experimentation is not easy, and it is not cheap, but there is no adequate alternative."

The hazards of small studies with sparse data include low statistical power, model overfitting, lack of reproducibility, and a reduced likelihood that a statistically significant result represents a true effect that is generalizable (Brouder et al. 2019). For example, a 2-year, single location correlation study produced a statistically significant relationship between maize yield and soil test K (STK) that implied a critical level that was definitive (Fig. 1.8a). Further, access to desktop computing permitted easy exploration of multiple empirical models with selection of the model and associated critical level based on best professional judgment regarding farmer perceptions of risk. Analysis of the correlation data demonstrated that common statistical models may perform similarly in explaining the variation among observations but identify critical levels that vary almost twofold (77 vs. 120+ mg kg^{-1}). Regardless, analysis across all five locations in the study demonstrated that the modeled correlations were not generalizable across multiple locations because substantial variation in relative yield was present at STK values ranging from 80 to 180 mg kg^{-1} (Fig. 1.8b).

For STK correlation and calibration, the lack of generalizability of significant small studies to other locations and across multiple years is common (e.g., Navarrete-Ganchozo 2014). Historically, insufficient data and overly localized correlation/calibration research combined with a lack of coordination among researchers contributed to the development of conflicting recommendations for the same or similar crop-soil systems (Tisdale 1967). In the USA, arbitrary, geopolitical differences in recommendations based solely on STK persist, reflecting a dearth of resources to collect and rigorously analyze data including important covariate and metadata. Additionally, lack of funding has largely restricted the evaluation of new soil assays to laboratory comparisons with existing protocols. New tests are being implemented with no or minimal field evaluation, a practice expected to introduce more unexplained variation among results (Gartley et al. 2002). Thus, the use of hypothetically important categorical (e.g., soil series, subsurface soil K supplies; Kelling et al. 1998) and/or continuous (e.g., cation exchange capacity; Vitosh et al. 1995) covariates in recommendations currently lack scientific support despite the now almost universal availability of advanced computing to facilitate covariate exploration.

1.6 Opportunities Moving Forward

The minimal extent to which a mass balance approach has been successfully used in existing K recommendations—despite being the theoretical underpinning—became widely apparent with the advent of precision technologies. In early implementation of variable rate technologies, there was both a focus on K and an implicit expectation that collecting spatially dense soil samples and applying existing recommendations to soil test results would generate a soil- and crop-specific rate recommendation (Mulla and Schepers 1997; Wollenhaupt et al. 1997) that would be more profitable than a whole-field, uniform application. Subsequent research has been sufficient to demonstrate the fallacy in assuming the existing, generic, tabular recommendations can be disaggregated in a meaningful way. We have learned that key assumptions, such as linkages between K and CEC and standard values for anticipated STK changes with fertilization and crop removal, are either not universal or are generally incorrect (e.g., Navarrete-Ganchozo 2014; Fulford and Culman 2018; Chap. 10).

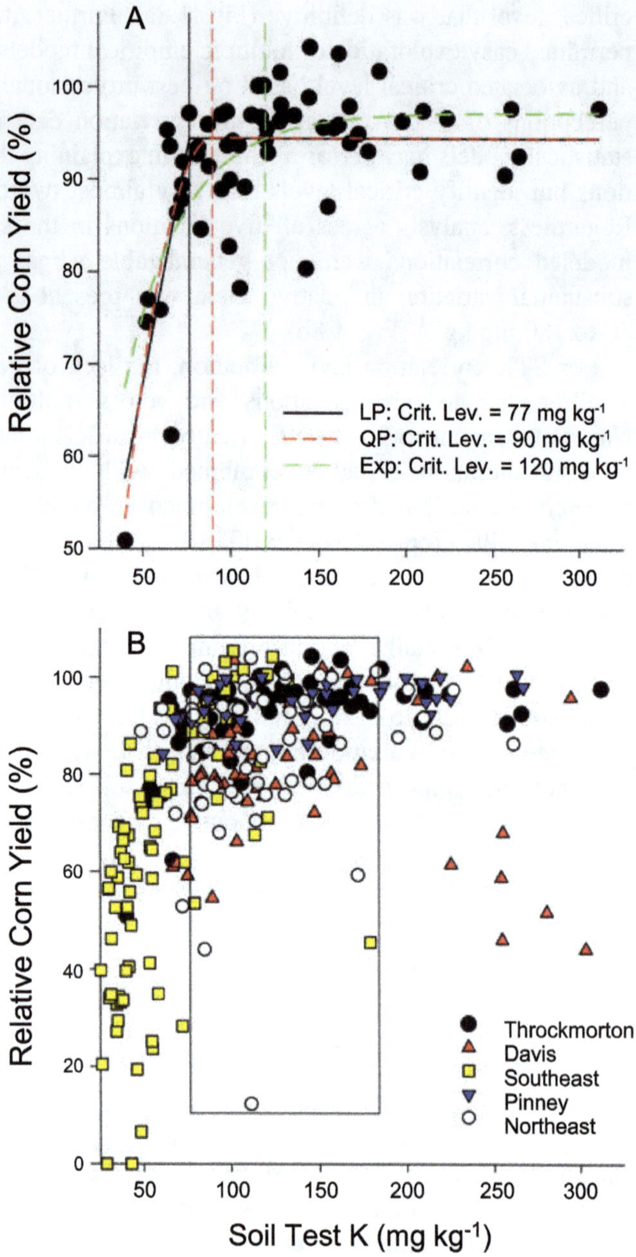

Fig. 1.8 Relationship between relative maize (*Zea mays* L.) yield and soil test K in surface soils (0–20 cm; NH_4OAc-Ext) in 2005 and 2005 at (**a**) one location (Throckmorton) and (**b**) five locations in Indiana, USA. (adapted from Navarrete-Ganchozo 2014)

1.6.1 Mechanistic Modeling

Mechanistic simulation models are intended to represent all system processes and attributes relevant to crop growth and development, including nutrient cycling and losses. To date, mechanistic models have been viewed primarily as research tools for hypothesis testing and predicting outcomes for the complex, system-level interactions that may occur when individual or suites of system parameters vary over their known biogeochemical and physiological ranges. As such, the rigorously verified mechanistic model has the potential to (1) identify system parameters that drive outcomes (e.g., via sensitivity analysis), (2) serve as a foundation for a simplification that provides for limited input of the critical on-farm data necessary to a site- or soil-specific recommendation, and (3) identify knowledge gaps via model failure (Brouder 1999). Models could serve these purposes in developing improved K management strategies, provided the K cycling processes have been fully and explicitly tested.

Modeling nutrient uptake has evolved from the seminal work of Nye and Spiers (1964) who developed mathematical expressions describing mass flow and diffusion of nutrients to a root segment, to an array of computational tools that help explain and predict K behavior in agroecosystems (Table 1.2). These models vary in focus, ranging from mechanistic models to more general integrative models with inference space at the ecosystem level. These models also differ in their spatial (nanometer to watershed) and temporal (sub-second to year) scales. Some models were purpose-built, with the primary goal of understanding nutrient uptake, including K, from the onset. This includes POTAS (Barnes et al. 1976; Zhang et al. 2007) and the Barber-Cushman model (Classen and Barber 1976). Other K models were adapted from computational tools originally developed to study other ecosystem processes. For example, PROFILE (Holmqvist et al. 2003) was initially developed to calculate critical loads for acid deposition in forest soils but later was modified to predict K release as soil minerals weather. Similarly, SWAT-K, used to estimate environmental K losses at the watershed scale (Wang et al. 2017), was developed by altering the Soil Water Assessment Tool (SWAT) hydrology model. Many models have undergone continuous improvement while retaining the same name (e.g., DSSAT). Improvements in others have resulted in model rebranding (Barber-Cushman to NST 3.0; COMP8 to SSAND to PCATS) (Lin and Kelly 2010). Many of the most successful models were created by teams including expertise from both the physical and agricultural/biological science domains (e.g., Barber-Cushman; DSSAT). At least some of the early models are no longer practically available for use by practitioners (e.g., Barber-Cushman and NST 3.0).

Critical to successful model development is access to data needed for calibration and validation. Barnes et al. (1976) highlighted the need for additional data from more cultivars, crop species, sites and soils to improve the accuracy and precision of their K and N model. Decades later, constraints imposed by limited availability of quality data unfortunately persist (Janssen et al. 1990; Boote et al. 1996). Rosenzweig et al. (2013) indicated that experimental data for most agricultural models tends to suffer from several common problems including: aggregation across sites and/or experiments, making it difficult to assign variation in agronomic performance to local climate and soil properties; absence of site-specific management information (e.g., metadata like planting date, pest control, tillage, soil characterization, or cultivars); and inexplicable yield results that cannot be readily attributed to environment or management. Additional K datasets are still needed to validate models, improving their accuracy and precision and extending their inference space (Lin and Kelly 2010; Wang et al. 2017). Specific examples of knowledge

Table 1.2 Summary of some models that could be or are used to understand potassium dynamics in agroecosystems

Model type	Fundamental/ mechanistic: soil and plant component	Process-based: soil and plant process	Whole plant and whole soil: Crop growth models	Ecosystem scale: Seasonal timeframe	Ecosystem scale: Multi-year timeframe
Example models:	HYDRUS 1-D; PROFILE	Barber-Cushman; HYDRUS 2-D; FUSSIM2; continuum/ multiscale; NST 3.0; SSAND; PCATS	APSIM; QUEFTS; DSSAT; EPIC; POTAS	APSIM; DSSAT; EPIC	AGNPS; SWAT-K; OVERSEER
Spatial scales:	nm–μm	mm–cm	dm–m	m–ha–field	Field—watershed
Temporal scales:	μs–min	min–hours	days–months	1–12 months	>12 months
Examples of processes:	Cation exchange processes; adsorption-desorption of ions; cell membrane transport of nutrients	Water and nutrient uptake by roots; root growth and morphology; nutrient concentration impacts on nutrient uptake; genotype and phenotypic expression	Crop growth, nutrient uptake and yield; cultivar effects on nutrient uptake	Seasonal weather patterns; single crop systems, double cropping systems; spatial and temporal variability of soil properties	Multi-year weather and climate; multi-year rotations; spatial and temporal variability of soil properties
Examples of parameters:	Soil properties: pH, organic matter, particle size distribution, aggregation, clay mineral distributions; genetic and physiological control on nutrient use efficiency		Field-scale phenotypic and genotypic variation; soil test correlations; genotype × environment × management; linkage to econometric crop production models		
References:	Holmqvist et al. (2003), da Silva Santos et al. (2015)	Classen and Barber (1976), Satpute and Singh (2017), Heinen (2001), Mai et al. (2019), Lin and Kelly (2010)	Scanlan et al. (2015a, b), de Barros et al. (2004), Smaling and Janssen (1993), Janssen et al. (1990), Jones et al. (2003), Zhang et al. (2007), Barnes et al. (1976), Wang et al. (2017), Wheeler et al. (2003), Young et al. (1989)		

Models are grouped by type and vary in relevant temporal and spatial scales. Typical model outputs and inference spaces also are indicated and data needed to improve K model calibration and/or validation include regulation of K luxury consumption, K adsorption/desorption and weathering of soil minerals, root morphology and mass impact on K uptake, soil moisture effects, transpiration rate impact on K uptake, and partitioning of K between shoots and roots, among others (Greenwood and Karpinets 1997; Holmqvist et al. 2003; Zhang et al. 2007; Scanlan et al. 2015a, b). It is hoped that ongoing national and international initiatives in open science and open access of publications and data will improve data availability for future modeling efforts.

1.6.2 Knowledge Gaps

Moving forward requires a careful assessment of the knowledge gaps and prioritization of investments with the highest probabilities of significantly advancing the science supporting K management. As discussed above (Sect. 1.2), a robust K recommendation must be complex enough in its components to represent important differences in biophysical and socioeconomic contexts but simple and cost-effective enough to be useful and efficiently used by the practitioner. Hence, explicit consideration must be given to whether or not K cycle parameters deemed scientifically important for an evidence-based K recommendation are practically assessable and whether cost-effective measurements of proxy variables have been fully explored. For example, the use of in-field elevation maps or telethermometry may prove more practical than soil sensors for determining localized moisture stress influencing K flux. Likewise, priorities of scientists must be balanced by those of practitioners. Important research foci include:

- *Correlation and calibration studies that explicitly test for covariates and important biophysical influencers of the K cycle*: What is clearly not needed are more primarily two-factor correlation calibration studies that focus on a putative, universal diagnostic (e.g., STK) and yield. As demonstrated above (Fig. 1.8b), covariate and metadata must be sufficient to explain variation in yield response across locations and yields to create a relationship to yield that is generalizable across space and time. While more expensive, long-term, multivariate studies not only improve the understanding of the crop-soil K cycle but would facilitate inclusion of important secondary biophysical influencers into a K recommendation (Fig. 1.2).
- *A better understanding of soil assays by mode of action*: Chap. 8 highlights the theoretical design features of different soil tests and proposes the choice of a test be based upon the expected duration of the meaning of a test result to crops and cropping systems of differing duration (short annual versus multi-harvest perennial). Such an approach is currently not used in most recommendation systems. In the USA, for example, recommended soil tests for routine testing are almost all assays of K_{Exch} with variations in extractant chemistry associated with geopolitical borders and not cropping system attributes (Nathan and Gelderman 2015).

- *Alternative approaches that are not grounded in soil testing and require development of large soil databases and routine, on-farm soil testing*: For regions and crops that have not historically relied on soil testing, developing soil tests and testing programs may be cost-prohibitive. Additionally, for some major production systems, K_{Soil} shows little relationship to quick chemical assays. For example, Dobermann et al. (2003) used omission plots in irrigated rice to characterize K_{Soil} and hypothesized that the approach held more promise for efficient development of site-specific recommendations.
- *Mechanistic and novel, empirical modeling of the K cycle*: The research needs for improved mechanistic modeling of the K cycle are highlighted above (Sect. 1.6.1). Additionally, machine learning and artificial intelligence approaches hold great promise for synthesizing K cycle data. Methods such as machine learning are tolerant to complex data characteristics (e.g., nonlinearity and outliers) (Brouder et al. 2019) and could be used to both detect patterns to underpin decision trees and to identify important proxy variables that are easy to assess. They can also automatically incorporate new information, creating opportunities for recommendations that are self-improving, potentially with on-farm data. Artificial neural networks are receiving increased interest as tools to explore input-output relationships in complex, agricultural systems where understanding remains incomplete (Yang et al. 2018; Liakos et al. 2018; Welikhe 2020).
- *K synthesis science*: Systematic review (SR), with or without application of meta-analysis statistics, is a well-recognized framework for translating knowledge into evidence-based policy and management practice. Cumulative systematic reviews permit efficient reanalysis of a relationship, effect, or efficacy of a practice as new data accrues. SR has long been considered a discrete field of research in medicine (Sackett and Rosenberg 1995; Cucherat et al. 1997, Jadad et al. 1998) and the gold standard for synthesizing research into a clinical practice and decision-making (Mulrow et al. 1997). Reports in the agricultural literature have highlighted the potential for routine use of systematic reviews to improve the quality of the primary literature and its use in evidenced-based decision-making (Philibert et al. 2012; Eagle et al. 2017; Brouder et al. 2019). The concept of developing minimum dataset guidelines to ensure small research studies can be synthesized (Brouder and Gomez-Macpherson 2014) has emerged as an important component of multi-investigator research networks and collaboratives. Proof-of-concept research is also needed regarding the value of integrating on-farm data with research data in developing site- and soil-specific K recommendations.
- *Probability, risk, and economics*: Although K recommendation frameworks have frequently associated an STK interpretation category (e.g., low, medium, high, very high) with the probability of response to fertilizer (50–0%) (Havlin et al. 2014), the probability of the response being of a particular magnitude was not included, as data typically were insufficient. Indeed, response probabilities for STK categories do not appear to have been rigorously evaluated despite their importance to farmer decision-making. The socioeconomic aspects of K management are beyond the scope of this chapter and book, but evaluation of any

improved framework should include a partial budgeting analysis to evaluate the changes in revenues and costs associated with applying the new recommendation compared to prior farmer practice.

1.6.3 Tools and Strategies, Data, and e-Infrastructure

Given the costs of conducting the system-level K cycle research required for strengthening our understanding, coordination among researchers, rigorous synthesis of new with existing studies, and application of a broad array of cyber-tools will all be paramount. Likewise, the planning for and sharing of structured data must become the norm, as most promising analytical strategies are data intensive. Scientists have reached near universal agreement that data sharing has value and advances research toward solutions to complex problems (e.g., Kim and Stanton 2016). In their analysis of the opportunities afforded to agriculture by data sharing, Brouder et al. (2019) argue that data sharing and the e-infrastructure needed to facilitate it will enhance the reliability of results and increase public trust in and use of agricultural science. Important data resources not currently available for synthesis and key attributes of a system to facilitate sharing and science synthesis to improve K recommendations are described here in brief.

1.6.3.1 Underutilized Data Sources with Potential

At present, the peer-reviewed literature is the primary source of data that is synthesized in systematic reviews and used to develop and verify the components of mechanistic models. Yet, for a variety of reasons, better characterization of study weights and effect sizes in statistical meta-analyses can be achieved with the use of the original study data than with extracted treatment means and their reported variance statistics (Cooper and Patall 2000). Likewise, mechanistic models can be more rigorously assessed with original data. Peer-reviewed journals tend to emphasize novel results (Fanelli 2012), thereby distorting the foundations of evidence-based practice by excluding confirmatory, null, or negative results. While numerous factors can contribute to irreproducible or non-generalizable results, solutions for improving the quality of science consistently stress complete reporting inclusive of data access (Begley and Ioannidis 2015; Button et al. 2013; Goodman et al. 2016). Furthermore, anecdotal evidence suggests that much of the data collected in previous and ongoing STK correlation and calibration research is inaccessible today—even in synthesized form—as publication in the less rigorous grey literature (e.g., newsletters, local- or state-level reports, conference proceedings) was and remains common when the goal is to test or adapt a peer-reviewed result for local constraints or conditions and farmer preferences. Such grey literature can be difficult to even identify (Debachere 1995). Mechanisms to ensure visibility and validity for the grey literature and data not associated with peer-reviewed publications along with recovery of legacy data associated with journal publications could greatly increase quality data available for synthesis.

Beyond research, other data that have the potential for improving site and soil specificity of a K recommendation include private research and on-farm data. Nutrient management research in the private sector typically remains in-house, although entities from that sector may be willing to share a portion of their research data to find new approaches with promising business opportunities or to bring additional credibility to their products through independent scientific evaluation. Additionally, combining on-farm data with research data could extend the inference space of research results. Few expect funding allocations to agricultural research to increase dramatically in the coming years (see USDA ERS 2018 and EPAR 2017 for trends in public and private support) making it even more imperative to improve efficiency of data collection and the use and reuse of existing data. However, the quantity of on-farm data generated by crop producers is projected to increase exponentially (BI Intelligence 2015). Harnessing these data for research on management recommendations is widely considered an untapped opportunity to leverage public research investments. Still, privacy, security, ownership, and intellectual property concerns need to be addressed when accessing and using privately generated data from any source.

1.6.3.2 FAIR Data

Moving K research from its present culture of small research studies and limited data sharing to one where data are collected with the anticipation that they will be reused in syntheses and modeling requires development and implementation of best practices that ensure readability over time and across an array of agronomic disciplines. In 2016, Wilkinson et al. articulated the basic principles of data sharing: data must be FAIR (findable, accessible, interoperable, and reusable). The principles emphasize consistent use of appropriate metadata, and machines must be able to assist in finding, obtaining, and subsequently reusing relevant data. In the absence of metadata, natural language processing can be used to search for relevant data resources that are inconsistently described (Joseph et al. 2016). Application programming interfaces can be written to convert dissimilarly structured data into a uniform structure for integration, provided the data are annotated sufficiently for a secondary user to understand. Moving toward common metadata and data standards will accelerate data reuse, and agricultural standards are under active development as are tools to facilitate FAIR data workflows.

Tools include workflows for consistent use of data and metadata standards, controlled vocabularies, and agricultural ontologies (Brouder et al. 2019). Baker et al. (2016) have described a global multilingual concept scheme, an attempt to combine the most useful terms from three broadly used sources. An ontology is an organized, typically hierarchical, set of concepts and categories with explicit properties and interrelationships. Ontology terms have uniform resource identifiers and, when used appropriately, contribute to interoperability. Aubert et al. (2017) used several existing ontologies, agronomic expertise, and the data standards from the International Consortium for Agricultural Systems Applications (White et al. 2013) to build an Agronomy Ontology (AgrO). AgrO describes agronomic practices, techniques, and variables used in agronomic experiments; associated field and desktop tools permit real-time development of FAIR data during experimentation as well as rectification of existing data into FAIR formats. Finally, because practitioners often use different terminologies than the scientists doing the research to inform practice, Ingram and Gaskell (2019) have proposed a methodology for co-production of ontologies with disparate stakeholders to make smarter search engines for agriculture.

1.6.3.3 Repositories and Data Publications, Catalogues, Registries, Knowledgebases

Potassium research will likely continue to be pursued by individuals or small teams of researchers dispersed across the globe, but an array of e-infrastructures is emerging to support a more data-driven, integrated approach to K recommendations. An ecosystem of repositories is emerging with the intent to foster FAIR data. General-purpose publishing repositories (e.g., Dryad [2020]) and institutional repositories (e.g., P.U.R.R n.d.) are non-specific, but they can ensure that data are appropriately annotated with metadata and exposed to relevant search engines. Potassium data are beginning to populate these resources. For example, Berg et al. (2020) published an 8-year study focused on the impact of P and K nutrition in alfalfa yield, quality, and persistence. These data are fully open access, have been downloaded to date over 600 times annually, and can be accessed, downloaded, reanalyzed, and formally cited when used in a novel analysis.

Some repositories can publish data as publications and assign a digital object identifier (DOI) ensuring their persistence in the scholarly record. Van Tuyl and Whitmire (2016) consider standalone data publications that follow citation conventions to be essential for incentivizing data preparation and ensuring wide accessibility. Several major repositories also offer data catalogues and registries. Catalogues facilitate locating similar datasets whose location is dispersed. Registries can expose metadata and content summaries to search engines but typically require additional steps to gain access to the data, thereby allowing concerns about privacy needs and embargoing to be addressed. Another form of registration could directly incentivize researchers to follow through with preparing all data for reuse and publication irrespective of outcomes. Advocates for improved reproducibility in science propose registering projects with a funding organization or a peer-reviewed journal prior to data collection as a way to prevent negative, null, and replicative results from being relegated to the grey literature (Nosek and Lakens 2014; Kupferschmidt 2018).

Finally, with sufficient investment and careful consideration for sustainability, the creation of a K knowledgebase would be a robust and innovative way to foster collaboration among researchers and advance K cycle science. As described by Gabella et al. (2017), knowledgebases are "organized and dynamic collections of information about a particular subject" and differ from repositories in that multiple data sources are not just archived but curated for a purpose with review, distillation, and manual annotation by experts. A K knowledgebase could create the consistent and reliable source of scientific knowledge to effectively deliver to practitioners credible, salient, and legitimate scientific information. Identifying space for K research activities within general agronomic knowledgebases such as the Ag Data Commons (USDA-NAL n.d.) or GARDIAN (CGIAR Platform for Big Data in Agriculture n.d.) would be an efficient way to leverage investments in data tools that would otherwise be well beyond the resources currently allocated to agronomic K research.

References

Amos B, Walters DT (2006) Maize root biomass and net rhizodeposited carbon. Soil Sci Soc Am J 70(5):1489–1503. https://doi.org/10.2136/sssaj2005.0216

Aubert C, Buttigieg PL, Laporte M-A, Devare M, Arnaud E (2017) CGIAR agronomy ontology. [WWW document]. http://purl.obolibrary.org/obo/agro.owl. Licensed under CC-BY-4.0. Accessed 4 June, 2020

Baker TC, Caracciolo C, Arnaud E (2016) Global agricultural concept scheme (GACS) a hub for agricultural vocabularies. In 7th international conference on biomedical ontologies, ICBO (vol 16, No. 2). https://pdfs.semanticscholar.org/9903/8eb708a26c1075f21952dad009bf299f3116.pdf. Accessed 4 June 2020

Barber SA (1995) Soil nutrient bioavailability: a mechanistic approach, 2nd edn. Wiley, New York

Barber TE, Matthews BC (1962) Release of non-exchangeable soil potassium by resin-equilibration and its significance for crop growth. Can J Soil Sci 42:266–272. https://doi.org/10.4141/cjss62-035

Barnes A, Greenwood DJ, Cleaver TJ (1976) A dynamic model for the effects of potassium and nitrogen fertilizers on the growth and nutrient uptake of crops. J Agric Sci 86:225–244. https://doi.org/10.1017/S002185960005468X

Begley CG, Ioannidis JP (2015) Reproducibility in science: improving the standard for basic and preclinical research. Circ Res 116:116–126. https://doi.org/10.1161/CIRCRESAHA.114.303819

Berg WK, Lissbrant S, Volenec JJ, Brouder SM, Joern BC, Johnson KD, Cunningham SM (2020) Phosphorus and potassium influence on alfalfa nutrition. (Version 2.0). Purdue University Research Repository. https://doi.org/10.4231/PPKB-VK18

Bertol I, Engel FL, Mafra AL, Bertol OJ, Ritter SR (2007) Phosphorus, potassium and organic carbon concentrations in runoff water and sediments under different soil tillage systems during soybean growth. Soil Tillage Res 94:142–150. https://doi.org/10.1016/j.still.2006.07.008

Bertsch PM, Thomas GW (1985) Potassium status of temperate region soils. In: Munson RD (ed) Potassium in agriculture. American Society of Agronomy, Madison, pp 131–162. https://doi.org/10.2134/1985.potassium.c7

BI Intelligence (2015) Seed by seed, acre by acre, big data is taking over the farm. https://www.businessinsider.com/big-data-and-farming-2015-8?r=UK&IR=T. Accessed 30 May 2020

Bolt GH, Sumner ME, Kamphorst A (1963) A study of the equilibria between three categories of potassium in an illitic soil. Soil Soc Am J 27(3):294–299. https://doi.org/10.2136/sssaj1963.03615995002700030024x

Bolton EF, Aylesworth JW, Hore FR (1970) Nutrient losses through tile drains under three cropping systems and two fertility levels on a Brookston clay soil. Can J Soil Sci 50(3):275–279. https://doi.org/10.4141/cjss70-038

Boote KJ, Jones JW, Pickering NB (1996) Potential uses and limitations of crop models. Agron J 88:704–716. https://doi.org/10.2134/agronj1996.00021962008800050005x

Bray RH (1932) A test for replaceable and water-soluble potassium in soils. Agron J 24 (4):312–316. https://doi.org/10.2134/agronj1932.00021962002400040007x

Bray RH (1944) Soil-plant relations: I. The quantitative relation of exchangeable potassium to crop yields and to crop response to potash additions. Soil Sci 58(4):305–324

Brouder (1999) Modeling soil-plant potassium relations. In: Oosterhuis DM, Berkowitz (eds) Frontiers in potassium nutrition: new perspectives on the effects of potassium on physiology of plants. Potash & Phosphate Institute/Potash & Phosphate Institute of Canada, Norcross, pp143–154

Brouder SM (2011) Potassium cycling. In: Hatfield JL, Sauer TJ (eds) Soil management: building a stable base for agriculture. American Society of Agronomy and Soil Science Society of America, Madison, pp 79–102

Brouder SM, Cassman KG (1994) Evaluation of a mechanistic model of potassium uptake by cotton in vermiculitic soil. Soil Sci Soc Am J 58(4):1174–1183. https://doi.org/10.2136/sssaj1994.03615995005800040024x

Brouder SM, Gomez-Macpherson H (2014) The impact of conservation agriculture on small agricultural yields: a scoping review of evidence. Agric Ecosyst Environ 187:11–32. https://doi.org/10.1016/j.agee.2013.08.010

Brouder SM, Volenec JJ (2008) Impact of climate change on crop nutrient and water use efficiencies. Physiol Plant 133(4):705–724. https://doi.org/10.1111/j.1399-3054.2008.01136.x

Brouder S, Eagle A, McNamara NKFJ, Murray S, Parr C, Tremblay N, Lyon D (2019) Enabling open-source data networks in public agricultural research. Council for Agricultural Science and Technology (CAST). CAST Commentary QTA2019-1. CAST, Ames, Iowa

Brown JR (1998) Recommended chemical soil test procedures for the North Central Region. No. 1001. Missouri Agricultural Experiment Station, University of Missouri, Columbia

Brown DA, Scott HD (1984) Dependence of crop growth and yield on root development and activity. In: Barber SA, Bouldin DR, Kral DM, Hawkins SL (eds) Roots, nutrient and water influx, and plant growth. Soil Science Society of America, Crop Science Society of America, and Soil Science Society of America, Madison, pp 101–136

Burke RH, Moore KJ, Shipitalo MJ, Miguez FE, Heaton EA (2017) All washed out? Foliar nutrient resorption and leaching in senescing switchgrass. Bioenergy Res 10:305–316. https://doi.org/10.1007/s12155-017-9819-6

Burks JL (2013) Eco-physiology of three perennial bioenergy systems. Dissertation, Purdue University. https://docs.lib.purdue.edu/dissertations/. Accessed 9 May 2020

Button KS, Ioannidis JP, Mokrysz C, Nosek BA, Flint J, Robinson ES, Munafò MR (2013) Power failure: why small sample size undermines the reliability of neuroscience. Nat Rev Neurosci 14 (5):365–376. https://doi.org/10.1038/nrn3475 [Erratum: Nat Rev Neurosci 14:451. https://doi.org/10.1038/nrn3502]

Byju G, Nedunchezhiyan M, Ravindran CS, Santhosh Mithra VS, Ravi V, Naskar SK (2012) Modeling the response of cassava to fertilizers: a site-specific nutrient management approach for greater tuberous root yield. Commun Soil Sci Plant 43(8):1149–1162. https://doi.org/10.1080/00103624.2012.662563

Cakmak I, Horst WJ (1991) Effect of aluminum on net efflux of nitrate and potassium from root tips of soybean (*Glycine max* L.). J Plant Physiol 138:400–403. https://doi.org/10.1016/S0176-1617(11)80513-4

Cash DW, Clark WC, Alcock F, Dickson NM, Eckley N, Guston DH, Jäger J, Mitchell RB (2003) Knowledge systems for sustainable development. Proc Natl Acad Sci USA 100(14):8086–8091. https://doi.org/10.1073/pnas.1231332100

Cassman KG, Bryant DC, Higashi SL, Roberts BA, Kerby TA (1989) Soil potassium balance and cumulative cotton response to annual potassium additions on a vermiculitic soil. Soil Sci Soc Am J 53(3):805–812. https://doi.org/10.2136/sssaj1989.03615995005300030030x

Cassman KG, Bryant DC, Roberts BA (1990) Comparison of soil test methods for predicting cotton response to soil and fertilizer potassium on potassium fixing soils. Commun Soil Sci Plan 21 (13–16):1727–1743. https://doi.org/10.1080/00103629009368336

Cassman KG, Bryant DC, Roberts BA (1992) Cotton response to residual fertilizer potassium on vermiculitic soil: organic matter and sodium effects. Soil Sci Soc Am J 56(3):823–830. https://doi.org/10.2136/sssaj1992.03615995005600030025x

Cassman KG, Dobermann A, Walters DT (2002) Agroecosystems, nitrogen-use efficiency, and nitrogen management. Ambio 31(2):132–140. https://doi.org/10.1579/0044-7447-31.2.132

Cate RB Jr, Nelson LA (1971) A simple statistical procedure for partitioning soil test correlation data into two classes. Soil Sci Soc Am Proc 35(4):658–660. https://doi.org/10.2136/sssaj1971.03615995003500040048x

Chuan L, He P, Jin J, Li S, Grant C, Xu X, Qiu S, Zhao S, Zhou W (2013) Estimating nutrient uptake requirements for wheat in China. Field Crops Res 146:96–104. https://doi.org/10.1016/j.fcr.2013.02.015

Ciampitti IA, Camberato JJ, Murrell ST, Vyn TJ (2013) Maize nutrient accumulation and partitioning in response to plant density and nitrogen rate: I. Macronutrients. Agron J 105 (3):783–795. https://doi.org/10.2134/agronj2012.0467

Classen N, Barber SA (1976) Simulation model for nutrient uptake from soil by a growing plant root system. Agron J 68:961–964. https://doi.org/10.2134/agronj1976.00021962006800060030x

Colwell JD (1967) The calibration of soil tests. J Aust Inst Agric Sci 33:321–323

Cong R, Li H, Zhang Z, Ren T, Li X, Lu J (2016) Evaluate regional potassium strategy of winter oilseed rape under intensive cropping systems: large-scale field experiment analysis. Field Crops Res 193:34–42. https://doi.org/10.1016/j.fcr.2016.03.004

Cooper H, Patall EA (2000) The relative benefits of meta-analysis conducted with individual participant data versus aggregated data. Pyschol Methods 14(2):165–176. https://doi.org/10.1037/a0015565

Cope JT Jr, Rouse RD (1973) Interpretation of soil test results. In: Walsh LM, Beaton JB (eds) Soil testing and plant analysis. Soil Science Society of America, Madison, pp 35–54

Cox AE, Joern BC, Brouder SM, Gao D (1999) Plant-available potassium assessment with a modified sodium tetraphenylboron method. Soil Sci Soc Am J 63(4):902–911. https://doi.org/10.2136/sssaj1999.634902x

Cucherat M, Boissel JP, Leizorovicz A (1997) La méta-analyse des essais thérapeutiques. Masson, Paris

da Silva Santos RS, Miranda JH, van Genuchten MT, Cooke RC (2015) HYDRUS-1D simulations of potassium transport in a saline tropical soil. In: 2015 ASABE annual international meeting. American Society of Agricultural and Biological Engineers, p 1. https://doi.org/10.13031/aim.20152189276

Dahnke WC, Olson RA (1990) Soil test correlation, calibration and recommendation. In: Westerman RL (ed) Soil testing and plant analysis, 3rd ed, soil science Society of America Book Ser 3. Soil Science Society of America, Madison, pp 45–71

de Barros I, Williams JR, Gaiser T (2004) Modeling soil nutrient limitations to crop production in semiarid NE of Brazil with a modified EPIC version I. Changes in the source code of the model. Ecol Model 178:441–456. https://doi.org/10.1016/j.ecolmodel.2004.04.015

Debachere MC (1995) Problems in obtaining grey literature. IFLA J 21(2):94–98. https://doi.org/10.1177/034003529502100205

Dobermann A, Cruz PS, Cassman KG (1996) Fertilizer inputs, nutrient balance, and soil nutrient-supplying power in intensive, irrigated rice systems. I. Potassium uptake and K balance. Nutr Cycling Agroecosyst 46(1):1–10. https://doi.org/10.1007/BF00210219

Dobermann A, Witt C, Abdulrachman S, Gines HC, Nagarajan R, Son TT, Tan PS, Wang GH, Chien NV, Thoa VTK, Phung CV (2003) Estimating indigenous nutrient supplies for site-specific nutrient management in irrigated rice. Agron J 95(4):924–935. https://doi.org/10.2134/agronj2003.9240

Doll EC, Lucas RE (1973) Testing soils for potassium, calcium and magnesium. In: Walsh LM, Beaton JB (eds) Soil testing and plant analysis. Soil Science Society of America, Madison, pp 133–151

Eagle AJ, Christianson LE, Cook RL, Harmel RD, Miguez FE, Qian SS, Ruiz Diaz DA (2017) Meta-analysis constrained by data: recommendations to improve relevance of nutrient management research. Agron J 109:2441–2449. https://doi.org/10.2134/agronj2017.04.0215

Eaton JS, Likens GE, Bormann FH (1973) Throughfall and stemflow chemistry in a northern hardwood forest. J Ecol 61(2):495–508. https://doi.org/10.2307/2259041

Evans School Policy Analysis and Research Group (EPAR) (2017) Private, public, and philanthropic funding for global agricultural and health research and development. https://evans.uw.edu/policy-impact/epar/blog/private-public-and-philanthropic-funding-global-agricultural-and-health. Accessed 7 Jun 2020

Fanelli D (2012) Negative results are disappearing from most disciplines and countries. Scientometrics 90:891–904. https://doi.org/10.1007/s11192-011-0494-7

Fernández FG, Brouder SM, Beyrouty CA, Volenec JJ, Hoyum R (2008) Assessment of plant-available potassium for no-till, rainfed soybean. Soil Sci Soc Am J 72(4):1085–1095. https://doi.org/10.2136/sssaj2007.0345

Fernández FG, Brouder SM, Volenec JJ, Beyrouty CA, Hoyum R (2009) Root and shoot growth, seed composition, and yield components of no-till rainfed soybean under variable potassium. Plant Soil 322(1–2):125–138. https://doi.org/10.1007/s11104-009-9900-9

Fulford AM, Culman SW (2018) Over-fertilization does not build soil test phosphorus and potassium in Ohio. Agron J 110(1):56–65. https://doi.org/10.2134/agronj2016.12.0701

Gabella C, Durinx C, Appel R (2017) Funding knowledgebases: towards a sustainable funding model for the UniProt use case. Version 2. F1000Res. 6:ELIXIR-2051. https://doi.org/10.12688/f1000research.12989.2

Gartley KL, Sims JT, Olsen CT, Chu P (2002) Comparison of soil test extractants used in mid-Atlantic United States. Commun Soil Sci Plan 33(5–6):873–895. https://doi.org/10.1081/CSS-120003072

Gaspar AP, Laboski CAM, Naeve SL, Conley SP (2017) Phosphorus and potassium uptake, partitioning, and removal across a wide range of soybean seed yield levels. Crop Sci 57 (4):2193–2204. https://doi.org/10.2135/cropsci2016.05.0378

Global Agricultural Research Data Innovation & Acceleration Network (GARDIAN) (n.d.) CGIAR platform for big data in agriculture. https://bigdata.cgiar.org/resources/gardian/. Accessed 29 May 2020

Goodman SN, Fanelli D, Ioannidis JP (2016) What does research reproducibility mean? Science Transl Med 8(341):341ps12-341ps12. https://doi.org/10.1126/scitranslmed.aaf5027

Greenwood DJ, Karpinets TV (1997) Dynamic model for the effects of K fertilizer on crop growth, K-uptake, and soil-K in arable cropping. 2. Field tests of the model. Soil Use Manag 13:184–189. https://doi.org/10.1111/j.1475-2743.1997.tb00583.x

Greenwood DJ, Cleaver TJ, Turner MK, Hunt J, Niendorf KB, Loquens SMH (1980) Comparison of the effects of potassium fertilizer on the yield, potassium content and quality of 22 different vegetable and agricultural crops. J Agric Sci Camb 95(2):441–456. https://doi.org/10.1017/S0021859600039496

Guiking FCT, Braakhekke WG, Dohme PAE (1995) Quantitative evaluation of soil fertility and the response to fertilizers. Wageningen Agricultural University, Wageningen

Hanway JJ (1973) Experimental methods for correlating and calibrating soil tests. In: Walsh LM, Beaton JD (eds) Soil testing and plant analysis, Rev. edn. Soil Science Society of America, Madison, pp 55–66

Havlin JH, Tisdale SL, Nelson WL, Beaton JD (2014) Soil fertility and fertilizers: an introduction to nutrient management, 8th edn. Pearson, New York

Heinen M (2001) FUSSIM2: brief description of the simulation model and application to fertigation scenarios. Agronomie 21:285–296. https://doi.org/10.1051/agro:2001124

Hocking PJ, Meyer CP (1991) Effects of CO_2 enrichment and nitrogen stress on growth, and partitioning of dry matter and nitrogen in wheat and maize. Funct Plant Biol 18(4):339–356. https://doi.org/10.1071/PP9910339

Holmqvist J, Ogaard AF, Oborn I, Edwards AC, Mattsson L, Sverdrup H (2003) Application of the PROFILE model to estimate potassium release from mineral weathering in Northern European agricultural soils. Eur J Agron 20:149–163. https://doi.org/10.1016/S1161-0301(03)00064-9

Horton ML (1959) Influence of soil type on potassium fixation. Dissertation. Purdue University

Ingram J, Gaskell P (2019) Searching for meaning: co-constructing ontologies with stakeholders for smarter search engines in agriculture. NJAS-Wageningen J Life Sci 90:100300. https://doi.org/10.1016/j.njas.2019.04.006

Jadad AR, Cook DJ, Jones A, Klassen TP, Tugwell P, Moher M, Moher D (1998) Methodology and reports of systematic reviews and meta-analyses: a comparison of Cochrane reviews with articles published in paper-based journals. J Am Med Assoc 280(3):278–280. https://doi.org/10.1001/jama.280.3.278

Janssen BH, Guiking FCT, Van der Eijk D, Smaling EMA, Wolf J, Van Reuler H (1990) A system for quantitative evaluation of the fertility of tropical soils (QUEFTS). Geoderma 46:299–318. https://doi.org/10.1016/0016-7061(90)90021-Z

Jiang W, Liu X, Wang X, Yin Y (2019) Characteristics of yield and harvest index, and evaluation of balanced nutrient uptake of soybean in northeast China. Agronomy 9:310–319. https://doi.org/10.3390/agronomy9060310

Jones JW, Hoogenboom G, Porter CH, Boote KJ, Batchelor WD, Hunt LA, Wilkens PW, Singh U, Gijsman AJ, Ritchie JT (2003) The DSSAT cropping system model. Eur J Agron 18:235–265. https://doi.org/10.1016/S1161-0301(02)00107-7

Joseph S, Sedimo K, Kaniwa F, Hlomani H, Letsholo K (2016) Natural language processing: a review. Nat Lang Proc Rev 6:207–210

Kaiser D, Fernandez F, Coulter J (2018) Fertilizing corn in Minnesota. https://extension.umn.edu/crop-specific-needs/fertilizing-corn-minnesota. Accessed 20 May 2020

Kelling KA, Bundy LG, Combs SM, Peters JB (1998) Soil test recommendations for field, vegetable and fruit crops. University of Wisconsin Extension Guide A2809, Madison

Kim Y, Stanton JM (2016) Institutional and individual factors affecting scientists' data-sharing behaviors: a multilevel analysis. J Assoc Inform Sci Technol 67:776–799. https://doi.org/10.1002/asi.23424

Knudsen D, Peterson GA, Pratt PF (1982) Lithium, sodium, and potassium. In: Page AL, Miller RH, Keeney DR (eds) Methods of soil analysis. Part 2. Chemical and microbiological properties, 2nd edn. American Society of Agronomy, Soil Science Society of America, Madison, pp 225–246

Kumar P, Byju G, Singh BP, Minhas JS, Dua VK (2016) Application of QUEFTS model for site-specific nutrient management of NPK in sweet potato (*Ipomoea batatas* L. Lam). Commun Soil Sci Plan 47(13–14):1599–1611. https://doi.org/10.1080/00103624.2016.1194989

Kumar P, Dua VK, Sharma J, Byju G, Minhas JS, Chakrabarti SK (2018) Site-specific nutrient requirements of NPK for potato (*Solanum tuberosum* L.) in western Indo-gangetic plains of India based on QUEFTS. J Plant Nutr 41(15):1988–2000. https://doi.org/10.1080/01904167.2018.1484135

Kupferschmidt K (2018) A recipe for rigor. Science 361(6408):1192–1193. https://doi.org/10.1126/science.361.6408.1192

Liakos KG, Busato P, Moshou D, Pearson S, Bochtis D (2018) Machine learning in agriculture: a review. Sensors 18(8):2674. https://doi.org/10.3390/s18082674

Lin W, Kelly JM (2010) Nutrient uptake estimates for woody species as described by the NST 3.0, SSAND, and PCATS mechanistic nutrient uptake models. Plant Soil 335:199–212. https://doi.org/10.1007/s11104-010-0407-1

Lindsay WL, Frazier AW, Stephenson HF (1962) Identification of reaction products from phosphate fertilizers in soils. Soil Sci Soc Am J 26:446–452. doi:0.2136/sssaj1962.03615995002600050013x

Lumbanraja J, Evangelou VP (1992) Potassium quantity-intensity relationships in the presence and absence of NH_4 for three Kentucky soils. Soil Sci 154(5):366–376

Mai TH, Schnepf A, Vereecken H, Vanderborght J (2019) Continuum multiscale model of root water and nutrient uptake from soil with explicit consideration of the 3D root architecture and the rhizosphere gradients. Plant Soil 439:273–292. https://doi.org/10.1007/s11104-018-3890-4

Malavolta E (1985) Potassium status of tropical and subtropical region soils. In: Munson RD (ed) Potassium in agriculture. American Society of Agronomy, Madison, pp 163–200. https://doi.org/10.2134/1985.potassium.c8

Mallarino AP, Sawyer JE, Barnhart SK (2013) A general guide for crop nutrient and limestone recommendations in Iowa. Iowa State University, Ames. https://store.extension.iastate.edu/Product/pm1688-pdf. Accessed 20 May 2020

Marschner P (2012) Mineral nutrition of higher plants, 3rd edn. Academic, London

Martin HW, Sparks DL (1983) Kinetics of nonexchangeable potassium release from two coastal plain soils. Soil Sci Soc Am J 47:883–887. https://doi.org/10.2136/sssaj1983.03615995004700050008x

Mayaki WC, Teare ID, Stone LR (1976) Top and root growth of irrigated and nonirrigated soybeans. Crop Sci 16(1):92–94. https://doi.org/10.2135/cropsci1976.0011183X001600010023x

McLean EO, Watson ME (1985) Soil measurements of plant-available potassium. In: Munson RD (ed) Potassium in agriculture. American Society of Agronomy, Madison, pp 131–162. https://doi.org/10.2134/1985.potassium.c10

Melsted SW, Peck TR (1973) The principles of soil testing. In: Walsh LM, Beaton JB (eds) Soil testing and plant analysis. Soil Science Society of America, Madison, pp 13–21

Morris TF, Murrell TS, Beegle DB, Camberato JJ, Ferguson RB, Grove J, Ketterings Q, Kyveryga PM, Laboski CA, McGrath JM, Meisinger JJ, Melkonian J, Moebius-Clune BN, Nafziger ED, Osmond D, Sawyer JE, Scharf PC, Smith W, Spargo JT, van Es HM, Yang H (2018) Strengths and limitations of nitrogen rate recommendations for corn and opportunities for improvement. Agron J 110(1):1–37. https://doi.org/10.2134/agronj2017.02.0112

Mulla DJ, Schepers JS (1997) Key processes and properties for site-specific soil and crop management. In: Pierce FJ, Sadler EJ (eds) The state of site specific management for agriculture. ASA/CSSA/SSSA, Madison, pp 1–18. https://doi.org/10.2134/1997.stateofsitespecific.c1

Mulrow CD, Cook DJ, Davidoff F (1997) Systematic reviews: critical links in the great chain of evidence. Ann Intern Med 126(5):389–391. https://doi.org/10.7326/0003-4819-126-5-199703010-00008

Munson RD, Nelson WL (1963) Movement of applied potassium in soils. J Agric Food Chem 11(3):193–201. https://doi.org/10.1021/jf60127a015

Murashkina MA, Southard RJ, Pettygrove GS (2007) Potassium fixation in San Joaquin Valley soils derived from granitic and nongranitic alluvium. Soil Sci Soc Am J 71(1):125–132. https://doi.org/10.2136/sssaj2006.0060

Nathan M, Gelderman R (2015) Recommended chemical soil test procedures for the North Central region. North Central Regional Research Publication No 221 Rev. https://agcrops.osu.edu/sites/agcrops/files/imce/fertility/Recommended%20Chemical%20Soil%20Test%20Procedures.pdf. Accessed 30 May 2020

Navarrete-Ganchozo RJ (2014) Quantification of plant-available potassium (K) in a corn-soybean rotation: a long-term evaluation of K rates and crop K removal effects. Dissertation. Purdue University. https://docs.lib.purdue.edu/dissertations/. Accessed 9 May 2020

Ning P, Li S, Li X, Li C (2014) New maize hybrids had larger and deeper post-silking root than old ones. Field Crops Res 166:66–71. https://doi.org/10.1016/j.fcr.2014.06.009

Nosek BA, Lakens D (2014) Registered reports: a method to increase the credibility of published results. Soc Psychol 45:137–141. https://doi.org/10.1027/1864-9335/a000192

Nye PH, Spiers JA (1964) Simultaneous diffusion and mass flow to plant roots. In: 8th International. Congress of Soil Science (Bucharest) August 1964, pp 535–542

Nyombi K, van Asten PJA, Corbeels M, Taulya G, Leffelaar PA, Giller KE (2010) Mineral fertilizer response and nutrient use efficiencies of East African highland banana (*Musa* spp., AAA-EAHB, cv. Kisansa). Field Crops Res 117:28–50. https://doi.org/10.1016/j.fcr.2010.01.011

Olk DC, Cassman KG (1995) Reduction of potassium fixation by two humic acid fractions in vermiculitic soils. Soil Sci Soc Am J 59(5):1250–1258. https://doi.org/10.2136/sssaj1995.03615995005900050007x

Olk DC, Cassman KG, Carlson RM (1995) Kinetics of potassium fixation in vermiculitic soils under different moisture regimes. Soil Sci Soc Am J 59(2):423–429. https://doi.org/10.2136/sssaj1995.03615995005900020022x

Olson RA, Dreier AF, Sorensen RC (1958) The significance of subsoil and soil series in Nebraska soil testing. Agron J 50(4):185–188. https://doi.org/10.2134/agronj1958.00021962005000040005x

Olson RA, Anderson FN, Frank KD, Grabouski PH, Rehm GW, Shapiro CA (1987) Soil test interpretations: sufficiency vs. build-up and maintenance. In: Brown JR (ed) Soil testing: sampling, correlation, calibration, and interpretation. SSSA Sp Pub No 21. Soil Science Society of America, Madison, pp 41–52. https://doi.org/10.2136/sssaspecpub21.c5

Oltmans RR, Mallarino AP (2015) Potassium uptake by corn and soybean, recycling to soil, and impact on soil test potassium. Soil Sci Soc Am J 79(1):314–327. https://doi.org/10.2136/sssaj2014.07.0272

Page AL, Ganje TJ (1964) The effect of pH on potassium fixed by an irreversible adsorption process. Soil Sci Soc Am J 28(2):199–202. https://doi.org/10.2136/sssaj1964.03615995002800020021x

Pampolino MF, Witt C, Pasquin JM, Johnston A, Fischer MJ (2012) Development approach and evaluation of the nutrient expert software for nutrient management in cereal crops. Comput Electron Agric 88:103–110. https://doi.org/10.1016/j.compag.2012.07.007

Philibert A, Loyce C, Makowski D (2012) Assessment of the quality of meta-analysis in agronomy. Agric Ecosyst Environ 148:72–82. https://doi.org/10.1016/j.agee.2011.12.003

Purdue University Research Repository (PURR) (n.d.) Research data management for Purdue. https://purr.purdue.edu. Accessed 29 May 2020

Raju J, Byju G (2019) Quantitative determination of NPK uptake requirements of taro (*Colocasia esculenta* (L.) Schott). J Plant Nutr 42(3):203–217. https://doi.org/10.1080/01904167.2018.1554070

Rhoades JD (1982) Soluble salts. In: Page AL et al (ed) Methods of soil analysis. Part 2. Chemical and microbiological properties, 2nd edn. Amer Soc Agron. Soil Science Society of America, Madison, pp 167–178

Rich CI, Black WR (1964) Potassium exchange as affected by cation size, pH, and mineral structure. Soil Sci 97(6):384–390

Rich CI, Obenshain SS (1955) Chemical and clay mineral properties of a red-yellow podzolic soil derived from muscovite schist. Soil Sci Soc Am J 19(3):334–339. https://doi.org/10.2136/sssaj1955.03615995001900030021x

Rosenzweig C, Jones JW, Hatfield JL, Ruane AC, Boote KJ, Thorburn P, Antle JM, Nelson GC, Porter C, Janssen S, Asseng S, Basso B, Ewert F, Wallach D, Baigorria G, Winter JM (2013) The agricultural model intercomparison and improvement project (AgMIP): protocols and pilot studies. Agric For Meteorol 170:166–182. https://doi.org/10.1016/j.agrformet.2012.09.011

Rouse RD (1967) Organizing data for soil test interpretations. In: Hardy GW (ed) Soil testing and plant analysis part I: soil testing. SSSA Sp Pub No. 2. Soil Science Society of America, Madison, pp 115–123. https://doi.org/10.2136/sssaspecpub2

Sackett DL, Rosenberg WM (1995) On the need for evidence-based medicine. J Public Health 17(3):330–334. https://doi.org/10.1093/oxfordjournals.pubmed.a043127

Satpute ST, Singh M (2017) Potassium and sulfur dynamics under surface drip fertigated onion crop. J Soil Salinity Water Qual 9:226–236

Scanlan CA, Huth NI, Bell RW (2015a) Simulating wheat growth response to potassium availability under field conditions with sandy soils. I. Model development. Field Crops Res 178:109–124. https://doi.org/10.1016/j.fcr.2015.03.022

Scanlan CA, Bell RW, Brennan RF (2015b) Simulating wheat growth response to potassium availability under field conditions in sandy soils. II. Effect of subsurface potassium on grain yield response to potassium fertilizer. Field Crops Res 178:125–134. https://doi.org/10.1016/j.fcr.2015.03.019

Shrader WD, Fuller WA, Cady FB (1966) Estimation of a common nitrogen response function for corn (*Zea mays*) in different crop rotations. Agron J 58(4):397–401. https://doi.org/10.2134/agronj1966.00021962005800040010x

Smaling EMA, Janssen BH (1993) Calibration of Quefts, a model predicting nutrient uptake and yields from chemical soil fertility indexes. Geoderma 59:21–44. https://doi.org/10.1016/0016-7061(93)90060-X

Soil Science Glossary Terms Committee (2008) Glossary of soil science terms. Soil Science Society of America, Madison

Stanford G (1973) Rationale for optimum nitrogen fertilization in corn production. J Environ Qual 2 (2):159–166. https://doi.org/10.2134/jeq1973.00472425000200020001x

Taiz L, Zeigler E, Moller IM, Murphy A (2018) Fundamentals of plant physiology, 1st edn. Oxford University Press, New York

Tisdale SL (1967) Problems and opportunities in soil testing. In: Hardy GW et al (eds) Soil testing and plant analysis: Part I Soil testing, vol 2. Soil Science Society of America, Madison, pp 1–11

U.S. Department of Agriculture–Economic Research Service (USDA–ERS) (2018) Agricultural research funding in the public and private sectors. https://www.ers.usda.gov/data-products/agricultural-research-funding-in-the-public-and-private-sectors. Accessed 29 May 2020

U.S. Department of Agriculture–National Agricultural Library (USDA–NAL) (n.d.) Ag data commons, https://data.nal.usda.gov. Accessed 29 May 2020

Van Tuyl S, Whitmire AL (2016) Water, water, everywhere: defining and assessing data sharing in academia. PLoS One 11(2):e0147942. https://doi.org/10.1371/journal.pone.0147942

Van Vuuren MM, Robinson D, Fitter AH, Chasalow SD, Williamson L, Raven JA (1997) Effects of elevated atmospheric CO_2 and soil water availability on root biomass, root length, and N, P and K uptake by wheat. New Phytol 135(3):455–465. https://doi.org/10.1046/j.1469-8137.1997.00682.x

Vitosh, ML, Johnson JW, Mengel DB (1995) Tri-state fertilizer recommendations for corn, soybeans, wheat and alfalfa. https://msu.edu/~warncke/E-2567%20Tri-State%20Fertilizer%20Recs.pdf. Accessed 9 May 2020

Wang CY, Boithias L, Ning ZG, Han YP, Sauvage S, Sanchez-Perez JM, Kuramochi K, Hatano R (2017) Comparison of Langmuir and Freundlich adsorption equations within the SWAT-K model for assessing potassium environmental losses at basin scale. Agric Water Manag 180:205–211. https://doi.org/10.1016/j.agwat.2016.08.001

Welch CD, Wiese RA (1973) Opportunities to improve soil testing programs. In: Walsh LM, Beaton JB (eds) Soil testing and plant analysis. Soil Science Society of America, Madison, pp 1–11

Welikhe PK (2020) Evaluating the effects of legacy phosphorus on dissolved reactive phosphorus losses in tile-drained systems. Dissertation, Purdue University. https://doi.org/10.25394/PGS.12253262.v1

Wheeler DM, Ledgard SF, de Klein CAM, Monaghan RM, Carey PL, McDowell RD, Johns KL (2003) OVERSEER® nutrient budgets – moving towards on-farm resource accounting. Proc New Zealand Grassl Assoc 65:191–194

White JA, Hunt LA, Boote KJ, Jones JW, Koo J, Kim S, Porter CH, Wilkens PW, Hoogenboom G (2013) Integrated description of agricultural field experiments and production: the ICASA version 2.0 data standards. Comp Electron Agric 96:1–12. https://doi.org/10.1016/j.compag.2013.04.003

Wilkinson MD, Dumontier M, Aalbersberg IJ, Appleton G, Axton M, Baak A, Blomberg N, Boiten JW, da Silva Santos LB, Bourne PE, Bouwman J (2016) The FAIR guiding principles for scientific data management and stewardship. Sci Data 3. https://doi.org/10.1038/sdata.2016.18 [Addendum: The FAIR guiding principles for scientific data management and stewardship. Sci Data 6(1):1–2. https://doi.org/10.1038/sdata.2016.18]

Witt C, Dobermann A, Abdulrachman S, Gines HC, Guanghuo W, Nagarajan R, Satawatananont S, Son TT, Tan PS, Tiem LV, Simbahan GC, Olk DC (1999) Internal nutrient efficiencies of irrigated lowland rice in tropical and subtropical Asia. Field Crops Res 63:113–138. https://doi.org/10.1016/S0378-4290(99)00031-3

Wollenhaupt NC, Mulla DJ Gotway Crawford CA (1997) Soil sampling and interpolation techniques for mapping spatial variability of soil properties. Pp. 19 to 54. In: F.J. Pierce and E.J. Sadler (Eds). The site-specific management of agricultural systems. ASA-CSSA-SSSA, Madison. doi:https://doi.org/10.2134/1997.stateofsitespecific.c2

Woodson P, Volenec JJ, Brouder SM (2013) Field-scale potassium and phosphorus fluxes in the bioenergy crop switchgrass: theoretical energy yields and management implications. J Plant Nutr Soil Sci 176(3):387–399. https://doi.org/10.1002/jpln.201200294

Wu L, Cui Z, Chen X, Zhao R, Si D, Sun Y, Yue S (2014) High-yield maize production in relation to potassium uptake requirements in China. Agron J 106(4):1153–1158. https://doi.org/10.2134/agronj13.0538

Wyn Jones RG, Brady CJ, Speirs J (1979) Ionic and osmotic relations in plant cells. In: Laidman DL, Wyn Jones RG (eds) Recent advances in the biochemistry of cereals. Academic, London, pp 63–103

Xie M, Wang Z, Xu X, Zheng X, Liu H, Shi P (2020) Quantitative estimation of the nutrient uptake requirements of peanut. Agron 10:119–130. https://doi.org/10.3390/agronomy10010119

Yang C, Na J, Li G, Li Y, Zhong J (2018) Neural network for complex systems: theory and applications. Complexity 2018:1–2. https://doi.org/10.1155/2018/3141805

Young RA, Onstad CA, Bosch DD, Anderson WP (1989) AGNPS: a nonpoint-source pollution model for evaluating agricultural watersheds. J Soil Water Cons 44:168–173

Zeng Q, Brown PH (2000) Soil potassium mobility and uptake by corn under differential soil moisture regimes. Plant Soil 221(2):121–134. https://doi.org/10.1023/A:1004738414847

Zhang KF, Greenwood DJ, White PJ, Burns IG (2007) A dynamic model for the combined effects of N, P and K fertilizers on yield and mineral composition; description and experimental test. Plant Soil 298:81–98. https://doi.org/10.1007/s11104-007-9342-1

Zhang Y, Hou P, Gao Q, Chen X, Zhang F, Cui Z (2012) On-farm estimation of nutrient requirements for spring corn in North China. Agron J 104(5):1436–1442. https://doi.org/10.2134/agronj2012.0125

Zhang J, He P, Ding W, Xu X, Ullah S, Abbas T, Ai C, Li M, Cui R, Jin C, Zhou W (2019) Estimating nutrient uptake requirements for radish in China based on QUEFTS model. Sci Rep 9:11663. https://doi.org/10.1038/s41598-019-48149-6

2

Inorganic and Organic Fertilizer Inputs: An Overview

Robert L. Mikkelsen and Terry L. Roberts

Abstract In the potassium (K) cycle, inputs encompass all K sources that move into a given volume of soil. These inputs may include atmospheric deposition, irrigation water, runoff, erosion, as well as seeds, cuttings, and transplants. Accounting for all inputs is seldom routinely done on the farm. Many K inputs have variable concentrations, making estimations difficult. Estimates for added K are provided in some planning documents and can be used where testing of on-farm inputs is not feasible, although testing is preferred. Standard commercial fertilizers have known concentrations of K and are concentrated enough to be economical to transport long distances. The global reserves for their production have an estimated lifetime of thousands of years. This chapter emphasizes considerations for using various commercial fertilizer sources.

2.1 Overview of Potassium Inputs

Potassium (K) inputs as a group is *the total quantity of K, originating outside a given volume of soil, that moves into that volume* (Fig. 2.1). Inputs include K in atmospheric deposition; irrigation water; K transported to the soil volume via runoff and erosion from other areas; K in seeds, cuttings, transplants, or residues; organic fertilizer applications; and commercial fertilizer additions. This pool is the sum of all these inputs. Inputs may occur directly to the soil or, as is the case with foliar fertilizer applications, directly to the plant.

For making improvements in K recommendations, this chapter emphasizes inorganic and organic fertilizer inputs. Of all the inputs, these are the ones that farmers have the greatest control over. This chapter defines organic inputs as K sources

R. L. Mikkelsen (✉)
African Plant Nutrition Institute and University Mohammed VI Polytechnic, Ben Guerir, Morocco
e-mail: r.mikkelsen@apni.net

T. L. Roberts
Formerly International Plant Nutrition Institute (IPNI), Peachtree Corners, GA, USA

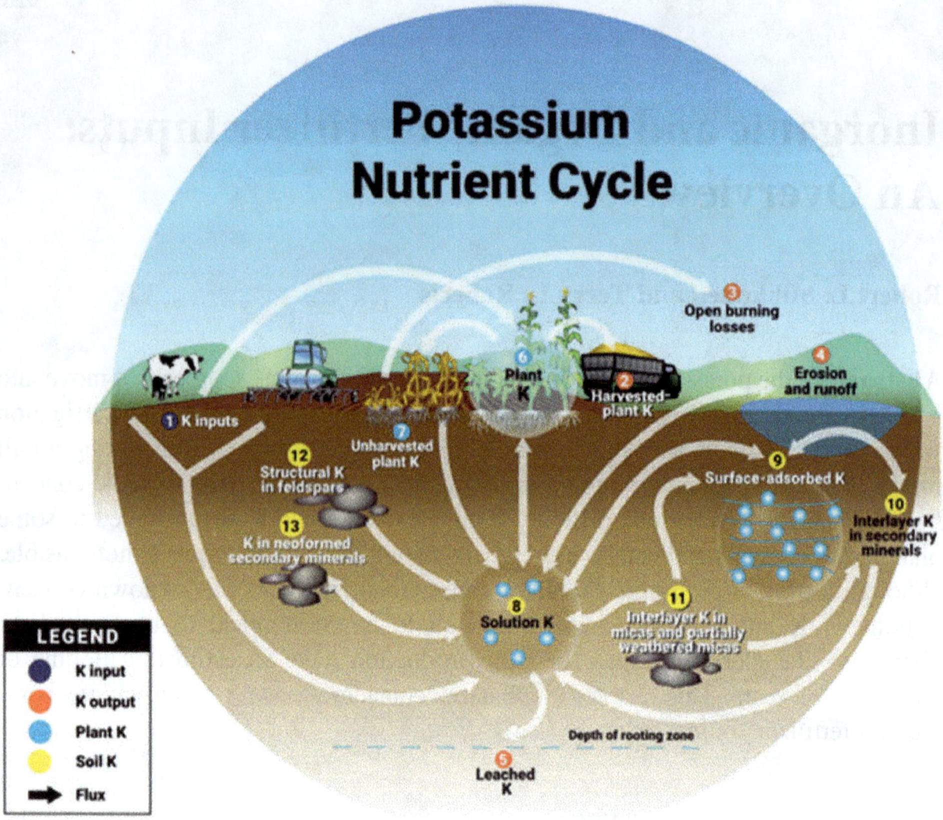

Fig. 2.1 The K cycle involves inputs, outputs, and transformations that occur in agricultural soils. Potassium biogeochemistry includes complex chemical, physical, geological, and biological processes within natural or managed ecosystems. This level of complexity has too often been overlooked for K. This figure outlines the major K pools and fluxes occurring within the rootzone during an annual cropping season. It is not comprehensive of all possible inputs, outputs, and transformations that can occur

derived from animal wastes (manure and biosolids) and plant residues. This definition is used to clarify that these inputs do not refer to K sources approved for the production of agricultural products labeled as "organic." Although the emphasis in this chapter is primarily on inorganic sources, we briefly mention other inputs.

2.2 Atmospheric Deposition

Atmospheric deposition is *the quantity of K transferred from the atmosphere to a given area of land* (adapted from National Agricultural Library 2019a). Atmospheric deposition is the sum of wet deposition and dry deposition. **Wet deposition** is *the quantity of K transferred from the atmosphere to a given area of land by rain, fog, or snow* (adapted from National Agricultural Library 2019b). **Dry deposition** is *the* *quantity of K in atmospheric particles transferred to a given area of land* (adapted from American Meteorological Society 2012). Dry deposition is the K in the deposited particles, while wet deposition is the K dissolved in precipitation. For both sources, K deposited on both soil and foliage are considered.

It is well documented that atmospheric transport takes place at very large scales (e.g., from Africa to South America; Prospero et al. 2014). However only a few studies have mapped K deposition at these large scales. In a study conducted in Northern China across ten sites and 3 years, Pan and Wang (2015) measured both wet and dry deposition of K. The average atmospheric deposition, the sum of wet and dry, ranged from 11 to 25 kg K ha^{-1} $year^{-1}$ across sites. At all but two sites, average dry deposition was several times greater than wet deposition, indicating that substantially more K was added in particles than in precipitation. Fluxes varied by time of year. The largest fluxes of dry deposition occurred in the spring, which the authors attributed to long-range transport of dust from deserts and loess deposits by 16 dust storms over the 3-year study period. The largest fluxes of wet deposition occurred in summer, which was the rainy season. Deposition also varied spatially, with the greatest atmospheric deposition in industrial areas and the smallest in agricultural areas.

Deposition of K across the contiguous 48 states of the United States was evaluated by Mikhailova et al. (2019) based on data from the US Atmospheric Deposition Program. Total inputs, the sum of both wet and dry deposition, ranged from 0 to 2.5 kg ha^{-1} $year^{-1}$ and varied markedly across states.

In power plants, the K composition of particulates in ash depends on the source of fuel. Ruscio et al. (2016) compared three sources of biomass fuel (olive residue, maize residue, and torrefied pine sawdust) to three sources of coal (bituminous, sub-bituminous, and lignite). In both coarse (560–1000 nm) and fine (100–180 nm) submicron particles of ash, the K concentrations in maize and olive waste were over 30%, several times higher than in any other fuel source. Ashes from coal sources were all less than 7% K across both particle size ranges. In the future, power plants relying solely on biomass feedstocks or power plants that cofire coal and biomass will have greater quantities of K in ash than traditional coal-only power plants. The high K content of ash from crop residues (e.g., 6% K in rice (*Oryza sativa* L.) straw ash; Hung et al. 2020) suggests that in areas where residues are burned in fields as a management practice, some K will be lost from fields and redeposited elsewhere in the environment, including nearby fields.

2.3 Irrigation Water

The quantity of K input by irrigation water is determined by the K concentration in the water source, the quantity of water used, the quantity of sediment transported with the water, and the K content in the sediment. For example, Hoa et al. (2006) examined K inputs into flooded rice fields in the Mekong Delta of Vietnam and found, on average, K input by irrigation water was 14–18 kg K ha^{-1} $year^{-1}$.

They also found that large amounts of K were added by the sediment in the water. The total quantity of K added ranged from 320 to 1892 kg K ha^{-1} $year^{-1}$, but only 3–10 kg ha^{-1} $year^{-1}$ was plant available, based on commonly used interpretations of the soil test used in the study. Over time, a portion of the remaining K in the sediment was expected to become available for plant uptake, based on transformations in the soil and the mechanisms used by plants to access K in various soil pools. The considerations of this study extend more generally to K input by runoff and erosion.

The waterborne K input can be rather large in some irrigated fields. For example, a common water source used for irrigating cotton (*Gossypium hirsutum* L.) in California contains ~0.3 mmol K L^{-1}. During the growing season, adding 7.5 ML ha^{-1} of irrigation water supplies approximately 90 kg K ha^{-1} $year^{-1}$ to the soil.

2.4 Runoff and Erosion

The loss of K with soil materials leaving agricultural fields is covered extensively in Chap. 3. Soil materials are most commonly transported by wind, by flowing water, and through slow hillside creep. These transport processes are complex and are the subject of detailed investigations (Williams 2012). However, these erosional materials are subsequently deposited again in another part of the landscape.

Particle deposition may be transitory, or it may persist for geologic periods (such as chalk and apatite). The eroded materials are deposited off-site from their source, too often resulting in clogged drainage ways, silted reservoirs, and a destruction of aquatic habitat. The loss of eroded soil directly leads to soil degradation through nutrient depletion and removal of valuable soil organic matter.

How much of the eroded material is carried back to the sea or is stored in river terraces, flood plans, or reservoirs is site specific. Sediments stored in reservoirs show an accumulation in nitrogen (N), phosphorus (P), and K that corresponds to watershed fertilizer application rates (Junakova and Balintova 2012).

The distribution of nutrients in reservoir sediments is not uniform, as the K-rich materials associated with the fine particles tend to settle more slowly than the larger-sized eroded material. Sediments can supply significant amounts of bioavailable K when it is dredged and added back to agricultural land (e.g., Woodard 1999; Darmody and Ruiz 2017).

2.5 Seeds, Cuttings, Transplants, and Residues

Very small amounts of K are added to the field during planting operations. Seeds and seedlings contain K in quantities that are commonly overlooked while calculating nutrient budgets. The mineral content of agronomic seeds will vary somewhat [e.g., maize seed can contain ~3 mg K g^{-1}, while tomato seeds contain 5 mg K g^{-1} (Liptay and Arevalo 2000)]. As an example, a typical maize field planted with 60,000 seeds ha^{-1} will add approximately 25 g K in the process. This seed-borne K provides nutrition during the germination process and the initial growth of the embryo. Once the seedling begins growing, the K demand from the soil quickly increases.

When seedlings are used for transplanting, a small amount of K is brought to the field. Using tomatoes as an example, transplanting 2000–3000 plants ha^{-1} will add between 600 and 900 g K to the field (Liptay and Arevalo 2000). When vineyards and orchards are established, small amounts of K will also be moved to the site in the woody tissue of the planting material.

Crop residues that remain in the field do not contribute to the K balance of the field, but when residues are imported from other fields or farms, they can contribute significant amounts of additional K. These residue inputs are discussed later.

2.6 Organic Fertilizer

The use of approved nutrient sources for organic crop production is governed by a variety of oversight organizations. Unfortunately, each of these organizations maintains somewhat different standards and allows different materials to be used in their organic production systems because they individually interpret the intent of organic agricultural principles. As a result, a grower seeking advice on permissible K materials should first know where the agricultural produce will be sold in order to meet the requirements of that market.

In general, regulations for mined K sources specify that they must not be processed, purified, or altered from their original form. However, there is disagreement among different certifying bodies over what specific materials can be used. Unfortunately, some of these restrictions on certain nutrient materials do not have solid scientific justification, and their inclusion or exclusion on various lists should not be viewed as one material being more or less "safe" than another fertilizer material (Mikkelsen 2007). Certain wastes (e.g., ash materials) and unprocessed minerals (e.g., glauconite) may also be permitted for organic crop production in certain conditions.

In the mixed livestock/crop systems, the nutrition of the animals generally takes first priority, and the residual manure is returned to surrounding cropland. In these cases, K imported to the farm in feed and bedding frequently exceeds the output in milk and meat products, sometimes leading to an accumulation of K in the surrounding fields that receive manure. Large losses of K often occur on these farms during manure storage and composting. Because excreted K is mostly expelled as urine, if this fraction is not effectively recovered in confined animal operations, most of the K will not be returned to the field with the solid portion of the manure.

The nutrient value of K in animal manures is generally equivalent to soluble K fertilizers. Since K is not a structural component of plant or animal cells and remains soluble in animal manure and urine, there is no true "organic" K.

Table 2.1 Approximate dry matter and K_2O content of selected animal manures. (IPNI 2012)

Livestock type	Waste handling system	Dry matter %	K_2O content kg t^{-1}
Solid handling systems			
Swine	Without bedding	18	4
	With bedding	18	3.5
Beef cattle	Without bedding	15	5
	With bedding	50	13
Dairy cattle	Without bedding	18	5
	With bedding	21	5
Poultry	Without litter	45	5
	With litter	75	17
	Deep pit (compost)	76	22.5
Liquid handling systems			
Swine	Liquid pit	4	9.5
	Oxidation ditch	2.5	9.5
	Lagoon	1	0.2
Beef cattle	Liquid pit	11	17
	Oxidation ditch	3	14.5
Dairy cattle	Lagoon	1	2.5
	Liquid pit	8	14.5
Poultry	Lagoon	1	2.5
	Liquid pit	13	48

The chemical composition of manures should be determined through laboratory analysis in order to apply material at rates that avoid either excessive or insufficient application of K to the field. Solid manures frequently contain between 5 and 25 kg K_2O Mg^{-1}, while liquid pit manures typically contain 1–4 kg K_2O 1000 L^{-1} (Table 2.1). Lagoon liquids have an even lower K concentration. When repeated applications of animal manures are used as a primary source of N for plant nutrition (such as on organic farms), the accumulation of excessive concentrations of K and P in the soil is common (Mikkelsen 2000a; Arienzo et al. 2009). Potassium concentration can change from load to load and also throughout the year (O'Dell et al. 1995). Agitation of manure pits prior to loading can reduce variability because it creates a more uniform distribution of manure solids (Duo et al. 2001).

In addition to recoverable manure, there can be significant amounts of K returned directly to the soil via animal urination and defecation. For example, at the localized site of grazing dairy cattle urination, effective application rates can reach 1000 kg K ha^{-1} in this small patch (Williams et al. 1990).

Nutrient concentrations in manure are generally low enough that it is uneconomical to transport manure long distances. Therefore, manure use is primarily local, often restricted to a single farm or nearby farms. There are also many cases where manure has not been distributed across an entire farm but instead spread only on fields nearest where animals were kept (Mikkelsen 2000b). Often overlooked, animal manure has additional benefits beyond just its mineral nutrient content and can aid in building and remediating soil (Mikha et al. 2017).

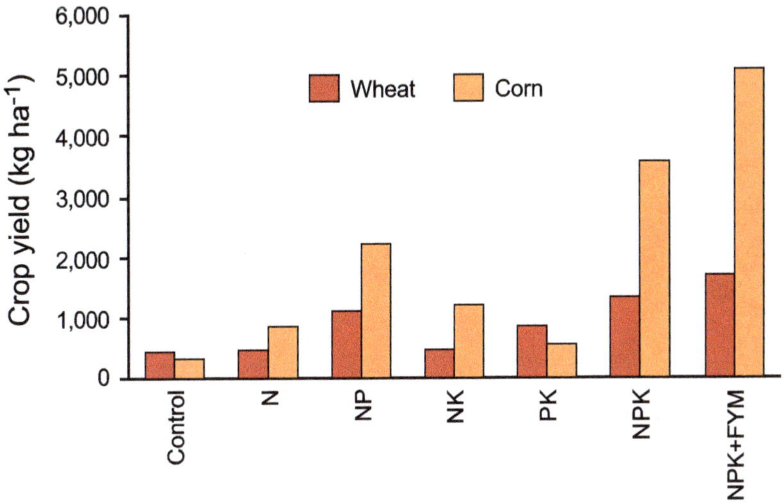

Fig. 2.2 Average wheat and corn yields for a 15-year study in Hunan, China, receiving combinations of N, P, K, and farmyard manure. (adapted from Zhang et al. 2009)

Evidence from long-term experiments show that combinations of inorganic and organic nutrients often achieve higher crop yields, improve soil quality and fertility, increase nutrient use efficiency, and lead to more sustainable nutrient management systems than either source alone (Miao et al. 2011). For example, Zhang et al. (2009) reported on a 15-year maize (*Zea mays* L.)—wheat (*Triticum aestivum* L.) rotational study in China where significant yield responses to N occurred for the first few years, but adding both N and P together increased yields fivefold, although yields decreased subsequently. Applying K along with N and P further increased yields in some years, but high yields were only sustained when farmyard manure was added with the N, P, and K (Fig. 2.2). The addition of farmyard manure prevented yields from declining over time and helped maintain the fertility and buffering capacity of the soil.

2.7 Commercial Fertilizer

Commercial fertilizer is the K input farmers can manage with the greatest accuracy and precision. In countries where the compositions of fertilizers are regulated, minimum concentrations of K are guaranteed by the fertilizer manufacturer and do not vary from the stated label. Inaccuracies in applying the correct amount arise from problems with distributing the products properly over the field, rather than from the composition of the products themselves. However, fertilizer composition integrity is not always assured in some parts of the world, and adulterated products find their way to the marketplace. Modern K fertilizers are manufactured with sufficiently high nutrient concentrations that they are economical to transport long distances, allowing them to be affordably shipped anywhere in the world where logistics permit.

Of all the sources enumerated above, regulated commercial fertilizers are the only inputs farmers do not need to test to know the K concentrations. The need to test other inputs prior to application means that an accurate accounting of all K inputs is rarely performed routinely on the farm. Where testing is feasible and economical, many farmers do test manure and irrigation water, especially if incentives or regulations are in place. Where testing is not practical, published estimates are used, but their inaccuracies for local conditions must be acknowledged.

2.7.1 Resources and Reserves

Potassium minerals and salts mined for use as fertilizers are generally referred to as potash, a term dating back to the nineteenth century when K used for fertilizer came from "potashes." Potash is a general term encompassing many different individual K fertilizers. Among these sources are potassium chloride (KCl), also called muriate of potash (MOP); potassium sulfate (K_2SO_4), also called sulfate of potash (SOP); potassium magnesium sulfate ($K_2SO_4 \cdot MgSO_4$), sometimes referred to as sulfate of potash magnesia (MgSOP or SOPM); potassium nitrate (KNO_3), also called nitrate of potash (NOP) or saltpeter; and mixed sodium-potassium nitrate ($NaNO_3 + KNO_3$), also called Chilean saltpeter. The K concentration of commercial fertilizers is reported either on an oxide basis (K_2O) or on an elemental basis (K). For a given concentration of elemental K, the reported oxide concentration is 1.2 times that of the reported elemental concentration: $K_2O = 1.2 \times K$.

Potassium is the seventh most abundant element in the Earth's crust (Fountain and Christensen 1989) but is never found in its elemental form in nature because it is highly reactive unless chemically bound to other elements. Feldspars, micas, and other silicate minerals contain significant amounts of K, but they are not currently commercial sources of potash.

Global potash resources most commonly occur as large, deeply buried deposits associated with marine evaporite sequences and, less commonly, with non-marine evaporites generally formed in arid climates (Sheldrick 1985; IFDC and UNIDO 1998). Sylvite, the mineral form of potassium chloride, is the most abundant mineral in commercial deposits (Table 2.2). Sylvinite, a physical mixture of sylvite and halite (NaCl), is the most common K-bearing ore. The second most common ore is carnallitite, a mixture of primarily carnallite ($KCl \cdot MgCl_2 \cdot 6H_2O$) and other salts. Other less common ores include kainite ($KCl \cdot MgSO_4 \cdot 3H_2O$), langbeinite ($K_2SO_4 \cdot 2MgSO_4$), polyhalite ($K_2SO_4 \cdot MgSO_4 \cdot 2CaSO_4 \cdot 2H_2O$), and hartsalz, a physical mixture of sylvite, halite, kieserite, and/or anhydrite. Potassium nitrate (niter) is a unique ore, with deposits occurring only in the Atacama Desert of northern Chile.

The world has abundant potash resources. There are approximately 980 deposits on five continents (Orris et al. 2014). The US Bureau of Mines and the US Geological Survey (1980) define a **resource** as *a concentration of naturally*

Table 2.2 Names, chemical formulas, and K concentrations of some common K-bearing minerals used in production of potash fertilizers

Mineral	Chemical name	Chemical formula	Weight % K_2O
Carnallite	Potassium magnesium chloride	$KCl{\cdot}MgCl_2{\cdot}6H_2O$	17
Carnallitite (ore)	Mixture of carnallite, halite, and others	$KCl{\cdot}MgCl_2{\cdot}6H_2O$, $NaCl$, $MgSO_4{\cdot}H_2O$, $CaSO_4$	Variable
Hartsalz (ore)	Mixture of sylvite, halite, and kieserite	KCl, $NaCl$, $MgSO_4{\cdot}H_2O$, $CaSO_4$	Variable
Kainite	Potassium magnesium sulfate chloride	$KCl{\cdot}MgSO_4{\cdot}3H_2O$	19
Langbeinite	Potassium magnesium sulfate	$K_2SO_4{\cdot}2MgSO_4$	23
Niter	Potassium nitrate	KNO_3	46
Polyhalite	Potassium magnesium calcium sulfate	$K_2SO_4{\cdot}MgSO_4{\cdot}2CaSO_4{\cdot}2H_2O$	16
Schoenite	Potassium magnesium sulfate	$K_2SO_4{\cdot}MgSO_4{\cdot}6H_2O$	23
Sylvinite (ore)	Mixture of sylvite and halite	KCl, $NaCl$	Variable
Sylvite	Potassium chloride	KCl	63

occurring solid, liquid, or gaseous material in or on the Earth's crust in such a form and amount that economic extraction of a commodity from the concentration is currently or potentially feasible. Feasibility depends on factors such as product prices, capital costs, and current and potential mining and processing technologies. The **reserve base** is *that part of an identified resource that meets specified minimum physical and chemical criteria related to current mining and production practices, including those for grade, quality, thickness, and depth. A* ***reserve*** *is that part of the reserve base which could be economically extracted or produced at the time of determination.* The quantity of reserves depends on factors such as expected product prices, capital costs, and current mining and processing technologies. Global potash reserves are estimated to be 250 billion metric tonnes (USGS 2020). That is sufficient potash for thousands of years, even if fertilizer production was to double.

The global K fertilizer industry is relatively mature and stable. However, there are many commercial endeavors underway to further develop geologic K resources around the world. In Africa, for example, the Danakil region of Ethiopia and Eritrea is being developed to be a major K producer. Resources in the Khemisset region of northern Morocco and the Sintoukola region in the west of the Republic of the Congo are under development as K fertilizer sources. Additionally, geologic accumulations of K in Western Australia and the Western United States are being developed as new K sources. The deep mine in North Yorkshire, UK, is also expected to bring new supplies of K to the commercial fertilizer market.

The percent minable K_2O ore reserves and resources vary widely in potash deposits. Mining costs depend on the depth to the ore, thickness, and uniformity of the potash bed, strength, and uniformity of the overlying strata, flooding risks, and other factors. Even so, potash mining and refining is a much simpler process than is

Table 2.3 Estimated global K_2O reserves and 2019 mine production. All values are in thousands of metric tonnes of K_2O (kt K_2O). (USGS 2020)

Country	Reserves kt K_2O	% of total reserves %	Mine production (2019) kt K_2O	% of total production %
Canada	1,000,000	28	13,300	32
Belarus	750,000	21	7000	17
Russia	600,000	17	6800	17
China	350,000	10	5510	13
Germany	150,000	4	3000	7
Israel	Large	(~5)	2000	5
Jordan	Large	(~5)	1500	3
Chile	100,000	3	950	2
Spain	68,000	2	600	1
Brazil	24,000	1	200	1
World total (rounded)	>3,600,000		41,000	

required for the manufacture of nitrogen and phosphorus fertilizers (Rahm 2017). Unlike the Haber-Bosh process for ammonia synthesis or wet process phosphoric acid production, potash processing does not involve complicated chemical reactions; the K is most commonly separated from the other compounds in the ore by flotation and selective crystallization.

Buried potash deposits can range from a few hundred to more than 3000 m deep and are obtained by conventional shaft mining using continuous mining machines that cut into the face of the deposit. Flexible conveyors are attached to the mining machines that convey the ore to the shaft for transport to the surface for processing. Most underground mines utilize room and pillar mining techniques to maintain structural integrity. Solution mining is used for deeper deposits. Hot water is first pumped into the deposit through bore holes to dissolve the ore, and then brine is injected to selectively dissolve KCl, which is withdrawn through another nearby well. Solution mining is energy intensive but is better suited to extremely deep deposits and to some ores such as carnallitite. For K-rich surface brines, potash is obtained by solar evaporation in shallow ponds and selective separation of the salts. A floating dredge or other heavy equipment is used to harvest the pond minerals. Underground deposits account for about 75% of global output: 70% from shaft mining operations and 5% from solution mining (Rahm 2017).

Although there are approximately 980 deposits of potash distributed around the world, potash is mined in less than 20 countries (Table 2.3). Russia, Canada, Belarus, and China account for three-quarters of world production. Since 2000, global production has increased at a compound annual growth rate of about 3.1%: from 24.5 Mt. K_2O in 2000 to 41.0 Mt in 2019 (Fig. 2.3). Erratic production has followed volatile prices. Potash prices spiked to record highs in 2008 which decreased demand and production the following year during the global recession and financial crisis.

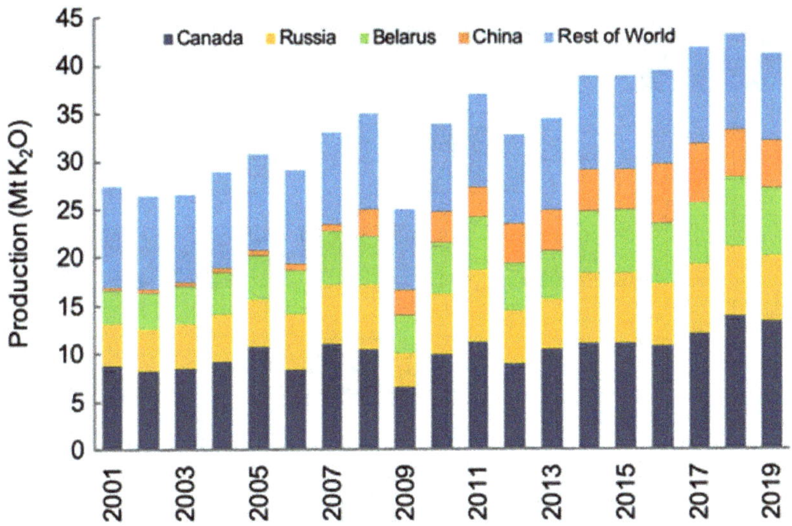

Fig. 2.3 Global potash mine production between 2000 and 2019. (Jasinksi 2020)

2.7.2 *Materials and Use*

Over 95% of global potash fertilizer production originates with MOP obtained from sylvinite, carnallitite, or hartsalz (Fig. 2.4). In 2015, approximately 74% of the KCl produced was used directly for plant nutrition. Of the remaining 26%, about 2% was reacted with nitric acid or nitrate salts to produce NOP, 5% was reacted with sulfuric acid to produce SOP, and 19% was incorporated into various N and P sources to produce N-P-K or P-K complex/compound fertilizers. The next largest quantity of fertilizer produced was primary SOP originating from kainite, hartsalz, or polyhalite, comprising about 4% of 2015 global fertilizer production. The remaining 1% of production was SOPM originating from langbeinite. Polyhalite is expected to become more common as new mines open. About 10% of global K production was used for industrial purposes or as an ingredient in animal feed. The remainder was used for crop nutrition.

The plant K nutritional value is identical for all K sources; however, their properties and solubility vary considerably (Table 2.4). All common commercial K fertilizers are sufficiently soluble to provide adequate K to plants growing in moist soil. Differences in water solubility become important when solid K sources are dissolved for use in foliar sprays or fluid fertilizers. When solid K fertilizers are used for these purposes, both the solubility and the time required for dissolution need to be considered when making a suitable liquid fertilizer. There are additional restrictions to consider when selecting a K source for organic crop production (Mikkelsen 2007).

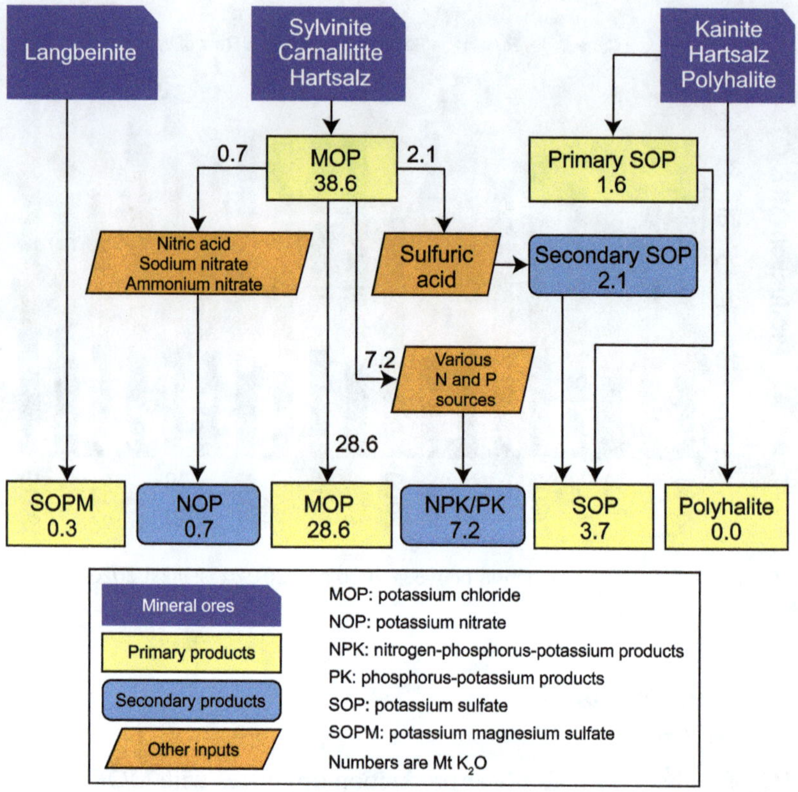

Fig. 2.4 The process for making final products of common K fertilizers, beginning with the mined ores; quantities are for 2015. (adapted from Rahm 2017)

Table 2.4 Selected properties of common K fertilizers

		K content	Solubility
Fertilizer name	Chemical formula	Weight % K_2O	g L^{-1} at 20° C
MOP	KCl	60–63	344
SOP	K_2SO_4	50	111
NOP	KNO_3	45	316
Potassium thiosulfate (KTS)	$K_2S_2O_3$	25	Fluid
Langbeinite (SOPM)	$K_2SO_4 \cdot 2MgSO_4$	21	240
Polyhalite	$K_2SO_4 \cdot MgSO_4 \cdot 2CaSO_4 \cdot 2H_2O$	13	12–18 (25 °C)
Potassium hydroxide	KOH	83	1120
Potassium phosphate	KH_2PO_4	34	226
Potassium chloride and kieserite	$KCl \cdot MgSO_4 \cdot H_2O$	40	

2.7.2.1 Potassium Chloride (MOP)

Potassium chloride is by far the most widely used K fertilizer for crop nutrition due to its relatively low cost and highest percentage of K_2O, 60–63% (Table 2.4). This high concentration means handling, storage, transport, and application costs are lower for KCl than for other sources. Most KCl fertilizer is produced by separating sylvinite ore into sylvite (KCl) and halite (NaCl), based on either their differing specific gravities or differing solubilities. The resulting purified sylvite-based fertilizer may have a reddish color arising from trace concentrations of iron oxide that are naturally occurring in the ore. Most of the MOP produced is used directly as a fertilizer; however, MOP can also be used to formulate other K fertilizer products (Fig. 2.4).

Potassium chloride is often spread onto the soil surface prior to tillage and planting or applied in a concentrated band near the seed or plant. Since all K fertilizers will increase the soluble salt concentration in the soil as they dissolve, banded KCl is typically placed to the side of the seed to avoid potential osmotic damage during germination. White-colored KCl is sometimes a more purified grade that can be solubilized as a component of fluid fertilizers and foliar sprays or for application through irrigation systems. Potassium chloride, as are most K fertilizers, is available in several particle sizes (from coarse to fine) to match the intended purpose.

2.7.2.2 Potassium Sulfate (SOP)

Potassium sulfate can be produced in three ways. Primary SOP originates from decomposition of ores containing sulfate (Fig. 2.4). As an example, water is added to kainite, producing schoenite as an intermediate product (for the chemical formulas, see Table 2.2). Potassium chloride is then added to the schoenite to produce K_2SO_4. A second method is to evaporate naturally occurring brines to crystallize K_2SO_4. Third, the Mannheim process reacts KCl with sulfuric acid, producing K_2SO_4, as indicated in Fig. 2.4. Potassium sulfate is available in either a crystalline form or as granules that are generally preferred for blending and uniform spreading. The K_2O content is approximately 50% (Table 2.4). In 2015, the primary production of K_2SO_4 accounted for about 9% of global K fertilizers.

Potassium sulfate provides both K and sulfur (S) in forms that are readily soluble and available to plants (SOPIB 2015). There may be certain soils and crops where K is needed but the addition of chloride (Cl^-) should be avoided. In these situations, K_2SO_4 makes a very useful K fertilizer source. Potassium sulfate is only one-third as soluble as KCl, so it is not as commonly used for application through irrigation water unless there is a need for additional S. Foliar sprays of K_2SO_4 are a convenient way to apply additional K and S to plants, supplementing the nutrients taken up from the soil by plant roots. Fine particles are used for making solutions for fertigation or foliar sprays since the small particles dissolve more rapidly. Due to the extra costs associated with producing K_2SO_4, the price is generally greater than KCl per unit of K.

Atmospheric deposition of S reduces the need for S fertilization. Most deposition originates from coal and petroleum combustion (Smith et al. 2011). In China, S deposition has been increasing (Smith et al. 2011), and bulk deposition in southeast China has been as high as 71 kg ha^{-1} $year^{-1}$ (Liu et al. (2016). In contrast, since the Clean Air Act was enacted in the United States, S deposition in that country has been declining (Smith et al. 2011), with areas that historically had the highest quantities of deposition now receiving only 15 kg S ha^{-1} $year^{-1}$ (Zhang et al. 2018). Because these changes occur over decades they can go unnoticed, leading to unknowingly outdated fertilizer recommendations for S. The use of SOP can help alleviate this increasing extent of S deficiency.

2.7.2.3 Potassium Nitrate (NOP)

There are several ways to produce potassium nitrate (Fig. 2.4). In two of them, KCl is mixed with nitrate salts, either sodium nitrate or ammonium nitrate. It is also commonly manufactured by the reaction of KCl with nitric acid. A relatively small amount of KNO_3 is produced from natural caliche deposits from the Chilean desert. Potassium nitrate has a K content of about 45% (Table 2.4).

The agronomic use of KNO_3 is often desirable in conditions where a highly soluble, chloride-free source of both K and N is needed. Potassium nitrate contains a relatively high proportion of K, with a N:K ratio of approximately 1:3 by weight. Applications of KNO_3 are typically made to the soil as prills, dissolved in a solution sprayed on plant foliage, or dissolved and applied through fertigation. Its use is most common for fertilizing high-value specialty crops, including fruits, vegetables, and tobacco (*Nicotiana tabacum* L.).

2.7.2.4 Potassium Thiosulfate (KTS)

Potassium thiosulfate is a clear fluid fertilizer. It is produced by reacting potassium hydroxide with aqueous ammonia, sulfur dioxide, and elemental S. It is used for direct soil application, in irrigation water, or as a foliar fertilizer. The thiosulfate portion of the molecule ($S_2O_3^{2-}$) oxidizes in soil to form sulfate (SO_4^{2-}) during an acid-forming process (Goos and Johnson 2001). The thiosulfate molecule has been shown to delay nitrification (Cai et al. 2018) and to sequester metals, particularly iron (Nayak and Dash 2006).

2.7.2.5 Langbeinite (SOPM)

Langbeinite (Table 2.2) has K, magnesium (Mg), and S all contained within a single geologic mineral. Its composition provides a uniform distribution of these nutrients when applied to the soil. The major source of langbeinite is from deposits in the

Southwest United States. Langbeinite is totally water soluble, but is slower to dissolve than some other common K fertilizers because of the higher particle density (2.8 g cm^{-3} compared with 2 g cm^{-3} for sylvite). This slower dissolution is an advantage in purification, as simple washing removes other impurities, leaving the less-soluble langbeinite (Harley and Atwood 1947). SOPM is frequently used where a low Cl^- source of K is desirable for crop nutrition or where additional Mg may be desired in the farming system, such as on acid soils or on soils where forage is produced for dairies. Langbeinite is a nutrient-rich fertilizer with a relatively low overall salt index. It is available in either a crystalline or granulated form. Langbeinite accounts for less than 1% of global production of primary K fertilizer.

2.7.2.6 Polyhalite

This soluble geologic mineral contains four essential plant nutrients: K, Mg, S, and calcium (Ca) (Table 2.2). The exact chemical formula will vary depending on its geologic origins. Polyhalite is frequently used in situations where a K fertilizer with a low Cl concentration is desirable. The lower K concentration in polyhalite (~13% K_2O) relative to other K fertilizer sources can in cases be compensated by the value of the additional Ca, Mg, and S in each particle (Yermiyahu et al. 2017). The lower solubility may also cause polyhalite to perform as a slightly slow-release source of nutrients, depending on the particle size (Barbier et al. 2017). Commercial supplies of polyhalite come from deep deposits in the Yorkshire region of the United Kingdom and in New Mexico, USA. In practice, both polyhalite and langbeinite are often blended with other less expensive K sources to obtain the desired blend of nutrients.

2.7.2.7 Potassium Hydroxide (KOH)

Potassium hydroxide (KOH) is a strongly alkaline liquid. It is commonly used as a component of fluid fertilizers. It has a low salt index, is free of Cl^-, and contains 83% K_2O (Table 2.4). It is commonly used to neutralize excess acidity in liquid fertilizer blends, irrigation water, and soil. Extreme safety measures must be used when handling this caustic material. Its use as a K source is generally limited to situations where the strongly alkaline properties are desired rather than its inherent K concentration.

2.7.2.8 Potassium Phosphate

Potassium phosphate (KH_2PO_4) is produced from the reaction of KCl with phosphoric acid (Fig. 2.4). It contains 34% K_2O (Table 2.4). This highly soluble product is commonly used in fertigation and for foliar applications where a source of both P and K is desired without additional N.

2.7.2.9 Mineral/Silicate K

Many geologic minerals contain abundant K, but their solubility is generally too low for agronomic use (Table 2.5). For example, potassium feldspar may contain almost 17% K_2O, but the dissolution is 20 million times slower than nepheline, a less abundant K-bearing mineral with 15% K_2O (Palandri and Kharaka 2004). The rate of K dissolution is key to its value as a source of plant nutrition (Manning 2018).

Considerable research has been conducted on developing various K-bearing minerals as fertilizers by utilizing a range of techniques to accelerate their dissolution through chemical or biological processes. The costs of transporting relatively low K concentration minerals often restricts their use to agricultural fields close to where they are mined.

The idea of using high K-content silicate rocks for fertilizer has been explored for many years (Ciceri et al. 2015), but most of the proposed techniques require significant energy or chemical inputs to accomplish a partial or complete dissolution of the rock. Based on dissolution rates of mineral K sources, rocks containing nepheline (including nepheline syenites, phonolites, and trachytes) may have the most commercial potential as agricultural potash sources (Manning 2010). Recent activity has focused on using K-rich feldspar as a nutrient source by milling, pH adjustment, and heating to accelerate natural weathering processes (Ciceri et al. 2016).

Glauconite (greensand) is a relatively insoluble silicate-based marine sediment (5–8% K_2O) that has been used with limited success as a rock-based K source that dissolves over multiple years. Similarly, a variety of K-rich micas and feldspars have been evaluated for direct soil application as a K source with limited commercial success due to their low solubility. A fine particle size ("rock dust") is required to allow dissolution at a rate that might provide a nutritional benefit for plants.

Table 2.5 Chemical composition and solubility in pure water for common K-bearing minerals. (adapted from Palandri and Kharaka 2004; Manning 2010)

	Mineral		Composition		Relative
Mineral	Family	Formula	% K	% K_2O[a]	Solubility
K feldspar	Feldspar	$KAlSi_3O_8$	14	16	Very low
Leucite	Feldspathoid	$KAlSi_2O_6$	17	21	Medium
Nepheline	Feldspathoid	$(K, Na)AlSiO_4$	13	15	Medium
Muscovite	Mica	$KAl_3Si_3O_{10}(OH)_2$	9	10	Very low
Biotite	Mica	$K_2Fe_6Si_6Al_2O_{20}(OH)_4$	7	9	Low
Phlogopite	Mica	$K_2Mg_6Si_6Al_2O_{20}(OH)_4$	9	11	Low
Glauconite	Mica	$(K, Na)\ (Mg, Fe^{2+})\ (Fe^{3+}, Al)\ (Si, Al)_4O_{10}(OH)_2$	1–4	1–6	Medium

[a]All values rounded down to the nearest whole number

Other sources were used to verify, but not cited; there are several ways to express the formulas for each mineral, etc.

Interest in biological additives to solubilize K from insoluble minerals has grown. Addition of various bacteria and fungi have been evaluated as a means for accelerating dissolution of K-bearing minerals or chelating silicon ions to enhance K solubility (Meena et al. 2016). Many of these chemical and biological approaches of facilitating K release from relatively insoluble minerals have attracted considerable attention, especially in regions where the cost of soluble K fertilizers poses a barrier to use.

The long-term impacts of using native minerals as sources of K for plant growth are not yet known. As an example, the release of K from the interlayers of micas can result in the formation of vermiculites in very short time periods (Hinsinger et al. 1992). Vermiculites have a high selectivity for K (Evangelou and Lumbanraja 2002), resulting in a significant portion of applied K ending up in interlayer positions where it is unavailable to plants, reducing its apparent recovery efficiency (RE_K). To overcome the reduction in RE_K, larger quantities of K fertilizer must be applied to achieve the same quantity of plant-available K (Chap. 1; Cassman et al. 1989). Future investigations will need to consider how the removal of K from minerals impacts the effectiveness of future K fertilizer applications.

2.7.2.10 Other Potassium Sources

Other K-containing materials have long been used as plant nutrients. Seaweed kelp (2–4% K_2O) has a long history as a K fertilizer. Wood ash (2–9% K_2O) has also been traditionally applied to supply additional K to crops. The alkaline nature of wood ash (containing $CaCO_3$ or CaO) needs to be considered if ash is added to soil at high application rates or repeatedly over long periods. The nature of the fuel and the combustion conditions will influence the K bioavailability in the ash. Biochar is typically applied to soils for agronomic and environmental benefits. Biederman and Harpole (2013) conducted a meta-analysis of 114 published manuscripts and found that biochar application significantly increased plant K concentration as well as soil test K levels.

A variety of harvested crop materials are returned to the field after processing (such as pomace and bagasse) to recycle organic material and nutrients, including K. If crop residues are allowed to remain in the field after harvest or are returned to the soil after processing, they quickly release any remaining cellular K, where it is recycled into the soil with rain and becomes available for uptake by succeeding crops.

Many wastewaters and food processing residuals contain K. Applications of large quantities of K-containing waste should consider the potential effects on soil mineralogy, soil cation ratios, soil physical properties (Oster et al. 2016), potential K leaching losses, and the overall nutrition of crops growing on the site (Arienzo et al. 2009).

2.7.3 *Forms of Potassium Fertilizer*

Commercial fertilizer comes in many forms. These include bulk blends of individual solid fertilizer products, complex granules, and fluids.

2.7.3.1 Bulk Blends

Most K fertilizer is mechanically blended with other solid fertilizers to make a mix of desired nutrients. This approach of blending separate components not only has the advantage of allowing various single materials to be selected based primarily on the cost of the separate components, but also achieves specific physical, nutritional, and chemical properties desired by the farmer. Freight costs associated with transporting individual solid fertilizers are usually less than transporting a variety of bagged materials or fluid fertilizers. A degree of flexibility is available to adjust the precise nutrient composition of the fertilizer blend according to specific crop and soil conditions.

When separate fertilizer materials are blended together, care needs to be taken to match the size of the raw materials to minimize separation (segregation) during transportation and field application. Proper blending of solid fertilizer materials takes experience and understanding of the individual components to be successful. This also involves selecting appropriate granular, crystalline, or prilled fertilizer materials with the proper particle size. The critical relative humidity needs to be considered as the fertilizer will absorb atmospheric moisture which will affect the storage life of blends. The critical relative humidity for common K fertilizers is relatively high, making them less susceptible to caking and clumping than many other solid fertilizer materials.

2.7.3.2 Complex (Compound) Granules

Potassium is commonly used in many complex fertilizers that are mixed with N, P, and other plant nutrients all within a single granule. There are a variety of processes for making these homogeneous fertilizers, but they have the advantage that they will not segregate during transport or application, and every granule delivers the same quantity and ratio of nutrients. This can be particularly important when very low rates of nutrients, for instance, micronutrients, need to be applied uniformly across the field or in a band. The K portion of the granules is most commonly derived from KCl, which readily dissolves in the soil and quickly becomes available for plant uptake.

2.7.3.3 Fluid Fertilizers

Solution fertilizers are preferred by some farmers. They are relatively easy to blend, are homogeneous, and can be applied in a variety of ways. When fertigating, fluid fertilizers provide a convenient way to introduce nutrients directly into the irrigation water for delivery to crops. Solid K fertilizers can also be dissolved to create fluids and then subsequently used in similar ways.

2.7.4 Potassium for Fertigation

Nutrients added to a pressurized irrigation system should be fully dissolved before adding them into the water stream. Although most common K fertilizers are relatively soluble, users should be aware of the differences between materials. The presence of impurities, fertilizer coatings, and conditioners can cause problems with plugging of the irrigation system, so these materials must be removed by a filter or avoided by using high-purity fertilizer products. The most common sources for fertigation are KCl, K_2SO_4, KNO_3, $K_2S_2O_3$, and KH_2PO_4. The selection of a particular K source is generally based on price, solubility, and the accompanying anion.

Choosing a specific K fertilizer for fertigation should account for potential chemical reactions between mixed fertilizers and the quality of the irrigation water. For example, the precipitation of calcium or phosphate salts in the irrigation lines can be minimized by selecting appropriate K fertilizer sources. Potassium itself is not generally a problem during fertigation, but the potential reactions of the accompanying anions need to be considered to avoid plugging the irrigation system.

2.7.5 Salt Index

Any soluble fertilizer will act as a salt when dissolved, thereby increasing the osmotic potential of the soil solution. The concept of "salt index" was first developed to predict the safety of a fertilizer placed in the proximity to a seedling (Rader et al. 1943). As various protocols have been developed to measure the safety of fertilizers placed near seeds and roots, it has been shown that other factors such as the crop and soil type, soil temperature and moisture, and fertilizer application rate have an equally or more important impact on potential seedling damage than the salinity developed while the fertilizer dissolves (Mortvedt 2001).

The original method for measuring salt index (Rader et al. 1943) was modified by Jackson (1958) and then again by Murray and Clapp (2004). Unfortunately, the different methods give inconsistent rankings in their prediction of potential salt damage. Additionally, the ranking of salt index reported for K fertilizers also varies in commonly used reference books. The use of salt index values does not predict the amount of fertilizer that will cause plant damage, but instead is most useful for providing a relative ranking of materials. In cases such as fertigation and foliar applications of K, the dilution of nutrients in relatively large quantities of water makes the salt index values less applicable. For example, Jifon and Lester (2009) used six K sources (KCl, KNO_3, KH_2PO_4, K_2SO_4, $K_2S_2O_3$, and a glycine amino-acid complexed K) as a foliar spray on musk melon (*Cucumis melo* L.) and found no differences in plant damage from any source or concentrations, with the solution buffered between pH 6.5 and 7.7.

The salt index of fertilizer is usually greatest for soluble N and K sources, indicating that it is usually best to avoid placement of large quantities close to the newly planted seed. A decision tool has been developed to help calculate the maximum amount of fertilizer that can be safely placed near seeds (SDSU and IPNI 2019). There is frequently little advantage for plant growth derived from placing fertilizer K very near the seed (unlike N and P) compared to the bulk soil. Damage to emerging seeds and seedlings from furrow-placed fertilizer can be minimized by avoiding high application rates of K.

2.7.6 *Chloride Considerations*

As mentioned previously, the selection of a particular source of K fertilizer is sometimes based on whether it contains Cl. There is a wide range of Cl sensitivity among plant species and cultivars. As a broad classification, many woody plant species, vegetables, and beans are more susceptible to Cl toxicity, whereas many non-woody crops tolerate higher concentrations of Cl^- in the root zone (Maas 1986).

There is a large body of literature related to the effects of Cl^- on crop performance (Xu et al. 2000). Chloride is an essential plant nutrient, and it is routinely added in some environments to enhance plant growth and disease resistance (Chen et al. 2010). However, excessively high Cl^- concentrations are linked to decreased crop growth and quality due to osmotic effects and specific ion toxicity.

To illustrate the difficulty in generalizing chloride's effects, we highlight some of the research on potatoes (*Solanum tuberosum* L.). Potatoes are an important crop that are rated as moderately sensitive to root zone salinity (Maas 1986). The effect of K fertilizer source on potato growth and quality has likely received more attention than any other crop. Potatoes accumulate large amounts of K during the growing season (>600 kg K ha^{-1} $year^{-1}$; Horneck and Rosen 2008) and frequently receive K fertilizer to sustain high yields. There are reports of an undesirable reduction in specific gravity (% dry matter) in potatoes receiving various K fertilizers. Some researchers have observed no differences in potato specific gravity when comparing various K fertilizer sources, while others have measured a decrease in specific gravity when using KCl compared with K_2SO_4 (e.g., Davenport and Bentley 2001). This reduction in specific gravity may not be directly due to the presence of

Cl^- but to greater K uptake from KCl and a higher salt index. These factors may cause the tubers to absorb more water than when fertilized with another K source.

Other research shows that the total K application rate has a greater impact on reducing specific gravity of tubers than the individual K fertilizer source (Westerman et al. 1994). Additionally, when KCl is split into multiple applications, or if KCl is applied in the autumn (with adequate winter rain) for a spring-planted potato crop, any negative impact on specific gravity is eliminated. Factors such as the climate and the potato variety can also influence the effect of K fertilization on potato specific gravity (Hütsch et al. 2018). These many interacting factors illustrate the difficulty in generalizing about selecting the proper K source for all conditions.

Differences in crop tolerance to Cl^- clearly exist. Although general osmotic damage can occur to germinating seeds and growing plants, there are also specific ion effects that can harm plants. For example, Tinker et al. (1977) reported that KCl fertilizer caused more damage to germinating seedlings than KNO_3, although the damage varied widely among the five plant species tested. They also reported that the fertilizer concentration, soil water content, and temperature also impacted the degree of seedling damage. Differences in the ability of species and cultivars to tolerate Cl^- stress are often related to the ability to restrict Cl uptake to the plant shoot (White and Broadley 2001).

The sensitivity of crops to salinity will change during the growing season, with young and newly developing seedlings generally most susceptible to salt damage. Once established, most crops become increasingly tolerant to salinity. In general practice, when application rates of KCl are not excessive or KCl is blended with another K source, no plant damage from excessive Cl^- is usually observed. In many situations, the amount of Cl added to a field as KCl is often small compared with the total amount of Cl already present in the soil and added as irrigation water. If KCl is applied in a localized band or zone in the soil, the initial soluble Cl concentration will spike and then decrease as Cl moves from the band into the surrounding soil by diffusion and mass flow.

2.7.7 Foliar Potassium Nutrition

Most crops have a relatively high demand for K throughout the growing season. If K uptake from the soil does not meet the plant demand, then growth, yield, and quality will suffer (Mikkelsen 2017). In the case where K uptake is insufficient, spraying an aqueous K-containing solution directly onto the plant foliage often overcomes this deficiency.

Applications of K-containing sprays directly onto the foliage of annual and perennial crops are common. Fernandez et al. have made several insightful reviews on this topic (Fernández and Brown 2013; Fernández et al. 2013). This practice is frequently done on high-value vegetables (e.g., Gunadi 2009; Jifon and Lester 2009; Salim et al. 2014) and perennial crops (e.g., Ben Yahmed and Ben Mimoun 2019; Shen et al. 2016; Solhjoo et al. 2017).

A foliar application of K during the growing season has been shown to improve crop yield and quality in many situations. Some plants have a high physiological K demand during specific stages of their growth cycle. For example, cotton (*Gossypium hirsutum* L.) is routinely sprayed with a foliar K solution in order to boost lint yield and quality during the late stage of crop development (Oosterhuis et al. 2014).

Crops frequently benefit from foliar K sprays if the soil-supplying capacity is too low (e.g., lack of proper fertilization) or if soil conditions do not permit uptake of soluble K by the roots (e.g., cold temperature, drought, water logging, nematodes, etc.). In these circumstances, a relatively small amount of K applied directly to the foliage can make a significant improvement to crop health and growth.

Foliar fertilization is common when tissue testing or observation of visual symptoms identifies an emerging K deficiency. Prompt correction of the K deficiency can halt additional damage, since foliar-applied K is rapidly taken up through the leaves and quickly utilized in the plant (Fageria et al. 2009). Application of the dilute K fertilizer solution can be done using a variety of spray equipment (such as an airblast-type sprayer) to deliver K onto crop leaves to supplement the soil supply. Potential benefits achieved from an intervention with a foliar spray need to be weighed against the financial costs of the field operation. Foliar K applications often occur while other chemicals are being sprayed on the plants to minimize the number of trips through the field.

Although foliar K sprays are beneficial in many situations, the majority of plant K is acquired and taken up by the roots. A foliar K fertilization program should be viewed as a supplement to maintaining an adequate concentration of soluble K in the soil. For example, Gordon and Niederholzer (2019) explain that a typical application to almond foliage of 27 kg KNO_3 ha^{-1} (~10 kg K ha^{-1}) in 1000 L will likely result in an increased leaf K concentration. However, this single application will only provide about 3% of the total crop K requirement (assuming a spray application efficiency of 75%). Clearly this relatively small amount of K added through foliar fertilization is merely a supplement to the soil supply, even when multiple applications are made through the growing season.

Repeated applications of foliar K solutions with a high concentration of fertilizer salt often leads to leaf damage ("salt burn"). Therefore, advice should be sought before beginning a foliar fertilization program in order to identify the appropriate 4R-based practices (Right Source, Right Rate, Right Time, and Right Place) suited to a specific crop and agro-environment.

Although KCl is the most common K fertilizer applied to soil, it has a high salt index (osmotic potential) and point of deliquescence (POD; crystallization) that may limit its use in foliar sprays. There are several excellent K fertilizer sources that are used for foliar nutrition. The most common of these are K_2SO_4, KNO_3, KH_2PO_4, and organic-based formulations.

2.8 Summary

Accounting for all K inputs is difficult to do accurately. Most inputs, except commercial fertilizers, have variable K contents that are not routinely measured on the farm. Testing those inputs for their content is preferred, but standardized values can be used as a substitute. The emphasis of this chapter has been on commercial fertilizers. These inputs are produced from K-bearing minerals found around the world. Reserves have an expected lifetime of thousands of years. When a need for additional K for plant growth has been identified, there are many sources of fertilizer that can be used to meet nutritional requirements. The selection of a specific potash fertilizer source largely depends on the nutrients that accompany the K, the fertilizer's availability in the market, and its price. As new technology is introduced, the importance of less-soluble K-bearing minerals as a plant nutrient source may increase, especially in regions of the world where soluble K fertilizers are not accessible or are too expensive for farmers. There are many factors to consider in selecting the most appropriate K source. Whatever source is selected, the soluble K^+ is the same for nutrition of the plant. It is essential that an adequate concentration of plant-available K is maintained in the rootzone to produce abundant and high-quality crops.

References

American Meteorological Society (2012) Dry deposition. Glossary of meteorology. http://glossary.ametsoc.org/wiki/climatology. Accessed 08 May 2020

Arienzo M, Christen EW, Quayle W, Kumar A (2009) A review of the fate of potassium in the soil-plant system after land application of wastewaters. J Hazard Mater 164:415–422. https://doi.org/10.1016/j.jhazmat.2008.08.095

Barbier M, Li YC, Liu G, He Z, Mylavarapu R, Zhang S (2017) Characterizing polyhalite plant nutritional properties. Agric Res Technol 6(3). https://doi.org/10.19080/ARTOAJ.2017.06.555690

Ben Yahmed J, Ben Mimoun M (2019) Effects of foliar application and fertigation of potassium on yield and fruit quality of 'Superior Seedless' grapevine. Acta Hortic (1253):367–372. https://doi.org/10.17660/ActaHortic.2019.1253.48

Biederman LA, Harpole WS (2013) Biochar and its effects on plant productivity and nutrient cycling: a meta-analysis. BCB Bioenergy 5:202–214. https://doi.org/10.1111/gcbb.12037

Cai Z, Gao S, Xu M, Hanson BD (2018) Evaluation of potassium thiosulfate as a nitrification inhibitor to reduce nitrous oxide emissions. Sci Total Environ 618:243–249. https://doi.org/10.1016/j.scitotenv.2017.10.274

Cassman KG, Roberts BA, Kerby TA, Bryant DC, Higashi SL (1989) Soil potassium balance and cumulative cotton response to annual potassium additions on a vermiculitic soil. Soil Sci Soc Am J 53:805–812. https://doi.org/10.2136/sssaj1989.03615995005300030030x

Chen W, He ZL, Yang XE, Mishra S, Stoffella PJ (2010) Chlorine nutrition of higher plants: progress and perspectives. J Plant Nutr 33(7):943–952. https://doi.org/10.1080/01904160903242417

Ciceri D, Manning DAC, Allanore A (2015) Historical and technical developments of potassium resources. Sci Total Environ 502:590–601. https://doi.org/10.1016/j.scitotenv.2014.09.013

Ciceri DT, Skorina C, Gadois K, Li, Allanore A (2016) Processing of potassium silicates for K-release. In: de Melo Benites V, de Oliveira Junior A, Pavinato PS, Teixeira PC, Moraes MF, de Campos Leite RMVB, de Oliveira RP (eds) Proceedings of 16th world fertilizer congress of CIEC, pp 102–104

Darmody RG, Ruiz D (2017) Dredged sediment: application as an agricultural amendment on sandy soils. Illinois Sustainable Technol Center TR-066. http://hdl.handle.net/2142/97824. Accessed 08 May 2020

Davenport JR, Bentley EM (2001) Does potassium fertilizer form, source, and time of application influence potato yield and quality in the Columbia Basin? Am J Potato Res 78:311–318. https://doi.org/10.1007/BF02875696

Duo Z, Galligan DT, Allshouse RD, Toth JD, Ramberg CF, Ferguson JD (2001) Manure sampling for nutrient analysis. J Environ Qual 30(4):1432–1437. https://doi.org/10.2134/jeq2001.3041432x

Evangelou VP, Lumbanraja J (2002) Ammonium-potassium-calcium exchange on vermiculite and hydroxy-aluminum vermiculite. Soil Sci Soc Am J 66:445–455. https://doi.org/10.2136/sssaj2002.4450

Fageria NK, Barbosa Filho MP, Moreira A, Guimaraes CM (2009) Foliar fertilization of crop plants. J Plant Nutr 32:1044–1064., 2009. https://doi.org/10.1080/01904160902872826

Fernández V, Brown PH (2013) From plant surface to plant metabolism: the uncertain fate of foliar-applied nutrients. Front Plant Sci 4:289. https://doi.org/10.3389/fpls.2013.00289

Fernández V, Sotiropoulos T, Brown P (2013) Foliar fertilization: scientific principles and field practices. IFA, Paris

Fountain DM, Christensen NI (1989) Composition of the continental crust and upper mantle; A review. In: Pakiser LC, Mooney WD (eds) Geophysical framework of the continental United States. Geol Soc Am Mem 172:711–742. https://doi.org/10.1130/MEM172-p711

Goos RJ, Johnson BE (2001) Thiosulfate oxidation by three soils as influenced by temperature. Commun Soil Sci Plan 32(17–18):2841–2849. https://doi.org/10.1081/CSS-120000966

Gordon P, Niederholzer FJA (2019) Potassium in nut crops? Plant uses and field applications. https://www.sjvtandv.com/blog/kgkt2owv6lxnrb1o4i15lmfipj9y4w. Accessed 01 June 2020

Gunadi N (2009) Response of potato to potassium fertilizer sources and application methods in Andisols of West Java. Indones J Agric Sci 10(2):65–72

Harley GT, Atwood GE (1947) Langbeinite. . .mining and processing. Ind Eng Chem 39(1):43–47. https://doi.org/10.1021/ie50445a020

Hinsinger P, Jaillard B, Dufey JE (1992) Rapid weathering of a trioctahedral mica by the roots of ryegrass. Soil Sci Soc Am J 56:977–982. https://doi.org/10.2136/sssaj1992.03615995005600030049x

Hoa NM, Janssen GH, Oenema O, Dobermann A (2006) Comparison of partial and complete soil K budgets under intensive rice cropping in the Mekong Delta, Vietnam. Agric Ecosyst Environ 116:121–131. https://doi.org/10.1016/j.agee.2006.03.020

Horneck D, Rosen C (2008) Measuring nutrient accumulation rates of potatoes – tools for better management. Better Crops 92(1):4–6

Hung NV, Maguyon-Detras MC, Migo MV, Quilloy R, Balingbing C, Chivenge P, Gummert M (2020) Rice straw overview: availability, properties, and management practices. In: Gummert M, Hung NV, Chivenge P, Douthwaite B (eds) Sustainable rice straw management. Springer, New York. https://doi.org/10.1007/978-3-030-32373-8_1

Hütsch BW, Keipp K, Glaser A-K, Schubert S (2018) Potato plants (*Solanum tuberosum* L.) are chloride-sensitive: is this dogma valid? J Sci Food Agr 98(8):3161–3168. https://doi.org/10.1002/jsfa.8819

IFDC, UNIDO (eds) (1998) Fertilizer manual. United Nations Industrial Development Organization and International Fertilizer Development Center, Kluwer Academic, Dordrecht

IPNI (2012) 4R Plant nutrition manual: a manual for improving the management of plant nutrition. Metric version. In: Bruulsema TW, Fixen PE, Sulewski GD (eds) International Plant Nutrition Institute, Norcross

Jackson WL (1958) Soil chemical analysis. Prentice Hall, Englewood Cliffs, NJ

Jifon JL, Lester GE (2009) Foliar potassium fertilization improves fruit quality of field-grown muskmelon on calcareous soils in south Texas. J Sci Food Agr 89:2452–2460. https://doi.org/10.1002/jsfa.3745

Junakova N, Balintova M (2012) Assessment of nutrient concentration in reservoir bottom sediments. Procedia Eng 42:165–170. https://doi.org/10.1016/j.proeng.2012.07.407

Liptay A, Arevalo AE (2000) Plant mineral accumulation, use and transport during the life cycle of plants: a review. Can J Plant Sci 80:29–38. https://doi.org/10.4141/P99-014

Liu L, Zhang X, Wang S, Zhang W, Lu X (2016) Bulk sulfur (S) deposition in China. Atmos Environ 135:41–49. https://doi.org/10.1016/j.atmosenv.2016.04.003

Maas EV (1986) Salt tolerance of plants. Appl Agric Res 1:12–26

Manning DAC (2010) Mineral sources of potassium for plant nutrition. A review. Agron Sustain Dev 30(2):281–294. https://doi.org/10.1051/agro/2009023

Manning DAC (2018) Innovation in resourcing geological materials as crop nutrients. Nat Resour Res 27(2):217–227. https://doi.org/10.1007/s11053-017-9347-2

Meena VS, Maurya BR, Verma P, Meena RS (eds) (2016) Potassium solubilizing microorganisms for sustainable agriculture. Springer, New Delhi. https://doi.org/10.1007/978-81-322-2776-2

Miao Y, Stewart BA, Zhang F (2011) Long-term experiments for sustainable nutrient management in China. A review. Agron Sustain Dev 31(2):397–414. https://doi.org/10.1051/agro/2010034

Mikha MM, Benjamin JG, Vigil MF, Poss DJ (2017) Manure use and tillage use in remediation of eroded land and impacts on soil chemical properties. PLoS One 12(4):e0175533. https://doi.org/10.1371/journal.pone.0175533

Mikhailova EA, Post GC, Cope MP, Post CJ, Schlautman MA, Zhang L (2019) Quantifying and mapping atmospheric potassium deposition for soil ecosystem services assessment in the United States. Front Environ Sci 7:Article 74. https://doi.org/10.3389/fenvs.2019.00074

Mikkelsen RL (2000a) Nutrient management for organic farming: a case study. J Nat Resour Life Sci Ed 29:88–92. https://doi.org/10.2134/jnrlse.2000.0088

Mikkelsen RL (2000b) Beneficial use of swine by-products: Opportunities for the future. In: Powers J et al (eds) Beneficial uses of agricultural, industrial, and municipal by-products. Soil Science Society of America, Madison, pp 451–480. https://doi.org/10.2136/sssabookser6.c16

Mikkelsen RL (2007) Managing potassium for organic crop production. HortTechn 17(4):455–460. https://doi.org/10.21273/HORTTECH.17.4.455

Mikkelsen R (2017) The importance of potassium management for horticultural crops. Indian J Fert 13(11):82–86

Mortvedt JJ (2001) Calculating salt index. Fluid J Spring 2001:1–3. https://fluidfertilizer.org/33p8-11/. Accessed 20 May 2020

Murray TP, Clapp JG (2004) Current fertilizer salt index tables are misleading. Commun Soil Sci Plan 35(19–20):2867–2873. https://doi.org/10.1081/CSS-200036474

National Agricultural Library (2019a) Atmospheric deposition. Term no. 2714, Agricultural thesaurus. https://agclass.nal.usda.gov/agt.shtml. Accessed 20 May 2020

National Agricultural Library (2019b) Wet deposition. Term no. 68066, Agricultural thesaurus. https://agclass.nal.usda.gov/agt.shtml. Accessed 20 May 2020

Nayak S, Dash A (2006) Oxidation of the thiosulfate ion by some iron(III) complexes with phenolate – amide-amine coordination: a comparative kinetic study. Transit Metal Chem 31:930–937. https://doi.org/10.1007/s11243-006-0086-1

O'Dell JD, Essington ME, Howard DD (1995) Surface application of liquid swine manure: chemical variability. Commun Soil Sci Plan 26(19–20):3113–3120. https://doi.org/10.1080/00103629509369513

Oosterhuis DM, Loka DA, Kawakami EM, Pettigrew WT (2014) The physiology of potassium in crop production. Adv Agron 126:203–233. https://doi.org/10.1016/B978-0-12-800132-5.00003-1

Orris GJ, Cocker MD, Dunlap P, Wynn J, Spanski GT, Briggs DA, Gass L, with contributions from Bliss JD, Bolm KS, Yang C, Lipin BR, Ludington S, Miller RJ, Slowakiewicz M (2014)

Potash—a global overview of evaporate-related potash resources, including spatial databases of deposits, occurrences, and permissive tracts. U.S. Geological Survey scientific investigations report 2010–5090-S. https://doi.org/10.3133/sir20105090S

Oster JD, Sposito G, Smith CJ (2016) Accounting for potassium and magnesium in irrigation water quality assessment. Calif Agr 70(2):71–76. https://doi.org/10.3733/ca.v070n02p71

Palandri JL, Kharaka YK (2004) A compilation of rate parameters of water-mineral interaction kinetics for application to geochemical modeling. U.S. Geological Survey, open file report 2004–1068. https://pubs.usgs.gov/of/2004/1068. Accessed 8 May 2020

Pan YP, Wang YS (2015) Atmospheric wet and dry deposition of trace elements at 10 sites in Northern China. Atmos Chem Phys 15:951–972. https://doi.org/10.5194/acp-15-951-2015

Prospero JM, Collard FX, Molinie J, Jeannot A (2014) Characterizing the annual cycle of African dust transport to the Caribbean Basin and South America and its impact on the environment and air quality. Glob Biogeo Cycl 28:757–773. https://doi.org/10.1002/2013GB004802

Rader LF Jr, White LM, Whittaker CW (1943) The salt index – a measure of the effect of fertilizers on the concentration of the soil solution. Soil Sci 55(3):201–218. https://journals.lww.com/soilsci/toc/1943/03000. Accessed 8 May 2020

Rahm M (2017) The global potassium market. In: Murrell TS, Mikkelsen RL (eds) Proceedings of frontiers potassium science conference, Rome, 25–27 Jan 2017. International Plant Nutrition Institute, Peachtree Corners. https://www.apni.net/k-frontier.org. Accessed 8 May 2020

Ruscio A, Kazanc F, Levendis YA (2016) Comparison of fine ash emissions generated from biomass and coal combustion and valuation of predictive furnace deposition indices: a review. J Energ Eng 142:E4015007. https://doi.org/10.1061/(ASCE)EY.1943-7897.0000310

Salim BBM, Abd El-Gawad HG, Abou El-Yazied A (2014) Effect of foliar spray of different potassium sources on growth, yield and mineral composition of potato (*Solanum tuberosum* L.). Middle East J Appl Sci 4:1197–1204

SDSU, IPNI (2019) Seed damage calculator. South Dakota Cooperative Extension Service and the International Plant Nutrition Institute. http://seed-damage-calculator.herokuapp.com. Accessed 08 May 2020

Sheldrick WF (1985) World potassium reserves. In: Munson RD (ed) Potassium in agriculture. American Society of Agronomy, Crop Science Society of America, Soil Science Society of America, Madison, pp 3–28. https://doi.org/10.2134/1985.potassium.c1

Shen C, Ding Y, Lei X, Zhao P, Wang S, Xu Y, Dong C (2016) Effects of foliar potassium fertilization on fruit growth rate, potassium accumulation, yield, and quality of 'Kousui' Japanese Pear. HortTechnology 26:270–277. https://doi.org/10.21273/HORTTECH.26.3.270

Smith SJ, van Aardenne J, Klimont Z, Andres RJ, Volke A, Arias SD (2011) Anthropogenic sulfur dioxide emissions: 1850-2005. Atmos Chem Phys 11:1101–1116. https://doi.org/10.5194/acp-11-1101-2011

Solhjoo S, Gharaghani A, Fallahi E (2017) Calcium and potassium foliar sprays affect fruit skin color, quality attributes, and mineral nutrient concentrations of 'red delicious' apples. Int J Fruit Sci 17:358–373. https://doi.org/10.1080/15538362.2017.1318734

SOPIB (Sulfate of Potash Information Board) (2015) The use of potassium sulfate fertilizer: principles and practices

Tinker PB, Reed L, Legg C, Højer-Pederson S (1977) The effects of chloride in fertiliser salts on crop seed germination. J Sci Food Agric 28(12):1045–1051. https://doi.org/10.1002/jsfa.2740281202

U.S. Bureau of Mines, U.S. Geological Survey (1980) Principles of a resource/reserve classification for minerals. Geological Survey Circular 831. https://pubs.usgs.gov/circ/1980/0831/report.pdf. Accessed 08 May 2020

U.S. Geological Survey (USGS) (2020) Mineral commodity summaries 2020: U.S. Geological Survey. https://doi.org/10.3133/mcs2020

Westerman DT, Tindall TA, James DW, Hurst RL (1994) Nitrogen and potassium fertilization of potatoes: yield and specific gravity. Am Potato J 71(7):417–431. https://doi.org/10.1007/BF02849097

White PJ, Broadley MR (2001) Chloride in soils and its uptake and movement within the plant: a review. Ann Bot-Lond 88(6):967–988. https://doi.org/10.1006/anbo.2001.1540

Williams M (2012) River sediments. Philos Trans R Soc A 370:2093–2122. https://doi.org/10.1098/rsta.2011.0504

Williams PH, Gregg PEH, Hedley GM (1990) Fate of potassium in dairy cow urine applied to intact soil cores. New Zealand J Agric Res 33:151–158. https://doi.org/10.1080/00288233.1990.10430672

Woodard HJ (1999) Plant growth on soils mixed with dredged lake sediment. J Environ Sci Health A 34:1229–1252. https://doi.org/10.1080/10934529909376893

Xu G, Magen H, Tarchitzky J, Kafkafi U (2000) Advances in chloride nutrition of plants. Adv Agron 68:97–150. https://doi.org/10.1016/S0065-2113(08)60844-5

Yermiyahu U, Zipori I, Faingold I, Yusopov L, Faust N, Bar-Tal A (2017) Polyhalite as a multi nutrient fertilizer potassium, magnesium, calcium and sulfate. Israel J Plant Sci 64 (3–4):145–157. https://doi.org/10.1163/22238980-06401001

Zhang H, Wang B, Xu M, Fan T (2009) Crop yield and soil responses to long-term fertilization on a red soil in southern China. Pedosphere 19:199–207

Zhang Y, Mathur R, Bash JO, Hogrefe C, Xing J, Roselle SJ (2018) Long-term trends in total inorganic nitrogen and sulfur deposition in the US from 1990 to 2010. Atmos Chem Phys 18:9091–9106. https://doi.org/10.5194/acp-18-9091-2018

3

Factors Affecting Potassium (K) Removals in Crops and Potassium (K) Losses

Keith Goulding, T. Scott Murrell, Robert L. Mikkelsen, Ciro Rosolem, Johnny Johnston, Huoyan Wang, and Marta A. Alfaro

Abstract Potassium (K) outputs comprise removals in harvested crops and losses via a number of pathways. No specific environmental issues arise from K losses to the wider environment, and so they have received little attention. Nevertheless, K is very soluble and so can be leached to depth or to surface waters. Also, because K is bound to clays and organic materials, and adsorbed K is mostly associated with fine soil particles, it can be eroded with particulate material in runoff water and by strong winds. It can also be lost when crop residues are burned in the open. Losses represent a potential economic cost to farmers and reduce soil nutritional status for plant growth. The pathways of loss and their relative importance can be related to: (a) the general characteristics of the agricultural ecosystem (tropical or temperate regions,

K. Goulding (✉) · J. Johnston
Rothamsted Research, Hertfordshire, UK
e-mail: keith.goulding@rothamsted.ac.uk; johnny.johnston@rothamsted.ac.uk

T. S. Murrell
African Plant Nutrition Institute and Mohammed VI Polytechnic University, Ben Guerir, Morocco

Department of Agronomy, Purdue University, West Lafayette, IN, USA
e-mail: s.murrell@apni.net

R. L. Mikkelsen
African Plant Nutrition Institute and Mohammed VI Polytechnic University, Ben Guerir, Morocco
e-mail: r.mikkelsen@apni.net

C. Rosolem
Faculdade de Ciências Agronômicas de Botucatu, Departamento de Produção e Melhoramento Vegetal, Universidade Estadual Paulista Júlio de Mesquita Filho, Botucatu, São Paulo, Brazil
e-mail: rosolem@fca.unesp.br

H. Wang
Institute of Soil Science, Chinese Academy of Sciences, Nanjing, China
e-mail: hywang@issas.ac.cn

M. A. Alfaro
Instituto de Investigaciones Agropecuarias, Ministerio de Agricultura, Osorno, Chile
e-mail: malfaro@inia.cl

cropping or grazing, tillage management, interactions with other nutrients such as nitrogen); (b) the specific characteristics of the agricultural ecosystem such as soil mineralogy, texture, initial soil K status, sources of K applied (organic, inorganic), and rates and timing of fertilizer applications. This chapter provides an overview of the main factors affecting K removals in crops and losses through runoff, leaching, erosion, and open burning.

Potassium is removed in harvested crops, a necessary and important part of agriculture and food production, but it is also lost by erosion, leaching, and open burning. Unlike nitrogen (N), there are no gaseous losses of K. Management practices can reduce losses, sustain plant-available soil K supplies, and improve the recovery efficiency of K applications. Compared to N and phosphorus (P), K losses have received little attention because they have few, if any, environmental impacts. Nevertheless, they represent a potential economic cost to farmers. They also increase the risk of poor soil K fertility and unbalanced nutrition for plant growth.

Higher losses are expected in soils with a lower K holding capacity. We define **potassium holding capacity** as *the maximum quantity of K that can be retained by a given volume of soil.* A better understanding of the main pathways of K losses and the key factors controlling them can improve practical recommendations to ensure farmers optimize the K-holding capacity of their soils to increase or sustain productivity and economic returns.

3.1 Removal in Harvested Crops

Harvested plant K is *the quantity of K in plant material removed from a given area.* The rate of K removal per unit area increases when the total K accumulation in harvested plant organs increases. Such increases occur with an increased K concentration in plant tissue, higher yields, or a shift toward removing additional plant organs from the area, such as straw and grain rather than just grain.

An associated term is **Harvest Index** (HI), which is *the fraction of the harvested yield divided by the total amount of biomass produced* (Unkovich et al. 2010). Although the denominator is defined as the total biomass produced (shoots and roots), it is most common to only measure the above-ground biomass due to the difficulty of measuring root biomass. *The unharvested portion that remains on the field following harvest* is defined as **crop residue**. Knowing the proportion of biomass or nutrients (such as K) that will be removed during harvest and what proportion will remain as residue or be returned to the soil is essential for estimating K budgets.

The HI can vary depending on the time of measurement. For example, many legume crops begin to shed leaves prior to physiological maturity and harvest. To correctly calculate the K requirement, the HI for these crops should be calculated using maximum biomass dry matter, including the leaves that fall to the ground.

During maize (*Zea mays* L.) harvest, the grain may be the only product removed from the field or in some circumstances the stalks, leaves, and husks may be all removed for economic purposes (such as for silage or bioenergy). Additionally, the residue stalks and stems are cut at varying heights above the ground.

For determining the HI of horticultural crops, "maturity" is determined by market-driven parameters such as gelatinous mass filling of tomato fruit (*Solanum lycopersicum* L.), coloring of peppers (*Capsicum annuum* L.), head formation of lettuce (*Lactuca sativa* L.), plant size for spinach (*Spinacia oleracea* L.), tuber size for potatoes (*Solanum tuberosum* L.), or sugar accumulation of grapes (*Vitis vinifera* L.)

Indeterminate crops may be harvested multiple times. The unharvested portion of annual crops is generally returned to the soil. For perennial crops, the woody portion will continue to accumulate biomass and nutrients, with only the leaves dropping to the soil. In some cases, crop residues may not be returned to the same field from which they were harvested (e.g., pomace and bagasse).

Understanding both the rate of uptake and the total amount of K accumulated in the crop during the growing season, and in the harvested portion removed during harvest, is required for assessing the seasonal crop demand. For example, Bender et al. (2013) examined seasonal K uptake of six modern maize hybrids. They reported that K accumulation occurred in a sigmoidal pattern over the growing season, with most K uptake already completed by the time vegetative growth transitioned to reproductive growth. After this, a large amount of K was translocated from the vegetative tissue to the reproductive organs. At harvest, 30% of the total K was in the grain.

Rogers et al. (2019) measured total K accumulation of five barley (*Hordeum vulgare* L.) cultivars and found it peaked at the soft dough stage (253 kg K ha^{-1}) and then declined to 172 K kg ha^{-1} at physiological maturity. This loss of 81 kg K ha^{-1} from the biomass occurred at the same time as a small increase in K in the barley heads (from 37 to 42 kg K kg ha^{-1}). Sugarbeet (*Beta vulgaris* L.) had a sigmoidal pattern of K accumulation, with approximately half of the total accumulated K (>500 kg K ha^{-1}) in the vegetative tops and half in the roots by the time of harvest (De et al. 2019).

Perennial crops can accumulate and export a large quantity of K at harvest. For example, almonds (*Prunus dulcis* (Mill.) D. A. Webb) remove 400 kg K in a typical 5 t ha^{-1} yield in the harvested hulls, shells, and kernels (Muhammad et al. 2009), and bananas (*Musa* spp.) remove even more ~750 kg K in a typical 50 t ha^{-1} yield during the extended fruit harvest period (Lahav and Turner 1989).

Crops grown for hay production, such as alfalfa (*Medicago sativa* L.), also remove large amounts of nutrients from the soil, especially K. For example, a healthy alfalfa crop may remove 25 kg K t^{-1} in a typical 25–35 t ha^{-1} yield. Cultivating high-yielding hay crops therefore requires special attention to avoid depleting the soil nutrient supply.

The K concentration in the residue of agricultural crops varies widely, as does the rate of subsequent K release from the residue (e.g., Anguria et al. 2017). Estimating both the quantity of crop residue and its K concentration are necessary steps for measuring the K in residues and the potential for K loss or recycling from residues remaining in the field.

Although the general differences in K removal among grain crops are well documented (e.g., soybeans seeds *Glycine max* (L.) Merr. contain approximately 4 times more K per t than maize), there is important variation in nutrient content among species based on the growing environment, yield level, and crop genetics. For example, Nathan et al. (2009) analyzed maize grain samples ($n = 141$ in 2006 and $n = 214$ in 2007) and measured a mean K concentration of 3.4 mg K kg^{-1} and 2.7 mg K kg^{-1} in the 2 years, respectively (± 0.5 mg K kg^{-1} standard deviation). Even within the fairly small geographic region sampled, they found >25% difference in corn grain K concentrations in the 2 years.

This temporal difference in nutrient removal during crop harvest illustrates the challenge of using "average values" for estimating K offtake. A number of published databases exist that provide average nutrient removal coefficients for most harvested crops. However, many of these tables and databases do not use up-to-date measurements, do not properly cite where the information came from, or are not reliable for accounting for the significant spatial, cultural, and temporal variation in nutrient concentrations that arise during routine crop production practices. General nutrient removal databases can be useful for making nutrient offtake estimates but should not be used for more precise planning. Samples of harvested crops should be periodically analyzed in the laboratory in order to confirm the quantity of nutrients being removed from the field.

3.1.1 Whole-Plant Removal

The practice of removing straw from grain fields is common in many parts of the world and has important implications for both soil health and nutrient cycling. Residue removal may be locally useful for purposes such as animal feed or bedding, fuel, or for use in cellulose-based ethanol production. However, some level of organic matter input is required for maintaining the long-term ecological function and the agricultural productivity of soils.

Crop residues contain valuable plant nutrients, so removing them from the field will speed nutrient depletion and have economic impacts, especially for K. For example, small grain straw contains less P and N than the grain, but a higher proportion of K; i.e., the average straw:grain mass nutrient ratio in wheat is 0.47 for N, 0.26 for P, but 4.12 for K; the straw:grain nutrient ratio in barley is 0.49 for N, 0.35 for P, but 5.04 for K (Tarkalson et al. 2009). Therefore, when both grain and straw are removed from fields, soil K depletion is accelerated compared with harvesting only grain. The financial expense associated with purchasing K fertilizer to replace this harvested K should be accounted for in long-term nutrient budgets and decisions on residue removal.

Methods of handling the straw and crop residues also need to be considered when calculating potential K losses. Since K is readily leached from crop residues with rainfall and irrigation, the length of time the residue remains in the field before removal and how the residue is distributed before removal (e.g., windrows, piles, or broadcast) will significantly impact the amount of K ultimately removed in the biomass.

3.2 Erosion

Erosion loss is *K lost from the movement of soil particles out of a given volume of soil.* Losses can occur in both water and wind erosion. Soil particles eroded from the field carry adsorbed K with them. Water erosion occurs mostly across the soil surface or at shallow depths by runoff, but particles can also be transported to depth and lost through field drains, if the land is drained.

3.2.1 Water Erosion

Runoff loss arises from surface and subsurface movement of water. **Surface runoff** loss is K *in water moving laterally over the soil surface in the direction of the slope.* **Subsurface runoff** loss is K *in water that infiltrates the soil surface to shallow depths and then moves laterally in the direction of the slope.*

Korucu et al. (2018) used a collection pan at the soil surface to measure only surface runoff. They conducted their study on a site with 2% slope composed of loam and clay loam soils. A day after maize silage harvest, they planted cereal rye to test the effects of a cover crop on surface runoff. Approximately 1 month after planting rye, they broadcast 13 kg NH_4-N ha^{-1}, 27 kg P ha^{-1}, and 83 kg K ha^{-1} as monoammonium phosphate and potassium chloride. An hour later, they simulated a 10-year extreme rainfall event, using spray nozzles to deliver 65 mm of water in 60 min. Such conditions favored runoff losses of fertilizer P and K. On treatments with no rye cover crop, it took an average of 4.9 min for runoff to begin after the start of the rainfall simulation (Table 3.1). Runoff averaged 27.3 mm. Total suspended solids averaged 444 kg ha^{-1} and K loss averaged 12.42 kg K ha^{-1}. Average concentration of K in the runoff was 43.0 mg L^{-1}. The month-old rye cover crop doubled the time for runoff to begin and reduced the total runoff amount by 65%, the total suspended solids by 68%, the K concentration by 75%, and the total K loss by 91%. Thus, even when a simulated 10-year rainfall event occurred just 1 h after a surface application of K, the rye cover crop reduced the K runoff loss to an average of 1.08 kg K ha^{-1}. Where no cover crops were present, K losses across replications were 4–32% of the amount applied an hour before the start of the rainfall simulation. These results and others indicate that K losses in surface runoff depend on rainfall intensity, the timing of precipitation events, and the management of K applications (Alfaro et al. 2004b, 2008).

Table 3.1 Results of a 60-min, 65 mm simulated rainfall event starting 1 h after a broadcast application of 13 kg NH_4-N ha^{-1}, 27 kg P ha^{-1}, and 83 kg K ha^{-1} on an experimental area with 2% slope and dominant soil series Clarion loam and Nicollett clay loam. (Korucu et al. 2018)

Cover crop	Rep. no.	Time to runoff min	Runoff mm	Total suspended solids kg ha^{-1}	K concentration mg L^{-1}	K loss kg ha^{-1}
None	1	5.0	10.6	134	43.9	4.66
None	2	4.0	42.6	797	39.6	16.83
None	3	5.3	31.7	491	33.5	10.62
None	4	7.5	17.9	407	20.0	3.57
None	5	2.7	33.9	392	77.9	26.42
Mean		**4.9b**	**27.3a**	**444a**	**43.0a**	**12.42a**
Rye	1	8.0	13.6	179	10.5	1.43
Rye	2	10.0	13.2	142	13.7	1.81
Rye	3	13.7	5.7	54	9.2	0.52
Rye	4	17.3	5.5	143	6.2	0.34
Rye	5	4.0	9.5	191	13.5	1.28
Mean		**10.6a**	**9.5b**	**142b**	**10.6b**	**1.08b**

Means separated by different letters within a column are significantly different at $p \leq 0.05$
Results are given for each of the five replications within each cover crop treatment

In temperate systems, favorable conditions for the development of surface runoff are mainly found during winter months. Although no mineral fertilizers are usually applied at that time, livestock farmers may apply substantial amounts of organic manures and slurries because there is more labor capacity available and the application does not interfere with grazing and cropping. Spreading in winter also helps to reduce storage requirements for manure or slurry. Some countries ban the spreading of manures and slurries in winter to reduce losses of nitrate and P to waters. However, the practice, as well as fertilizer applications in early spring, is still common and increases the risk of K losses in runoff (Alfaro et al. 2004a). The effect is likely to be greater in grazing areas because of the excessive trampling by animals that modifies soil structure, puddling the soil surface and reducing soil porosity (Heathwaite et al. 1996). In this case, the control of stocking rates at critical points during the grazing season is a key factor to reduce K losses from grazed paddocks (Alfaro et al. 2004a, b).

Rain reduces the porosity of the soil over time, increasing the likelihood of K runoff losses. Raindrop impact destroys soil aggregates and increases the thickness of the compacted surface layer (Rousseva et al. 2002), especially at high rainfall intensities, such as those in tropical regions (e.g., Acharya et al. 2007). This risk increases in soils with poor drainage. Under no-till, surface-applied K fertilizer increases the K concentration in runoff; however, because surface crop residue reduces the force of raindrop impact, sediment loss in the runoff is reduced, resulting in an overall decrease in total K loss (Bertol et al. 2005).

Zöbisch et al. (1995) measured total K loss from water erosion and the impact of cropping on losses in a soil with 8% slope at the Kabete Steep Lands Research Station in Nairobi, Kenya. The four treatments were maize (*Zea mays* L.), common

Table 3.2 Erosion and runoff losses of K from a soil with 8% slope with four cropping treatments: bare fallow, maize, bean, and maize intercropped with bean. (Zöbisch et al. 1995)

	Total K loss	Soil loss	Erosion K loss[a]	Runoff loss	Runoff (dissolved) K loss[b]
Cropping[c]	kg K ha^{-1}	Mg ha^{-1}	% of total	m^3 ha^{-1}	% of total
Bare fallow	52.3	25.4a	97.7	246a	2.3
Maize	14.7	6.17b	97.1	140b	2.9
Bean	8.1	3.39b	94.1	114b	5.9
Maize and bean	5.5	1.89b	92.1	46c	7.9

[a]Bare soil lost significantly more K (kg ha^{-1}) from erosion than the other cropping treatments
[b]Quantities of runoff K were not statistically different across cropping treatments
[c]Maize (*Zea mays* L.) and common bean (*Phaseolus vulgaris* L.)

bean (*Phaseolus vulgaris* L.), maize intercropped with common bean, and bare fallow. Sediment was collected after each rainfall event during the rainy season. There were 22 rainfall events during that period, eight of which, totaling 189.6 mm, produced runoff and erosion. Table 3.2 shows that bare fallow lost the most total K. Losses from the cropped treatments did not differ significantly, although losses from maize tended to be higher. Erosion contributed most to K loss—over 90% for all treatments; K dissolved in runoff comprised less than 10% of the total across all treatments.

Bertoluzzi et al. (2013) observed that the composition of suspended sediment differed significantly from that of the soil from which it was lost. During a 39-mm rainfall event with a peak intensity of 3.25 mm min^{-1}, they sampled suspended solids from a stream monitoring point at the outlet of a 36-farm, 480 ha watershed in the Rio Grande do Sul State in southern Brazil. They divided stream flow occurring 120 min after the start of rainfall into three periods: phase A (15-min duration) was the initial period when streamflow was still near background levels and suspended solid concentrations were low; phase B (70-min duration) was characterized by high suspended solid concentrations and a rapid increase in flow in response to the rainfall event, followed by a slow decrease; and phase C (35-min duration) was when flow rate returned to background levels but contained low concentrations of suspended solids composed of fine particles. The total transported sediment was 29.2 Mg. Clay-sized particles dominated the sediment composition (Table 3.3) even though clay contents of the soils in the watershed were all less than 21%. Smectite comprised more than 90% of the clay, with most of the remainder being kaolinite. Illite, present in quantities up to 25% in some of the soils in the watershed, was not detected in the sediment. Potassium, defined as "labile K," was lost in both phases. Bertoluzzi et al. (2013) defined **labile K** as *the quantity of K in the soil solution plus the quantity of K most readily desorbed into solution from particle surfaces* (i.e., the most soluble K in soil). To quantify labile K, they extracted K from soil with a cation/anion exchange resin for 16 h and then measured the K adsorbed to the resin. They repeated the extraction on the same sample of soil for a total of four successive extractions but considered labile K to be the K desorbed from the soil in only the first extraction. The

Table 3.3 Composition of sediment lost from a 480-ha watershed during a single rainfall event totaling 39 mm with a peak intensity of 3.25 mm min^{-1}. (Bertoluzzi et al. 2013)

Streamflow phase	K lost from watershed[a]	Labile K	Additional bioavailable K	Clay in sediment[b]	Silt in sediment	Sand in sediment
	kg K	% of K lost	% of K lost	%	%	%
A	201	65	32	49	36	14
B	123	22	18	53	34	13
C	189	37	26	72	19	8

[a]Values calculated from sediment K concentration (mg kg^{-1}) and a total sediment load of 29.2 Mg
[b]The sum of clay, silt, and sand contents may not total 100% due to rounding errors

sum of all the K desorbed during the second through fourth extractions was considered to be additional, potentially bioavailable K desorbed more slowly than the labile K (Table 3.3). They found that labile K was greatest in phase A and second highest in phase C, even though phase C had the highest clay content and most persistent suspension of fine particulates. Phases A and C also contained the highest quantities of total bioavailable K (labile plus additional, more slowly desorbed bioavailable K). Both of these phases were associated with slower streamflow.

3.2.2 *Wind Erosion*

Potassium is also lost through wind erosion. The greater the velocity of wind, the more soil is eroded (Wang et al. 2018). The smallest dry particle sizes are most susceptible to wind erosion (Yan et al. 2018). Dry particle size can be measured by air-drying soil samples and then passing them through a series of sieves of progressively smaller mesh sizes. (Dry particle size is not the same as soil texture. Soil texture is determined by using a dispersing solution, typically sodium hexametaphosphate, to break up aggregates into sand-, silt-, and clay-sized fractions.) Depending on the soil and its management, dry particles of a given size can be made up of a variety of percentages of clay, silt, and sand. Yan et al. (2018) observed that wind-eroded soils lost more fine dry particles (<0.2 mm in diameter), than larger dry particles; however, the sand, silt, and clay composition did not change. They also found that, compared to the composition of the bulk soil, a disproportionate amount of K was lost with the fine dry particles.

These few studies elucidate key points about erosion. First, erosion losses of K, though they have not been studied to the extent that leaching losses have, can be a dominant form of loss. Second, the composition of eroded soil can be very different from the bulk soil. Smaller-sized particles are more subject to erosion, and those smaller particles contain a significant portion of the bulk soil's K supply.

3.3 Leaching

Leaching is *the displacement of K below the rhizosphere volume by water percolating down the soil profile*. The **rhizosphere volume** is *the volume of soil adjacent to and influenced by plant roots*. Leaching losses can be expected in the presence of drainage when K inputs exceed the sum of K holding capacity and plant uptake (Johnston 2003). Leaching losses can be as low as 0.2 kg K ha^{-1} $year^{-1}$ in the prairies of northern America (Brye and Norman 2004) and as high as 185 kg K ha^{-1} $year^{-1}$ under urine patches in a silt loam soil in New Zealand (Di and Cameron 2004). These losses are influenced by the rate of K applied, the timing of fertilizer or manure application, soil type and land use, and the amount and pathways of drainage.

It has been proposed that K leaching losses follow a two-phase pattern (Fig. 3.1). Phase A (fast) arises from macropore flow and the presence of K in solution. **Macropore flow**, or **preferential flow** is *the rapid movement of water and solutes through large pores*. These large pores may be channels left by roots or worms, cracks in the soil, or other larger voids formed from biological, geological, or anthropogenic causes. The presence of K in solution at the beginning of the drainage season may result from any one or more of the following: release of K from soil particles upon rewetting; applications of K as fertilizer or manure; leaching from crop residues; or soil biological activity (Alfaro et al. 2004b; Askegaard et al. 2003). Phase B (slow) is dependent on the amount and intensity of rainfall and the associated development of matrix flow later in the drainage season (Alfaro et al. 2004b). **Matrix flow** is *the slow movement of water and solutes from soil volumes of higher total soil water potential to soil volumes of lower total soil water potential.*

Fig. 3.1 Simplified diagram showing the initial rapid leaching phase arising from macropore flow and the presence of K in solution (Phase A), and the subsequent slow leaching phase (Phase B) caused by matrix flow. (Alfaro et al. 2004a)

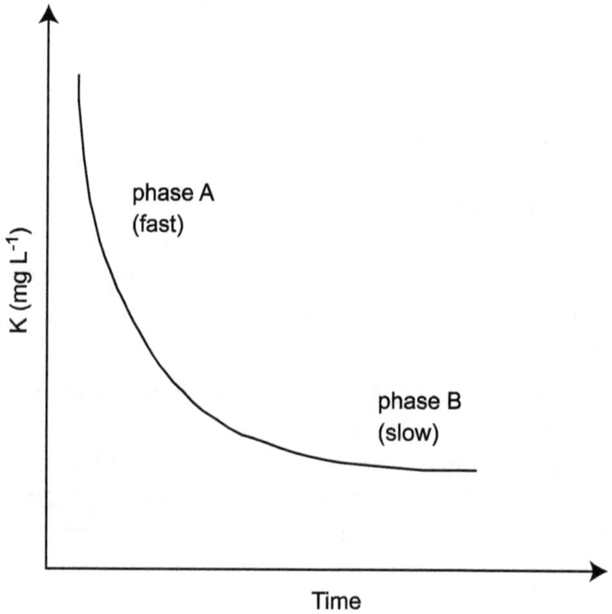

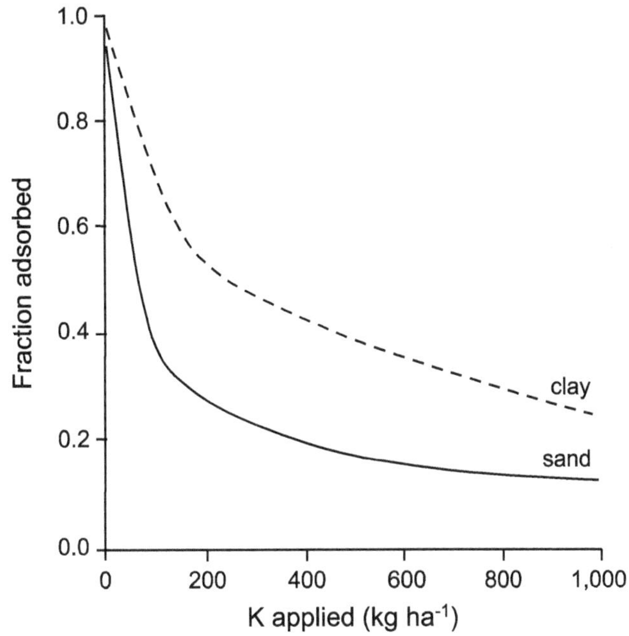

Fig. 3.2 Modelled relationships between the fraction of K adsorbed to soil particle surfaces and the rate of K applied for a soil layer 1-cm thick. (Johnston and Goulding 1992)

Potassium-holding capacity is determined by the presence of soil particles with adsorption sites that hold K on the planar, edge, wedge, and interlayer sites of phyllosilicate minerals. These sites bind K with sufficient energy to keep it from readily re-entering the soil solution. The largest number of these sites is usually present in the clay fraction where the smallest particle sizes maximize surface area. Fewer sites are available in the silt fraction and fewer still are in the sand fraction. Thus, the expectation is that K leaching is least in clay soils, greatest in sandy soils, and somewhere between in loamy soils. Figure 3.2 shows this conceptual relationship for sand and clay for a layer of soil 1-cm thick. As progressively more K is added, an exponentially lower proportion of it is adsorbed as adsorption sites are saturated with K, leaving a greater proportion of the added K in the soil solution. This soil solution K is free to move downward with the wetting front to the next soil layer.

For a given K-holding capacity, a history of higher application rates of K reduces the quantity of additional K that can be retained. Rosolem et al. (2010) observed greater movement of K down the soil profile with successively greater rates of previously applied K. High application rates of manure can have the same effect. Not accounting for K when using organic sources (Askegaard and Eriksen 2002; Bernal et al. 1993) or wastewater (Arienzo et al. 2009) can result in overapplications of K, saturating adsorption sites and exceeding the K holding capacity.

Proper manure management should require farmers to account for K in manures, which is much more variable than in commercial fertilizers. Ideally, each load of manure should be tested for its K content to allow the farmer to back calculate how much K was applied. While some testing is done, in practice, it is more common to use standardized estimates of K concentration and plant-availability. Software

Table 3.4 Results from leachate collected from four different soils placed intact into lysimeters 80 cm in diameter and 135 cm deep and then buried in soil in the field. (Alfaro et al. 2004b)

		Flow	Average K concentration	Total K loss
Soil series	Texture	L	mg L^{-1}	kg ha^{-1}
Radyr	Loam	385a	0.2b	1b
Frilsham	Loam over chalk	371a	0.6b	4b
Newport	Sand	404a	2.4b	19a
Hallsworth	Clay	197b	6.0a	39a

Means separated by different letters within a column are significantly different ($p \leq 0.05$)

decision support tools like MANNER-NPK estimate plant-available nutrients in manures and other organic materials (Nicholson et al. 2013) to help farmers account for the K applied. Another decision support tool is OVERSEER (Wheeler et al. 2003), a farm-scale nutrient budget model used by farmers and consultants throughout New Zealand. The model strives to optimize nutrient input (both inorganic and organic) to maximize production while minimizing nutrient losses to water (Wheeler et al. 2008).

Soils with higher clay contents may not necessarily have greater K-holding capacities. Macropore flow has such a dominant effect on K leaching that it can override differences in soil texture. As an example, Alfaro et al. (2004b) conducted leaching studies on monolith lysimeters of four soils: one sand, two loams, and one clay. Each lysimeter was 80 cm in diameter and 135 cm deep. The study used four lysimeters of each soil as replicates, buried in a field so that the top of the lysimeter was level with the soil surface. The excavated monolith columns preserved the original structure of each soil. Dairy slurry containing 5.7% K (dry basis) was applied at a rate of 24 L per lysimeter, split evenly across four applications during the year. Table 3.4 presents the results for leachate collected between October 2000 and April 2001. As expected, total K leaching losses from the two loam soils were both less than those from the sandy soil, even though flow from all three soils was the same. Unexpectedly, the clay soil lost as much K as the sand. Leachate from the clay soil had the lowest flow but the greatest concentration of K. The dominance of macropore flow in two of the four lysimeters containing the clay soil prevented K in the percolating water from diffusing into clay aggregates and being adsorbed. In those two lysimeters, the average K concentrations across the sampling period were 25.5 and 36.3 mg L^{-1}. In the other two clay soil lysimeters, matrix flow dominated, resulting in average K concentrations of 1.6 and 0.9 mg L^{-1}. When all four replicates of the clay soil were averaged it had the highest losses of K, but with very large variation. These results show how important macropore flow can be for determining the quantity of K leaching losses and the likely spatial variation of leaching in clay soils.

Macropore flow has been shown to occur immediately after a cattle urination event. The quantity of K in the event is large, and is deposited to a small volume of soil. Also, the K supply is usually in excess of the short-term requirements of the plants growing in the urine patch. Consequently, K penetrates to depth in the profile

(Williams and Haynes 1992). As with the lysimeters above, percolation occurs too quickly for any significant sorption reactions between the soil and solutes in the urine (Williams and Haynes 1992). In dairy systems, urine from dairy cattle is responsible for 74 to 92% of total K losses (Williams et al. 1990), accounting for 3–29 kg K ha^{-1} $year^{-1}$ in grazing areas of Chile and New Zealand (Alfaro et al. 2006; Williams et al. 1990), and up to 185 kg K ha^{-1} $year^{-1}$ under urine patches in New Zealand soils (Di and Cameron 2004).

While practitioners in the field often associate greater K retention with higher cation exchange capacity (CEC), CEC has not proven to be a good predictor of K loss from soils (Quémener 1986). An important confounding factor is organic matter, which has a high CEC but does not bind K as strongly as mineral adsorption sites (Quémener 1986; Thomas and Hipp 1968). In high organic matter soils, heavy rains can seriously deplete the amount of soluble K in a matter of days (Thomas and Hipp 1968). In addition, the presence of clay-sized phyllosilicate minerals with high CEC does not limit K leaching losses if macropore flow is present. Thus, CEC alone is not a good predictor of K leaching losses; however, it could be a useful factor in prediction models that incorporate additional factors.

When N increases crop K uptake and yield, lower K leaching losses may follow. For instance, the use of nitrification inhibitors such as dicyandiamide in grassland soils has been found to reduce K leaching losses by up to 65%, probably as an indirect effect of its increasing yield (Di and Cameron 2004). However, N applications may also result in larger quantities of leached K, even when system productivity is increased. In a study conducted in southwest England (Alfaro et al. 2003), larger K leaching losses occurred even though N applications increased system productivity so much that K outputs exceeded K inputs, resulting in a net negative K budget.

In intensively managed agricultural systems with nutrient surpluses, greater K leaching losses are usually linked to greater N leaching losses (Brye and Norman 2004). Nitrate leached through the soil profile forms ion pairs with other solution cations to balance charge. When K^+ is part of the ion pair, it will move with NO_3^- down the soil profile. This effect has also been observed after liming. In acidic soils, liming promotes nitrification and increases nitrate concentration in the uppermost soil layers, resulting in higher K leaching losses (Crusciol et al. 2011). Additionally, calcium (Ca^{2+}) and magnesium (Mg^{2+}) in lime can exchange with K^+ on soil particle surfaces, moving it into the soil solution. Greater K leaching can also occur when competing cations are present in irrigation water (Kolahchi and Jalali 2007; Sekhon 1982).

3.4 Modeling Potassium Losses

Several models, ranging from conceptual to computational, have been developed to estimate K behavior. This section highlights some of these past efforts to model K losses from soils.

3.4.1 *Conceptual Model of Leaching*

Alfaro et al. (2004a) proposed a conceptual model that combined surface runoff, macropore flow, and matrix flow. Their model was based on research conducted on field drainage plots (Armstrong and Garwood 1991). Each of those plots collected surface runoff and subsurface runoff downslope from the upper 30 cm of soil. Subsurface runoff from this layer was associated with macropore flow. On drained plots, mole channels placed below 30 cm were added along with an associated second drainage collection point downslope. Flow collected at that second point was associated with matrix flow and classified as leaching. The conceptual model is presented in Fig. 3.3. The solid black line represents the combination of surface runoff, macropore flow through the upper 30 cm of soil, and subsurface runoff from that upper layer. The dashed line represents matrix flow. At the start of the time period considered, the soil is dry, and there is little runoff or matrix flow. When rainfall starts, runoff flow increases, but matrix flow remains low. Once enough time has passed for water to infiltrate the soil, matrix flow subsequently increases. According to these principles, higher intensity rainfall events favor larger losses by runoff, because larger quantities of water fall before matrix flow become significant.

3.4.2 *EPIC*

De Barros et al. (2004) modified the EPIC model to estimate K losses through leaching, surface runoff, and subsurface runoff. They adjusted the model to match the climatic (semiarid), pedological, and cropping system conditions in Brazil. Soil is modeled as a stack of soil layers. For only the first layer, K lost in surface runoff is subtracted and not added to any remaining soil layers. For all layers, K lost in subsurface runoff and leaching are subtracted from a given layer and added to the next layer until the bottom of the soil profile is reached.

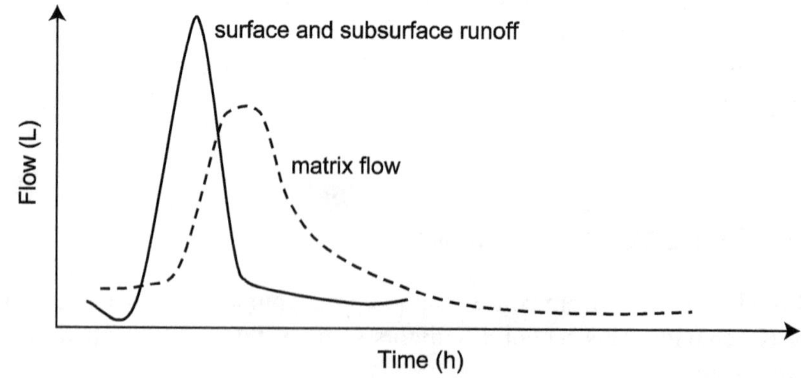

Fig. 3.3 Idealized hydrograph of water flow after rainfall on a dry soil. (Alfaro et al. 2004a)

3.4.3 KLEACH

Johnston and Goulding (1992) used relationships similar to Fig. 3.2 as the basis for the model KLEACH. This model considers soil to be a series of consecutively stacked 1-cm thick layers. Added K not adsorbed by one layer moves to the adjacent layer below it. The model estimates K adsorbed by each successive layer as K moves down the soil profile. Figure 3.4 shows the results of two simulations for the cultivated layer (20 cm) of soil. The first (Fig. 3.4a) is for 200 kg K ha^{-1} applied as KCl to a clay loam soil. The second (Fig. 3.4b) models 800 kg K ha^{-1} applied as manure slurry to a sandy soil. In both cases, KLEACH models 100% adsorption of

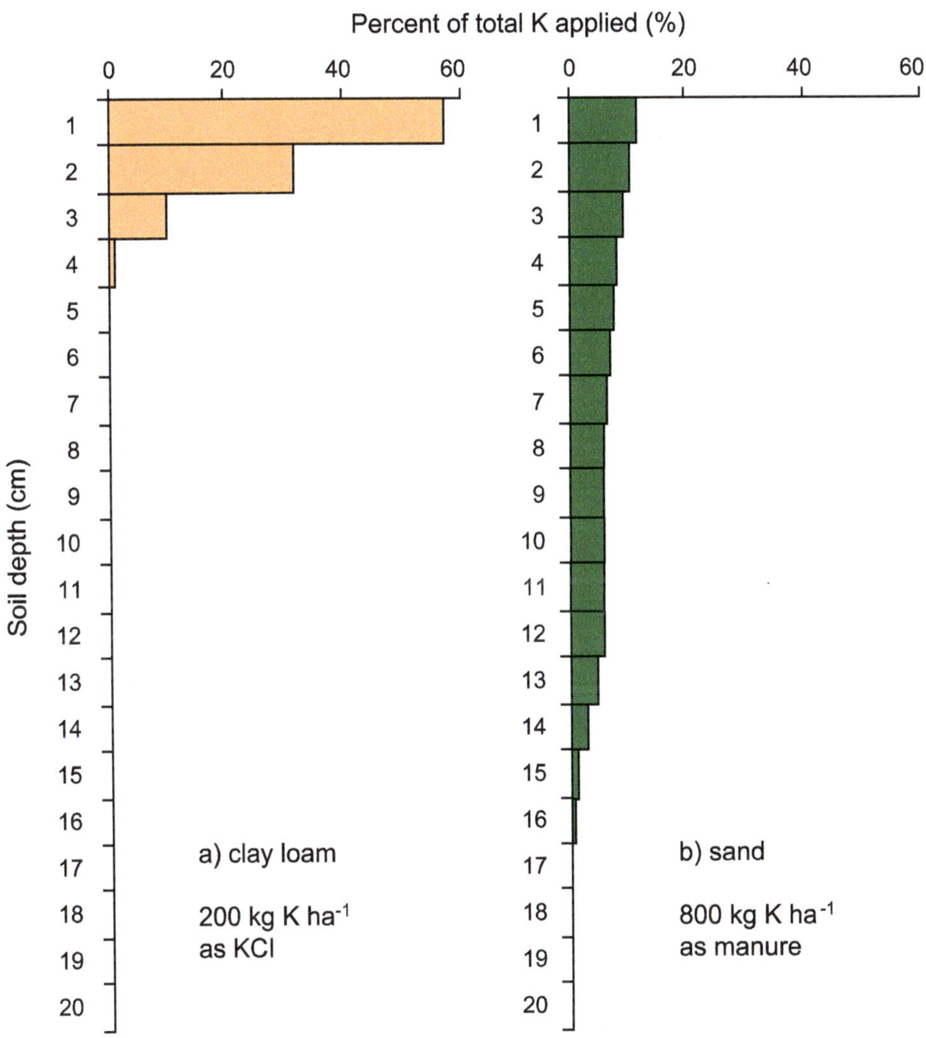

Fig. 3.4 Predicted distribution of K applied (**a**) at a rate of 200 kg K ha^{-1} as KCl on a clay loam soil, and (**b**) 800 kg K ha^{-1} as manure slurry on a sandy soil in the cultivated layer (20 cm) of each soil. (Johnston and Goulding 1992)

the applied K when summed over all soil depths. Compared to the clay loam, the sand is predicted to adsorb less K in each 1-cm layer, resulting in deeper penetration of K down the soil profile. The model predicts the movement of K through the soil profile and so the risk of leaching, rather than actual leaching loss.

3.4.4 NUTMON

Smaling and Fresco (1993) developed the multilevel decision support model for monitoring the soil nutrient balance (NUTMON). Data collected from the Kisii District in Kenya were used to develop the model (Smaling et al. 1993), and the authors stated that the regional level was most appropriate for operating it.

In NUTMON, loss of K by erosion is based on estimates of soil erosion, using the universal soil loss equation (Smaling et al. 1993). Soil loss is multiplied by estimated total K concentration in the 0–20 cm layer of soil, which varies by clay content and exchangeable K. To account for the enrichment of K in eroded sediment, compared to the soil in situ, the estimated total K loss is multiplied by an enrichment factor. The result is multiplied by 0.75 to quantify net K loss. In a related effort, although not formally part of NUTMON, Lesschen et al. (2007) used the LAPSUS model (Schoorl et al. 2000, 2002) to estimate sediment transport rates, quantifying soil erosion as well as sedimentation.

NUTMON estimates the amount of K leached as a percentage of the K applied as fertilizer and manure (Smaling and Fresco 1993; Smaling et al. 1993). These percentages range from 0.5 to 1.0% and are adjusted for annual rainfall and clay content, with higher annual rainfall adjusting percentages toward 1.0% and higher clay content adjusting percentages the opposite direction, toward 0.5%. Lesschen et al. (2007) revised this methodology based on new algorithms developed for nitrogen (N) in NUTMON. They developed the following multiple regression model (Eq. 3.1) to estimate K leaching for a wide range of soils and climates:

$$\text{K leached} = -6.87 + 0.0117P + 0.173F_{\text{K}} - 0.265\text{CEC} \tag{3.1}$$

where:

P annual precipitation (mm)
F_{K} mineral and organic fertilizer addition (kg K ha^{-1})
CEC cation exchange capacity (cmol(+) kg^{-1}).

This equation was used to develop the K-leaching estimates used to calculate the K balances of sub-Saharan African countries presented in Chap. 11.

3.4.5 SVMLEACH-NK POTATO

Fortin et al. (2015) developed SVMLEACH-NK POTATO to simulate the daily dynamics of N and K leaching losses during the potato-growing season in Quebec, Canada. The model uses the least squares support vector machines (LS-SVMs) method, a machine learning technique that looks for patterns in data and performs regression in a high-dimension feature space (Fortin et al. 2014). The best-fit empirical model built from the training data set used the following variables: seasonal precipitation, seasonal temperature, N rate, day of year, and percent clay. Consequently, model users enter planting date, N application rate, percent clay, daily mean temperature, and daily total precipitation. The model relates nitrate leaching to K leaching, consistent with K^+ being one cation forming an ion pair with NO_3^- in the soil solution during leaching.

3.4.6 SWAT-K

Wang et al. (2016) modified the Soil and Water Assessment Tool (SWAT) to simulate stream K load and K budgets of the Shibetwu River Watershed, an area of dairy farming in Japan where 95% of the agricultural land is under pasture. The SWAT-K model is an example of a more integrative K management approach. The model considers K losses in surface runoff, subsurface runoff (lateral flow), leaching, and erosion and simulates total K load at the outlet of the watershed. Wang et al. (2016) calibrated the model to measured levels of streamflow, suspended sediment load, and dissolved K.

3.5 Open Burning

Open burning is *the unenclosed combustion of materials in an ambient environment* (Lemieux et al. 2004). We limit our discussion to the burning of plant material in place at or near the soil surface, including wildfires, prescribed burning, burning for land clearing, and stubble burning (the burning of crop residues on agricultural land). Emissions from burning include particulate matter, nutrients, water, carbon dioxide, carbon monoxide, methane, volatile and semi-volatile organics, acid aerosols, metals, polychlorinated dibenzo-*p*-dioxins, polychlorinated dibenzofurans, and polychlorinated biphenyls (Lemieux et al. 2004). Potassium is among the nutrients lost in burning.

Gaudichet et al. (1995) identified three K-containing particles emitted from biomass burning: carbonaceous, drop-shaped microsoot particles ranging from 0.2 to 0.5μm contain K associated with sulfur (S); potassium salt condensed as cubic crystals of KCl less than 1μm in size and occurring either separately or embedded in other constituents associated with open flames; larger vegetation relics (greater than 1µm) containing K in partially combusted material. Long-distance transport is possible for all of these particles.

Potassium is the most abundant of the nutrients in particles emitted from burning vegetation. This is in contrast to emissions from fossil fuels that contain little K (Ruscio et al. 2016). Gaudichet et al. (1995) suggested that the presence of K could be used to differentiate biomass emissions from fossil fuel emissions. Support for such differentiation was provided by Amici et al. (2011) who used hyperspectral imaging to confirm that K emission was a characteristic of open flames from biomass burning.

We know relatively little about the quantity of K lost by open burning. The fact that K can be lost in this way is contrary to the belief that burning simply leaves K in the ash on the field. While ash does contain K, it clearly does not contain all the K that was in the plant material prior to burning. Further research will need to determine just how much K is lost during open burning.

3.6 Considerations for Potassium Recommendations

The development and implementation of Best Management Practices (BMP) for fertilizer use, with a focus on source, rate, time, and placement (i.e., the "4R" approach: right source, right rate, right time and right place), is necessary in the short-term to increase productivity and economic returns and in the long-term to provide more efficient ways of using non-renewable resources upon which food, feed, fiber, and fuel production depend (Fixen and Johnston 2012).

Source of K can be a factor when adjusting recommendations to limit losses. Most K sources (fertilizers, manures, composts, crop residues, and wastewaters) contain K in the simple cationic form of K^+ (Arienzo et al. 2009; Stockdale et al. 2002), and inorganic fertilizers and organic manures are equally effective for meeting the K requirements of crops (Johnston and Goulding 1990). Organic sources, however, may sustain higher concentrations of solution K (Addiscott and Johnston 1975). To what extent these higher levels contribute to greater leaching or more efficient plant utilization has yet to be clarified in research. Enhanced-efficiency fertilizers (for example, Di and Cameron 2004; Gillman and Noble 2005; Yang et al. 2016) are a technological approach to increasing K use efficiency. The main advantage of these is that the K release may more closely match plant requirements through the season, reducing the risk of losses. However, in a greenhouse experiment, Bley et al. (2017) found that using slow-release (polymer coated) KCl reduced leaching compared to traditional KCl, but the slow rate of release was not sufficient to meet crop K demand during the initial phase of growth. The rate of adoption of technologies such as these is limited because of: the cost of the materials in comparison to traditional sources; the existence of regulatory policies (Gillman and Noble 2005); health and safety issues associated with their application

(Timilsena et al. 2015); and sometimes a lack of information available on its impacts on productivity at the farm level.

The rate of K fertilizer recommended to farmers should account for K losses. While some decision tools exist, the focus has been on calculating more accurate nutrient budgets to eliminate overapplications. Adjusting K inputs in relation to estimated outputs to avoid a surplus at risk to leaching is a key factor for reducing K leaching losses. In tropical, coarse-textured soils managed under no-till, K leaching is high when fertilizer application exceeds plant demand (Rosolem and Steiner 2017). Process models that simulate K loss by water and wind erosion have yet to be developed, and even those that exist and estimate runoff and leaching losses for a given set of conditions have yet to be incorporated into algorithms that adjust K fertilizer rates recommended to farmers.

Timing of fertilizer applications is another approach to managing K losses. High K application rates may generate high K leaching losses, especially when drainage exceeds 500 mm (Bolton et al. 1970; Thomas and Hipp 1968). In fact, rainfall distribution and intensity are often more important than total precipitation (Quémener 1986) because of their impacts on surface runoff relative to the amount of matrix flow (Heathwaite et al. 1996). Potassium leaching losses may significantly increase when K fertilizer is applied to drained soils because of preferential flow, as discussed previously. In these situations, a key aspect to reduce K losses is the time interval between K application and the rainfall event: the longer this interval, the lower the losses (Alfaro et al. 2004a). Splitting a large application into two or more smaller applications is recommended when the risk of loss is high. On organic soils, K should be applied close to the time of active uptake by crops to avoid leaching, since these soils do not bind K tightly, even though they have a high CEC.

Fertilizer placement is also likely to be an effective way to reduce K loss, but almost no research has examined how various K placement methods affect it. Work on P has shown that subsurface banding of P fertilizers reduces runoff P losses compared to broadcast applications (Kimmel et al. 2001). The work cited earlier by Korucu (2018) showed that when K fertilizer was applied by surface broadcasting, a cover crop reduced K runoff losses. Sato et al. (2009), examining K placement in raised beds with seepage irrigation, suggested that evaluating K losses will require information about the height and seasonal dynamics of the water table. At the time of writing, it appears that no general guidance exists for placement strategies to minimize K losses.

3.7 Conclusions

Relative to other nutrients, K losses and transfers have not been well researched. Leaching has received the most attention, but erosion appears to be equally important in terms of the quantity of K lost. Dissolved K in runoff may contribute less to total loss than leaching and erosion. How much K is lost from open burning is still not well known. Potassium losses are associated particularly with losses of smaller soil particulates, which in turn are associated with loss of clay minerals. Not only is this loss of fertility detrimental in the short term, but it also appears to lead to reductions in K-holding capacity in the long term. Improved K management strategies must go beyond considering only fertilizer source, rate, time, and placement and be developed to incorporate strategies to maintain soil cover so that nutrients can be recycled more effectively. Building better decision support tools that incorporate process models will better inform farmers and help them make decisions that achieve the desired outcomes of efficient K use, including minimal losses.

Acknowledgments Marta A. Alfaro, Deputy Director of National Research and Development, Instituto de Investigaciones Agropecuarias, Ministerio de Agricultura, Osorno, Chile, led the discussions on this topic at the workshop in Rome and contributed to an early version of the chapter.

References

Acharya GP, McDonald MA, Tripathi BP, Gardner RM, Mawdesley KJ (2007) Nutrient losses from rain-fed bench terraced cultivation systems in high rainfall areas of the mid-hills of Nepal. Land Degrad Dev 18(5):486–499. https://doi.org/10.1002/ldr.792

Addiscott TM, Johnston AE (1975) Potassium in soils under different cropping systems: 3. Nonexchangeable potassium in soils from long-term experiments at Rothamsted and Woburn. J Agric Sci 84(3):513–524. https://doi.org/10.1017/S0021859600052734

Alfaro MA, Jarvis SC, Gregory PJ (2003) Potassium budgets in grassland systems as affected by nitrogen and drainage. Soil Use Manag 19(2):89–95. https://doi.org/10.1111/j.1475-2743.2003.tb00286.x

Alfaro MA, Gregory PJ, Jarvis SC (2004a) Dynamics of potassium leaching on a hillslope grassland soil. J Environ Qual 33(1):192–200. https://doi.org/10.2134/jeq2004.1920

Alfaro MA, Jarvis SC, Gregory PJ (2004b) Factors affecting potassium leaching in different soils. Soil Use Manag 20(2):182–189. https://doi.org/10.1111/j.1475-2743.2004.tb00355.x

Alfaro M, Salazar F, Teuber N (2006) Potassium surface runoff and leaching losses in a beef cattle grazing system on volcanic soil. Better Crops 90(4):20–22

Alfaro M, Salazar F, Iraira S, Teuber N, Villarroel D, Ramírez L (2008) Nitrogen, phosphorus and potassium losses in a grazing system with different stocking rates in a volcanic soil. Chil J Agric Res 68(2):146–155

Amici S, Wooster MJ, Piscini A (2011) Multi-resolution spectral analysis of wildfire potassium emission signatures using laboratory, airborne and spaceborn remote sensing. Remote Sens Environ 115:1811–1823. https://doi.org/10.1016/j.rse.2011.02.022

Anguria P, Chemining'wa GN, Onwonga RN, Ugen MA (2017) Decomposition and nutrient release of selected cereal and legume crop residues. J Agric Sci 9(6):108–119. https://doi.org/10.5539/jas.v9n6p108

Arienzo M, Christen EW, Quayle W, Kumar A (2009) A review of the fate of potassium in the soil–plant system after land application of wastewaters. J Hazard Mater 164(2–3):415–422. https://doi.org/10.1016/j.jhazmat.2008.08.095

Armstrong AC, Garwood EA (1991) Hydrological consequences of artificial drainage of grassland. Hydrol Process 5:157–174. https://doi.org/10.1002/hyp.3360050204

Askegaard M, Eriksen J (2002) Exchangeable potassium in soil as indicator of potassium status in an organic crop rotation on loamy sand. Soil Use Manag 18(2):84–90. https://doi.org/10.1111/j.1475-2743.2002.tb00224.x

Askegaard M, Eriksen J, Olesen JE (2003) Exchangeable potassium and potassium balances in organic crop rotations on a coarse sand. Soil Use Manag 19(2):96–103. https://doi.org/10.1111/j.1475-2743.2003.tb00287.x

Bender RR, Haegele JW, Ruffo ML, Below FE (2013) Nutrient uptake, partitioning, and remobilization in modern, transgenic insect-protected maize hybrids. Agron J 105:161–170. https://doi.org/10.2134/agronj2012.0352

Bernal MP, Roig A, García D (1993) Nutrient balances in calcareous soils after application of different rates of pig slurry. Soil Use Manag 9(1):9–11. https://doi.org/10.1111/j.1475-2743.1993.tb00920.x

Bertol I, Guadagnin JC, González AP, do Amaral AJ, Brignoni LF (2005) Soil tillage, water erosion, and calcium, magnesium and organic carbon losses. Sci Agric 62(6):578–584

Bertoluzzi EC, dos Santos DR, Santanna MA, Caner L (2013) Mineralogy and nutrient desorption of suspended sediments during a storm event. J Soils Sediments 13:1093–1105. https://doi.org/10.1007/s11368-013-0692-4

Bley H, Gianello C, Santos LS, Selau LPR (2017) Nutrient release, plant nutrition, and potassium leaching from polymer-coated fertilizer. Rev Bras Ciênc Solo 41:e0160142. https://doi.org/10.1590/18069657rbcs20160142

Bolton EF, Aylesworth JW, Hore FR (1970) Nutrient losses through tile drains under three cropping systems and two fertility levels on a Brookston clay soil. Can J Soil Sci 50(3):275–279. https://doi.org/10.4141/cjss70-038

Brye KR, Norman JM (2004) Land-use effects on anion-associated cation leaching in response to above-normal precipitation. Acta Hydrochim Hydrobiol 32(3):235–248. https://doi.org/10.1002/aheh.200300534

Crusciol CAC, Garcia RA, Castro GSA, Rosolem CA (2011) Nitrate role in basic cation leaching under no-till. Rev Bras Ciênc Solo 35(6):1975–1984

de Barros I, Williams JR, Gaiser T (2004) Modeling soil nutrient limitations to crop production in semiarid NE of Brazil with a modified EPIC version. I. Changes in the source code of the model. Ecol Model 178:441–456. https://doi.org/10.1016/j.ecolmodel.2004.04.015

De M, Moore AD, Mikkelsen RL (2019) In-season accumulation and partitioning of macronutrients and micronutrients in irrigated sugar beet production. J Sugar Beet Res 56(3–4):54–78. https://doi.org/10.5274/Jsbr.56.3.56

Di HJ, Cameron KC (2004) Effects of the nitrification inhibitor dicyandiamide on potassium, magnesium and calcium leaching in grazed grassland. Soil Use Manag 20(1):2–7. https://doi.org/10.1111/j.1475-2743.2004.tb00330.x

Fixen PE, Johnston AM (2012) World fertilizer nutrient reserves: a view to the future. J Sci Food Agr 92(5):1001–1005. https://doi.org/10.1002/jsfa.4532

Fortin JG, Morais A, Anctil F, Parent LE (2014) Comparison of machine learning regression methods to simulate NO3 flux in soil solution under potato crops. Appl Math 5:832–841. https://doi.org/10.4236/am.2014.55079

Fortin JG, Morais A, Anctil F, Parent LE (2015) SVMLEACH-NK POTATO: a simple software tool to simulate nitrate and potassium co-leaching under potato crop. Comput Electron Agric 110:259–266. https://doi.org/10.1016/j.compag.2014.11.025

Gaudichet A, Echalar F, Chatenet B, Quisefit JP, Malingre G, Cachier H, Buat-Menard P, Artaxo P, Maenhaut W (1995) Trace elements in tropical African savanna biomass burning aerosols. J Atmos Chem 22:19–39. https://doi.org/10.1007/BF00708179

Gillman G, Noble A (2005) Environmentally manageable fertilizers: a new approach. Environ Qual Manag 15(2):59–70. https://doi.org/10.1002/tqem.20081

Heathwaite AL, Johnes PJ, Peters NE (1996) Trends in nutrients. Hydrol Process 10(2):263–293. https://doi.org/10.1002/(SICI)1099-1085(199602)10:2<263::AID-HYP441>3.0.CO;2-K

Johnston AE (2003) Understanding potassium and its use in agriculture. European Fertilizer Manufacturers Association, Brussels. http://www.pda.org.uk/wp/wp-content/uploads/2015/08/EFMA_Potassium_booklet_2003.pdf. Accessed 14 May 2020

Johnston AE, Goulding KWT (1990) The use of plant and soil analyses to predict the potassium supplying capacity of soil. In: Proceedings of the 22nd colloquium of the international potash institute: development of K-fertilizer recommendations, International Potash Institute, Basel, pp 177–204. https://ipipotash.org/en/. Accessed 14 May 2020

Johnston AE, Goulding KWT (1992) Potassium concentrations in surface and groundwaters and the loss of potassium in relation to land use. In: Proceedings of the 23rd colloquium of the International Potash Institute: potassium in ecosystems: biogeochemical fluxes of cations in agro- and forest-systems. International Potash Institute, Basel, pp 177–204. https://ipipotash.org/en/. Accessed 14 May 2020

Kimmel RJ, Pierzynski GM, Janssen KA, Barnes PL (2001) Effects of tillage and phosphorus placement on phosphorus runoff losses in a grain sorghum-soybean rotation. J Environ Qual 30:1324–1330. https://doi.org/10.2134/jeq2001.3041324x

Kolahchi Z, Jalali M (2007) Effect of water quality on the leaching of potassium from sandy soil. J Arid Environ 68(4):624–639. https://doi.org/10.1016/j.jaridenv.2006.06.010

Korucu T, Shipitalo MJ, Kaspar TC (2018) Rye cover crop increases earthworm populations and reduces losses of broadcast, fall-applied, fertilizers in surface runoff. Soil Till Res 180:99–106. https://doi.org/10.1016/j.still.2018.03.004

Lahav E, Turner DW (1989) Banana nutrition. International Potash Institute Bulletin 7. Berne

Lemieux PM, Lutes CC, Santoianni DA (2004) Emissions of organic air toxics from open burning: a comprehensive review. Prog Energ Combust 30:1–32. https://doi.org/10.1016/j.pecs.2003.08.001

Lesschen JP, Stoorvogel JJ, Smaling EMA, Heuvelink GBM, Veldkamp A (2007) A spatially explicit methodology to quantify soil nutrient balances and their uncertainties at the national level. Nutr Cycl Agroecosyst 78(2):111–131. https://doi.org/10.1007/s10705-006-9078-y

Muhammad S, Luedeling E, Brown PH (2009) A nutrient budget approach to nutrient management in almond. Proc Inter Plant Nutr Coll XVI. http://ucanr.org/sites/nm/files/76672.pdf. Accessed 14 May 2020

Nathan MV, Sun Y, Dunn D (2009) Nutrient removal values for major agronomic crops in Missouri report for 2006–2007. https://www.researchgate.net/publication/268082826. Accessed 14 May 2020

Nicholson FA, Bhogal A, Chadwick D, Gill E, Gooday RD, Lord E, Misselbrook T, Rollett AJ, Sagoo E, Smith KA, Thorman RE, Williams JR, Chambers BJ (2013) An enhanced software tool to support better use of manure nutrients: MANNER-*NPK*. Soil Use Manag 29(4):473–484. https://doi.org/10.1111/sum.12078

Quémener J (1986) Nutrient balances and need for potassium. In: Proceedings of the 13th international potash institute congress. International Potash Institute, Basel, pp 41–69. https://ipipotash.org/en/. Accessed 14 May 2020

Rogers CW, Dari B, Hu G, Mikkelsen R (2019) Dry matter production, nutrient accumulation, and nutrient partitioning of barley. J Plant Nutr Soil Sci 182:367–373. https://doi.org/10.1002/jpln.201800336

Rosolem CA, Steiner F (2017) Effects of soil texture and rates of K input on potassium balance in tropical soil. Eur J Soil Sci 68:658–666. https://doi.org/10.1111/ejss.12460

Rosolem CA, Sgariboldi T, Garcia RA, Calonego JC (2010) Potassium leaching as affected by soil texture and residual fertilization in tropical soils. Commun Soil Sci Plan 41(16):1934–1943. https://doi.org/10.1080/00103624.2010.495804

Rousseva S, Torri D, Pagliai M (2002) Effect of rain on the macroporosity at the soil surface. Eur J Soil Sci 53(1):83–94. https://doi.org/10.1046/j.1365-2389.2002.00426.x

Ruscio A, Kazanc F, Levendis YA (2016) Comparison of fine ash emissions generated from biomass and coal combustion and valuation of predictive furnace deposition indices: a review. J Energ Eng 142:E4015007. https://doi.org/10.1061/(ASCE)EY.1943-7897.0000310

Sato S, Morgan KT, Ozores-Hampton M, Simonne EH (2009) Spatial and temporal distributions in sandy soils with seepage irrigation: II. Phosphorus and potassium. Soil Sci Soc Am J 73:1053–1060. https://doi.org/10.2136/sssaj2008.0114

Schoorl JM, Sonneveld MPW, Veldkamp A (2000) Three-dimensional landscape process modelling: the effect of DEM resolution. Earth Surf Proc Land 25(9):1025–1034. https://doi.org/10.1002/1096-9837(200008)25%3A9<1025%3A%3AAID-ESP116>3.0.CO%3B2-Z

Schoorl JM, Veldkamp A, Bouma J (2002) Modeling water and soil redistribution in a dynamic landscape context. Soil Sci Soc Am J 66(5):1610–1619. https://doi.org/10.2136/sssaj2002.1610

Sekhon GS (1982) Potassium recycling in agriculture. In: Food and Fertilizer Technology Centre (ed) Recycling of potassium and phosphorus in agriculture, Tech Bull 69. Food and Fertilizer Technology Centre, Taiwan, pp 1–15

Smaling EMA, Fresco LO (1993) A decision-support model for monitoring nutrient balances under agricultural land use (NUTMON). Geoderma 60(1–4):235–256. https://doi.org/10.1016/0016-7061(93)90029-K

Smaling EMA, Stoorvogel JJ, Sindmeijer PN (1993) Calculating soil nutrient balances in Africa at different scales. II District scale. Fert Res 35:237–250. https://doi.org/10.1007/BF00750642

Stockdale EA, Shepherd MA, Fortune S, Cuttle SP (2002) Soil fertility in organic farming systems – fundamentally different? Soil Use Manag 18.(S1:301–308. https://doi.org/10.1111/j.1475-2743.2002.tb00272.x

Tarkalson DD, Brown B, Kok H, Bjorneberg DL (2009) Impact of removing straw from wheat and barley fields: a literature review. Better Crops 93(3):17–19

Thomas GW, Hipp BW (1968) Soil factors affecting potassium availability. In: Kilmer VJ (ed) The role of potassium in agriculture. American Society of Agronomy, Crop Science Society of America, Soil Science Society of America, Madison, pp 269–291. https://doi.org/10.2134/1968.roleofpotassium.c13

Timilsena YP, Adhikari R, Casey P, Muster T, Gill H, Adhikari B (2015) Enhanced efficiency fertilizers: a review of formulation and nutrient release patterns. J Sci Food Agr 95 (6):1131–1142. https://doi.org/10.1002/jsfa.6812

Unkovich M, Baldock J, Forbes M (2010) Variability in harvest index of grain crops and potential significance for carbon accounting: examples from Australian agriculture. Adv Agron 105:173–219. https://doi.org/10.1016/S0065-2113(10)05005-4

Wang C, Jiang R, Boithias L, Sauvage S, Sánchez-Pérez J-M, Mao X, Han Y, Hayakawa A, Kuramochi K, Hatano R (2016) Assessing potassium environmental losses from a dairy farming watershed with the modified SWAT model. Agric Water Manage 175:91–104. https://doi.org/10.1016/j.agwat.2016.02.007

Wang X, Lang L, Hua T, Li H, Zhang C, Ma W (2018) Effects of aeolian processes on soil nutrient loss in the Gonghe Basin, Qinghai-Tibet Plateau: an experimental study. J Soils Sediments 18:229–238. https://doi.org/10.1007/s11368-017-1734-0

Wheeler DM, Ledgard SF, DeKlein CAM, Monaghan RM, Carey PL, McDowell RW, Johns KL (2003) OVERSEER® nutrient budgets – moving towards on-farm resource accounting. Proc NZ Grassland Assoc 65:191–194. https://www.grassland.org.nz/proceedings.php. Accessed 14 May 2020

Wheeler DM, Ledgard SF, DeKlein CAM (2008) Using the OVERSEER nutrient budget model to estimate on-farm greenhouse gas emissions. Aust J Exp Agric 48(1–2):99–103. https://doi.org/10.1071/EA07250

Williams PH, Haynes RJ (1992) Balance sheet of phosphorus, sulphur and potassium in a long-term grazed pasture supplied with superphosphate. Fert Res 31(1):51–60. https://doi.org/10.1007/BF01064227

Williams PH, Gregg PEH, Hedley MJ (1990) Mass balance modelling of potassium losses from grazed dairy pasture. New Zeal J Agric Res 33(4):661–668. https://doi.org/10.1080/00288233.1990.10428470

Yan Y, Wang X, Guo Z, Chen J, Xin X, Xu D, Yan R, Chen B, Xu L (2018) Influence of wind erosion on dry aggregate size distribution and nutrients in three steppe soils in northern China. Catena 170:159–168. https://doi.org/10.1016/j.catena.2018.06.013

Yang X, Geng J, Li C, Zhang M, Chen B, Tian X, Zheng W, Liu Z, Wang C (2016) Combined application of polymer coated potassium chloride and urea improved fertilizer use efficiencies, yield and leaf photosynthesis of cotton on saline soil. Field Crops Res 197:63–73. https://doi.org/10.1016/j.fcr.2016.08.009

Zöbisch MA, Richter C, Heiligtag B, Schlott R (1995) Nutrient losses from cropland in the Central Highlands of Kenya due to surface runoff and soil erosion. Soil Till Res 33:109–116. https://doi.org/10.1016/0167-1987(94)00441-G

4

Root Traits and Rhizosphere Characteristics Determining Potassium (K) Acquisition from Soils

Philippe Hinsinger, Michael J. Bell, John L. Kovar, and Philip J. White

Abstract Plants acquire K^+ ions from the soil solution, and this small and dynamic pool needs to be quickly replenished via desorption of surface-adsorbed K from clay minerals and organic matter, by release of interlayer K from micaceous clay minerals and micas, or structural K from feldspars. Because of these chemical interactions with soil solid phases, solution K^+ concentration is kept low and its mobility is restricted. In response, plants have evolved efficient strategies of root foraging. Root traits related to root system architecture (root angle and branching), root length and growth, together with root hairs and mycorrhiza-related traits help to determine the capacity of plants to cope with the poor mobility of soil K. Rooting depth is also important, given the potentially significant contribution of subsoil K in many soils. Root-induced depletion of K^+ shifts the exchange equilibria, enhancing desorption of K, as well as the release of nonexchangeable, interlayer K from minerals in the rhizosphere. Both these pools can be bioavailable if plant roots can take up significant amounts of K at low concentrations in the soil solution (in the micromolar range). In addition, roots can significantly acidify their environment or release large amounts of organic compounds (exudates). These two processes ultimately promote the dissolution of micas and feldspars in the rhizosphere, contributing to the mining strategy evolved by plants. There are thus several root or rhizosphere-related traits

P. Hinsinger (✉)
Eco&Sols, University of Montpellier, CIRAD, INRAE, Institut Agro, IRD, Montpellier, France
e-mail: philippe.hinsinger@inrae.fr

M. J. Bell
School of Agriculture and Food Science, The University of Queensland, Brisbane, Australia
e-mail: m.bell4@uq.edu.au

J. L. Kovar
USDA-ARS, National Laboratory for Agriculture and the Environment, Ames, IA, USA
e-mail: john.kovar@usda.gov

P. J. White
Ecological Sciences, The James Hutton Institute, Dundee, UK

King Saud University, Riyadh, Saudi Arabia
e-mail: philip.white@hutton.ac.uk

(morphological, physiological, or biochemical) that determine the acquisition of K by crop species and genotypes.

4.1 Soil Properties and Processes Determining the Acquisition of Potassium by Plants

A number of soil characteristics determine K mobility,[1] availability, and bioavailability to plants. These properties, together with the actual distribution of the various pools of K in the soil profile and horizons, ultimately determine the most desirable root and rhizosphere-related traits to search for in order to improve K acquisition efficiency in crops.

4.1.1 Potassium Mobility: Mass Flow Versus Diffusion in the Rhizosphere

Potassium is present in the soil solution as K^+ ions, which experience rather strong interactions (adsorption/desorption) with the many soil constituents contributing to cation exchange capacity, notably clay minerals and organic matter (Sparks and Huang 1985; Sparks 1987; Chap. 7). The consequences of such interactions are twofold. First, they buffer the concentration of K^+ in the soil solution to values that commonly range from one to several hundred micromoles per dm^3 (Asher and Ozanne 1967; Hinsinger 2006), i.e., concentrations that are significantly greater than those of phosphate, but less than those of nitrate. Second, they limit K mobility in the soil. Thus, compared with nitrate, K leaching occurs in significant amounts only in fertilized, light-textured soils. In addition, while mass flow can contribute significantly to the transport of nitrate toward the root surface as a consequence of transpiration-driven water uptake and corresponding solute movement, its contribution to the supply of K^+ and phosphate ions is small (Barber 1995). Hence most K^+ is transported to the root surface by diffusion, as a consequence of the concentration gradients that develop in the rhizosphere (Tinker and Nye 2000; Jungk 2001, 2002). Barber (1995) estimated that diffusion contributed about 80% of the K delivered to maize (*Zea mays* L.) roots in a Chalmers silt loam (Mollisol) soil (Table 4.1).

[1]Mobility is used here to describe the ability of K^+ ions to move in soils, either vertically through leaching or laterally, through mass flow and diffusion (e.g., Hinsinger 2004).

Table 4.1 Estimated contributions of diffusion and mass flow to the acquisition of major nutrient ions in maize grown in field conditions in a Chalmers silt loam (Mollisol) soil and yielding 9500 kg grain ha^{-1}. (adapted from Barber 1995)

Nutrient ion	Diffusion	Mass flow	Acquisition
	kg ha^{-1}	kg ha^{-1}	kg ha^{-1}
Potassium	156	35	195
Phosphate	37	2	40
Nitrate	38	150	190

4.1.2 Potassium Availability and Bioavailability: Exchangeable Versus Nonexchangeable Pools in the Rhizosphere

The *availability* of a nutrient is an intrinsic property of the soil that is usually assessed by chemical methods designed to extract the fraction of the nutrient that is likely to replenish the soil solution in response to depletion by nutrient uptake (Harmsen et al. 2005; Harmsen 2007). While it is usually expressed as a concentration, the *bioavailability* is best defined as the actual flux of a nutrient into a living organism (Harmsen et al. 2005; Harmsen 2007), which means that it varies for a given soil, nutrient, and set of environmental conditions, as well as the organism of interest (e.g., the plant species or genotype). This is due to both differences in uptake capacities and abilities to alter the availability in the bio-influenced zone (Harmsen et al. 2005), which corresponds to the rhizosphere for plants (Hinsinger et al. 2011). For K, it was long assumed that the only bioavailable pools were K^+ ions in the soil solution and surface-adsorbed K, i.e., K^+ ions adsorbed onto negatively charged soil constituents (Sparks and Huang 1985; Sparks 1987), which are often assessed via an extraction with ammonium salts. These correspond to the so-called exchangeable K pool that represents typically about 1–2% of total soil K (Chap. 7). It has been well documented that plants can exploit this pool, which is therefore bioavailable.

Most soil K is, however, nonexchangeable in the sense that it cannot be extracted by an ammonium salt. There are two main nonexchangeable pools, corresponding to either K^+ contained in the interlayers of micas, partially weathered micas and secondary layer silicates (Chap. 7, Fig. 7.1, pools 10 and 11) or in the structure of other K-bearing silicates (Chap. 7, Fig. 7.1, pool 12), feldspars being the most abundant ones (Sparks and Huang 1985; Sparks 1987; Chap. 7). These have been referred to as interlayer K and structural K, respectively, and were thought to be poorly or not bioavailable. However, there is growing evidence that the K in these pools is bioavailable to some plants, as further explained below (Hinsinger 2006, 2013).

4.1.3 *Soil Profile Distribution: Topsoil Versus Subsoil Potassium Availability and Bioavailability*

It is often observed that the topsoil is enriched in nutrients compared to the subsoil, or at least exhibits greater nutrient availability. This occurs in many natural ecosystems due to the role of vegetation in the rapid recycling of nutrients through uptake and litterfall as well as throughfall, the latter being especially important for K. Nutrients accumulate in the topsoil, and whenever uptake occurs at greater depth, from subsoil layers, this ultimately contributes to nutrient accumulation in the topsoil (Jobbagy and Jackson 2001). In agroecosystems, the topsoil can also be enriched by K fertilization (Obrycki et al. 2018), but there is also some evidence for significant uptake of K occurring from the subsoil with redistribution to the soil surface in residues (Barré et al. 2009). This overlooked component of the soil, namely the subsoil and the potential reservoir of bioavailable nutrients it can represent, has been reviewed by Kautz et al. (2013), who stressed the need to assess its contribution to plant nutrition further. Kuhlmann (1990) provided some quantitative assessment of the contribution of the subsoil to wheat (*Triticum aestivum* L.) K nutrition, which ranged from 7 to 70%, with an average of 34% in Luvisols of Northern Germany. The contribution may be less in deeply weathered soils such as Oxisols and Ultisols, which contain less exchangeable and nonexchangeable K stocks. However, for deep-rooted plants such as eucalypt (*Eucalyptus grandis*) in deep Oxisols in Brazil, it has been shown that significant root–soil interactions occur at considerable depths, affecting the fate of K to at least 4 m (Pradier et al. 2017). This reinforces the need to take subsoil K into consideration in future research and in K-fertilizer recommendations, as well as when designing more K-efficient ideotypes of crops in breeding programs (Thorup-Kristensen et al. 2020).

4.2 Root Morphological Traits Determining the Acquisition of Potassium by Plants

Because of the restricted mobility of various nutrient ions, including K^+, ammonium, and phosphate in soils, plants have evolved a range of foraging strategies in order to increase the volume of their rhizosphere, i.e., the actual volume of soil from which they can acquire these poorly mobile nutrients (Hinsinger 2004; Lynch 2007; Hinsinger et al. 2011).

4.2.1 *Root System Architecture and Plasticity*

Plant species can differ considerably in root system architecture (RSA), with the tap-rooted systems and fibrous systems found in crops being good examples

(Kutschera et al. 2009). Tap-rooted systems usually enable plants to access deeper horizons, while colonizing the topsoil less densely than fibrous systems. Witter and Johansson (2001) compared forage species and estimated that the tap- and deep-rooted alfalfa (*Medicago sativa* L.) obtained about 67% of its K from the subsoil, while ryegrass (*Lolium* spp.), with its fibrous root system, obtained only 42% of its K from the subsoil under the same conditions. While the type of RSA (e.g., tap-rooted vs. fibrous) is genetically determined, it has been shown that there is considerable variation in RSA within a given species, which is a promising avenue for selecting more efficient crop genotypes of a broad range of species (Lynch 2007, 2015; Hammond et al. 2009; White et al. 2013; Mi et al. 2016; Thomas et al. 2016; Jin et al. 2017). While most of the work done so far has focused on N or phosphorus (P), some of these results could be easily extended to K. The root angle and distance between lateral roots are traits that will largely determine inter-root competition and the overlapping of the rhizosphere of neighboring roots, which is of greater concern for mobile resources, such as water and nitrate, than for poorly mobile nutrients, such as K or P (Ge et al. 2000).

Lynch and co-workers have shown in common bean (*Phaseolus vulgaris* L.) and maize that shallow-rooted genotypes may perform better than deep-rooted genotypes whenever there is a strong vertical gradient of fertility, with much greater nutrient availability in the topsoil than in the subsoil (Ge et al. 2000) and that past breeding schemes have resulted in selecting more shallow root systems in maize in the USA (York et al. 2015). Conversely, in soils exhibiting significant resources of K at depth, which is common in temperate conditions and even more so when accounting for nonexchangeable K pools, crop species that invest in deeper roots may derive more K from the subsoil, as shown by Kuhlmann (1990) in a loess soil in Germany (Fig. 4.1). The work of York et al. (2015) has also shown considerable plasticity of RSA traits, some of which vary substantially with sowing density, for instance. Such plasticity is an intrinsic property of root systems, which further complicates their study and phenotyping, but which plays a major role in the adaptive strategy of plants to acquire mineral nutrients. In contrast to nitrogen (N) and P, plants do not seem to respond to K-rich patches by enhanced root proliferation (Drew 1975; Hermans et al. 2006), which may restrict the options for effective fertilizer K placement, unless co-located with N or P. Nevertheless, we would argue that there is considerable progress to be expected from integrating RSA-related traits into breeding programs, and this is urgently needed to obtain genotypes that can better cope with spatially restricted availability of nutrients such as N, P, or K (Lynch 2015; Thorup-Kristensen et al. 2020).

4.2.2 Root Length and Growth

For poorly mobile nutrients, it has long been known that root length or root surface area is among the most relevant traits determining their acquisition (Barber 1995). In their sensitivity analysis, Silberbush and Barber (1983) showed that the predicted K

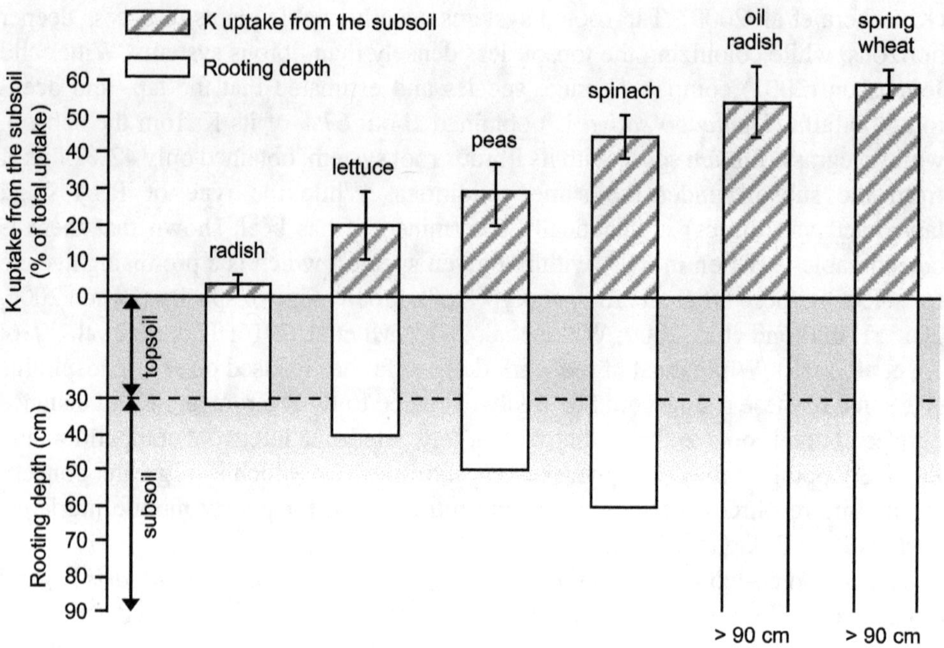

Fig. 4.1 Relative contribution of the subsoil to K acquisition as a function of rooting depth of different crop species grown in field conditions in a loess-derived, deep soil (Cambisol) with topsoil containing 90 mg kg^{-1} exchangeable K and subsoil containing 160 mg kg^{-1} exchangeable K. Deep-rooted crops such as oil radish (*Raphanus sativus* L.) and spring wheat acquired more than 50% of their K from the subsoil. (adapted from Kuhlmann 1990)

acquisition by soybean was increased more by an increase in root surface area than by the same relative increase in any other parameter in the classic Barber-Cushman model. More recently, Wissuwa (2003) predicted that a 22% increase in the root surface area of rice (*Oryza sativa* L.) was enough to give a threefold increase in P acquisition under P-limiting conditions. Given that the mobility of K^+ is greater than that of phosphate in soils, it is likely that an even smaller change of root surface area would have a significant impact on K acquisition. In this respect, the modelling work by Pagès (2011), conducted with a more realistic distribution of roots based on RSA, rather than an evenly distributed root system as in the Barber–Cushman-derived models, revealed that greater root length was ecologically relevant for the acquisition of poorly mobile nutrients, such as P, but not for mobile ions, such as nitrate (due to large overlapping of nitrate depletion zones). The results for K^+ and ammonium ions were intermediate.

Genotypes within a species can exhibit considerable variation in root length, as has been shown for potato (Wishart et al. 2013) and maize (Erel et al. 2017). While the relationship between crop performance (growth or yield) under nutrient-limiting conditions and root length was not consistently significant or positive, these studies suggest that the impact of root length variation is worthy of more detailed investigation. While root length was weakly correlated with K uptake in a range of lentil (*Lens culinaris* Medik) genotypes, Gahoonia et al. (2006) showed that root hair length was an even more relevant root trait for poorly mobile nutrients such as K or P. As for RSA, root length and related traits such as specific root length or root surface area are not just genetically determined, but are also highly plastic, responding to many environmental factors and biological stimuli. Plant growth-promoting microorganisms are an example of the latter, with some of these directly altering root growth or proliferation (Vacheron et al. 2013).

4.2.3 Root Hairs and Mycorrhizae

It has been well documented for the least mobile nutrients, such as P, that morphological or anatomical features other than RSA and root length-related traits can play a major role in extending the rhizosphere volume, and hence the actual amounts of nutrients acquired. These include root hairs that can extend up to several millimeters from the root surface (e.g., Gahoonia et al. 1997), and mycorrhizal hyphae which can access even greater volumes, extending up to several centimeters away from the root surface (e.g., Jakobsen et al. 1992; Thonar et al. 2011). Their direct implication for the acquisition of K^+ is less well documented than for P, but there are a number of reports on the potential role of root hairs and mycorrhiza-related traits for improving the foraging capacity of plants for soil K. In their modelling of K uptake, Samal et al. (2010) showed that, assuming root hair surface areas ranging from 0.38 to 0.47 $cm^2\ cm^{-2}$ root, root hairs contributed slightly less to K uptake than the roots alone (without their root hairs) in wheat and maize, but more than the roots alone in sugar beet (*Beta vulgaris* L.). The significant role of root hairs in K acquisition is also supported by the strong correlation between root hair length and K acquisition found among crop species in decreasing order of efficiency (Fig. 4.2): oilseed rape (*Brassica napus oliefera* L.), tomato (*Solanum lycopersicum* Mill.), ryegrass, maize, onion (*Allium* spp.) (Claassen and Jungk 1984; Jungk 2001) or rye (*Secale cerale* L.), ryegrass (*Lolium perenne*), oilseed rape, alfalfa, barley (*Hordeum vulgare*), pea (*Pisum sativum* L.), and red clover (*Trifolium pretense* L.) (Høgh-Jensen and Pedersen 2003). Høgh-Jensen and Pedersen (2003) also reported some plasticity for this trait, as root hairs exhibited greater length at lower K supply, suggesting that investment in the length of root hairs is an adaptive strategy for improving K acquisition. Mycorrhizal hyphae can access a much greater volume of soil than roots, and thereby increase the effective radius of the rhizosphere. While their quantitative impact on K acquisition has been studied much less than for P acquisition, the K uptake transport systems involved in the mycorrhizal symbiosis are now well documented (Garcia and Zimmermann 2014). Additional research is needed on the functional side of this symbiosis before identifying relevant traits worthy of being pursued for improving K acquisition efficiency.

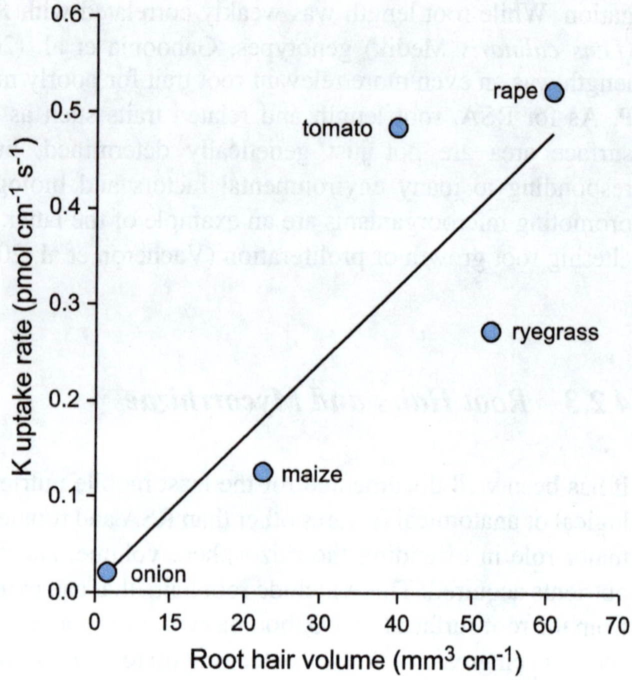

Fig. 4.2 Relationship between K acquisition (K uptake rate) and the volume of the root hair cylinder in a range of crops. (adapted from Claassen and Jungk 1984)

4.3 Root Physiological Traits Determining the Acquisition of Potassium by Plants

As potassium is present only as K^+ ions in soils, which interact strongly with negatively charged soil constituents or are part of the crystal structures of silicate minerals, K acquisition by plants is dependent on the mobilization of K from these sources. Depletion of K^+ in the rhizosphere soil solution and the excretion of protons and other K-mobilizing exudates (a so-called "mining" strategy) can increase the availability of poorly available forms of soil K in the vicinity of roots and contribute to improved plant nutrition (Hinsinger et al. 2011).

4.3.1 Traits Related to Potassium Uptake and Depletion in the Rhizosphere

For poorly mobile nutrients such as P or K, the uptake capacity of root cells, determined by the rate of transport of ions across the plasma membrane, is not the limiting step for their acquisition. This contrasts markedly with the situation for more mobile nutrients, such as nitrate, as confirmed by sensitivity analysis of Barber-Cushman and other plant nutrition models (Rengel 1993). Nevertheless, in the case of K, the uptake of K^+ is a driving process for accessing both the exchangeable pool and even a significant part of the nonexchangeable pool (Hinsinger 2006; Hinsinger

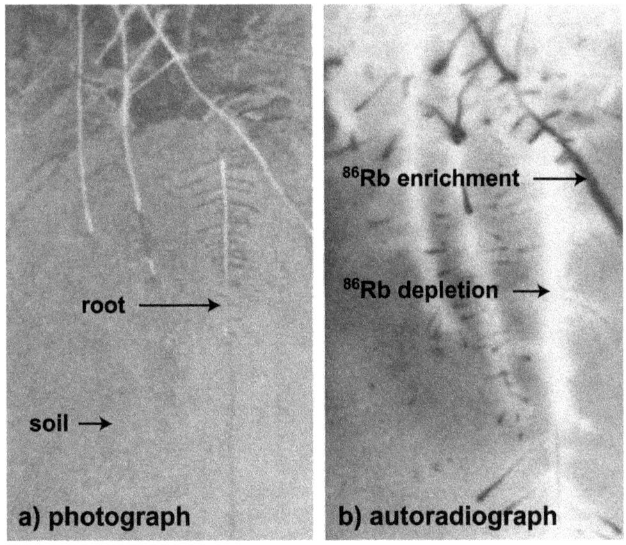

Fig. 4.3 Depletion of ^{86}Rb in the rhizosphere of 13-day-old maize roots, as revealed by autoradiography, using radioactive Rb as a proxy for K. The white areas around roots correspond to depletion zones, while the black areas correspond to zones of ^{86}Rb accumulation inside the roots, especially the root apices, where the uptake presumably occurs at a higher flux. (adapted from Walker and Barber 1962)

et al. 2011; White et al. 2013), namely interlayer K from micas and micaceous clay minerals (Chap. 7, Fig. 7.1, pools 10 and 11). When roots take up K^+ from the soil solution, a rapid depletion of K^+ occurs in the rhizosphere. This was first observed in the early 1960s using autoradiography of a radioactive analogue of K, ^{86}Rb, which demonstrated the occurrence of a depletion zone extending a few millimeters from the surface of maize roots (Fig. 4.3) (Walker and Barber 1962), and was later confirmed by a range of approaches, based on direct measurements or modelling (e.g., Kuchenbuch and Jungk 1982; Claassen et al. 1986). The latter studies also demonstrated that K^+ uptake by roots can deplete the exchangeable pool of K by causing a shift in the cation exchange equilibria toward enhanced desorption of K^+ ions from the surface-adsorbed K pool (Chap. 7, Fig. 7.1, pool 9).

As plant roots are capable of decreasing the concentration of K^+ from several hundreds of micromoles per dm^{-3} in the bulk soil down to concentrations in the micromolar range at the root surface (Fig. 4.4), they can even shift the equilibria determining the release of interlayer K in micaceous phyllosilicate minerals (e.g., micas, illite, illite interstratified with smectite and vermiculite, i.e., pools 10 and 11 in Fig. 7.1, Chap. 7), ultimately depleting the large pool of nonexchangeable K contained in soils (Kuchenbuch and Jungk 1982; Niebes et al. 1993; Moritsuka et al. 2004) and altering soil mineralogy (Kodama et al. 1994; Barré et al. 2007, 2008). This mechanism has been demonstrated to occur in the rhizosphere of ryegrass, using a phlogopite mica as the sole source of (almost exclusively interlayer) K, which released significant amounts of interlayer K and was transformed into a vermiculite clay mineral within only a few days of growth (Hinsinger et al. 1992; Hinsinger and Jaillard 1993). Barré et al. (2007) further confirmed this alteration of soil mineralogy for illitic clay minerals in the rhizosphere of ryegrass in a pot experiment, and it was also shown in a field experiment with maize (Adamo et al. 2016).

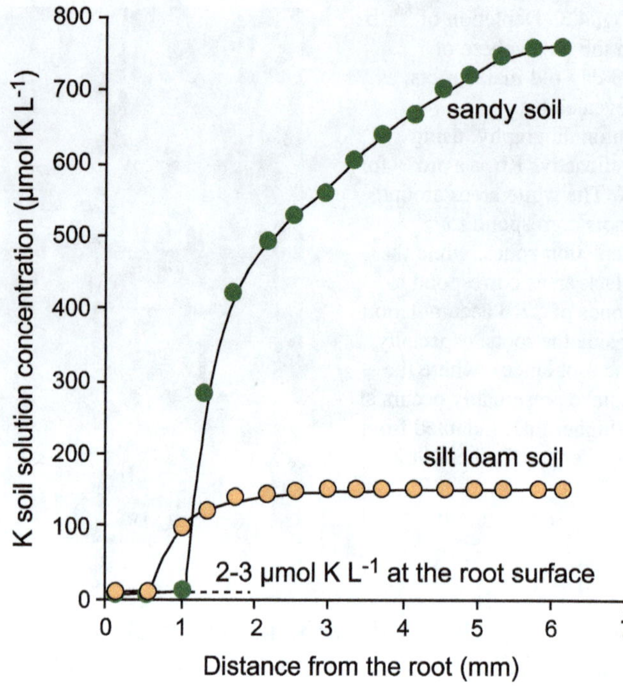

Fig. 4.4 Depletion of soil solution K in the rhizosphere of maize grown in two different soils exhibiting different K availabilities. (adapted from Claassen and Jungk 1982)

Springob and Richter (1998) have shown that the rate of release of nonexchangeable K in soils can be considerably enhanced below a threshold concentration of about 2–3 micromoles K dm^{-3}, which approximates the K concentration occurring close to the root surface as a consequence of K^+ uptake and subsequent depletion of K in the rhizosphere (Claassen and Jungk 1982; Hinsinger 2006). This steep decrease of K^+ concentration in the soil solution in the vicinity of roots thus drives the substantial and rapid depletion of exchangeable K, but also the release of K from the nonexchangeable, interlayer pool (Fig. 4.5). The substantial release that can contribute from 20 up to 80 or 90% of the actual amount of K acquired by plants over rather short periods (a few days) shows that, in the peculiar conditions of the rhizosphere, and especially its low solution K^+ concentration (Fig. 4.6), the rates of this normally slow process can be much faster than expected (Claassen and Jungk 1982; Kuchenbuch and Jungk 1982; Hinsinger 2006; Niebes et al. 1993; Samal et al. 2010).

In this respect, the three important uptake characteristics that need to be considered are the C_{min} value (minimal solution K concentration below which plants cannot take up K) and the capacity to achieve a large K uptake rate at low K^+ ion concentrations, which depends on the K_m and V_{max} parameters of the Michaelis–Menten equation. A low C_{min} value is achieved by coupling the proton gradient generated by a plasma membrane H^+-ATPase to K^+ influx to root cells via a H^+/K^+ coupled symporter (White and Karley 2010; White 2013). This transporter has a low K_m for K^+, and the required rate of K^+ influx is achieved by regulating its abundance and activity in response to plant K status (White and Karley 2010; White 2013). Plant species or genotypes showing very low C_{min} values (Asher and Ozanne 1967),

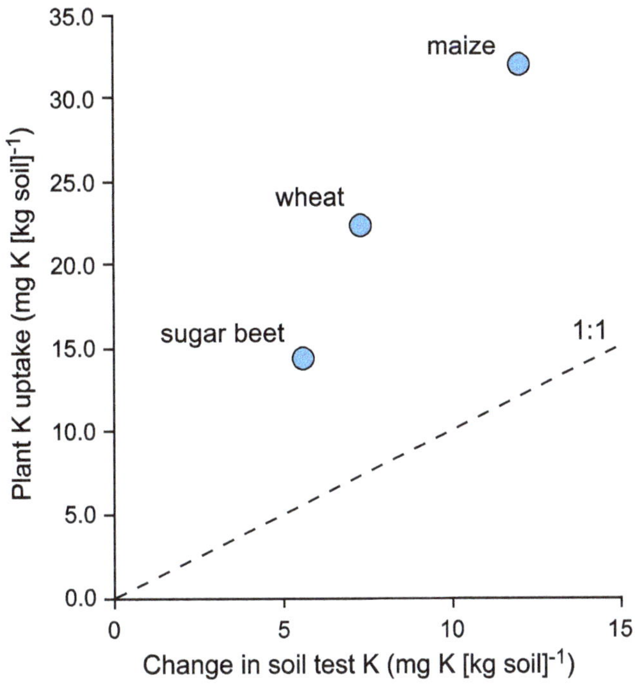

Fig. 4.5 Relative contribution of exchangeable K to the actual K acquisition for different crops, showing significant contributions of nonexchangeable K. (adapted from Samal et al. 2010)

and high affinity (low K_m) and capacity (V_{max}) of their K transporters, would be better equipped for depleting K^+ ions to concentrations low enough to induce a significant release of interlayer K and thus access the large pool of nonexchangeable K that would be otherwise unavailable. Such K uptake traits differ among and within plant species, thus explaining differences in K bioavailability in a given soil, and might be worth considering for screening K-efficient crop genotypes (White 2013; White et al. 2016).

4.3.2 *Traits Related to pH Modification in the Rhizosphere*

It should be noted that a number of field experiments have reported an increase of some K pools in the rhizosphere, notably the exchangeable K pool, instead of depletion. This has been shown mostly in perennial tree species, e.g., by Courchesne and Gobran (1997) in a Norway spruce (*Picea abies* (L.) H. Karst.) forest, Bourbia et al. (2013) in an olive (*Olea europaea* L.) grove and Pradier et al. (2017) in a eucalypt (*Eucalyptus grandis*) plantation. These observations strongly suggest the occurrence of root-induced weathering processes, resulting in an increase of exchangeable K at the expense of the nonexchangeable K pool through processes other than those mentioned above. For instance, Pradier et al. (2017) showed that root-induced acidification of the rhizosphere of eucalypt trees may have partly contributed to the increase of exchangeable K that was observed in the rhizosphere throughout the soil profile to a depth of 4 m. Plant roots can modify rhizosphere pH

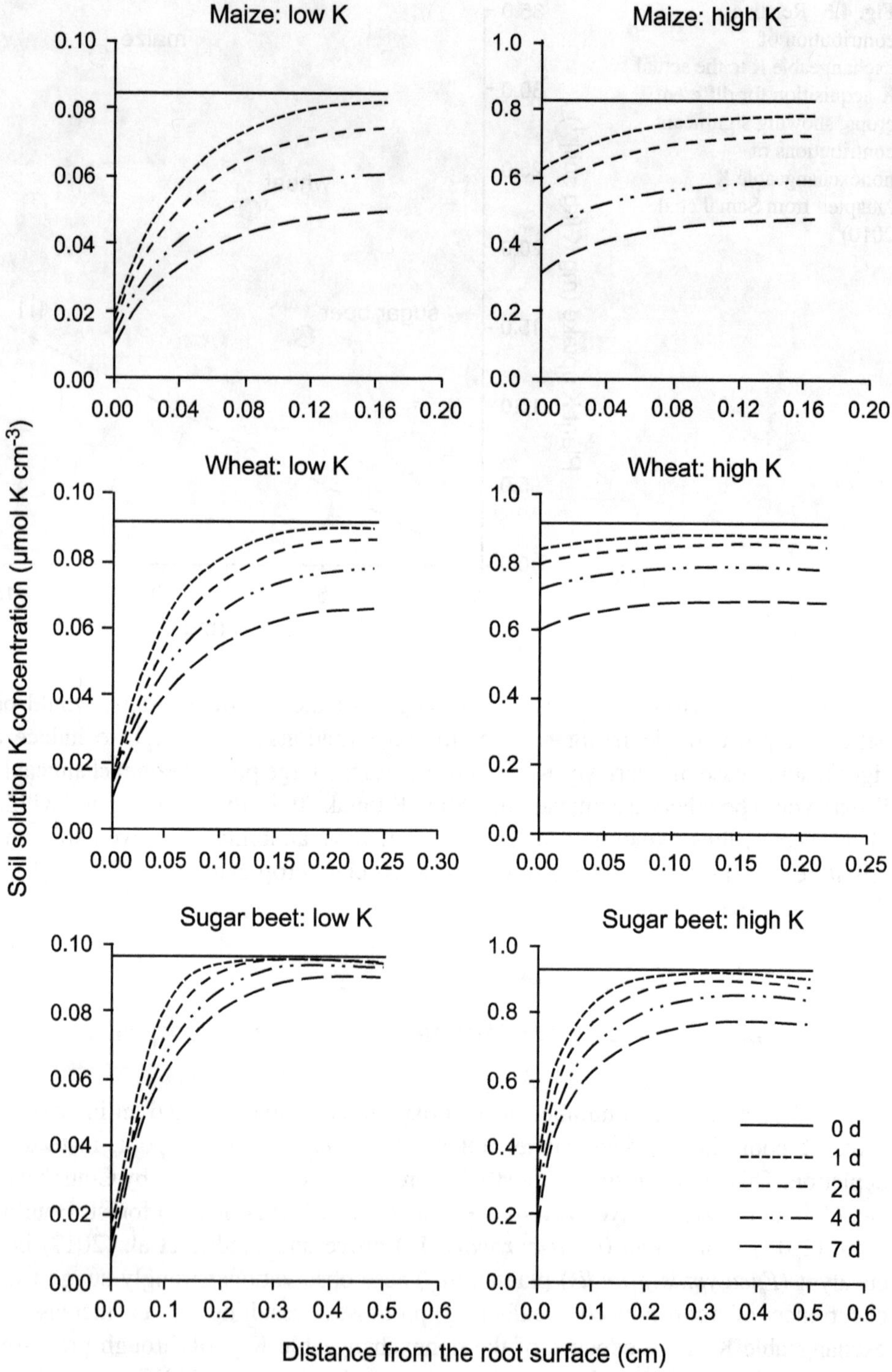

Fig. 4.6 Simulated depletion of K in the rhizosphere as a function of time for different crops. (adapted from Samal et al. 2010)

considerably and root-mediated acidification of up to 2–3 pH units has been observed repeatedly (e.g., Römheld 1986; Hinsinger et al. 2003; Blossfeld et al. 2013). Such a decrease in pH can have a dramatic effect on the weathering rate of minerals such as K-bearing silicates, through proton-promoted dissolution (Berner et al. 2003; Taylor et al. 2009; Hinsinger 2013).

The ability to modify rhizosphere pH varies among plant species. For instance, faba bean (*Vicia faba* L.) reduced rhizosphere pH more effectively than maize (Liu et al. 2016) and oilseed rape induced a dissolution of the phlogopite mica as a consequence of rhizosphere acidification, while ryegrass did not (Hinsinger et al. 1993). Such root-induced dissolution of K-bearing minerals has not been widely studied in crops to our knowledge, but a number of studies have been conducted in the context of forest trees. For example, Arocena et al. (1999) showed that the acidification occurring around ectomycorrhizas could be involved in the dissolution of micas and feldspars. The capacity of roots to change the rhizosphere pH is not a simple trait to target, however, as root-mediated pH changes are essentially the consequence of an imbalanced uptake of major cations and anions (Hinsinger et al. 2003). Rhizosphere acidification occurs when a net surplus of major cations (K^+ being often predominant in the cation budget of plants) are taken up relative to the sum of major anions. Thus, there are many ways to increase rhizosphere acidification, and it is not expected to be related to a single trait, as the underlying mechanisms are largely determined by the environmental context. Nevertheless, plant species or genotypes could be screened quite easily for their capacity to acidify their rhizosphere in a given context (Gahoonia et al. 2006). This has been shown in common bean by Yan et al. (2004), who found significant heritability of a measured trait called total acid exudation, which was also able to account for a significant part of genotypic variation in P uptake.

4.3.3 Traits Related to Exudates in the Rhizosphere

Besides protons, roots can release large quantities of diverse exudates, including some that can also promote the dissolution of K-bearing silicates and thus the release of nonexchangeable K, i.e., interlayer or structural K (pools 10–11 or 12, Fig. 7.1, Chap. 7). Examples of these include carboxylates, such as citrate, oxalate, and malate, which are able to complex cations when they are released from the crystal structure of K-bearing minerals (Jones 1998). Such exudates can thus be involved in the dissolution of feldspars and micas (Razzaghe and Robert 1979; Robert and Berthelin 1986; Song and Huang 1988; Barman et al. 1992; Lawrence et al. 2014), and enhance the release of nonexchangeable K, i.e., structural or interlayer K, together with the other metal cations contained in these silicates. Screening wheat, maize, and sorghum (*Sorghum bicolor* (L.) Moench) genotypes for their release of malate and citrate to detoxify Al has been successful, which shows that such a trait can vary substantially within some plant species (e.g., Ryan et al. 2011). However, variation is certainly greater between species, with some crops such as white lupin

(*Lupinus albus* L.) or chickpea (*Cicer arietinum* L.) being well known for their large carboxylate exudation capacity (Jones 1998). Even minute amounts of malate or citrate exuded at the root tip may provide sufficient protection against Al toxicity (Ryan et al. 2011), but much greater concentrations would be needed to induce significant release of K through such ligand-promoted dissolution of K-bearing silicates. There is no direct evidence to our knowledge that such exudation traits are worth pursuing for improving K acquisition efficiency in crops.

In addition to the exudation of such ligands, a considerable range and amount of other C-compounds can be released by roots (Jones 1998; Jones et al. 2009). Exudation is thus an important but complex process occurring in the rhizosphere, which stimulates the microbial communities and has potential implications for the dissolution of nutrient-bearing minerals (e.g., Philippot et al. 2013), as reviewed for fungi by Hoffland et al. (2004).

4.4 Summary and Conclusions

A number of root or rhizosphere-related traits determine the K acquisition efficiency of crops, influencing both their foraging and mining strategies. To deal with the restricted mobility of K^+ ions in soils, the foraging strategy of plants is based on a number of root traits, including densely branched root system architectures with substantial portions exploring the topsoil and the subsoil, with a large root surface area or length. In addition, roots can expand the volume of exploited soil and K-depletion zones considerably by developing long root hairs or supporting strong mycorrhizal symbioses and extensive hyphal networks.

In addition to this foraging strategy, which enlarges the volume of the rhizosphere, plants have also evolved various mining strategies to increase the bioavailability of all the K pools in the rhizosphere. The corresponding traits are related first to the ability of roots to sustain high fluxes of K at very low concentrations in the soil solution. This induces a shift in the exchange equilibria and an enhanced desorption of exchangeable K (surface-adsorbed K from clay minerals and organic matter), as well as an enhanced release of nonexchangeable, interlayer K contained in micaceous minerals. Second, roots can promote the dissolution of K-bearing silicates such as micas and feldspars through rhizosphere acidification and/or exudation of complexing ligands, such as some carboxylates. How to make better use of these traits in the context of a sustainable intensification of agroecosystems is not obvious though, and breeders have not yet fully integrated belowground traits in their breeding schemes. To do so may introduce additional challenges, such as potential trade-offs with those traits required for the acquisition of other belowground resources (e.g., water, N, P, and micronutrients).

References

Adamo P, Barré P, Cozzolino V, Di Meo V, Velde B (2016) Short term clay mineral release and re-capture of potassium in a *Zea mays* field experiment. Geoderma 264:54–60. https://doi.org/10.1016/j.geoderma.2015.10.005

Arocena JM, Glowa KR, Massicotte HB, Lavkulich L (1999) Chemical and mineral composition of ectomycorrhizosphere soils of subalpine fir (*Abies lasiocarpa* (Hook.) Nutt.) in the Ae horizon of a Luvisol. Can J Soil Sci 79(1):25–35. https://doi.org/10.4141/S98-037

Asher CJ, Ozanne PG (1967) Growth and potassium content of plants in solution cultures maintained at constant potassium concentrations. Soil Sci 103:155–161. https://doi.org/10.1097/00010694-196703000-00002

Barber SA (1995) Soil nutrient bioavailability: a mechanistic approach, 2nd edn. Wiley, New York

Barman AK, Varadachari C, Ghosh K (1992) Weathering of silicate minerals by organic acids. I. Nature of cation solubilisation. Geoderma 53:45–63. https://doi.org/10.1016/0016-7061(92)90020-8

Barré P, Velde B, Catel N, Abbadie L (2007) Soil-plant potassium transfer: impact of plant activity on clay minerals as seen from X-ray diffraction. Plant Soil 292:137–146. https://doi.org/10.1007/s11104-007-9208-6

Barré P, Montagnier C, Chenu C, Abbadie L, Velde B (2008) Clay minerals as a soil potassium reservoir: observation and quantification through X-ray diffraction. Plant Soil 302:213–220. https://doi.org/10.1007/s11104-007-9471-6

Barré P, Berger G, Velde B (2009) How element translocation by plants may stabilize illitic clays in the surface of temperate soils. Geoderma 151:22–30. https://doi.org/10.1016/j.geoderma.2009.03.004

Berner EK, Berner RA, Moulton KL (2003) Plants and mineral weathering: present and past. In: Holland HD, Turekian KK (eds) Treatise on geochemistry, vol 5, pp 169–188. https://doi.org/10.1016/B0-08-043751-6/05175-6

Blossfeld S, Schreiber CM, Liebsch G, Kühn AJ, Hinsinger P (2013) Quantitative imaging of rhizosphere pH and CO_2 dynamics with planar optodes. Ann Bot 112:267–276. https://doi.org/10.1093/aob/mct047

Bourbia SM, Barré P, Kaci MBN, Derridj A, Velde B (2013) Potassium status in bulk and rhizospheric soils of olive groves in North Algeria. Geoderma 197:161–168. https://doi.org/10.1016/j.geoderma.2013.01.007

Claassen N, Jungk A (1982) Kaliumdynamik im wurzelnahen Boden in Beziehung zur Kaliumaufnahme von Maispflanzen. Z Pflanzenernähr Bodenkd 145:513–525

Claassen N, Jungk A (1984) Bedeutung von Kaliumaufnahmerate, Wurzelwachstum und Wurzelhaaren für das Kaliumaneignungsvermögen verschiedener Pflanzenarten. Z Pflanzenernähr Bodenkd 147:276–289

Claassen N, Syring KM, Jungk A (1986) Verification of a mathematical model by simulating potassium uptake from soil. Plant Soil 95:209–220. https://doi.org/10.1007/BF02375073

Courchesne F, Gobran GR (1997) Mineralogical variation of bulk and rhizosphere soils from a Norway spruce stand. Soil Sci Soc Am J 61:1245–1249. https://doi.org/10.2136/sssaj1997.03615995006100040034x

Drew MC (1975) Comparison of effects of a localized supply of phosphate, nitrate, ammonium and potassium on growth of seminal root system, and the shoot, in barley. New Phytol 75:479–490. JSTOR. www.jstor.org/stable/2431588. Accessed 29 May 2020

Erel R, Bérard A, Capowiez L, Doussan C, Arnal D, Souche G, Gavaland A, Fritz C, Visser EJW, Salvi S, Le Marié C, Hund A, Hinsinger P (2017) Soil type determines how root and rhizosphere traits relate to phosphorus acquisition in field-grown maize genotypes. Plant Soil 412:115–132. https://doi.org/10.1007/s11104-016-3127-3

Gahoonia TS, Care D, Nielsen NE (1997) Root hairs and phosphorus acquisition of wheat and barley cultivars. Plant Soil 191:181–188. https://doi.org/10.1023/A:1004270201418

Gahoonia TS, Ali O, Sarker A, Nielsen NE, Rahman MM (2006) Genetic variation in root traits and nutrient acquisition of lentil genotypes. J Plant Nutr 29:643–655. https://doi.org/10.1080/01904160600564378

Garcia K, Zimmermann SD (2014) The role of mycorrhizal associations in plant potassium nutrition. Front Plant Sci 5:1–9. https://doi.org/10.3389/fpls.2014.00337

Ge ZY, Rubio G, Lynch JP (2000) The importance of root gravitropism for inter-root competition and phosphorus acquisition efficiency: results from a geometric simulation model. Plant Soil 218:159–171. https://doi.org/10.1023/A:1014987710937

Hammond JP, Broadley MR, White PJ, King GJ, Bowen HC, Hayden R, Meacham MC, Mead A, Overs T, Spracklen WP, Greenwood DJ (2009) Shoot yield drives phosphorus use efficiency in *Brassica oleracea* and correlates with root architecture traits. J Exp Bot 60:1953–1968. https://doi.org/10.1093/jxb/erp083

Harmsen J (2007) Measuring bioavailability: from a scientific approach to standard methods. J Environ Qual 36:1420–1428. https://doi.org/10.2134/jeq2006.0492

Harmsen J, Rulkens W, Eijsackers H (2005) Bioavailability: concept for understanding or tool for predicting? Land Contam Reclam 13:161–171

Hermans C, Hammond JP, White PJ, Verbruggen N (2006) How do plants respond to nutrient shortage by biomass allocation? Trends Plant Sci 11:610–617. https://doi.org/10.1016/j.tplants.2006.10.007

Hinsinger P (2004) Nutrient availability and transport in the rhizosphere. In: Goodman RM (ed) Encyclopedia of plant and crop science. Taylor and Francis, London, pp 1094–1097. https://doi.org/10.1201/9780203757604

Hinsinger P (2006) Potassium. In: Lal R (ed) Encyclopedia of soil science, 2nd edn. Taylor and Francis, London. https://doi.org/10.4324/9781315161860

Hinsinger P (2013) Plant-induced changes of soil processes and properties. In: Gregory PJ, Nortcliff S (eds) Soil conditions and plant growth. Wiley-Blackwell, London, pp 323–365. https://doi.org/10.1002/9781118337295

Hinsinger P, Jaillard B (1993) Root-induced release of interlayer potassium and vermiculitization of phlogopite as related to potassium depletion in the rhizosphere of ryegrass. J Soil Sci 44:525–534. https://doi.org/10.1111/j.1365-2389.1993.tb00474.x

Hinsinger P, Jaillard B, Dufey JE (1992) Rapid weathering of a trioctahedral mica by the roots of ryegrass. Soil Sci Soc Am J 56:977–982. https://doi.org/10.2136/sssaj1992.03615995005600030049x

Hinsinger P, Elsass F, Jaillard B, Robert M (1993) Root-induced irreversible transformation of a trioctahedral mica in the rhizosphere of rape. J Soil Sci 44:535–545. https://doi.org/10.1111/j.1365-2389.1993.tb00475.x

Hinsinger P, Plassard C, Tang C, Jaillard B (2003) Origins of root-meditated pH changes in the rhizosphere and their responses to environmental constraints: a review. Plant Soil 248:43–59. https://doi.org/10.1023/A:1022371130939

Hinsinger P, Brauman A, Devau N, Gérard F, Jourdan C, Laclau JP, Le Cadre E, Jaillard B, Plassard C (2011) Acquisition of phosphorus and other poorly mobile nutrients by roots. Where do plant nutrition models fail? Plant Soil 348:29–61. https://doi.org/10.1007/s11104-011-0903-y

Hoffland E, Kuyper TW, Wallander H, Plassard C, Gorbushina A, Haselwandter K, Holmström S, Landeweert R, Lundström US, Rosling A, Sen R, Smits MM, Van Hees PAW, Van Breemen N (2004) The role of fungi in weathering. Front Ecol Environ 2:258–264. https://doi.org/10.1890/1540-9295(2004)002[0258:TROFIW]2.0.CO;2

Høgh-Jensen H, Pedersen MB (2003) Morphological plasticity by crop plants and their potassium use efficiency. J Plant Nutr 26:969–984. https://doi.org/10.1081/PLN-120020069

Jakobsen I, Abbott LK, Robson AD (1992) External hyphae of vesicular arbuscular mycorrhizal fungi associated with *Trifolium subterraneum* L. 2. Hyphal transport of ^{32}P over defined distances. New Phytol 120:509–516. www.jstor.org/stable/2557412. Accessed 29 May 2020

Jin K, White PJ, Whalley WR, Shen J, Shi L (2017) Shaping an optimal soil by root-soil interaction. Trends Plant Sci 22:823–829. https://doi.org/10.1016/j.tplants.2017.07.008

Jobbagy EG, Jackson RB (2001) The distribution of soil nutrients with depth: global patterns and the imprint of plants. Biogeochemistry 53:51–77. https://doi.org/10.1023/A:1010760720215

Jones DL (1998) Organic acids in the rhizosphere – a critical review. Plant Soil 205:25–44. https://doi.org/10.1023/A:1004356007312

Jones DL, Nguyen C, Finlay RD (2009) Carbon flow in the rhizosphere: carbon trading at the soil-root interface. Plant Soil 321:5–33. https://doi.org/10.1007/s11104-009-9925-0

Jungk A (2001) Root hairs and the acquisition of plant nutrients from soil. J Plant Nutr Soil Sci 164:121–129. https://doi.org/10.1002/1522-2624(200104)164:2%3C121::AID-JPLN121%3E3.0.CO;2-6

Jungk A (2002) Dynamics of nutrient movement at the soil-root interface. In: Waisel Y, Eshel A, Kafkafi U (eds) Plant roots: the hidden half, 3rd edn. CRC Press, Boca Raton, pp 587–616. https://doi.org/10.1201/9780203909423

Kautz T, Amelung W, Ewert F, Gaiser T, Horn R, Jahn R, Javaux M, Kemna A, Kuzyzakov Y, Munch JC, Pätzold S, Peth S, Scherer HW, Schloter M, Schneider H, Vanderborght J, Vetterlein D, Walter A, Wiensenberg GLB, Köpke U (2013) Nutrient acquisition from arable subsoils in temperate climates: a review. Soil Biol Biochem 57:1003–1022. https://doi.org/10.1016/j.soilbio.2012.09.014

Kodama H, Nelson S, Yang F, Kohyama N (1994) Mineralogy of rhizospheric and non-rhizospheric soils in corn fields. Clay Clay Miner 42:755–763. https://doi.org/10.1346/CCMN.1994.0420612

Kuchenbuch R, Jungk A (1982) A method for determining concentration profiles at the soil-root interface by thin slicing rhizospheric soil. Plant Soil 68:391–394. https://doi.org/10.1007/BF02197944

Kuhlmann H (1990) Importance of the subsoil for the K-nutrition of crops. Plant Soil 127:129–136. https://doi.org/10.1007/BF00010845

Kutschera L, Lichtenegger E, Sobotik M (2009) Wurzelatlas der Kulturpflanzen gemäßigter Gebiete mit Arten des Feldgemüsebaues, 7 Band der Wurzelatlas Reihe. DLG, Frankfurt/Main

Lawrence C, Harden J, Maher K (2014) Modeling the influence of organic acids on soil weathering. Geochim Cosmochim Acta 139:487–507. https://doi.org/10.1016/j.gca.2014.05.003

Liu H, White PJ, Li C (2016) Biomass partitioning and rhizosphere responses of maize and faba bean to phosphorus deficiency. Crop Pasture Sci 67:847–856. https://doi.org/10.1071/CP16015

Lynch JP (2007) Roots of the second green revolution. Aust J Bot 55:493–512. https://doi.org/10.1071/BT06118

Lynch JP (2015) Root phenes that reduce the metabolic costs of soil exploration: opportunities for 21st century agriculture. Plant Cell Environ 38:1775–1784. https://doi.org/10.1111/pce.12451

Mi G, Chen F, Yuan L, Zhang F (2016) Ideotype root system architecture for maize to achieve high yield and resource use efficiency in intensive cropping systems. Adv Agron 139:73–97. https://doi.org/10.1016/bs.agron.2016.05.002

Moritsuka N, Yanai J, Kosaki T (2004) Possible processes releasing nonexchangeable potassium from the rhizosphere of maize. Plant Soil 258:261–268. https://doi.org/10.1023/B:PLSO.0000016556.79278.7f

Niebes JF, Hinsinger P, Jaillard B, Dufey JE (1993) Release of nonexchangeable potassium from different size fractions of two highly K-fertilized soils in the rhizosphere of rape (*Brassica napus* cv Drakkar). Plant Soil 155:403–406. https://doi.org/10.1007/BF00025068

Obrycki JF, Kovar JL, Karlen DL (2018) Subsoil potassium in Central Iowa soils: status and future challenges. Agrosyst Geosci Environ 1:1–8. https://doi.org/10.2134/age2018.07.0018

Pagès L (2011) Links between root developmental traits and foraging performance. Plant Cell Environ 34:1749–1760. https://doi.org/10.1111/j.1365-3040.2011.02371.x

Philippot L, Raaijmakers JM, Lemanceau P, Van Der Putten WH (2013) Going back to the roots: the microbial ecology of the rhizosphere. Nat Rev Microbiol 11:789–799. https://doi.org/10.1038/nrmicro3109

Pradier C, Hinsinger P, Laclau JP, Pouillet JP, Guerrini IA, Goncalves JLM, Asensio V, Abreu-Junior CH, Jourdan C (2017) Rainfall reduction impacts rhizosphere biogeochemistry in eucalypts grown in a deep Ferralsol in Brazil. Plant Soil 414:339–354. https://doi.org/10.1007/s11104-016-3107-7

Razzaghe MK, Robert M (1979) Géochimie des éléments majeurs des micas en milieu organique: mécanismes de l'altération des silicates. Ann Agron 30:493–512

Rengel Z (1993) Mechanistic simulation models of nutrient uptake: a review. Plant Soil 152:161–173. https://doi.org/10.1007/BF00029086

Robert M, Berthelin J (1986) Role of biological and biochemical factors in soil mineral weathering. In: Huang PM, Schnitzer M (eds) Interactions of soil minerals with natural organics and microbes, Spec Pub 17. Soil Science Society of America, Madison, pp 453–495. https://doi.org/10.2136/sssaspecpub17.c12

Römheld V (1986) pH-Veränderungen in der Rhizosphäre verschiedener Kulturpflanzenarten in Abhängigkeit vom Nährstoffangebot. Potash Rev 55:1–8

Ryan PR, Tyerman SD, Sasaki T, Furuichi T, Yamamoto Y, Zhang WH, Delhaize E (2011) The identification of aluminium-resistance genes provides opportunities for enhancing crop production on acid soils. J Exp Bot 62:9–20. https://doi.org/10.1093/jxb/erq272

Samal D, Kovar JL, Steingrobe B, Sadana US, Bhadoria PS, Claassen N (2010) Potassium uptake efficiency and dynamics in the rhizosphere of maize (*Zea mays* L.), wheat (*Triticum aestivum* L.), and sugar beet (*Beta vulgaris* L.) evaluated with a mechanistic model. Plant Soil 332:105–121. https://doi.org/10.1007/s11104-009-0277-6

Silberbush M, Barber SA (1983) Sensitivity analysis of parameters used in simulating K uptake with a mechanistic mathematical model. Agron J 75:851–854. https://doi.org/10.2134/agronj1983.00021962007500060002x

Song SK, Huang PM (1988) Dynamics of potassium release from potassium-bearing minerals as influenced by oxalic and citric acids. Soil Sci Soc Am J 52:383–390. https://doi.org/10.2136/sssaj1988.03615995005200020015x

Sparks DL (1987) Potassium dynamics in soils. Adv Soil Sci 6:1–63. https://doi.org/10.1007/978-1-4612-4682-4_1

Sparks DL, Huang PM (1985) Physical chemistry of soil potassium. In: Munson RD (ed) Potassium in agriculture. Am Soc Agron, Madison, pp 201–276. https://doi.org/10.2134/1985.potassium.c9

Springob G, Richter J (1998) Measuring interlayer potassium release rates from soil materials. II. A percolation procedure to study the influence of the variable 'solute' K in the <1...10 μM range. J Plant Nutr Soil Sci 161:323–329. https://doi.org/10.1002/jpln.1998.3581610321

Taylor LL, Leake JR, Quirk J, Hardy K, Banwarts SA, Beerling DJ (2009) Biological weathering and the long-term carbon cycle: integrating mycorrhizal evolution and function into the current paradigm. Geobiology 7:171–191. https://doi.org/10.1111/j.1472-4669.2009.00194.x

Thomas CL, Graham NS, Hayden R, Meacham MC, Neugebauer K, Nightingale M, Dupuy LX, Hammond JP, White PJ, Broadley MR (2016) High-throughput phenotyping (HTP) identifies seedling root traits linked to variation in seed yield and nutrient capture in field-grown oilseed rape (*Brassica napus* L.). Ann Bot – London 118:655–665. https://doi.org/10.1093/aob/mcw046

Thonar C, Schnepf A, Frossard E, Roose T, Jansa J (2011) Traits related to differences in function among three arbuscular mycorrhizal fungi. Plant Soil 339:231–245. https://doi.org/10.1007/s11104-010-0571-3

Thorup-Kristensen K, Halberg N, Nicolaisen M, Olesen JE, Crews TE, Hinsinger P, Kirkegaard J, Pierret A, Dresbøll DB (2020) Digging deeper for agricultural resources, the value of deep rooting. Trends Plant Sci 25:406–417. https://doi.org/10.1016/j.tplants.2019.12.007

Tinker PB, Nye PH (2000) Solute movement in the rhizosphere. Oxford University Press, New York

Vacheron J, Desbrosses G, Bouffaud ML, Touraine B, Moënne-Loccoz Y, Muller D, Legendre L, Wisniewski-Dyé F, Prigent-Combaret C (2013) Plant growth-promoting rhizobacteria and root system functioning. Front Plant Sci 4. https://doi.org/10.3389/fpls.2013.00356

Walker JM, Barber SA (1962) Absorption of potassium and rubidium from the soil by corn roots. Plant Soil 17:243–259. https://doi.org/10.1007/BF01376227

White PJ (2013) Improving potassium acquisition and utilisation by crop plants. J Plant Nutr Soil Sci 176:305–316. https://doi.org/10.1002/jpln.201200121

White PJ, Karley AJ (2010) Potassium. In: Hell R, Mendel R-R (eds) Cell biology of metals and nutrients. Springer, Berlin, pp 199–224

White PJ, George TS, Gregory PJ, Bengough AG, Hallett PD, Mckenzie BM (2013) Matching roots to their environment. Ann Bot – Lond 112:207–222. https://doi.org/10.1093/aob/mct123

White PJ, Kawachi T, Thompson JA, Wright G, Dupuy LX (2016) Minimizing the treatments required to determine the responses of different crop genotypes to potassium supply. Commun Soil Sci Plant Anal 47:104–111. https://doi.org/10.1080/00103624.2016.1232103

Wishart J, George TS, Brown LK, Ramsay G, Bradshaw JE, White PJ, Gregory PJ (2013) Measuring variation in potato roots in both field and glasshouse: the search for useful yield predictors and a simple screen for root traits. Plant Soil 368:231–249. https://doi.org/10.1007/s11104-012-1483-1

Wissuwa M (2003) How do plants achieve tolerance to phosphorus deficiency? Small causes with big effects. Plant Physiol 133:1947–1958. www.jstor.org/stable/4281510. Accessed 29 May 2020

Witter E, Johansson G (2001) Potassium uptake from the subsoil by green manure crops. Biol Agric Hortic 19:127–141. https://doi.org/10.1080/01448765.2001.9754917

Yan X, Liao H, Beebe SE, Blair MW, Lynch JP (2004) QTL mapping of root hair and acid exudation traits and their relationship to phosphorus uptake in common bean. Plant Soil 265:17–29. https://doi.org/10.1007/s11104-005-0693-1

York LM, Galindo-Castaneda T, Schussler JR, Lynch JP (2015) Evolution of US maize (*Zea mays* L.) root architectural and anatomical phenes over the past 100 years corresponds to increased tolerance of nitrogen stress. J Exp Bot 66:2347–2358. https://doi.org/10.1093/jxb/erv074

5

Understanding Potassium Use, Uptake and Utilization Efficiency

Philip J. White, Michael J. Bell, Ivica Djalovic, Philippe Hinsinger, and Zed Rengel

Abstract There are many terms used to define aspects of potassium (K) use efficiency of plants. The terms used most frequently in an agricultural context are (1) agronomic K use efficiency (KUE), which is defined as yield per unit K available to a crop and is numerically equal to the product of (2) the K uptake efficiency (KUpE) of the crop, which is defined as crop K content per unit K available and (3) its K utilization efficiency (KUtE), which is defined as yield per unit crop K content. There is considerable genetic variation between and within plant species in KUE, KUpE, and KUtE. Root systems of genotypes with greatest KUpE often have an ability (1) to exploit the soil volume effectively, (2) to manipulate the rhizosphere to release nonexchangeable K from soil, and (3) to take up K at low rhizosphere K concentrations. Genotypes with greatest KUtE have the ability (1) to redistribute K from older to younger tissues to maintain growth and photosynthesis and (2) to reduce vacuolar K concentration, while maintaining an appropriate K concentration in metabolically active subcellular compartments, either by anatomical adaptation or

P. J. White (✉)
Ecological Sciences, The James Hutton Institute, Dundee, UK

King Saud University, Riyadh, Saudi Arabia
e-mail: philip.white@hutton.ac.uk

M. J. Bell
School of Agriculture and Food Sciences, The University of Queensland, Brisbane, QLD, Australia
e-mail: m.bell4@uq.edu.au

I. Djalovic
Institute of Field and Vegetable Crops, Novi Sad, Serbia
e-mail: ivica.djalovic@ifvcns.ns.ac.rs

P. Hinsinger
Eco&Sols, University of Montpellier, CIRAD, INRAE, Institut Agro, IRD, Montpellier, France
e-mail: philippe.hinsinger@inrae.fr

Z. Rengel
Soil Science and Plant Nutrition M087, School of Agriculture and Environmental Science, University of Western Australia, Crawley, WA, Australia
e-mail: zed.rengel@uwa.edu.au

by greater substitution of K with other solutes in the vacuole. Genetic variation in traits related to KUpE and KUtE might be exploited in breeding crop genotypes that require less K fertilizer. This could reduce fertilizer costs, protect the environment, and slow the exhaustion of nonrenewable resources.

5.1 Metrics of Potassium Use Efficiency and Their Relationships

There are many terms defining aspects of the potassium (K) use efficiency of plants (Table 5.1; White 2013). The terms used most frequently in an agricultural context are (1) agronomic K use efficiency (KUE), which is defined as crop yield (Y) per unit K available (Ka) from the soil plus fertilizer (g Y g^{-1} Ka) and is numerically equal to the product of (2) the K uptake efficiency (KUpE) of a crop, which is defined as crop K content (K_{crop}) per unit K available in the soil plus fertilizer (g K_{crop} g^{-1} Ka) and (3) its K utilization efficiency (KUtE), which is defined as yield per unit crop K content (g Y g^{-1} K_{crop}). These are often complemented by measurements of (4) the

Table 5.1 Mathematical definitions of aspects of potassium (K) use efficiency in crops

	Name	Abbreviation	Calculation	Units
1	Agronomic K use efficiency	KUE	Y/Ka	g DM g^{-1} K
2	K uptake efficiency	KUpE	K_{crop}/Ka	g K g^{-1} K
3	K utilization efficiency	KUtE	Y/K_{crop}	g DM g^{-1} K
4	Yield response to K supply		$Y = Y_{max} \times (Ka/(Km_{Ka} + Ka))$	
5	Response of plant K content to K supply		Derived from Eqs. (4) and (6)	
6	Yield response to plant K content		$Y = Y_{max} \times (K_{crop}/(Km_{Kcrop} + K_{crop}))$	
7	Apparent fertilizer recovery efficiency	ARE	$((K_{crop(Kf)} - K_{crop(Ks)})/Kf) \times 100$	%
8	Agronomic efficiency of K fertilizer	AE	$(Y_{Kf} - Y_{Ks})/Kf$	g DM g^{-1} K
9	Root uptake capacity		K_{crop}/R	g K g^{-1} DM
10	Apparent remobilization efficiency	AKR	$((K_{tissue(o)} - K_{tissue(t)})/K_{tissue(0)}) \times 100$	%

Abbreviations: DM = dry matter, Ka = K available from both soil and fertilizer, K_{crop} = crop K content, $K_{crop(Kf)}$ = crop K content when fertilizer is applied, $K_{crop(Ks)}$ = crop K content without fertilizer, $K_{crop(max)}$ = maximum crop K content, $K_{tissue(o)}$ = original tissue K content, $K_{tissue(t)}$ = tissue K content after remobilization, Km_{Ka} = Ka at which Y equals $Y_{max}/2$, Km_{Kcrop} = K_{crop} at which Y equals $Y_{max}/2$, Kf = K supplied as fertilizer, Ks = available K in soil with no fertilizer applied, R = root DM, Y = yield, Y_{Kf} = yield with fertilizer applied, Y_{Ks} = yield without fertilizer applied, Y_{max} = maximum yield

For further information see Fageria (2009), White (2013), Maillard et al. (2015) and White et al. (2016)

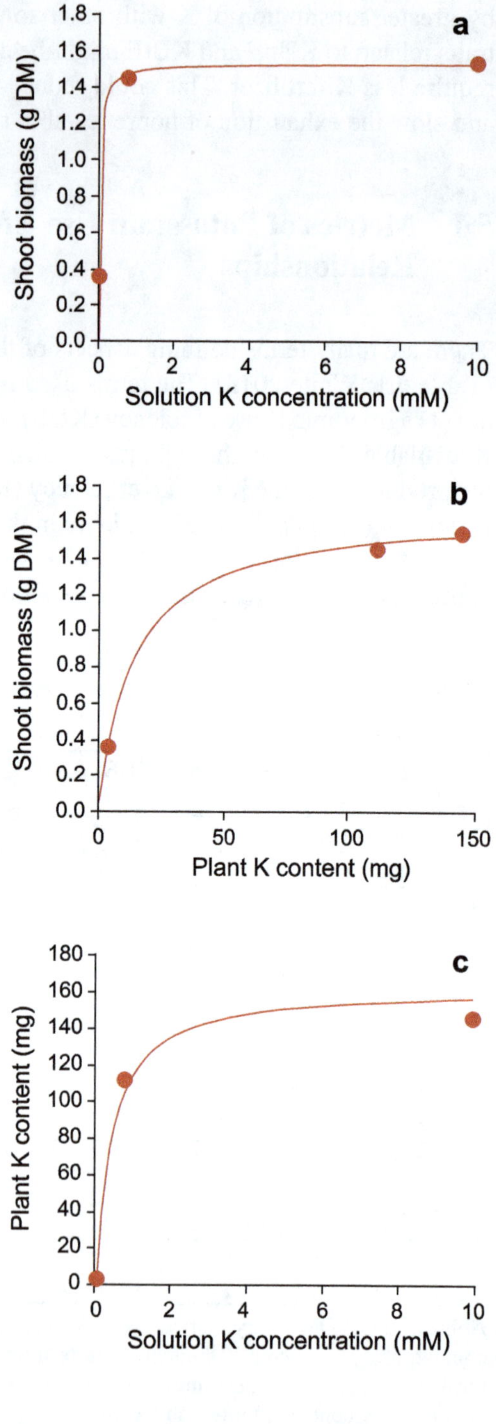

Fig. 5.1 Relationships between (**a**) shoot dry biomass and the K concentration in the nutrient solution, (**b**) shoot dry biomass and plant K content, and (**c**) plant K content and the K concentration in the nutrient solution for seedlings of spring barley "Prisma" grown hydroponically for 21 days in complete nutrient solutions containing 10 μM, 0.75 mM, or 10 mM K^+. Lines show regressions to the data assuming Michaelis–Menten relationships with (**a**) $Km_{Ka} = 0.032$ mM and $Y_{max} = 1.53$ g DM, (**b**) $Km_{Kcrop} = 13.9$ mg K and $Y_{max} = 1.66$ g DM, and (**c**) the relationship between shoot K content and the K concentration in the nutrient solution predicted using these regressions. (data from White et al. 2016)

response of crop yield to K availability, (5) the response of crop K content, or tissue K concentration, to K availability, and (6) the relationship between crop yield and crop K content or tissue K concentration (Figs. 5.1 and 5.2). In practice, these

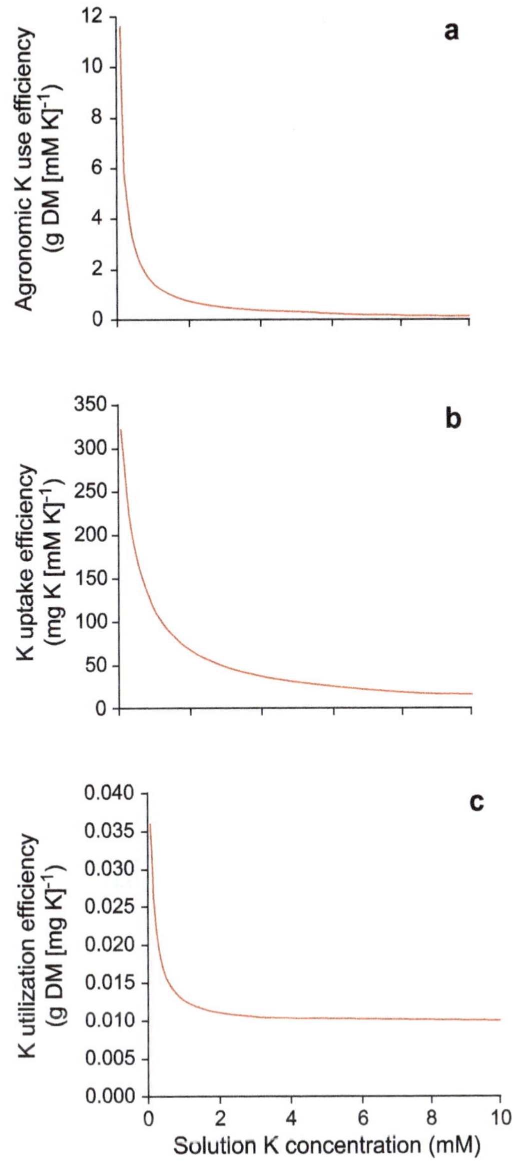

Fig. 5.2 Relationships between (**a**) agronomic K use efficiency (KUE) and the K concentration in the nutrient solution, (**b**) K uptake efficiency (KUpE) and the K concentration in the nutrient solution, and (**c**) K utilization efficiency (KUtE) and the K concentration in the nutrient solution for seedlings of spring barley "Prisma" grown hydroponically for 21 days in complete nutrient solutions containing various K concentrations. Lines were calculated from the data shown in Fig. 5.1. (White et al. 2016)

relationships are difficult to determine accurately even when data are obtained at many K availabilities and depend upon many environmental factors.

Other frequent assessments include (7) the apparent recovery (acquisition) of applied K fertilizer, which is numerically equal to KUpE when there is no available K in the unfertilized soil but is proportionally decreased as the available K in the unfertilized soil increases, and (8) the increased crop yield resulting from the application of K fertilizers relative to the amount of K fertilizer applied (Fageria 2009). The latter is often referred to as K fertilizer use efficiency or agronomic efficiency (AE). It can be determined relatively simply in field experiments, but the values obtained depend upon a variety of environmental factors, including the K availability in the unfertilized soil and factors affecting K acquisition, plant growth rates, and harvest index. The ability of a plant to tolerate low K availability can be expressed as the proportion of yield potential that it achieves without the application of K fertilizer (Rengel and Damon 2008). There are differences in all these aspects of K use efficiency both between and within plant species. This chapter describes plant traits affecting these characteristics and highlights those that commonly account for differences in KUE, KUpE, and KUtE between and within plant species.

5.2 Differences in Potassium Uptake and Utilization Between Plant Species

Plant species differ in their growth response to K supply either because of differences in their ability to acquire K from the soil (KUpE) or their ability to utilize K physiologically (KUtE) for vegetative and reproductive growth (Fageria 2009; Römheld and Kirkby 2010; White 2013; White and Bell 2017). Plant roots can acquire sufficient K for maximal growth from solutions containing micromolar K concentrations, provided the K supply to the roots matches the minimal K demand of the plant and the concentration of ammonium, which competes with K^+ for transport and inhibits the expression of genes encoding the dominant high-affinity H^+-coupled K^+ transporter in roots (e.g., *AtHAK5* in arabidopsis, *Arabidopsis thaliana* (L.) Heynh.; Qi et al. 2008), in the rhizosphere is small (Asher and Ozanne 1967; Wild et al. 1974; Spear et al. 1978a; Siddiqi and Glass 1983a; White 1993). The minimum tissue K concentration that can be tolerated without impacting plant growth and development must be sufficient to maintain about 100 mM K^+ in metabolically active compartments including the cytosol, mitochondria, and plastids (White and Karley 2010). This requires a minimal vacuolar K^+ concentration in living cells of 10–20 mM, which corresponds to a tissue K concentration of 5–40 mg g^{-1} dry weight (White and Karley 2010; White 2013).

Species from the Poales and Brassicales generally achieve their growth potential at a lower K supply than many other angiosperms and compete best in K-limited environments (Asher and Ozanne 1967; Hoveland et al. 1976; Grant et al. 2007; Hafsi et al. 2011; White et al. 2012). Species from these orders are, therefore, considered to be tolerant to K deficiency (i.e., K-efficient; Rengel and Damon 2008). Similarly, cereal and brassica crops generally require less K fertilizer than most vegetable, solanaceous, or beet (*Beta vulgaris* L.) crops to achieve maximum yields (Greenwood et al. 1980; Pretty and Stangel 1985; Steingrobe and Claassen 2000; Brennan and Bolland 2004; Trehan 2005; Fageria 2009; Kuchenbuch and Buczko 2011; Brennan and Bell 2013; Trehan and Singh 2013; White 2013; Schilling et al. 2016). Other crops that have a large demand for K fertilizer include oil palm (*Elaeis guineensis* Jacq.) and banana (*Musa acuminata* Colla/*Musa balbisiana* Colla) grown in plantations (Mengel et al. 2001; White 2020).

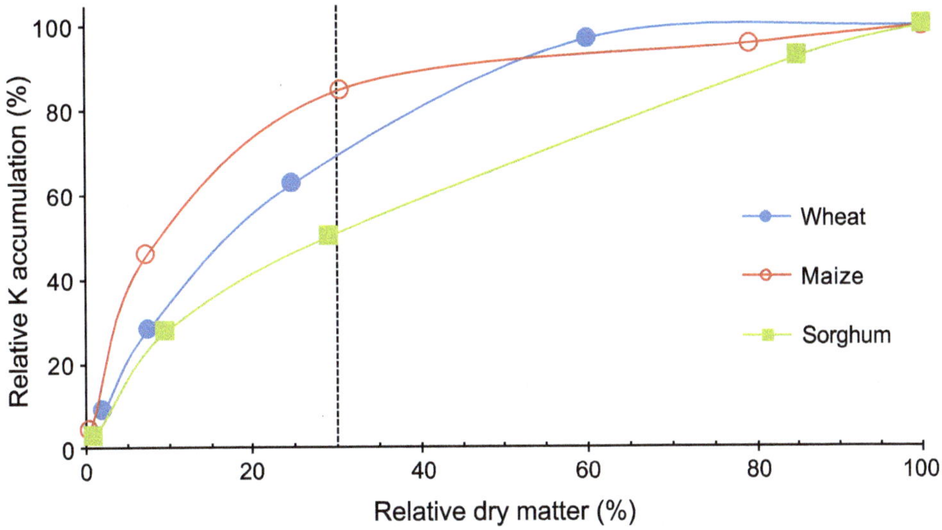

Fig. 5.3 Relative accumulation of dry matter and potassium (K) in wheat, maize, and sorghum grown to maturity in an Oxisol soil under controlled conditions in the glasshouse (Bell et al. unpublished). Maize shows the classic K accumulation curve that is well in advance of biomass in relative terms, with >80% of total K uptake occurring in the first third of the growing season (when relative biomass accumulation is only ~30%). Wheat shows a similar tendency, although relative K accumulation occurs less rapidly than in maize, while grain sorghum shows accumulation that more closely reflects the pattern of dry matter accumulation

Crops also differ in their temporal demand for K, which is related to their individual phenology, and K supply must be synchronized with their K demand to achieve maximal yields (White 2013). For example, both maize (*Zea mays* L.) and wheat (*Triticum aestivum* L.) accumulate K during early growth, while grain sorghum (*Sorghum bicolor* [L.] Moench) accumulates K roughly in proportion to its biomass accumulation (Fig. 5.3). One explanation for the temporal difference in K accumulation between these species might be tillering: The main stems of wheat and sorghum show an almost identical pattern of relative accumulation of K and DM as the uniculm maize, but the subsequent production of tillers requires continued K accumulation in new vegetative structures. While tillering in wheat occurs at a similar time to the development of the main stem, tillering in sorghum continues until much later in crop development.

5.2.1 Differences in KUpE Between Plant Species

Differences between plant species in their ability to acquire K from the soil has been attributed to differences in (1) the capacity of their root cells to take up K^+ at low rhizosphere K^+ concentrations, (2) the ability of their root systems to proliferate and exploit the soil volume effectively, and (3) their ability to acquire nonexchangeable K from the soil (Greenwood et al. 1980; Steingrobe and Claassen 2000; Wang et al.

2000, 2011; Jungk 2001; Rengel and Damon 2008; El Dessougi et al. 2010; Römheld and Kirkby 2010; Samal et al. 2010; White 2013; White et al. 2017).

5.2.1.1 Kinetics of Potassium Uptake

The uptake of K and its movement within plants are dynamic processes involving many transport proteins in many cellular membranes (White and Karley 2010; White and Bell 2017). These transporters are regulated precisely to ensure K homeostasis in metabolic compartments (White and Karley 2010; Véry et al. 2014; Nieves-Cordones et al. 2016). Thus, the relationship between K uptake by plant roots and the K concentration in the rhizosphere solution can vary markedly, both spatially and temporally, as the plant matches K supply and K demand through its K transport systems. When plants lack sufficient K, either because of low substrate K supply or high plant K demand for growth, there is an induction of genes encoding high-affinity K^+ transporters (Hermans et al. 2006; White and Karley 2010; Véry et al. 2014; Nieves-Cordones et al. 2016; White and Bell 2017), which not only increases cellular capacity for K uptake, but also increases the affinity for K in the rhizosphere solution. This reduces the K^+ concentration in the rhizosphere solution at K flux equilibrium. Indeed, the K^+ concentration at the root surface can decline to <2–3 μM, which not only accelerates K^+ diffusion to the root surface but also promotes the release of nonexchangeable K from soil minerals (Hinsinger 1998, 2013; Chap. 4).

When assayed under the same conditions, there are large differences between plant species in the maximal rate of K uptake, the solution K concentration at which K uptake is half maximal, and the minimal K concentration in the rhizosphere solution when there is K flux equilibrium. Plant species differ in both (1) the relationship between K uptake and the K concentration in the rhizosphere solution (e.g., Asher and Ozanne 1967; Wild et al. 1974; Spear et al. 1978a; Steingrobe and Claassen 2000; El Dessougi et al. 2002, 2010; Brennan and Bolland 2004; Wang et al. 2011; White 2013) and (2) the selectivity of monovalent cation accumulation (Broadley et al. 2004; Watanabe et al. 2007; White et al. 2012, 2017). This has been attributed to differences in both the capacity and complement of transport proteins catalyzing K^+ influx to root cells of different plant species (White 2013; Nieves-Cordones et al. 2016), although the molecular mechanisms, and evolutionary processes, underlying these differences are largely unknown. Roots of rapidly growing plant species with large shoot/root biomass quotients and a great K demand often have greater K uptake capacities than those of other plant species, and the roots of cereals and grasses generally have large K uptake capacities (Pettersson and Jensén 1983; Jungk and Claassen 1997; Steingrobe and Claassen 2000; Végh et al. 2008; Samal et al. 2010; Wang et al. 2011; Coskun et al. 2013). The ability of perennial ryegrass (*Lolium perenne* L.) to accumulate more K than grain amaranth (*Amaranthus* sp.) when, for example, phlogopite (1.6-fold difference) or vermiculite (12.8-fold difference) was the growth substrate was attributed to a greater K uptake capacity and a lower K concentration at which there was net K uptake in perennial ryegrass than in grain amaranth (Wang et al. 2011).

5.2.1.2 Root System Investment and Architecture

A larger root system generally allows greater access to soil K and increasing the density of roots in soil can help reduce the K concentration in the rhizosphere solution, which accelerates K diffusion to the root and promotes the release of nonexchangeable K (Zörb et al. 2014). In general, grasses and cereals invest more in root biomass than other plants, which often results in rapid and effective exploitation of the soil volume, greater root density throughout the soil volume, and potentially deeper rooting (Steingrobe and Claassen 2000; Høgh-Jensen and Pedersen 2003; Végh et al. 2008; Samal et al. 2010; White 2013; Thorup-Kristensen et al. 2020). This effect is enhanced by increasing the specific surface area ($m^2\ g^{-1}$ DM) of roots, for example by producing a finer, more densely branched root system, which increases the contact between roots and soil for a given biomass investment (White et al. 2013). Thus, it has been hypothesized that plants with greater KUpE might have a relatively larger proportion of thin roots in their root system than those with lower KUpE (Rengel and Marschner 2005; Végh et al. 2008). In addition to differences in the absolute biomass investment in the root system, the placement of roots in the soil profile also differs between plant species (Gregory 2006; Hinsinger 2013; Thorup-Kristensen et al. 2020). Kuhlmann (1990) showed that plant species with deeper roots were more reliant on K located in the subsoil than those with shallower roots, which could sometimes make a major contribution to K uptake. When growing on sandy soils that are susceptible to K leaching, it can benefit plants to have deeper root systems to acquire K at depth (Ehdaie et al. 2010; Maeght et al. 2013).

An abundance of long root hairs also facilitates K uptake by roots. It increases both the volume of soil that is explored and the surface area of the root in contact with the soil. This enhances K depletion in the rhizosphere solution and creates a steeper K^+ diffusion gradient within the bulk soil solution (Rengel and Marschner 2005). This trait also differs between plant species (White 2013). Jungk (2001) reported a linear relationship between the specific rate of K uptake (mg K cm^{-1} root) and the length of root hairs among onion (*Allium cepa* L.), maize, perennial ryegrass, tomato (*Solanum lycopersicum* L.), and canola (oilseed rape; *Brassica napus* L.). Høgh-Jensen and Pedersen (2003) reported a linear relationship between K accumulation and root hair length among red clover (*Trifolium pratense* L.), pea (*Pisum sativum* L.), barley (*Hordeum vulgare* L.), alfalfa (*Medicago sativa* L.), canola, perennial ryegrass, and rye (*Secale cereale* L.), illustrating the importance of this trait for K uptake.

5.2.1.3 Rhizosphere Acidification and Root Exudates

Root-induced acidification of the rhizosphere can lead to a significant release of exchangeable K in soils (Hinsinger 2013; Hinsinger et al. 2017). Plant species differ in their ability to acidify the rhizosphere and access nonexchangeable K in the soil. For example, legumes reduce rhizosphere pH more effectively than cereals (Liu et al. 2016; Giles et al. 2017) and oilseed rape can induce the dissolution of phlogopite mica, and the subsequent release of interlayer K, by rhizosphere acidification more effectively than Italian ryegrass (*Lolium multiflorum* Lam.; Hinsinger 2013).

Root exudates can also have a profound effect on the dissolution of feldspars and micas and, therefore, on the availability of nonexchangeable (structural and interlayer, respectively) soil K to plants. The composition of root exudates differs between plant species, which affects their ability to acquire nonexchangeable K (Hinsinger 2013; Giles et al. 2017; Hinsinger et al. 2017). Root exudates can also change during plant development and in response to environmental factors (Neumann and Römheld 2012; Kuijken et al. 2015; Giles et al. 2017). The exudation of carboxylates, such as citrate, malate, and oxalate, promotes the dissolution of feldspars and micas by complexing cations contained in their crystal lattice (Marchi et al. 2012; Chap. 4). Plant species vary greatly in the amounts and diversity of carboxylates their roots release into the rhizosphere (Hinsinger 2013; Zörb et al. 2014; Bell et al. 2017; Rengel and Djalovic 2017). Roots of Caryophyllales, including grain amaranths and beets, can access nonexchangeable K by exuding copious amounts of carboxylates (Wang et al. 2011). Roots of white lupin (*Lupinus albus* L.), and other species forming cluster roots, exude considerable quantities of both citrate and malate, as do many brassica crops (White et al. 2005; Hinsinger 2013). Greater acquisition of nonexchangeable K by *Cucurbita pepo* subsp. *pepo* than *C. pepo* subsp. *ovifera* was attributed to the greater citrate content in root exudates of subsp. *pepo* (Gent et al. 2005), while the dominant carboxylate in root exudates of K-deficient crested wheatgrass (*Agropyron cristatum* [L.] Gaertn.) appears to be malate (Henry et al. 2007). By contrast, solanaceous crops generally release carboxylates such as succinate, rather than citrate, into the rhizosphere and are relatively ineffective in acquiring nonexchangeable K from the soil (Steingrobe and Claassen 2000; White et al. 2005; White 2013). Legumes, such as alfalfa and pea, are also relatively ineffective in acquiring nonexchangeable K from the soil (Høgh-Jensen and Pedersen 2003). In addition to carboxylates, roots of different species exude a variety of amino acids and phytosiderophores, proteins, including enzymes, sugars, and polysaccharides (mucilage), flavonoids, and phenolic compounds (e.g., ferulic acid, p-coumaric acid, and cinnamic acid) into the rhizosphere (Neumann and Römheld 2012), although it is not yet known whether these compounds facilitate the acquisition of K by plants.

5.2.2 Differences in KUtE Between Plant Species

Plant species also differ in their ability to utilize the K they have acquired for growth and yield formation (White 2013). Most crops have a high K demand, which is ultimately set by their growth rate and, most often, by the nitrogen supply that generally determines their growth rate (Fageria 2009, 2015a; White and Greenwood 2013). The physiological K requirement of a plant is determined by its critical tissue K concentration, defined as the concentration at which the plant achieves 90% of its maximum growth, and its growth rate (White 2013). The tissue K concentration at which K deficiency symptoms appear in leaves is generally lower in cereals and grasses than in legumes and other eudicots, which reflects their lower physiological K requirements (Johnson 1973; Greenwood et al. 1980; Brennan and Bolland 2004, 2007; Römheld 2012; White 2013). Similarly, seed K concentrations are generally lower in cereals (3–5 g K kg^{-1} grain) than in oilseeds (5–10 g K kg^{-1} grain) and legumes (10–20 g K kg^{-1} grain; Fig. 5.4). Since crops generally have large harvest indices, achieving appropriate K concentrations in seed has significant implications for the agronomic use of K fertilizers in crop production.

In general, physiological K utilization efficiency can be improved by (1) reducing vacuolar K concentration while maintaining an appropriate cytoplasmic K concentration, either by anatomical adaptations or by greater substitution of K with other solutes in the vacuole, and (2) redistributing K from older to younger tissues to maintain growth and photosynthesis (Rengel and Damon 2008; Wakeel et al. 2011;

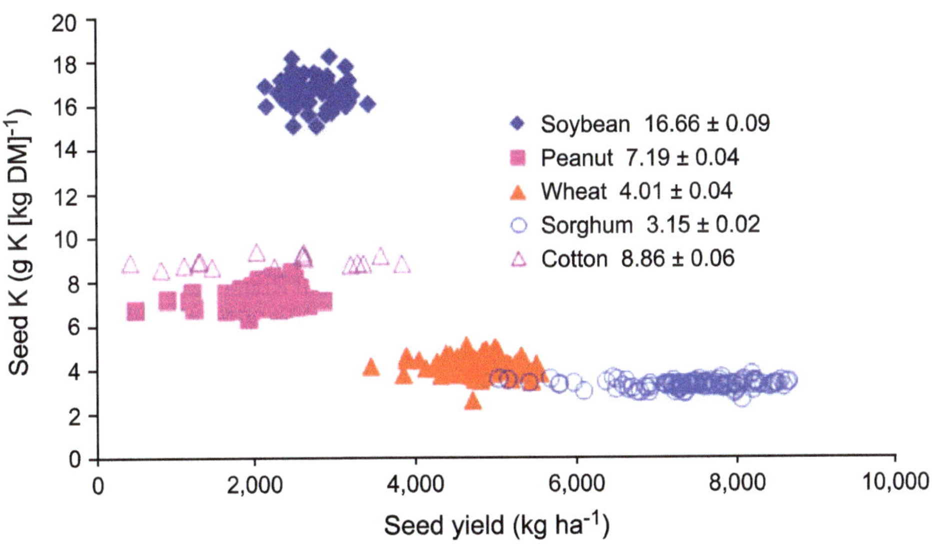

Fig. 5.4 Relationships between crop yield (as determined by variation in soil K status) and the K concentration in grains harvested from soybean, peanut, wheat, sorghum, and cotton crops grown on an Oxisol soil at Kingaroy, SE Queensland, Australia (Bell et al. unpublished). Data for each species were obtained over 2–3 growing seasons. The data illustrate the consistency of grain K concentration within each plant species irrespective of either yield or leaf K concentration (not shown)

White 2013; Maillard et al. 2015). The ability to substitute K with sodium (Na) in the vacuole is important for efficient K utilization in many, but not all, plant species and is particularly evident in species adapted to soils with low K availability and in natrophilic species, such as sugar beet (Wakeel et al. 2011; Gattward et al. 2012; Battie-Laclau et al. 2014; Erel et al. 2014; Zörb et al. 2014; White et al. 2017). About 60% of the K in cells of sugar beet can be replaced by Na, whereas less than 15% of the K in cells of wheat can be replaced (Zörb et al. 2014). The ability to retranslocate K from senescing tissues also differs between plant species (Hocking and Pate 1977; Milla et al. 2005; Maillard et al. 2015). In general, plant species with greater KUtE can maintain their water relations, photosynthetic activity, and harvest index when grown in environments with a low K supply (Rengel and Damon 2008; White 2013).

5.3 Differences in Potassium Uptake and Utilization Within Crop Species

Differences in growth and yield responses to K supply, KUE, KUpE, and KUtE have been reported among genotypes of many crop species (Baligar et al. 2001; Rengel and Damon 2008; Fageria 2009, 2015a; Römheld and Kirkby 2010; White 2013; Zörb et al. 2014; White and Bell 2017). Although variation in KUE has been correlated with variation in both KUpE and KUtE, depending upon plant species and growth conditions, it is most often correlated with KUpE in crop species (Rengel and Damon 2008; Fageria 2009; White 2013).

5.3.1 Differences in KUpE Within Plant Species

Variation in KUpE has been observed among genotypes of barley (Pettersson and Jensén 1983; Siddiqi and Glass 1983a; Wu et al. 2011; Kuzmanova et al. 2014; White et al. 2016), wheat (Zhang et al. 1999; Damon and Rengel 2007; Damon et al. 2011), wild oats (*Avena fatua* L.; Siddiqi et al. 1987), rice (*Oryza sativa* L.; Yang et al. 2004; Fageria 2009, 2015b; Liu et al. 2009; Fageria et al. 2010, 2013; Sanes et al. 2013; Fageria and dos Santos 2015), maize (Feil et al. 1992; Allan et al. 1998; Nawaz et al. 2006; Ning et al. 2013), common bean (*Phaseolus vulgaris* L.; Fageria et al. 2001, 2015; Fageria and Melo 2014), faba bean (*Vicia faba* L.; Stelling et al. 1996), soybean (*Glycine max* (L.) Merr.; Moreira et al. 2015), lupin (*Lupinus angustifolius* L.; Brennan and Bolland 2004), canola (Damon et al. 2007; Lu et al. 2016), *Brassica oleracea* L. (White et al. 2010), Indian mustard (*Brassica juncea* (L.) Czern.; Shi et al. 2004), cassava (*Manihot esculenta* Crantz; Spear et al. 1978b), sweet potato (*Ipomoea batatas* L.; George et al. 2002; Wang et al. 2015a), tomato (Chen and Gabelman 1995, 2000; Sánchez-Rodríguez et al. 2010), potato (*Solanum tuberosum* L.; Trehan 2005), cotton (*Gossypium hirsutum* L.; Ali et al. 2006; Zhang et al. 2007; Yang et al. 2011; Chen et al. 2014; Zia-ul-hassan et al. 2014; Rochester and Constable 2015) and watermelon (*Citrullus lanatus* (Thunb.) Matsum. and Nakai; Fan et al. 2013). The same traits that contribute to differences in KUpE between plant species also contribute to differences in KUpE among genotypes within plant species. These include differences in (1) the capacity of their root cells to take up K^+ at low rhizosphere K^+ concentrations, (2) the ability of their root systems to proliferate and exploit the soil volume effectively, and (3) the ability of their roots to induce the release of nonexchangeable K from the soil, depending upon the crop species and the environment in which they are grown.

5.3.1.1 Kinetics of Potassium Uptake

The rate of K uptake by roots is determined by both the cellular capacity for K uptake, the affinity for K in the rhizosphere solution, and the K concentration in the rhizosphere solution at K flux equilibrium (White 2013; Hinsinger et al. 2017). Differences in the capacity for K uptake of roots have been observed among genotypes of many crops (White 2013; Rengel and Djalovic 2017) and, when assayed at low K^+ concentrations in the rhizosphere solution, genotypes of, for example, barley (Siddiqi and Glass 1983b), Chinese cabbage (*Brassica rapa* L.; Li et al. 2015), tomato (Chen and Gabelman 1995, 2000) and potato (Trehan 2005) with greatest root K uptake capacities often having the greatest KUpE.

5.3.1.2 Root System Investment and Architecture

In general, the ability of a root system to forage the soil is related to its length and its direct interaction with the rhizosphere, which is conferred by its surface area (White 2013). There is considerable variation among genotypes of crop species in the length and architecture of their root system, the distribution of roots in the soil, the length/biomass quotients of root types, and the abundance, length, and longevity of root hairs (e.g., White et al. 2005; Gahoonia et al. 2006, 2007; Hammond et al. 2009; Wishart et al. 2013; Adu et al. 2014; Atkinson et al. 2015; Lynch 2015; Yu et al. 2015; Thomas et al. 2016; Chen et al. 2017; Erel et al. 2017).

Chromosomal loci (QTL) affecting these traits in seedlings have been identified (Lynch 2007; White et al. 2013; Atkinson et al. 2015; Kuijken et al. 2015). When compared at low K supply, maize (Minjian et al. 2007), rice (Jia et al. 2008; Sanes et al. 2013), wheat (Ehdaie et al. 2010), potato (Trehan 2005), tomato (Chen and Gabelman 1995, 2000), Chinese cabbage (Li et al. 2015), and cotton (Yang et al. 2011; Zia-ul-hassan and Arshad 2011) genotypes with larger roots have greater KUpE, and often faster growth and greater yields, than other genotypes. Similarly, enlarging the root system of rice by overexpressing the WUSCHEL-related homeobox gene WOX11 increased both K uptake and grain yield when K availability was low (Chen et al. 2015). Although there was a weak correlation between KUpE and root length among different genotypes of lentil (*Lens culinaris* Medikus), there was a stronger correlation between KUpE and the length of root hairs (Gahoonia et al. 2006). A strong correlation between KUpE and the abundance and length of root hairs was also observed among genotypes of chickpea (*Cicer arietinum* L.; Gahoonia et al. 2007) and cotton (Tao et al. 2012). Other aspects of root architecture can also contribute to differences in KUpE among genotypes of a particular species. For example, genotypes of ramie (*Boehmeria nivea* (L.) Gaudich.) whose root systems comprise a large proportion of thin roots often have greater KUpE than other genotypes (Cui and Li 2000), although this phenomenon was not observed in Chinese cabbage (Li et al. 2015).

5.3.1.3 Root Exudates

When the K uptake capacity of root cells exceeds the rate at which K is supplied to the root, K uptake is determined by the rate at which K can be replenished at the root surface. This is determined both by the movement of solution to the root surface, which is often governed by transpiration, and by the ability of the plant to mobilize nonexchangeable K from the soil, which is influenced by root exudates (White 2013).

There is considerable variation between genotypes within plant species in both the composition and quantity of root exudates that can induce the release of nonexchangeable K from the soil. For example, genotypes of barley, wheat, maize, and sorghum vary greatly in their exudation of malate and citrate into the rhizosphere (e.g., Ryan et al. 2011; Giles et al. 2017), root exudates of *Cucurbita pepo* subsp. *pepo* contain more citrate than those of *Cucurbita pepo* subsp. *ovifera* (Gent et al. 2005), canola genotypes differ in the quantity and diversity of carboxylates they release into the rhizosphere (Akhtar et al. 2006, 2008) and in their ability to acquire nonexchangeable K (Shi et al. 2004), and genotypes of potato with greater KUpE mobilize more nonexchangeable K than other genotypes (Trehan 2005).

5.3.2 *Differences in KUtE Within Crop Species*

Variation in KUtE has been observed among genotypes of barley (Pettersson and Jensén 1983; Wu et al. 2011; Kuzmanova et al. 2014; White et al. 2016), wheat (Woodend and Glass 1993; Zhang et al. 1999; Baligar et al. 2001; Damon and Rengel 2007; Damon et al. 2011; Moriconi et al. 2012), wild oats (Siddiqi et al. 1987), rice (Yang et al. 2003, 2004; Fageria 2009, 2015b; Liu et al. 2009; Fageria et al. 2010, 2013; Zhang et al. 2013; Fageria and dos Santos 2015), maize (Feil et al. 1992; Baligar et al. 2001; Nawaz et al. 2006), sorghum (Baligar et al. 2001), common bean (Fageria et al. 2001, 2015; Fageria and Melo 2014), faba bean (Stelling et al. 1996), soybean (Moreira et al. 2015), alfalfa (Baligar et al. 2001), lupin (Brennan and Bolland 2004), canola (Damon et al. 2007; Lu et al. 2016), *Brassica oleracea* (White et al. 2010), Chinese cabbage (Wu et al. 2008), Indian mustard (Shi et al. 2004), spinach (*Spinacia oleracea* L.; Grusak and Cakmak 2005), cassava (Spear et al. 1978a, b), sweet potato (George et al. 2002; Wang et al. 2015a), tomato (Chen and Gabelman 1995), potato (Trehan 2005), cotton (Ali et al. 2006; Zhang et al. 2007; Yang et al. 2011; Chen et al. 2014; Zia-ul-hassan et al. 2014; Rochester and Constable 2015) and watermelon (Fan et al. 2013). However, it is noteworthy that KUtE for vegetative growth does not always correlate with KUtE for crop yield. The same traits that contribute to differences in KUtE between plant species also contribute to differences in KUtE among genotypes of a particular species.

5.3.2.1 Partitioning of Potassium Within the Cell and Its Substitution with Other Ions

In metabolically active compartments, such as the cytosol, mitochondria, and plastids, K^+ concentrations must be maintained at about 100 mM to ensure protein function and provide charge balance (White and Karley 2010). When K is in limited supply, these compartments take precedence and cellular K can be reduced by substituting vacuolar K with other elements. Thus, it has been observed that genotypes of barley that are less susceptible to K deficiency symptoms partition K more effectively from the vacuole to the cytoplasm of root cells at low K supply (Memon et al. 1985), and the ability of tomato (Figdore et al. 1989) and maize (Moriconi et al. 2012) genotypes to grow in Na-rich, K-limiting conditions correlates with their ability to substitute Na for K as a vacuolar osmoticum.

5.3.2.2 Partitioning and Redistribution of Potassium Within the Plant

Potassium is required for stomatal opening, photosynthetic performance, and the movement of photosynthates to developing tissues (White and Karley 2010). The ability to maintain gas exchange, photosynthesis, and phloem translocation to developing tissues under conditions of restricted K supply requires effective redistribution of K from older to younger tissues. Thus, the redistribution of K within the plant can contribute significantly to KUtE. For example, the ability to redistribute K from older to younger leaves has been found to correlate with greater KUtE among genotypes of cassava (Spear et al. 1978b) and rice (Yang et al. 2004) and the ability to maintain photosynthesis at a low K supply correlates with better growth among soybean genotypes (Wang et al. 2015b). Differences in harvest index (the ability to translocate carbon into the harvested tissue), which is a component trait of KUtE, contribute to variation in yield among rice (Yang et al. 2003, 2004; Fageria et al. 2010; Zhang et al. 2013), wheat (Woodend and Glass 1993; Zhang et al. 1999; Damon and Rengel 2007), common bean (Fageria et al. 2001), faba bean (Stelling et al. 1996), canola (Rose et al. 2007), sweet potato (George et al. 2002) and cotton (Rochester and Constable 2015) genotypes, especially when grown with a low K supply.

5.3.2.3 Partitioning of Resources into the Economic Product

Potassium is required for electroneutrality in both the loading of sucrose and the transport of anions in the phloem (White and Karley 2010). Although there are considerable differences among genotypes of a crop species, the seed K concentration of a particular genotype is often relatively insensitive to plant K nutrition (Fig. 5.4). However, tuber K concentration does vary with plant K nutrition (White et al. 2009). The relationships between KUE, KUtE, and K partitioning to edible portions are currently unknown. However, given that K is essential for animal nutrition and there is substantial interest in the links between plant and human nutrition (White 2016), these relationships should be investigated.

5.4 Breeding Crops for Greater Agronomic Potassium Use Efficiency

Breeding for greater KUE relies upon (1) useful variation in component traits within germplasm resources, (2) the ability to identify beneficial traits in large germplasm collections, either through phenotypic or genetic analyses, and (3) the ability to incorporate beneficial traits into commercial varieties or locally adapted germplasm (Rengel and Damon 2008; White 2013; White and Bell 2017).

There appears to be sufficient, heritable genetic variation within crop species to breed for genotypes with greater KUE, KUpE, and KUtE (White 2013). However, these traits are controlled by multiple chromosomal loci (QTL) and strong interactions between genotype and environment can occur (e.g., White et al. 2010; Guo et al. 2012; Genc et al. 2013; Gong et al. 2015). This implies that breeding programs should incorporate beneficial alleles of several genes to improve KUE and consider carefully the conditions under which genotypes are screened and cultivated. Breeding programs have generally focused on increasing yield under current management practices, which, although resulting in greater KUE under current management practices, does not address the needs of reduced-input agriculture. This omission must be redressed in the future.

To breed for greater KUE, breeding programs must be able to screen many genotypes for variation in KUE, KUpE, or KUtE or to identify genetic variation linked to these traits (Rengel and Damon 2008; White and Bell 2017). A successful breeding program also requires the ability to characterize the relationships between K supply, plant K content, and yield formation in a variety of environments to reveal interactions between genotype, management, and environmental conditions. In principle, the required data can be obtained from simple measurements of the response of yield and K content to varying K fertilizer application at several well-chosen sites across several years (White and Bell 2017). This effort can be facilitated by reducing the number of treatments required to estimate the responses of KUE, KUpE, and KUtE to management and fertilizer practices using crop modelling approaches or theoretical considerations (Moriconi and Santa-María 2013; Santa-María et al. 2015; White et al. 2016) and developing techniques to estimate crop biomass and K content that are less costly and labor intensive than conventional mineral analyses (White and Bell 2017). An alternative approach is to screen for morphological, physiological, or biochemical traits associated with greater KUpE and KUtE using high-throughput laboratory or glasshouse systems (Downie et al. 2015; Kuijken et al. 2015).

Chromosomal loci influencing KUpE, KUtE, shoot K concentration, or biomass production at low K supply have been identified in a few model species, such as arabidopsis (e.g., Harada and Leigh 2006; Ghandilyan et al. 2009; Kanter et al. 2010; Prinzenberg et al. 2010), and in several crops, including rice (Wu et al. 1998; Koyama et al. 2001; Lin et al. 2004; Cheng et al. 2012; Wang et al. 2012; Miyamoto et al. 2012; Fang et al. 2015; Khan et al. 2015), wheat (Genc et al. 2010, 2013; Guo et al. 2012; Kong et al. 2013; Zhao et al. 2014; Gong et al. 2015), barley (Nguyen et al. 2013a, b), maize (Zdunić et al. 2014), miscanthus (*Miscanthus sinensis* Andersson; Atienza et al. 2003), tomato (Villalta et al. 2008; Asins et al. 2013), barrel medic (*Medicago truncatula* Gaertn.; Arraouadi et al. 2012), *Brassica oleracea* (White et al. 2010), apple (*Malus pumila* Miller; Fazio et al. 2013), and cotton (Liu et al. 2015). However, few genes underpinning these QTL have been identified. Nevertheless, it has been reported that genes encoding K^+ transporters, such as AtAKT1, AtHAK5, AtKUP9, AtTPK1, AtCNGC1, and AtSKOR, are located within QTL affecting shoot K concentration in arabidopsis (Harada and Leigh 2006; Kanter et al. 2010) and genes encoding homologs of the arabidopsis K^+ transporters AtKUP9, AtAKT2, AtKAT2, and AtTPK3 occur within a QTL affecting shoot K concentration in *Brassica oleracea* (White et al. 2010). Similarly, genes affecting shoot K concentration located within a QTL on chromosome 14 of cotton include numerous cation transporters, such as AKT2/3 and a Na^+/H^+-antiporter (Liu et al. 2015). In rice, the gene *OsHKT1;5* (*OsHKT8*), which encodes a Na^+ transporter expressed predominantly in the parenchyma cells surrounding the xylem, underpins the locus SKC1 that affects shoot K concentration under saline conditions (Ren et al. 2005). Similarly, HvHKT1;5, TmHKT1;5-A, and TaHKT1:5-D have been implicated in the control of shoot Na and K concentrations in barley and wheat (Munns et al. 2012; Nguyen et al. 2013a) and SlHKT1;1 and SlHKT1;2 have been implicated in the control of shoot Na and K concentrations in tomato (Asins et al. 2013).

5.5 Conclusions

Many terms have been used to define aspects of K use efficiency in plants (Table 5.1). Agronomic K use efficiency (KUE) is defined based on crop yield and is equal to the product of K uptake efficiency (KUpE) and K utilization efficiency (KUtE). Differences in KUE between plant species, and between genotypes within a species, reflect differences in their KUpE and KUtE. In crop species, KUE is most often correlated with KUpE.

Differences in KUpE have been attributed to differences in (1) the capacity of root cells to take up K^+ at low rhizosphere K^+ concentrations, (2) the ability of root systems to exploit the soil volume effectively, and (3) the release of exudates into the rhizosphere that promote the release of nonexchangeable K from the soil. Differences in KUtE have been attributed to differences in (1) the ability to reduce cellular K concentration while maintaining appropriate K concentrations in metabolically active compartments, either by anatomical adaptations or by greater substitution of K with other solutes in the vacuole, and (2) the ability to redistribute K from older to younger tissues and, thereby, maintain growth and photosynthetic capacity. There is sufficient heritable variation in both KUpE and KUtE to develop crops with greater KUE.

Given that KUpE and KUtE are polygenic and there are strong interactions between genotype and environment, breeding programs should include beneficial alleles of several genes and consider carefully the conditions under which genotypes are developed and deployed. It is likely that the full economic benefit of genotypes with greater KUE will require complementary agricultural management practices. Combining genetic and agronomic strategies to make better use of K fertilizers in agriculture would reduce fertilizer costs, protect the environment, and slow the exhaustion of nonrenewable resources.

Acknowledgment The contribution of PJW to this work was supported by the Rural and Environment Science and Analytical Services Division (RESAS) of the Scottish Government.

References

Adu MO, Chatot A, Wiesel L, Bennett MJ, Broadley MR, White PJ, Dupuy LX (2014) A scanner system for high-resolution quantification of variation in root growth dynamics of *Brassica rapa* genotypes. J Exp Bot 65:2039–2048. https://doi.org/10.1093/jxb/eru048

Akhtar MS, Oki Y, Adachi T, Murata Y, Khan MHR (2006) Phosphorus starvation induced root-mediated pH changes in solubilization and acquisition of sparingly soluble P sources and organic acids exudation by Brassica cultivars. Soil Sci Plant Nutr 52:623–633. https://doi.org/10.1111/j.1747-0765.2006.00082.x

Akhtar MS, Oki Y, Adachi T (2008) Genetic variability in phosphorus acquisition and utilization efficiency from sparingly soluble P-sources by *Brassica* cultivars under P-stress environment. J Agron Crop Sci 194:380–392. https://doi.org/10.1111/j.1439-037X.2008.00326.x

Ali L, Rahmatullah, Ranjha AM, Aziz T, Maqsood MA, Ashraf M (2006) Differential potassium requirement and its substitution by sodium in cotton genotypes. Pak J Agric Sci 43:108–113

Allan DL, Rehm GW, Oldham JL (1998) Root system interactions with potassium management in corn. In: Oosterhuis D, Berkowitz G (eds) Frontiers in potassium nutrition: new perspectives on the effects of potassium on physiology of plants. Potash and Phosphate Institute, Norcross, pp 111–116

Arraouadi S, Badri M, Abdelly C, Huguet T, Aouani ME (2012) QTL mapping of physiological traits associated with salt tolerance in *Medicago truncatula* recombinant inbred lines. Genomics 99:118–125. https://doi.org/10.1016/j.ygeno.2011.11.005

Asher CJ, Ozanne PG (1967) Growth and potassium content of plants in solution cultures maintained at constant potassium concentrations. Soil Sci 103:155–161

Asins MJ, Villalta I, Aly MM, Olias R, De Morales PA, Huertas R, Li J, Jaime-Perez N, Haro R, Raga V, Carbonell EA, Belver A (2013) Two closely linked tomato HKT coding genes are positional candidates for the major tomato QTL involved in Na^+/K^+ homeostasis. Plant Cell Environ 36:1171–1191. https://doi.org/10.1111/pce.12051

Atienza SG, Satovic Z, Petersen KK, Dolstra O, Martín A (2003) Identification of QTLs influencing combustion quality in *Miscanthus sinensis* Anderss. II. Chlorine and potassium content. Theor Appl Genet 107:857–863. https://doi.org/10.1007/s00122-003-1218-z

Atkinson JA, Wingen LU, Griffiths M, Pound MP, Gaju O, Foulkes MJ, Le Gouis J, Griffiths S, Bennett MJ, King J (2015) Phenotyping pipeline reveals major seedling root growth QTL in hexaploid wheat. J Exp Bot 66:2281–2292. https://doi.org/10.1093/jxb/erv006

Baligar VC, Fageria NK, He ZL (2001) Nutrient use efficiency in plants. Commun Soil Sci Plant Anal 32:921–950. https://doi.org/10.1007/978-81-322-2169-2_1

Battie-Laclau P, Laclau J-P, Beri C, Mietton L, Muniz MRA, Arenque BC, Piccolo MDC, Jordan-Meille L, Bouillet JP, Nouvellon Y (2014) Photosynthetic and anatomical responses of *Eucalyptus grandis* leaves to potassium and sodium supply in a field experiment. Plant Cell Environ 37:70–81. https://doi.org/10.1111/pce.12131

Bell MJ, Moody P, Thompson M, Guppy C, Mallarino AP, Goulding K (2017) Improving potassium rate recommendations by recognizing soil potassium pools with dissimilar bioavailability. In: Murrell TS, Mikkelsen RL (eds) Proceedings for the frontiers of potassium science conference, 25–27 January 2017, Rome, Italy. International Plant Nutrition Institute, Peachtree Corners, pp 239–248. https://www.apni.net/k-frontiers/. Accessed 29 May 2020

Brennan RF, Bell MJ (2013) Soil potassium-crop response calibration relationships and criteria for field crops grown in Australia. Crop Pasture Sci 64:514–522. https://doi.org/10.1071/CP13006

Brennan RF, Bolland MDA (2004) Lupin takes up less potassium but uses the potassium more effectively to produce shoots than canola and wheat. Aust J Exp Agric 44:309–319. https://doi.org/10.1071/EA02232

Brennan RF, Bolland MDA (2007) Comparing the potassium requirements of canola and wheat. Aust J Agric Res 58:359–366. https://doi.org/10.1071/AR06244

Broadley MR, Bowen HC, Cotterill HL, Hammond JP, Meacham MC, Mead A, White PJ (2004) Phylogenetic variation in the shoot mineral concentration of angiosperms. J Exp Bot 55:321–336. https://doi.org/10.1093/jxb/erh002

Chen J, Gabelman WH (1995) Isolation of tomato strains varying in potassium acquisition using a sand-zeolite culture system. Plant Soil 176:65–70. https://doi.org/10.1007/BF00017676

Chen J, Gabelman WH (2000) Morphological and physiological characteristics of tomato roots associated with potassium-acquisition efficiency. Sci Hortic 83:213–225. https://doi.org/10.1016/S0304-4238(99)00079-5

Chen Y, Wen Y, Wang J, Zhang X, Chen D (2014) Cotton potassium uptake and use efficiency vary with potassium application rates and soil potassium nutrition levels. J Food Agric Environ 12:221–227

Chen G, Feng H, Hu Q, Qu H, Chen A, Yu L, Xu G (2015) Improving rice tolerance to potassium deficiency by enhancing OsHAK16p:WOX11-controlled root development. Plant Biotechnol J 13:833–848. https://doi.org/10.1111/pbi.12320

Chen Y, Ghanem ME, Siddique KHM (2017) Characterising root trait variability in chickpea (*Cicer arietinum* L.) germplasm. J Exp Bot 68:1987–1999. https://doi.org/10.1093/jxb/erw368

Cheng L, Wang Y, Meng L, Hu X, Cui Y, Sun Y, Zhu L, Ali J, Xu J, Li Z (2012) Identification of salt-tolerant QTLs with strong genetic background effect using two sets of reciprocal introgression lines in rice. Genome 55:45–55. https://doi.org/10.1139/g11-075

Coskun D, Britto DT, Li M, Oh S, Kronzucker HJ (2013) Capacity and plasticity of potassium channels and high-affinity transporters in roots of barley and Arabidopsis. Plant Physiol 162:496–511. https://doi.org/10.1104/pp.113.215913

Cui G, Li Z (2000) Relationship between potassium absorption and root parameters of different genotypes of ramie. Res Agric Modernization 21:371–375

Damon PM, Rengel Z (2007) Wheat genotypes differ in potassium efficiency under glasshouse and field conditions. Aust J Agric Res 58:816–825. https://doi.org/10.1071/AR06402

Damon PM, Osborne LD, Rengel Z (2007) Canola genotypes differ in potassium efficiency during vegetative growth. Euphytica 156:387–397. https://doi.org/10.1007/s10681-007-9388-4

Damon PM, Ma QF, Rengel Z (2011) Wheat genotypes differ in potassium accumulation and osmotic adjustment under drought stress. Crop Pasture Sci 62:550–555. https://doi.org/10.1071/CP11071

Downie HF, Adu MO, Schmidt S, Otten W, Dupuy LX, White PJ, Valentine TA (2015) Challenges and opportunities for quantifying roots and rhizosphere interactions through imaging and image analysis. Plant Cell Environ 38:1213–1232. https://doi.org/10.1111/pce.12448

Ehdaie B, Merhaut DJ, Ahmadian S, Hoops AC, Khuong T, Layne AP, Waines JG (2010) Root system size influences water-nutrient uptake and nitrate leaching potential in wheat. J Agron Crop Sci 196:455–466. https://doi.org/10.1111/j.1439-037X.2010.00433.x

El Dessougi H, Claassen N, Steingrobe B (2002) Potassium efficiency mechanisms of wheat, barley, and sugar beet grown on a K fixing soil under controlled conditions. J Plant Nutr Soil Sci 165:732–737. https://doi.org/10.1002/jpln.200290011

El Dessougi HI, Claassen N, Steingrobe B (2010) Potassium efficiency of different crops grown on a sandy soil under controlled conditions. Univ Khartoum J Agric Sci 18:310–334

Erel R, Ben-Gal A, Dag A, Schwartz A, Yermiyahu U (2014) Sodium replacement of potassium in physiological processes of olive trees (var. Barnea) as affected by drought. Tree Physiol 34:1102–1117. https://doi.org/10.1093/treephys/tpu081

Erel R, Bérard A, Capowiez L, Doussan C, Arnal D, Souche G, Gavaland A, Fritz C, Visser EJW, Salvi S, Le Marié C, Hund A, Hinsinger P (2017) Soil type determines how root and rhizosphere traits relate to phosphorus acquisition in field-grown maize genotypes. Plant Soil 412:115–132. https://doi.org/10.1007/s11104-016-3127-3

Fageria NK (2009) The use of nutrients in crop plants. CRC Press, Boca Raton

Fageria NK (2015a) Potassium. In: Barker AV, Pilbeam DJ (eds) A handbook of plant nutrition, 2nd edn. CRC Press, Boca Raton, pp 127–163

Fageria NK (2015b) Lowland rice genotypes evaluation for potassium-use efficiency. Commun Soil Sci Plant Anal 46:1628–1635. https://doi.org/10.1081/PLN-120015539

Fageria NK, dos Santos AB (2015) Agronomic evaluation of lowland rice genotypes for potassium-use efficiency. Commun Soil Sci Plant Anal 46:1327–1344. https://doi.org/10.1080/01904167.2014.911889

Fageria NK, Melo LC (2014) Agronomic evaluation of dry bean genotypes for potassium use efficiency. J Plant Nutr 37:1899–1912. https://doi.org/10.1080/01904167.2014.911889

Fageria NK, Barbosa Filho MP, da Costa JGC (2001) Potassium-use efficiency in common bean genotypes. J Plant Nutr 24:1937–1945. https://doi.org/10.1081/PLN-100107605

Fageria NK, dos Santos AB, de Moraes MF (2010) Yield, potassium uptake, and use efficiency in upland rice genotypes. Commun Soil Sci Plant Anal 41:2676–2684. https://doi.org/10.1080/00103624.2010.517882

Fageria NK, Moreira A, Ferreira EPB, Knupp AM (2013) Potassium-use efficiency in upland rice genotypes. Commun Soil Sci Plant Anal 44:2656–2665. https://doi.org/10.1080/00103624.2013.813031

Fageria NK, Melo LC, Knupp AM (2015) Dry bean genotype evaluation for potassium-use efficiency. Commun Soil Sci Plant Anal 46:1061–1075. https://doi.org/10.1080/00103624.2014.981637

Fan M, Bie Z, Xie H, Zhang F, Zhao S, Zhang H (2013) Genotypic variation for potassium efficiency in wild and domesticated watermelons under ample and limited potassium supply. J Plant Nutr Soil Sci 176:466–473. https://doi.org/10.1002/jpln.201200007

Fang Y, Wu W, Zhang X, Jiang H, Lu W, Pan J, Hu J, Guo L, Zeng D, Xue D (2015) Identification of quantitative trait loci associated with tolerance to low potassium and related ions concentrations at seedling stage in rice (*Oryza sativa* L.). Plant Growth Regul 77:157–166. https://doi.org/10.1007/s10725-015-0047-9

Fazio G, Kviklys D, Grusak MA, Robinson T (2013) Phenotypic diversity and QTL mapping of absorption and translocation of nutrients by apple rootstocks. Asp Appl Biol 119:37–50

Feil B, Thiraporn R, Geisler G, Stamp P (1992) Yield, development and nutrient efficiency of temperate and tropical maize germplasm in the tropical lowlands. II. Uptake and redistribution of nitrogen, phosphorus and potassium. Maydica 37:199–207

Figdore SS, Gabelman WH, Gerloff GC (1989) Inheritance of potassium efficiency, sodium substitution capacity, and sodium accumulation in tomatoes grown under low-potassium stress. J Am Soc Hortic Sci 114:322–327

Gahoonia TS, Ali O, Sarker A, Nielsen NE, Rahman MM (2006) Genetic variation in root traits and nutrient acquisition of lentil genotypes. J Plant Nutr 29:643–655. https://doi.org/10.1080/01904160600564378

Gahoonia TS, Rawshan A, Malhotra RS, Jahoor A, Rahman MM (2007) Variation in root morphological and physiological traits and nutrient uptake of chickpea genotypes. J Plant Nutr 30:829–841. https://doi.org/10.1080/15226510701373213

Gattward JN, Almeida A-AF, Souza JO, Gomes FP, Kronzucker HJ (2012) Sodium-potassium synergism in *Theobroma cacao*: stimulation of photosynthesis, water-use efficiency and mineral nutrition. Physiol Plant 146:350–362. https://doi.org/10.1111/j.1399-3054.2012.01621.x

Genc Y, Oldach K, Verbyla AP, Lott G, Hassan M, Tester M, Wallworth H, McDonald GK (2010) Sodium exclusion QTL associated with improved seedling growth in bread wheat under salinity stress. Theor Appl Genet 121:877–894. https://doi.org/10.1007/s00122-010-1357-y

Genc Y, Oldach K, Gogel B, Wallwork H, McDonald GK, Smith AB (2013) Quantitative trait loci for agronomical and physiological traits for a bread wheat population grown in environments with a range of salinity levels. Mol Breed 32:39–59. https://doi.org/10.1007/s11032-013-9851-y

Gent MPN, Parrish ZD, White JC (2005) Nutrient uptake among subspecies of *Cucurbita pepo* L. is related to exudation of citric acid. J Am Soc Hortic Sci 130:782–788. https://doi.org/10.21273/JASHS.130.5.782

George MS, Lu G, Zhou W (2002) Genotypic variation for potassium uptake and utilization efficiency in sweet potato (*Ipomoea batatas* L.). Field Crops Res 77:7–15. https://doi.org/10.1016/S0378-4290(02)00043-6

Ghandilyan A, Ilk N, Hanhart C, Mbengue M, Barboza L, Schat H, Koornneef M, El-Lithy M, Vreugdenhil D, Reymond M, Aarts MGM (2009) A strong effect of growth medium and organ type on the identification of QTLs for phytate and mineral concentrations in three *Arabidopsis thaliana* RIL populations. J Exp Bot 60:1409–1425. https://doi.org/10.1093/jxb/erp084

Giles CD, Brown LK, Adu MO, Mezeli MM, Sandral GA, Simpson RJ, Wendler R, Shand CA, Menezes-Blackburn D, Darch T, Stutter MI, Lumsdon DG, Zhang H, Blackwell MSA, Wearing C, Cooper P, Haygarth PM, George TS (2017) Response-based selection of barley cultivars and legume species for complementarity: root morphology and exudation in relation to nutrient source. Plant Sci 255:12–28. https://doi.org/10.1016/j.plantsci.2016.11.002

Gong X-P, Liang X, Guo Y, Wu C-H, Zhao Y, Li X-H, Li S-S, Kong F-M (2015) Quantitative trait locus mapping for potassium use efficiency traits at the seedling stage in wheat under different nitrogen and phosphorus treatments. Crop Sci 55:2690–2700. https://doi.org/10.2135/cropsci2014.10.0711

Grant CA, Derksen DA, Blackshaw RE, Entz T, Janzen HH (2007) Differential response of weed and crop species to potassium and sulphur fertilizers. Can J Plant Sci 87:293–296. https://doi.org/10.4141/P06-138

Greenwood DJ, Cleaver TJ, Turner MK, Hunt J, Niendorf KB, Loquens SMH (1980) Comparison of the effects of potassium fertilizer on the yield, potassium content and quality of 22 different vegetables and agricultural crops. J Agric Sci 95:441–456. https://doi.org/10.1017/S0021859600039496

Gregory PJ (2006) Plant roots. Growth, activity and interaction with soils. Blackwell, Oxford. https://doi.org/10.1002/9780470995563.fmatter

Grusak MA, Cakmak I (2005) Methods to improve the crop delivery of minerals to humans and livestock. In: Broadley MR, White PJ (eds) Plant nutritional genomics. Blackwell, Oxford, pp 265–286

Guo Y, Kong F-M, Xu Y-F, Zhao Y, Liang X, Wang Y-Y, An D-G, Li S-S (2012) QTL mapping for seedling traits in wheat grown under varying concentrations of N, P and K nutrients. Theor Appl Genet 124:851–865. https://doi.org/10.1007/s00122-011-1749-7

Hafsi C, Atia A, Lakhdar A, Debez A, Abdelly C (2011) Differential responses in potassium absorption and use efficiencies in the halophytes *Catapodium rigidum* and *Hordeum maritimum* to various potassium concentrations in the medium. Plant Prod Sci 14:135–140

Hammond JP, Broadley MR, White PJ, King GJ, Bowen HC, Hayden R, Meacham MC, Mead A, Overs T, Spracklen WP, Greenwood DJ (2009) Shoot yield drives phosphorus use efficiency in *Brassica oleracea* and correlates with root architecture traits. J Exp Bot 60:1953–1968. https://doi.org/10.1093/jxb/erp083

Harada H, Leigh RA (2006) Genetic mapping of natural variation in potassium concentrations in shoots of *Arabidopsis thaliana*. J Exp Bot 57:953–960. https://doi.org/10.1093/jxb/erj081

Henry A, Doucette W, Norton J, Bugbee B (2007) Changes in crested wheatgrass root exudation caused by flood, drought, and nutrient stress. J Environ Qual 36:904–912. https://doi.org/10.2134/jeq2006.0425sc

Hermans C, Hammond JP, White PJ, Verbruggen N (2006) How do plants respond to nutrient shortage by biomass allocation? Trends Plant Sci 11:610–617. https://doi.org/10.1016/j.tplants.2006.10.007

Hinsinger P (1998) How do plant roots acquire mineral nutrients? Chemical processes involved in the rhizosphere. Adv Agron 64:225–265. https://doi.org/10.1016/S0065-2113(08)60506-4

Hinsinger P (2013) Plant-induced changes of soil processes and properties. In: Gregory PJ, Nortcliff S (eds) Soil conditions and plant growth. Blackwell, Oxford, pp 323–365. https://doi.org/10.1002/9781118337295.ch10

Hinsinger P, Bell M, White PJ (2017) Root traits and rhizosphere characteristics determining potassium acquisition from soils. In: Murrell TS, Mikkelsen RL (eds) Proceedings for the frontiers of potassium science conference, Rome, Italy, 25–27. January 2017. International Plant Nutrition Institute, Peachtree Corners, GA, USA, pp 289–299. https://www.apni.net/kfrontiers/. Accessed 28 Sept 2020

Hocking PJ, Pate JS (1977) Mobilization of minerals to developing seeds of legumes. Ann Bot 41:1259–1278. https://doi.org/10.1093/oxfordjournals.aob.a085415

Høgh-Jensen H, Pedersen MB (2003) Morphological plasticity by crop plants and their potassium use efficiency. J Plant Nutr 26:969–984. https://doi.org/10.1081/PLN-120020069

Hoveland CS, Buchanan GA, Harris MC (1976) Response of weeds to soil phosphorus and potassium. Weed Sci 24:194–201. https://doi.org/10.1017/S0043174500065747

Jia Y-B, Yang X-E, Feng Y, Jilani G (2008) Differential response of root morphology to potassium deficient stress among rice genotypes varying in potassium efficiency. J Zhejiang Univ Sci B 9:427–434. https://doi.org/10.1631/jzus.B0710636

Johnson CR (1973) Symptomatology and analyses of nutrient deficiencies produced on flowering annual plants. Commun Soil Sci Plant Anal 4:185–196. https://doi.org/10.1080/00103627309366436

Jungk A (2001) Root hairs and the acquisition of plant nutrients from soil. J Plant Nutr Soil Sci 164:121–129. https://doi.org/10.1002/1522-2624(200104)164:2<121::AID-JPLN121>3.0.CO;2-6

Jungk A, Claassen N (1997) Ion diffusion in the soil-root system. Adv Agron 61:53–110. https://doi.org/10.1016/S0065-2113(08)60662-8

Kanter U, Hauser A, Michalke B, Draexl S, Schaeffner AR (2010) Caesium and strontium accumulation in shoots of *Arabidopsis thaliana*: genetic and physiological aspects. J Exp Bot 61:3995–4009. https://doi.org/10.1093/jxb/erq213

Khan MSK, Saeed M, Iqbal J (2015) Identification of quantitative trait loci for Na^+, K^+ and Ca^{++} accumulation traits in rice grown under saline conditions using F_2 mapping population. J Braz Bot 38:555–565. https://doi.org/10.1007/s40415-015-0160-z

Kong F-M, Guo Y, Liang X, Wu C-H, Wang Y-Y, Zhao Y, Li S-S (2013) Potassium (K) effects and QTL mapping for K efficiency traits at seedling and adult stages in wheat. Plant Soil 373:877–892. https://doi.org/10.1007/s11104-013-1844-4

Koyama ML, Levesley A, Koebner RMD, Flowers TJ, Yeo AR (2001) Quantitative trait loci for component physiological traits determining salt tolerance in rice. Plant Physiol 125:406–422. https://doi.org/10.1104/pp.125.1.406

Kuchenbuch RO, Buczko U (2011) Re-visiting potassium- and phosphate-fertilizer responses in field experiments and soil-test interpretations by means of data mining. J Plant Nutr Soil Sci 174:171–185. https://doi.org/10.1002/jpln.200900162

Kuhlmann H (1990) Importance of the subsoil for the K-nutrition of crops. Plant Soil 127:129–136. https://doi.org/10.1007/BF00010845

Kuijken RCP, van Eeuwijk FA, Marcelis LFM, Bouwmeester HJ (2015) Root phenotyping: from component trait in the lab to breeding. J Exp Bot 66:5389–5401. https://doi.org/10.1093/jxb/erv239

Kuzmanova L, Kostadinova S, Ganusheva N (2014) Efficiency of potassium in barley genotypes. Turk J Agric Nat Sci S1:584–589

Li HY, Si DX, Lv FT (2015) Differential responses of six Chinese cabbage (*Brassica rapa* L. ssp. *pekinensis*) cultivars to potassium ion deficiency. J Hortic Sci Biotechnol 90:483–488

Lin HX, Zhu MZ, Yano M, Gao JP, Liang ZW, Su WA, Hu XH, Ren ZH, Chao DY (2004) QTLs for Na^+ and K^+ uptake of the shoots and roots controlling rice salt tolerance. Theor Appl Genet 108:253–260. https://doi.org/10.1007/s00122-003-1421-y

Liu G, Li Y, Porterfield DM (2009) Genotypic differences in potassium nutrition in lowland rice hybrids. Commun Soil Sci Plant Anal 40:1803–1821. https://doi.org/10.1080/00103620902896704

Liu S, Lacape J-M, Constable GA, Llewellyn DJ (2015) Inheritance and QTL mapping of leaf nutrient concentration in a cotton inter-specific derived RIL population. PLoS One 10(5): e0128100. https://doi.org/10.1371/journal.pone.0128100

Liu H, White PJ, Li C (2016) Biomass partitioning and rhizosphere responses of maize and faba bean to phosphorus deficiency. Crop Pasture Sci 67:847–856. https://doi.org/10.1071/CP16015

Lu ZF, Lu JW, Pan YH, Li XK, Cong RH, Ren T (2016) Genotypic variation in photosynthetic limitation responses to K deficiency of *Brassica napus* is associated with potassium utilisation efficiency. Funct Plant Biol 43:880–891. https://doi.org/10.1071/FP16098

Lynch JP (2007) Roots of the second green revolution. Aust J Bot 55:493–512. https://doi.org/10.1071/BT06118

Lynch JP (2015) Root phenes that reduce the metabolic costs of soil exploration: opportunities for 21st century agriculture. Plant Cell Environ 38:1775–1784. https://doi.org/10.1111/pce.12451

Maeght J-L, Rewald B, Pierret A (2013) How to study deep roots – and why it matters. Front Plant Sci 4:299. https://doi.org/10.3389/fpls.2013.00299

Maillard A, Diquélou S, Billard V, Laîné P, Garnica M, Prudent M, Garcia-Mina J-M, Yvin J-C, Ourry A (2015) Leaf mineral nutrient remobilization during leaf senescence and modulation by nutrient deficiency. Front Plant Sci 6:317. https://doi.org/10.3389/fpls.2015.00317

Marchi G, Silva VA, Guilherme LRG, Lima JM, Nogueira FD, Guimaraes PTG (2012) Potassium extractability from soils of Brazilian coffee regions. Biosci J 28:913–919

Memon AR, Saccomani M, Glass ADM (1985) Efficiency of potassium utilization by barley varieties: the role of subcellular compartmentation. J Exp Bot 36:1860–1876. https://doi.org/10.1093/jxb/36.12.1860

Mengel K, Kirkby EA, Kosegarten H, Appel T (2001) Principles of plant nutrition. Kluwer Academic, Dordrecht

Milla R, Castro-Díez P, Maestro-Martínez M, Montserrat-Martí G (2005) Relationships between phenology and the remobilization of nitrogen, phosphorus and potassium in branches of eight Mediterranean evergreens. New Phytol 168:167–178. https://doi.org/10.1111/j.1469-8137.2005.01477.x

Minjian C, Haiqiu Y, Hongkui Y, Chunji J (2007) Difference in tolerance to potassium deficiency between two maize inbred lines. Plant Prod Sci 10:42–46. https://doi.org/10.1626/pps.10.42

Miyamoto T, Ochiai K, Takeshita S, Matoh T (2012) Identification of quantitative trait loci associated with shoot sodium accumulation under low potassium conditions in rice plants. Soil Sci Plant Nutr 58:728–736. https://doi.org/10.1080/00380768.2012.745797

Moreira A, Moraes LAC, Fageria NK (2015) Variability on yield, nutritional status, soil fertility, and potassium-use efficiency by soybean cultivar in acidic soil. Commun Soil Sci Plant Anal 46:2490–2508. https://doi.org/10.1080/00103624.2015.1085555

Moriconi JI, Santa-María GE (2013) A theoretical framework to study potassium utilization efficiency in response to withdrawal of potassium. J Exp Bot 64:4289–4299. https://doi.org/10.1093/jxb/ert236

Moriconi JI, Buet A, Simontacchi M, Santa-María GE (2012) Near-isogenic wheat lines carrying altered function alleles of the *Rht-1* genes exhibit differential responses to potassium deprivation. Plant Sci 185/186:199–207. https://doi.org/10.1016/j.plantsci.2011.10.011

Munns R, James RA, Xu B, Athman A, Conn SJ, Jordans C, Byrt CS, Hare RA, Tyerman SD, Tester M, Plett D, Gilliham M (2012) Wheat grain yield on saline soils is improved by an ancestral Na^+ transporter gene. Nat Biotechnol 30:360–364. https://doi.org/10.1038/nbt.2120

Nawaz I, Zia-ul-hassan, Ranjha AM, Arshad M (2006) Exploiting genotypic variation among fifteen maize genotypes of Pakistan for potassium uptake and use efficiency in solution culture. Pak J Bot 38:1689–1696

Neumann G, Römheld V (2012) Rhizosphere chemistry in relation to plant nutrition. In: Marschner P (ed) Marschner's mineral nutrition of higher plants, 3rd edn. Academic, London, pp 347–368. https://doi.org/10.1016/B978-0-12-384905-2.00014-5

Nguyen VL, Dolstra O, Malosetti M, Kilian B, Graner A, Visser RGF, van der Linden CG (2013a) Association mapping of salt tolerance in barley (*Hordeum vulgare* L.). Theor Appl Genet 126:2335–2351. https://doi.org/10.1007/s00122-013-2139-0

Nguyen VL, Ribot SA, Dolstra O, Niks RE, Visser RGF, van der Linden CG (2013b) Identification of quantitative trait loci for ion homeostasis and salt tolerance in barley (*Hordeum vulgare* L). Mol Breed 31:137–152. https://doi.org/10.1007/s11032-012-9777-9

Nieves-Cordones M, Martinez V, Benito B, Rubio F (2016) Comparison between arabidopsis and rice for main pathways of K^+ and Na^+ uptake by roots. Front Plant Sci 7:992. https://doi.org/10.3389/fpls.2016.00992

Ning P, Li S, Yu P, Zhang Y, Li C (2013) Post-silking accumulation and partitioning of dry matter, nitrogen, phosphorus and potassium in maize varieties differing in leaf longevity. Field Crops Res 144:19–27. https://doi.org/10.1016/j.fcr.2013.01.020

Pettersson S, Jensén P (1983) Variation among species and varieties in uptake and utilization of potassium. Plant Soil 72:231–237. https://doi.org/10.1007/BF02181962

Pretty KM, Stangel PJ (1985) Current and future use of world potassium. In: Munson RD (ed) Potassium in agriculture. ASA, CSSA & SSSA, Madison, pp 99–128

Prinzenberg AE, Barbier H, Salt DE, Stich B, Matthieu R (2010) Relationships between growth, growth response to nutrient supply, and ion content using a recombinant inbred line population in Arabidopsis. Plant Physiol 154:1361–1371. https://doi.org/10.1104/pp.110.161398

Qi Z, Hampton CR, Shin R, Barkla BJ, White PJ, Schachtman DP (2008) The high affinity K^+ transporter AtHAK5 plays a physiological role *in planta* at very low K^+ concentrations and provides a caesium uptake pathway in Arabidopsis. J Exp Bot 59:595–607. https://doi.org/10.1093/jxb/erm330

Ren H-Z, Goa J-P, Li L-G, Cai X-L, Huang W, Chao D-Y, Zhu M-Z, Wang Z-Y, Luan S, Lin H-X (2005) A rice quantitative trait locus for salt tolerance encodes a sodium transporter. Nat Genet 37:1141–1146. https://doi.org/10.1038/ng1643

Rengel Z, Damon PM (2008) Crops and genotypes differ in efficiency of potassium uptake and use. Physiol Plant 133:624–636. https://doi.org/10.1111/j.1399-3054.2008.01079.x

Rengel Z, Djalovic I (2017) Differential potassium-use efficiency in crops and genotypes. In: Murrell TS, Mikkelsen RL (eds) Proceedings for the frontiers of potassium science conference, 25–27 January 2017, Rome, Italy. International Plant Nutrition Institute, Peachtree Corners, pp 65–73. https://www.apni.net/k-frontiers/. Accessed 29 May 2020

Rengel Z, Marschner P (2005) Nutrient availability and management in the rhizosphere: exploiting genotypic differences. New Phytol 168:305–312. https://doi.org/10.1111/j.1469-8137.2005.01558.x

Rochester IJ, Constable GA (2015) Improvements in nutrient uptake and nutrient use-efficiency in cotton cultivars released between 1973 and 2006. Field Crops Res 173:14–21. https://doi.org/10.1016/j.fcr.2015.01.001

Römheld V (2012) Diagnosis of deficiency and toxicity of nutrients. In: Marschner P (ed) Marschner's mineral nutrition of higher plants, 3rd edn. Academic, London, pp 299–312. https://doi.org/10.1016/B978-0-12-384905-2.00011-X

Römheld V, Kirkby EA (2010) Research on potassium in agriculture: needs and prospects. Plant Soil 335:155–180. https://doi.org/10.1007/s11104-010-0520-1

Rose TJ, Rengel Z, Ma Q, Bowden JW (2007) Differential accumulation patterns of phosphorus and potassium by canola cultivars compared to wheat. J Plant Nutr Soil Sci 170:404–411. https://doi.org/10.1002/jpln.200625163

Ryan PR, Tyerman SD, Sasaki T, Furuichi T, Yamamoto Y, Zhang WH, Delhaize E (2011) The identification of aluminium-resistance genes provides opportunities for enhancing crop production on acid soils. J Exp Bot 62:9–20. https://doi.org/10.1093/jxb/erq272

Samal D, Kovar JL, Steingrobe B, Sadana US, Bhadoria PS, Claassen N (2010) Potassium uptake efficiency and dynamics in the rhizosphere of maize (*Zea mays* L.), wheat (*Triticum aestivum* L.), and sugar beet (*Beta vulgaris* L.) evaluated with a mechanistic model. Plant Soil 332:105–121. https://doi.org/10.1007/s11104-009-0277-6

Sánchez-Rodríguez E, del Mar Rubio-Wilhelmi M, Cervilla LM, Blasco B, Rios JJ, Leyva R, Romero L, Ruiz JM (2010) Study of the ionome and uptake fluxes in cherry tomato plants under moderate water stress conditions. Plant Soil 335:339–347. https://doi.org/10.1007/s11104-010-0422-2

Sanes FSM, Castilhos RMV, Scivittaro WB, Vahl LC, de Morais JR (2013) Root morphology and potassium uptake kinetic parameters in irrigated rice genotypes. Rev Bras Ciênc Solo 37:688–697. https://doi.org/10.1590/S0100-06832013000300015

Santa-María GE, Moriconi JI, Oliferuk S (2015) Internal efficiency of nutrient utilization: what is it and how to measure it during vegetative plant growth? J Exp Bot 66:3011–3018. https://doi.org/10.1093/jxb/erv162

Schilling G, Eißner H, Schmidt L, Peiter E (2016) Yield formation of five crop species under water shortage and differential potassium supply. J Plant Nutr Soil Sci 179:234–243. https://doi.org/10.1002/jpln.201500407

Shi W, Wang X, Yan W (2004) Distribution patterns of available P and K in rape rhizosphere in relation to genotypic difference. Plant Soil 261:11–16. https://doi.org/10.1023/B:PLSO.0000035571.26352.99

Siddiqi MY, Glass ADM (1983a) Studies of the growth and mineral nutrition of barley varieties. I. Effect of potassium supply on the uptake of potassium and growth. Can J Bot 61:671–678. https://doi.org/10.1139/b83-076

Siddiqi MY, Glass ADM (1983b) Studies of the growth and mineral nutrition of barley varieties. II. Potassium uptake and its regulation. Can J Bot 61:1551–1558. https://doi.org/10.1139/b83-167

Siddiqi MY, Glass ADM, Hsiao AI, Minjas AN (1987) Genetic differences among wild oat lines in potassium uptake and growth in relation to potassium supply. Plant Soil 99:93–105. https://doi.org/10.1007/BF02370157

Spear SN, Asher CJ, Edwards DG (1978a) Response of cassava, sunflower, and maize to potassium concentration in solution. I. Growth and plant potassium concentration. Field Crops Res 1:347–361. https://doi.org/10.1016/0378-4290(78)90036-9

Spear SN, Asher CJ, Edwards DG (1978b) Response of cassava, sunflower, and maize to potassium concentration in solution. II. Potassium absorption and its relation to growth. Field Crops Res 1:363–373. https://doi.org/10.1016/0378-4290(78)90037-0

Steingrobe B, Claassen N (2000) Potassium dynamics in the rhizosphere and K efficiency of crops. J Plant Nutr Soil Sci 163:101–106. https://doi.org/10.1002/(SICI)1522-2624(200002)163:1<101::AID-JPLN101>3.0.CO;2-J

Stelling D, Wang S-H, Römer W (1996) Efficiency in the use of phosphorus, nitrogen and potassium in *topless* faba beans (*Vicia faba* L.) – variability and inheritance. Plant Breed 115:361–366. https://doi.org/10.1111/j.1439-0523.1996.tb00934.x

Tao Y, Wang L, Wang X, Xia Y, Wan K, Chen F (2012) Adaptive phenotypic differences to low potassium soil of two cotton genotypes with various potassium-use efficiencies. Commun Soil Sci Plant Anal 43:1984–1993. https://doi.org/10.1080/00103624.2012.693230

Thomas CL, Graham NS, Hayden R, Meacham MC, Neugebauer K, Nightingale M, Dupuy LX, Hammond JP, White PJ, Broadley MR (2016) High-throughput phenotyping (HTP) identifies seedling root traits linked to variation in seed yield and nutrient capture in field-grown oilseed rape (*Brassica napus* L.). Ann Bot 118:655–665. https://doi.org/10.1093/aob/mcw046

Thorup-Kristensen K, Halberg N, Nicolaisen M, Olesen JE, Crews TE, Hinsinger P, Kirkegaard J, Pierret A, Dresbøll DB (2020) Digging deeper for agricultural resources, the value of deep rooting. Trends Plant Sci 25:406–417. https://doi.org/10.1016/j.tplants.2019.12.007

Trehan SP (2005) Nutrient management by exploiting genetic diversity of potato – a review. Potato J 32:1–15

Trehan SP, Singh BP (2013) Nutrient efficiency of different crop species and potato varieties – in retrospect and prospect. Potato J 40:1–21

Végh KR, Köszegi B, Gupta SC (2008) Bioavailability of soil potassium for different crops. Cereal Res Commun 36:1883–1886

Véry AA, Nieves-Cordones M, Daly M, Khan I, Fizames C, Sentenac H (2014) Molecular biology of K^+ transport across the plant cell membrane: what do we learn from comparison between plant species? J Plant Physiol 171:748–769. https://doi.org/10.1016/j.jplph.2014.01.011

Villalta I, Reina-Sánchez A, Bolarín MC, Cuartero J, Belver A, Venema K, Carbonell EA, Asins MJ (2008) Genetic analysis of Na^+ and K^+ concentrations in leaf and stem as physiological components of salt tolerance in tomato. Theor Appl Genet 116:869–880. https://doi.org/10.1007/s00122-008-0720-8

Wakeel A, Farooq M, Qadir M, Schubert S (2011) Potassium substitution by sodium in plants. Crit Rev Plant Sci 30:401–413. https://doi.org/10.1080/07352689.2011.587728

Wang JG, Zhang FS, Cao YP, Zhang XL (2000) Effect of plant types on release of mineral potassium from gneiss. Nutr Cycl Agroecosyst 56:37–44. https://doi.org/10.1023/A:1009826111929

Wang H-Y, Shen Q-H, Zhou J-M, Wang J, Du C-W, Chen X-Q (2011) Plants use alternative strategies to utilize nonexchangeable potassium in minerals. Plant Soil 343:209–220. https://doi.org/10.1007/s11104-011-0726-x

Wang Z, Chen Z, Cheng J, Lai Y, Wang J, Bao Y, Huang J, Zhang H (2012) QTL analysis of Na^+ and K^+ concentrations in roots and shoots under different levels of NaCl stress in rice (*Oryza sativa* L.). PLoS One 7(12):e51202. https://doi.org/10.1371/journal.pone.0051202

Wang JD, Wang H, Zhang Y, Zhou J, Chen X (2015a) Intraspecific variation in potassium uptake and utilization among sweet potato (*Ipomoea batatas* L.) genotypes. Field Crops Res 170:76–82. https://doi.org/10.1016/j.fcr.2014.10.007

Wang X-G, Zhao X-H, Jiang C-J, Li C-H, Cong S, Wu D, Chen Y-Q, Yu H-Q, Wang C-Y (2015b) Effects of potassium deficiency on photosynthesis and photoprotection mechanisms in soybean (*Glycine max* (L.) Merr.). J Integr Agric 14:856–863. https://doi.org/10.1016/S2095-3119(14)60848-0

Watanabe T, Broadley MR, Jansen S, White PJ, Takada J, Satake K, Takamatsu T, Tuah SJ, Osaki M (2007) Evolutionary control of leaf element composition in plants. New Phytol 174:516–523. https://doi.org/10.1111/j.1469-8137.2007.02078.x

White PJ (1993) Relationship between the development and growth of rye (*Secale cereale* L.) and the potassium concentration in solution. Ann Bot 72:349–358. https://doi.org/10.1006/anbo.1993.1118

White PJ (2013) Improving potassium acquisition and utilisation by crop plants. J Plant Nutr Soil Sci 176:305–316. https://doi.org/10.1002/jpln.201200121

White PJ (2016) Biofortification of edible crops. In: eLS. Wiley, Chichester. https://doi.org/10.1002/9780470015902.a0023743

White PJ (2020) Potassium in crop physiology. In: Rengel Z (ed) Achieving sustainable crop nutrition. Burleigh Dodds, Cambridge, pp 213–236. https://doi.org/10.19103/AS.2019.0062.10

White PJ, Bell MJ (2017) The genetics of potassium uptake and utilization in plants. In: Murrell TS, Mikkelsen RL (eds) Proceedings for the frontiers of potassium science conference, 25–27 January 2017, Rome. International Plant Nutrition Institute, Peachtree Corners, pp 46–65. https://www.apni.net/k-frontiers/. Accessed 29 May 2020

White PJ, Greenwood DJ (2013) Properties and management of cationic elements for crop growth. In: Gregory PJ, Nortcliff S (eds) Soil conditions and plant growth. Blackwell, Oxford, pp 160–194. https://doi.org/10.1002/9781118337295.ch6

White PJ, Karley AJ (2010) Potassium. In: Hell R, Mendel R-R (eds) Cell biology of metals and nutrients. Springer, Berlin, pp 199–224. https://doi.org/10.1007/978-3-642-10613-2_9

White PJ, Broadley MR, Greenwood DJ, Hammond JP (2005) Genetic modifications to improve phosphorus acquisition by roots. In: Proceedings of the International Fertiliser Society 568. IFS, York. ISBN: 0853102058

White PJ, Bradshaw JE, Dale MFB, Ramsay G, Hammond JP, Broadley MR (2009) Relationships between yield and mineral concentrations in potato tubers. HortScience 44:6–11. https://doi.org/10.21273/HORTSCI.44.1.6

White PJ, Hammond JP, King GJ, Bowen HC, Hayden RM, Meacham MC, Spracklen WP, Broadley MR (2010) Genetic analysis of potassium use efficiency in *Brassica oleracea*. Ann Bot 105:1199–1210. https://doi.org/10.1093/aob/mcp253

White PJ, Broadley MR, Thompson JA, McNicol JW, Crawley MJ, Poulton PR, Johnston AE (2012) Testing the distinctness of shoot ionomes of angiosperm families using the Rothamsted Park Grass Continuous Hay experiment. New Phytol 196:101–109. https://doi.org/10.1111/j.1469-8137.2012.04228.x

White PJ, George TS, Gregory PJ, Bengough AG, Hallett PD, McKenzie BM (2013) Matching roots to their environment. Ann Bot 112:207–222. https://doi.org/10.1093/aob/mct123

White PJ, Kawachi T, Thompson JA, Wright G, Dupuy LX (2016) Minimising the treatments required to determine the responses of different crop genotypes to potassium supply. Commun Soil Sci Plant Anal 47.(S1:104–111. https://doi.org/10.1080/00103624.2016.1232103

White PJ, Bowen HC, Broadley MR, El-Serehy HA, Neugebauer K, Taylor A, Thompson JA, Wright G (2017) Evolutionary origins of abnormally large shoot sodium accumulation in non-saline environments within the Caryophyllales. New Phytol 214:284–293. https://doi.org/10.1111/nph.14370

Wild A, Skarlou V, Clement CR, Snaydon RW (1974) Comparison of potassium uptake by four plant species grown in sand and in flowing solution culture. J Appl Ecol 11:801–812

Wishart J, George TS, Brown LK, Ramsay G, Bradshaw JE, White PJ, Gregory PJ (2013) Measuring variation in potato roots in both field and glasshouse: the search for useful yield predictors and a simple screen for root traits. Plant Soil 368:231–249. https://doi.org/10.1007/s11104-012-1483-1

Woodend JJ, Glass ADM (1993) Genotype-environment interaction and correlation between vegetative and grain production measures of potassium use-efficiency in wheat (*T. aestivum* L.) grown under potassium stress. Plant Soil 151:39–44. https://doi.org/10.1007/BF00010784

Wu P, Ni JJ, Luo AC (1998) QTLs underlying rice tolerance to low-potassium stress in rice seedlings. Crop Sci 38:1458–1462

Wu J, Yuan Y-X, Zhang X-W, Zhao J, Song X, Li Y, Li X, Sun R, Koornneef M, Aarts MGM, Wang X-W (2008) Mapping QTLs for mineral accumulation and shoot dry biomass under different Zn nutritional conditions in Chinese cabbage (*Brassica rapa* L. ssp. *pekinensis*). Plant Soil 310:25–40. https://doi.org/10.1007/s11104-008-9625-1

Wu J-T, Zhang X-Z, Li T-X, Yu H-Y, Huang P (2011) Differences in the efficiency of potassium (K) uptake and use in barley varieties. Agric Sci China 10:101–108. https://doi.org/10.1016/S1671-2927(11)60312-X

Yang XE, Liu JX, Wang WM, Li H, Luo AC, Ye ZQ, Yang Y (2003) Genotypic differences and some associated plant traits in potassium internal use efficiency of lowland rice (*Oryza sativa* L.). Nutr Cycl Agroecosyst 67:273–282. https://doi.org/10.1023/B:FRES.0000003665.90952.0c

Yang XE, Liu JX, Wang WM, Ye ZQ, Luo AC (2004) Potassium internal use efficiency relative to growth vigor, potassium distribution, and carbohydrate allocation in rice genotypes. J Plant Nutr 27:837–852. https://doi.org/10.1081/PLN-120030674

Yang F, Wang G, Zhang Z, Eneji AE, Duan L, Li Z, Tian X (2011) Genotypic variations in potassium uptake and utilization in cotton. J Plant Nutr 34:83–97. https://doi.org/10.1080/01904167.2011.531361

Yu P, Li X, White PJ, Li C (2015) A large and deep root system underlies high nitrogen-use efficiency in maize production. PLoS One 10(5):e0126293. https://doi.org/10.1371/journal.pone.0126293

Zdunić Z, Grljušić S, Ledenčan T, Duvnjak T, Šimić D (2014) Quantitative trait loci mapping of metal concentrations in leaves of the maize IBM population. Hereditas 151:55–60. https://doi.org/10.1111/hrd2.00048

Zhang G, Chen J, Tirore EA (1999) Genotypic variation for potassium uptake and utilization efficiency in wheat. Nutr Cycl Agroecosyst 54:41–48. https://doi.org/10.1023/A:1009708012381

Zhang Z, Tian X, Duan L, Wang B, He Z, Li Z (2007) Differential responses of conventional and Bt-transgenic cotton to potassium deficiency. J Plant Nutr 30:659–670. https://doi.org/10.1080/01904160701289206

Zhang Y, Zhang C-C, Yan P, Chen X-P, Yang J-C, Zhang F-S, Cui Z-L (2013) Potassium requirement in relation to grain yield and genotypic improvement of irrigated lowland rice in China. J Plant Nutr Soil Sci 176:400–406. https://doi.org/10.1002/jpln.201200206

Zhao Y, Li X-Y, Zhang S-H, Wang J, Yang X-F, Tian J-C, Hai Y, Yang X-J (2014) Mapping QTLs for potassium-deficiency tolerance at the seedling stage in wheat (*Triticum aestivum* L.). Euphytica 198:185–198. https://doi.org/10.1007/s10681-014-1091-7

Zia-ul-hassan, Arshad M (2011) Relationship among root characteristics and differential potassium uptake and use efficiency of selected cotton genotypes under potassium deficiency. Pak J Bot 43:1831–1835

Zia-ul-hassan, Kubar KA, Rajpar I, Shah AN, Tunio SD, Shah JA, Maitlo AA (2014) Evaluating potassium-use-efficiency of five cotton genotypes of Pakistan. Pak J Bot 46:1237–1242

Zörb C, Senbayram M, Peiter E (2014) Potassium in agriculture – status and perspectives. J Plant Physiol 171:656–669. https://doi.org/10.1016/j.jplph.2013.08.008

6

Non-Harvested Plant Potassium

Ciro A. Rosolem, Antonio P. Mallarino, and Thiago A. R. Nogueira

Abstract Potassium (K) is found in plants as a free ion or in weak complexes. It is easily released from living or decomposing tissues, and it should be considered in fertilization programs. Several factors affect K cycling in agroecosystems, including soil and fertilizer K contributions, plant K content and exports, mineralization rates from residues, soil chemical reactions, rainfall, and time. Soil K^+ ions can be leached, remain as exchangeable K, or migrate to non-exchangeable forms. Crop rotations that include vigorous, deep-rooted cover crops capable of exploring non-exchangeable K in soil are an effective strategy for recycling K and can prevent leaching below the rooting zone in light-textured soils. The amount of K released by cover crops depends on biomass production. Potassium recycled with non-harvested components of crops also varies greatly. Research with maize, soybean, and wheat has shown that 50–60% of K accumulated in vegetative tissues is released within 40–45 days. A better understanding of K cycling would greatly improve the efficacy of K management for crop production. When studying K cycling in agricultural systems, it is important to consider: (1) K addition from fertilizers and organic amendments; (2) K left in residues; (3) K partitioning differences among species; (4) soil texture; (5) soil pools that act as temporary sources or sinks for K. In this chapter, the role of cash and cover crops and organic residues on K cycling are explored to better understand how these factors could be integrated into making K fertilizer recommendations.

C. A. Rosolem (✉)
School of Agricultural Sciences, São Paulo State University, Botucatu, São Paulo, Brazil
e-mail: ciro.rosolem@unesp.br

A. P. Mallarino
Department of Agronomy, Iowa State University, Ames, IA, USA
e-mail: apmallar@iastate.edu

T. A. R. Nogueira
School of Engineering, São Paulo State University, Ilha Solteira, São Paulo, Brazil
e-mail: tar.nogueira@unesp.br

6.1 The Crop Canopy as a Source of Potassium

After nitrogen (N), K is the nutrient required in largest amounts by plants. Its concentration in plants has been reported to range from 4 to 43 g K kg^{-1} (Askegaard et al. 2004), and it is affected mainly by plant species, site, year, tissue age, and fertilizer input. Most K uptake in annual species is observed as the shoot undergoes rapid growth (Gregory et al. 1979). For cereals, more than 70% of K remains in the straw after grain harvest, and the concentration is increased by fertilization; therefore, this is an important source of K for the next crop and should be considered and integrated in fertilizer recommendations.

Mobility of K in plants is high at all levels—within individual cells, tissues, and in long-distance transport via xylem and phloem. It is the most abundant cation in the cytoplasm, and except for cytosol and the vacuole, its subcellular distribution is largely uncharacterized. The concentration of K in the cytoplasm is kept relatively constant from 50 to 150 mM, while the concentration in the vacuole varies depending on supply status (Zörb et al. 2014). Potassium mineral salts are highly soluble, and K is not metabolized, forming weak complexes with organic molecules from which it is readily exchangeable (Marschner 1995). Therefore, K is prone to be easily leached from living or dead plant tissues irrespective of plant residue decomposition or mineralization.

Leached K is *the quantity of K removed from plants by the action of aqueous solutions, such as rain, dew, mist, and fog* (adapted from Tukey 1970). Potassium can be lost from living tissues by guttation at leaf margins and tips, or leached from damaged or old plant parts, such as senescent leaves or from intact plant tissues. Leaching is thought to be a passive process, but the driving forces and mechanisms of nutrient leaching from live tissues are unclear. The K leaching rate from living tissues is increased with leaf age, intensity, and duration of rainfall. This happens because the accumulation of substances in the apoplast of mature leaves may result in a steeper concentration gradient favoring leaching (Eichert and Fernández 2012). Furthermore, as annual plants complete their life cycle after flowering, the older tissues start to senesce and slowly cell membranes are disrupted favoring K leaching.

After harvest, or when cover crops are terminated, K leaches from the dead tissues. It has been shown that K leaching from dead tissues is proportional to the K concentration in the tissue, time after termination and rainfall (Grimes and Hanway 1967a; Schomberg and Steiner 1999; Rosolem et al. 2005).

6.2 Potential of Potassium Cycling by Crops and Cover Crops

The potential of plants for cycling K in cropping systems is defined by the capacity to access various soil K pools. This capacity depends on the quantity of K the plant accumulates, the size and depth of the root system, and the effectiveness of mechanisms the plants use to access K in both exchangeable and non-exchangeable pools. Potassium accumulated in plant tissues that are not harvested from the field will be later returned by leaching from plant residues. Therefore, in K-limited areas, species or varieties efficient in utilizing non-exchangeable forms of soil K have a great potential to increase K cycling and K use efficiency in the system (Zörb et al. 2014).

A variable amount of the K taken up by crops harvested for grain, grazed pastures, and cover crops will be recycled to the soil as an ion, highly soluble mineral forms, or weakly complexed in organic compounds. The cycled K will enrich the soil solution, be available to the next crop, lost with runoff or leached through the soil profile, or be transformed into less readily plant-available pools in the soil. While adequate K uptake is important in supplying K to plants, uptake of K beyond plant needs will compromise sustainability (Rosolem and Steiner 2017). Grimes and Hanway (1967a) and Oltmans and Mallarino (2015) showed that the soil K increase after harvesting maize in the fall to the following spring was directly related to the amount of K in the residue.

Cash grain crops play an important role in K cycling in agricultural cropping systems. At maturity, approximately 25–45% of the total aboveground plant K is found in the maize grain but more than 50% is in soybean (*Glycine max* (L.) Merr) grain; furthermore, there is a large variation across growing conditions, species, and cultivars (Bender et al. 2013; Ciampitti et al. 2013; Oltmans and Mallarino 2015). In soybean, K fertilization results in very little additional K accumulation in grain but markedly increases K accumulation in the mature stems, pods, leaves, and petioles (Hanway and Weber 1971; Rosolem and Nakagawa 1985; Farmaha et al. 2012). In Iowa, average K accumulation in soybean grain or residue at harvest was 68 and 34 kg ha^{-1} in a 14 site-year experiment. Maize averaged, in a 33 site-year experiment, 29 and 52 kg ha^{-1} in grains and residue, respectively (Oltmans and Mallarino 2015). In maize, 50% of the K accumulated in vegetative tissue at physiological maturity remained in the straw after 2 months, then decreased to 31% after 6 months. In soybean, 19% of accumulated K remained 2 months after physiological maturity, then decreased to 12% in 6 months (Fig. 6.1). The amount of K remaining in the crop residue decreased as precipitation increased, and soil test K increased from fall to spring. Despite a greater amount of K remaining in plant tissues with K fertilization, the decrease over time was similar. The greatest K leaching was observed between physiological maturity and grain harvest. Due to different plant structures of maize and soybean, mainly the maize stalks containing most of the K, much more rain was necessary to leach out a similar amount of K from maize than from soybean (Oltmans and Mallarino 2015).

Crop and cover crop residues may have a high amount of K potentially available for the next crop and may be able to supply K for the new crop early growth, depending on the release synchrony. The quantity of K released depends on the species or even the cultivar, as shown in Table 6.1. For example, forages of the genus *Brachiaria*, one of the main grasses used as a cover crop in Brazil, can accumulate over 400 kg K ha^{-1}. *Panicum* species accumulate up to 800 kg K ha^{-1} in an entire cycle. Cover crops such as brachiaria increase exchangeable K in the topsoil layers

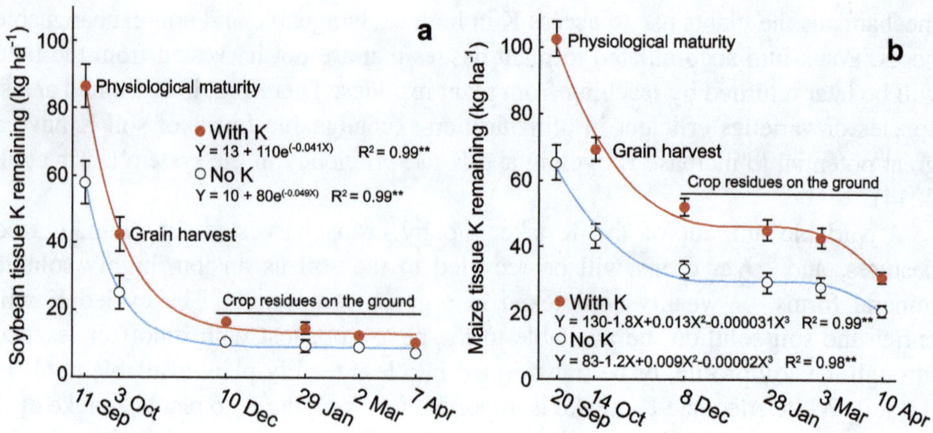

Fig. 6.1 Potassium accumulation in (**a**) soybean and (**b**) maize vegetative tissue or residue over time for two K treatments. Coefficients of determination (R^2) for all regression models are significant (**) at $P \leq 0.01$. Vertical lines indicate confidence intervals ($P = 0.10$). (adapted from Oltmans and Mallarino 2015)

by bringing it up from deeper soil layers (Eckert 1991; Garcia et al. 2008)—a process defined as nutrient uplift by Jobbágy and Jackson (2004).

When these species or others are used as cover crops in crop–livestock integrated systems or when the growing season between two main crops is short, K accumulation is not so high, as shown in research with cover crops in the northern areas of the Corn Belt of the United States, where a maize–soybean rotation system predominates. Cover crops reduce soil and nutrient loss, but because of a long period with very cold or frozen soils in some regions, cover crops have little effective time to grow and take up nutrients between two cash crops. Unpublished research with cereal rye (*Secale cereale* L.) across 12 site-years (A. P. Mallarino, Iowa State University personal communication), which is the most widely used cover crop in these conditions and is commonly terminated when it is 20–30 cm tall, shows that the aboveground K recycled at the spring termination time ranged from 7 to 84 kg K ha^{-1}, being greatly affected by soil test K concentration, the active growth period, and dry matter production.

6.3 Synchrony of Potassium Availability in Cropping Systems

Losses of K from plant residues are affected by several factors such as the species, rainfall (Fig. 6.2), and time after desiccation. Potassium is found in plants as a free cation or in weak complexes. It is easily leached from dead plant tissues independent of plant residue mineralization or decomposition.

Initial K loss ranges from 4.4 to to 29.3 g K kg^{-1} day^{-1} depending on the species and precipitation. Across crops, a 150-mm rainfall removed 500 g K kg^{-1} from plant residues (Schomberg and Steiner 1999). The maximum rate of K released from

Table 6.1 Amount of residue and average K accumulated and non-harvested in some crops and cover crops

Crop/cover crop	Scientific name	Amount of residue[a]	Non-harvested K[b]	References
		Mg ha^{-1}	kg t^{-1}	
Black mucuna	*Stizolobium aterrimum* Piper & Tracy	1–9	17	Borkert et al. (2003)
Black oat	*Avena strigosa* Schreb.	2–12	30	Crusciol et al. (2008)
Calopo	*Calopogonium mucunoides Desv.*	4–6	15	Teodoro et al. (2011)
Crambe	*Crambe abyssinica* Hochst.	1–3	32	Heinz et al. (2011)
Common vetch	*Vicia sativa* L.	2–6	23	Borkert et al. (2003), Rossato (2004)
Congo grass	*Urochloa ruziziensis*	2–15	28	Pereira et al. (2016)
Dwarf mucuna	*Stizolobium Deeringianum* Bort	2–4	10	Caceres and Alcarde (1995)
Finger millet	*Eleusine coracana* L. Gaertn	3–12	22	Francisco et al. (2007)
Forage sorghum	*Sorghum bicolor* L. Moench	5–16	16	Oliveira et al. (2002)
Forage turnip	*Raphanus sativus* L.	2–6	30	Crusciol et al. (2005)
Guinea grass	*Panicum maximum* cv. Tanzania	2–18	33	Pereira et al. (2016)
Guinea grass	*Panicum maximum* cv. Áries	2–20	42	Pereira et al. (2016)
Jack bean	*Canavalia ensiformis* L.	3–10	14	Caceres and Alcarde (1995)
Lablab	*Dolichos lablab* L.	3–8	14	Caceres and Alcarde (1995)
Lupin	*Lupinus albus* L.	3–8	19	Borkert et al. (2003)
Maize	*Zea mays* L.	6–12	18	Oliveira et al. (2002)
Oilseed radish	*Raphanus sativus* L.	2–9	42	Heinz et al. (2011)
Palisade grass	*Urochloa brizantha* cv. Marandu	3–23	27	Pereira et al. (2016)
Pearl millet	*Pennisetum glaucum* (L.) R.Br.	2–12	25	Braz et al. (2004)
Peanut	*Arachis hypogaea* L.	1–4	20	Teodoro et al. (2011), Crusciol (2016)
Perennial soybean	*Glycine wightii* L.	4–6	18	Teodoro et al. (2011)
Pigeon pea	*Cajanus cajan* L.	2–12	14	Borkert et al. (2003)
Soybean	*Glycine max* (L.) Merr	3–4	19	Kurihara et al. (2013)

(continued)

Table 6.1 (continued)

Crop/cover crop	Scientific name	Amount of residue[a]	Non-harvested K[b]	References
		Mg ha^{-1}	kg t^{-1}	
Showy rattlebox	*Crotalaria spectabilis* Roth	3–8	22	Caceres and Alcarde (1995)
Sunflower	*Helianthus annuus* L.	7–10	15	Ambrosano et al. (2013)
Sunn hemp	*Crotalaria juncea* L.	5–14	14	Caceres and Alcarde (1995)
Sugarcane	*Saccharum officinarum* L.	5–13	15	Oliveira et al. (1999)
Tropical kudzu	*Pueraria phaseoloides* L.	4–7	16	Teodoro et al. (2011)
Triticale	*X Triticosecale* Wittm	1–5	48	Rosolem et al. (2003)
Wheat	*Triticum aestivum* L.	1–5	18	Rossato (2004)
Upland rice	*Oryza sativa* L.	6–9	24	Crusciol (2016)

[a]These values (in dry weight) were reported in the literature and can vary with plant age, soil type, fertility, climate, season, and sowing density
[b]K non-harvested (dry matter) = K accumulated × average amount of residue

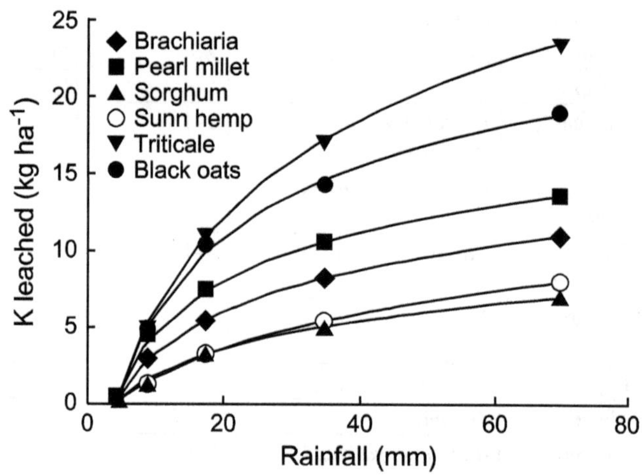

Fig. 6.2 Accumulated K leaching as affected by rainfall and cover crop species 8 days after desiccation (adapted from Rosolem et al. 2003); species were brachiaria, pearl millet, sorghum (*Sorghum bicolor* (L.) Moench), sunn hemp, triticale, and black oat

several plant species by rain soon after desiccation ranges from 200 to 650 g K ha^{-1} per mm of rain and is strongly correlated with the amount of nutrient accumulated in the crop residues (Rosolem et al. 2003), probably because a large proportion of this nutrient is present in the vacuole and not bound to organic compounds (Marschner 1995). Rosolem et al. (2005) found that K fertilization increased both K accumulation in pearl millet straw and K leached from the residue under simulated rainfall, and it was estimated that residue leaching could provide 24–64 kg K ha^{-1} to the next crop. Oltmans and Mallarino (2015) also reported that K fertilization increased the amount of K leached by natural rainfall from maize and soybean residues compared with non-fertilized treatments with or without grain yield response.

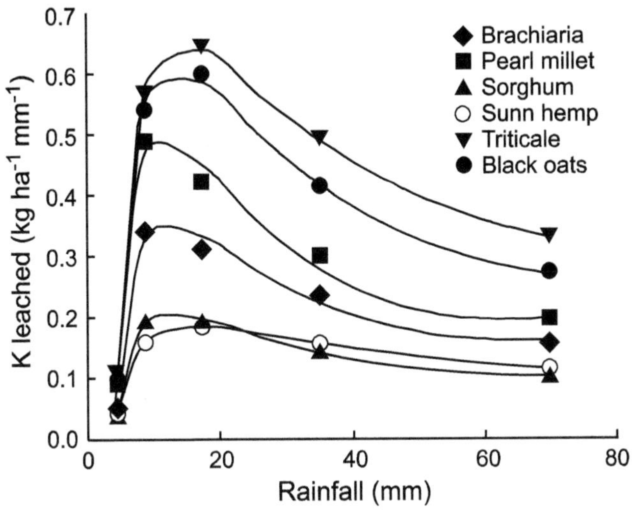

Fig. 6.3 Rate of K leaching as affected by plant species and rainfall, 5 days after cutting (adapted from Rosolem et al. 2003); species were brachiaria, pearl millet, sorghum, sunn hemp, triticale, and black oat

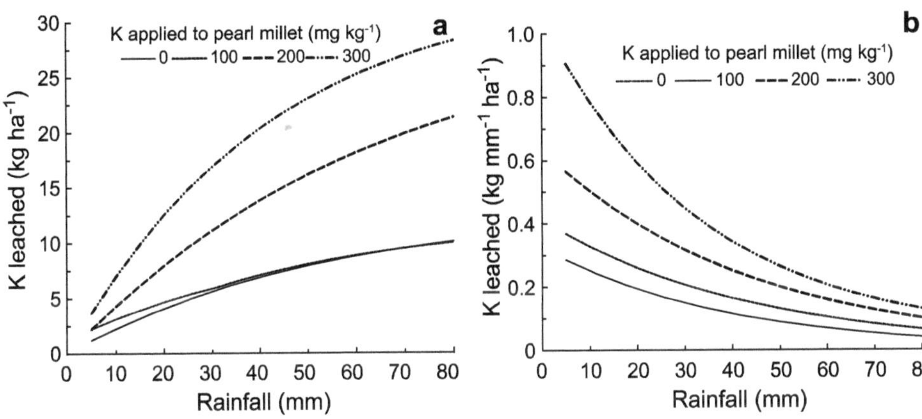

Fig. 6.4 Accumulated K leaching (**a**) and K leaching rate (**b**) from pearl millet residues as affected by simulated rainfall and tissue K contents. (from Rosolem et al. 2005)

After cover crop desiccation, some K is remobilized to roots as tissues die. Then, rainfall is the main driver of K release from plant residues. Potassium leaching rates, or the amount of K washed from plant residues per mm rain, is highly variable among species and is very low with rains up to 5 mm (Fig. 6.3) because this is barely enough to wet plant residues. Then, there is a steep increase with rains up to 20 mm, then decreasing exponentially, tending to a constant with rains >75–80 mm (Rosolem et al. 2003, 2005). The K concentration in the plant tissue also plays an important role not only in the amount of K released from plant residues, but also on the rate of K leaching (Fig. 6.4), as shown by Rosolem et al. (2005).

This occurs because, at first, all K in plant residues is potentially available to be leached, and the first rain will saturate the residues. From this point, additional rain can only wash out the K present at the superficial layers of the residue, with little leaching from deeper layers. As a result, to be leached, K has to diffuse to the straw surface. Therefore, heavier rains would have little effect on the process, and a greater nutrient release will be observed with successive drying–wetting cycles, which occurs in many agricultural areas.

Up to 50–60% of the K accumulated in the straw is washed within 40–45 days from plant desiccation in several species under field conditions (Giacomini et al. 2003; Calonego et al. 2012; Oltmans and Mallarino 2015), and within 130–140 days over 70% of the nutrient will return to the soil (Spain and Salinas 1985; Calonego et al. 2012). For grasses such as palisade grass (*Urochloa brizantha* cv. Marandu), pearl millet (*Pennisetum glaucum* L.), and Guinea grass (*Panicum maximum* L.), most of the K in the straw is washed back to the soil in less than 50 days. The K release from straw left on the soil surface is not related to straw mineralization rate, since K loss is faster than dry matter loss (Rosolem et al. 2003; Calonego et al. 2012). However, leaching is increased as the plant residues age, probably as a result of cellular disruption (Calonego et al. 2005). The amount of rainfall or irrigation that occurs also drives the K release processes.

The varied K release from different plant species in a cropping system defines its role in supplying K to the next crop. For instance, Sunn hemp, maize, and sorghum release K slowly for a long time compared to triticale, black oats, soybean, and cover crops terminated during early vegetative growth stages which can release considerable amounts of the nutrient very fast (Rosolem et al. 2003; Oltmans and Mallarino 2015). Therefore, to estimate the value of recycled K for a following crop, it is essential to consider the species from which the K is recycled, the time between desiccation and planting of the next crop, the rainfall amount and pattern within this period, and soil properties that influence the fate of recycled K. Tropical grasses used as cover crops in Brazil can release from 1.5 to 6.5 kg K ha^{-1} day^{-1}. Considering that Palisade grass and *Panicum* have between 90 and 100 kg K ha^{-1} and pearl millet has 200 kg K ha^{-1} accumulated in the residues, this is more than enough to supply the nutrient for subsequent crops, such as soybean, maize, or cotton (*Gossypium hirsutum* (L.)).

In the US Corn Belt, from 50 to over 70 kg K ha^{-1} were washed back to the soil from soybean residues up to the time of planting the next crop, depending on K fertilization. For maize, washed K ranged from 25 to 50 kg K ha^{-1} (Oltmans and Mallarino 2015). In Brazil, in a soybean–pearl millet rotation, around 70 kg K ha^{-1} was released from plant residues from the day of soybean planting up to 50 days after emergence. By this time soybean had taken up around 90 kg K ha^{-1} (Foloni and Rosolem 2004). These results show the importance of the nutrient accumulated in plant residues in supplying K to the next crop.

6.4 Residue Potassium as a Means of Reducing Potassium Losses from the System

Pal et al. (1999) showed that soil soluble K is negatively related to the proportion of coarse sand and positively related to the amounts of clay and silt. Thus, greater K leaching losses might be expected from sandy soils than from clayey soils (Malavolta 1985). Potassium leaching below the arable layer increases with K application rates, although the effect is less noticeable in clayey soils.

Potassium leaching in a sandy clay loam soil is related to the soil K content from prior fertilization. With no excess water and in the presence of soybean roots, the K distribution through the profile was significant in a light textured soil but was not observed on a heavy-textured soil (Rosolem et al. 2012). Furthermore, in sandy soils K leaching is proportional to K fertilizer application rates (Rosolem et al. 2012), and it strongly increases with annual applications >65 kg K ha^{-1}. The increase in K fertilizer application rates intensifies K leaching losses below 1.0 m in sandy clay loam soils, representing 16–52% of the applied fertilizer K (Rosolem and Steiner 2017). Therefore, due to the high potential of K leaching, splitting of K fertilizer applications and conserving K in residues are both important management strategies to minimize K leaching losses and to improve K use efficiency in tropical, low-clay soils.

6.5 Potassium from Agro-Industrial Residues

The application of agricultural waste to the soil to complement or substitute for K fertilization is an important alternative adopted in the agricultural sector. Such practice, besides decreasing production costs, is an appropriate way to dispose and utilize these materials. Many residues can be used as K sources in agricultural systems (Table 6.2). However, the decision to apply a residue to the soil is related not only to the K concentration, but also to its availability and ease of acquisition by farmers.

In sugar mills, filter cake is obtained from impurities removed during the flocculation process, decanting and filtering the sugarcane in a rotary filter. It is estimated that 30–40 kg of filter cake are produced for each ton of cane processed (Santos et al. 2011). This residue has a considerable amount of organic matter and nutrients (Almeida-Júnior et al. 2011). Filter cake can be applied to agricultural soils, increasing the plant availability of K and other nutrients, as well as decreasing exchangeable Al (Korndörfer and Anderson 1997).

Vinasse, a byproduct of biomass distillation, is the largest source of pollution in the ethanol industry (Santos et al. 2013), and its disposal has become a problem in sugarcane growing countries. Considering that 1 L of ethanol generates around 9–14 L of vinasse, it is forecast that about 6 trillion L (TL) of this material will need to be managed by 2023 (Carrilho et al. 2016). On the other hand, land

Table 6.2 Concentration of K and K_2O in some organic by-products and residues from vegetal, animal, and agro-industrial sources

Residue	Unit	K	K_2O	References
Vinasse Molasses (M)	mg L^{-1}	–[a]	3740–7830	Carrilho et al. (2016)
Juice (J)	mg L^{-1}		1200–2100	
M + J	mg L^{-1}		3340–4600	
Green sugarcane	kg m^{-3}	–	2.10–3.40	Korndörfer and Anderson (1997)
Filter cake	g kg^{-1}	–	0.2–0.4	Prado et al. (2013)
Boiler ash	g kg^{-1}	–	2.7	Vitti and Luz (2008)
Poultry litter	g kg^{-1}	–	25.7	Vitti and Luz (2008)
Chicken manure	g kg^{-1}	–	30.1	Vitti and Luz (2008)
Pig slurry	kg m^{-3}	–	1.0–1.25	Vitti and Luz (2008)
Cattle manure fresh	g kg^{-1}	6.0	–	Raij et al. (1997)
Tanned cattle manure	g kg^{-1}	21.0	–	Raij et al. (1997)
Castor cake	g kg^{-1}	11.0	–	Raij et al. (1997)
Natural coffee husk	g kg^{-1}	–	30.0	Matiello (2005)
Coffee cherry husk	g kg^{-1}	–	39.0	Matiello (2005)
Parchment of coffee beans	g kg^{-1}	–	3.7	Matiello (2005)

[a]Information is not available
[b]Dry matter ranging from 11.7 to 20.9 g kg^{-1}

application of vinasse and sugar industry effluents is gaining importance due to the presence of high quantities of mineral nutrients essential for plant growth and organic matter content. Land application not only improves crop yields but also addresses the problem of effluent disposal (Jiang et al. 2012). Vinasse has relatively high concentrations of K, calcium (Ca), and organic matter, as well as moderate amounts of N and other nutrients (Abreu-Junior et al. 2008). This residue can be profitably recycled to improve soil chemical and physical properties and is an alternative for supplying valuable crop nutrients.

It is estimated that 30 Mg ha^{-1} of filter cake and 150 m^3 ha^{-1} of vinasse are equivalent to 60 and 690 kg ha^{-1} of potassium chloride, respectively. Thus, vinasse is applied to provide 100% of the K required by sugarcane (Bataglia et al. 1986), typically applied in amounts from 60 to 350 m^3 ha^{-1}. Filter cake (wet) can be applied to the total area (80–100 Mg ha^{-1}) preplant, at planting (15–30 Mg ha^{-1}) or between cane lines (40–50 Mg ha^{-1}). The K added by such wastes is fully deducted from the mineral fertilizer recommendation (Raij et al. 1997). This practice has become so popular that vinasse use is now regulated by environmental agencies in Brazil to avoid over application.

Using coffee (*Coffea arábica* L.) as another example, 50% of the harvested fruit consists of beans and another 50% is husk (by weight). The large amount of coffee husk from processing has caused environmental concerns and alternative uses for these residues must be found. Depending on processing, various wastes are generated, such as husk, pulp, parchment, mucilage, and wastewater. Coffee processing residues are rich in several nutrients, especially K, although the K concentration depends on the type of coffee husk. Although coffee processing residues are considered a good source of organic fertilizer (Matiello 2005), little is known about the release and mineralization of the nutrients from these residues. It has been shown that K release from coffee husk is rapid, and it does not depend on the type of coffee processing; it can be used as a substitute for mineral fertilizers (Zoca et al. 2014).

6.6 Fertilizer Recommendations and Potassium Cycling

According to Mallarino et al. (2013), the Iowa State University fertilizer guidelines consider the amount of crop residue removed and average concentrations of 7.50 and 9.58 g K kg^{-1} for maize and soybean residue, respectively (150 and 100 g kg^{-1} basis). These average concentrations were determined for a variety of management conditions during the 1990s and 2000s. High soil-test K (STK) values and high K fertilization rates would lead to more K being removed in residues than the published numbers because of the large K increase in vegetative tissues in response to high K supply (Rosolem et al. 2010; Oltmans and Mallarino 2015).

Vitko et al. (2009) evaluated K fertilization for maize harvested for grain or for silage and the time of soil sampling on STK at five Wisconsin sites over 3 years. They reported that spring STK was consistently greater than fall STK (20–45% greater) at only one site for both harvest systems. Only in 1 year was the STK increase lower with silage harvest than with grain harvest. Grimes and Hanway (1967b) showed that maize and alfalfa (*Medicago sativa* (L.)) residues added to soil did not differ in K availability and were equal to K added with KCl fertilizer after 72 days of ryegrass (*Lolium multiflorum* Lam.) growth. Soil test K concentrations were usually higher in spring than in the previous fall. The STK difference was correlated with the amount of K lost for both crops, although there was greater unexplained variability in maize ($r^2 = 0.16$) than in soybean ($r^2 = 0.54$). Oltmans and Mallarino (2015) also reported that STK increased from fall to spring, that the increase was correlated to the K lost from maize and soybean residue, and that both crop type and rainfall strongly influenced the K recycled and the effect on STK temporal change. It is possible that unmeasured changes among soil K pools in these studies could further explain measured STK differences between fall and spring. Furthermore, the K supply to ruzigrass (*Brachiaria ruziziensis* Germ. & C.M. Evrard) has been shown to be more dependent on recently added K fertilizer than on the residual effect of previous fertilizations in a light-textured Cerrado soil from Brazil (Rosolem et al. 2012).

6.6.1 Modeling Potassium Release from Residues

Most of the studies on K release from plant residues report that the process fits a single exponential model (Wider and Lang 1982). This makes sense, because K leaching from live or dead tissues is practically independent of tissue decomposition. When tissue decomposition is important, a double exponential model would be more appropriate (Wider and Lang 1982).

A rather simple soil and plant K model was developed to be incorporated into EPIC model code, version 0160 (Barros et al. 2004). The modification takes into account the transfer between soil K pools, fertilizer addition, K losses, K transport by soil water evaporation, uptake by crops, effect on biomass production, and K release from crop residues. However, the proposed modification oversimplifies the soil K transformations as well as the contribution of the plant residues for K availability. The model considers the K concentration in plant residues, but rainfall and residue composition are ignored. The modified EPIC model was tested for a maize–cowpea intercropping in NE Brazil, with reasonably good accuracy and agreement between the measured and simulated values for 3 years (Barros et al. 2005). However, results could be improved if the K contributions of plant residues were better estimated.

The problem is that K leaching from plant residues is regulated by K concentration in the tissue, period of leaching, and rainfall; therefore, simple models are not likely to work in accurately predicting the amount of K available for the next crop.

6.6.2 Implications for Timing of Soil Sampling

According to Oltmans and Mallarino (2015), 43% of the K accumulated in vegetative soybean tissues at maturity remained in residue by early December (after harvest), and only 12% remained by early April (before the next growing season). In maize, however, 67% of the K accumulated in vegetative tissues at maturity remained in residue by early December, and 31% remained by early April. Increasing precipitation decreased K remaining in tissues exponentially to a minimum across all site years. Soil test K concentrations usually were higher in spring than the previous fall, and the soil test K difference was correlated with the amount of K lost for both crops. However, it is important to note that changes in soil K pools, both exchangeable and non-exchangeable forms, depend also on the rainfall (Rosolem et al. 2006). Therefore, the result of the K soil test will be dependent on the time after the previous crop maturity and harvest, desiccation of cover crops, the rainfall, the species, and the tissue K concentration.

6.7 Conclusion

It is not difficult to measure or estimate the amount of K to be released from crop or cover crop residues. However, there is uncertainty in estimating exactly how much K will be available in time for the next crop. One approach is to use soil testing as a monitoring tool and then estimate K fertilizer rates to be applied, considering the harvested K. In this case, soil samples must be always taken at the same time of the year. Fertilizer recommendations based on such sampling would not only promote an adequate K supply for the crop but would also contribute to system sustainability.

References

Abreu-Junior CH, Nogueira TAR, Oliveira FC et al (2008) Aproveitamento agrícola de resíduos no canavial. In: Marques MO, Mutton MA, Nogueira TAR et al (eds) Tecnologias na agroindústria canavieira. FCAV, Jaboticabal, pp 183–210. https://www.alice.cnptia.embrapa.br/bitstream/doc/16271/1/2008CL47.pdf. Accessed 21 May 2020

Almeida-Júnior AB, Nascimento CWA, Sobral MF et al (2011) Fertilidade do solo e absorção de nutrientes em cana-de-açúcar fertilizada com torta de filtro. Rev Brasil de Eng Agríc e Ambient 15(10):1004–1013. https://doi.org/10.1590/S1415-43662011001000003

Ambrosano EJ, Foltran DE, Camargo MS et al (2013) Mass and nutrient accumulation by green manures and sugarcane plant yield grown in succession, in two locations of Sao Paulo, Brazil. Rev Brasil de Agroecol 8(1):199–209. http://revistas.aba-agroecologia.org.br/index.php/rbagroecologia/article/view/12944. Accessed 21 May 2020

Askegaard M, Eriksen J, Johnston AE (2004) Sustainable management of potassium. In: Schjorring P, Elmholt S, Christensen BT (eds) Managing soil quality: challenges in modern agriculture. CABI, Wallingford, pp 85–102. https://www.cabi.org/bookshop/book/9780851996714. Accessed 21 May 2020

Barros I, Williams JR, Gaiser T (2004) Modeling soil nutrient limitations to crop production in semiarid NE of Brazil with a modified EPIC version. I. Changes in the source code of the model. Ecol Model 178:441–456. https://doi.org/10.1016/j.ecolmodel.2004.04.015

Barros I, Williams JR, Gaiser T (2005) Modeling soil nutrient limitations to crop production in semiarid NE of Brazil with a modified EPIC version. II: Field test of the model. Ecol Model 181:567–580. https://doi.org/10.1016/j.ecolmodel.2004.03.018

Bataglia OC, Camargo OA, Berton RS (1986) Emprego da vinhaça na cultura de citros. Laranja 7:277–289

Bender RR, Haegele JW, Ruffo ML et al (2013) Nutrient uptake, partitioning, and remobilization in modern, transgenic insect-protected maize hybrids. Agron J 105(1):161–170. https://doi.org/10.2134/agronj2012.0352

Borkert CM, Gaudêncio CA, Pereira JE et al (2003) Mineral nutrients in the shoot biomass of soil cover crops. Pesqui Agropecu Brasil 38(1):143–153. https://doi.org/10.1590/S0100-204X2003000100019

Braz AJBP, Silveira PM, Kliemann HJ et al (2004) Nutrient accumulation in leaves of millet, brachiaria and guineagrass. Pesqui Agropecu Trop 34(2):83–87. https://www.revistas.ufg.br/pat/article/view/2315. Accessed 11 May 2020

Caceres NT, Alcarde JC (1995) Adubação verde com leguminosas em rotação com cana-de-açúcar (Saccharum spp). Rev STAB 13(5):16–20

Calonego JC, Foloni JSS, Rosolem CA (2005) Potassium leaching from plant cover straw at different senescence stages after chemical desiccation. Rev Brasil de Ciênc do Solo 29 (1):99–108. https://doi.org/10.1590/S0100-06832005000100011

Calonego JC, Gil FC, Rocco VF et al (2012) Persistence and nutrient release from maize, brachiaria and lablab straw. Biosci J 28(5):770–781. http://www.seer.ufu.br/index.php/biosciencejournal/article/view/13885. Accessed 21 May 2020

Carrilho ENVM, Labuto G, Kamogawa MY (2016) Destination of vinasse, a residue from alcohol industry: resource recovery and prevention of pollution. In: Prasad MNV, Shih K (eds) Environmental materials and waste: resource recovery and pollution prevention. Elsevier, New York, pp 21–43. https://doi.org/10.1016/B978-0-12-803837-6.00002-0

Ciampitti IA, Camberato JJ, Murrell ST et al (2013) Maize nutrient accumulation and partitioning in response to plant density and nitrogen rate: I. Macronutrients. Agron J 105(3):783–795. https://doi.org/10.2134/agronj2012.0467

Crusciol CAC (2016) Absorção e remoção de nutrientesemcultivares de amendoim. COPERCANA, Botucatu. Technical report

Crusciol CAC, Cottica RL, Lima EV et al (2005) Persistence and nutrients release of forage turnip straw utilized as mulching in no-tillage crop system. Pesqui Agropec Brasil 40(2):161–168. https://doi.org/10.1590/S0100-204X2005000200009

Crusciol CAC, Moro E, Lima EV et al (2008) Decomposition rate and nutrient release of oat straw used as mulching in no-till system. Bragantia 67(2):481–489. https://doi.org/10.1590/S0006-87052008000200024

Eckert DJ (1991) Chemical attributes of soils subjected to no-till cropping with rye cover crops. Soil Sci Soc Am J 55:405–409. https://doi.org/10.2136/sssaj1991.03615995005500020019x

Eichert T, Fernández V (2012) Uptake and release of elements by leaves and other aerial plant parts. In: Marschner P (ed) Marchner's mineral nutrition of higher plants. Academic, San Diego, pp 71–84. https://doi.org/10.1016/b978-0-12-384905-2.00004-2

Farmaha BS, Fernandez FG, Nafziger ED (2012) Soybean seed composition, aboveground growth, and nutrient accumulation with phosphorus and potassium fertilization in no-till and strip-till. Agron J 104(4):1006–1015. https://doi.org/10.2134/agronj2012.0010

Foloni JSS, Rosolem CA (2004) Potassium balance in soybean grown under no-till. In: Proceedings of 4th international crop science congress, Brisbane, Proceedings, Brisbane, ICSS. Available from: CD-Rom. http://www.cropscience.org.au/icsc2004/poster/2/5/5/907_rosolemca.htm#TopOfPage. Accessed 21 May 2020

Francisco EAB, Câmara GMS, Segatelli CR (2007) Nutritional condition and yield of finger millet and soybean grown in succession in a system of anticipated fertilization. Bragantia 66 (2):259–266. https://doi.org/10.1590/S0006-87052007000200009

Garcia RA, Crusciol CAC, Calonego JC et al (2008) Potassium cycling in a corn-brachiaria cropping system. Eur J Agron 28:579–585. https://doi.org/10.1016/j.eja.2008.01.002

Giacomini SJ, Aita C, Vendruscolo ERO et al (2003) Matéria seca, relação C/N e acúmulo de nitrogênio, fósforo e potássio em misturas de plantas de cobertura de Solo. Rev Brasil deCiênc do Solo 27(2):325–334. https://doi.org/10.1590/S0100-06832003000200012

Gregory PJ, Crawford DV, McGowan M (1979) Nutrient relations of winter wheat. 1. Accumulation and distribution of Na, K, Ca, Mg, P, S and N. J Agric Sci 93(2):485–494. https://doi.org/10.1017/S0021859600038181

Grimes DW, Hanway JJ (1967a) Exchangeable soil potassium as influenced by seasonal cropping and potassium added in crop residues. Soil Sci Soc Am Proc 31:502–506. https://doi.org/10.2136/sssaj1967.03615995003100040024x

Grimes DW, Hanway JJ (1967b) An evaluation of the availability of K in crop residues. Soil Sci Soc Am Proc 31:705–706. https://doi.org/10.2136/sssaj1967.03615995003100050027x

Hanway JJ, Weber CR (1971) Accumulation of N, P, and K by soybean (*Glycine max* (L.) Merrill)' plants. Agron J 63:406–408. https://doi.org/10.2134/agronj1971.00021962006300030017x

Heinz R, Garbiate MV, ViegasNeto AL et al (2011) Decomposition and nutrient release of crambe and fodder radish residues. Ciênc Rural 41(9):1549–1555.v. https://doi.org/10.1590/S0103-84782011000900010

Jiang ZP, Li YR, Wei GP et al (2012) Effect of long-term vinasse application on physico-chemical properties of sugarcane field soils. Sugar Tech 14(4):412–417. https://doi.org/10.1007/s12355-012-0174-9

Jobbágy EG, Jackson RB (2004) The uplift of soil nutrients by plants: biogeochemical consequences across scales. Ecology 89(9):2380–2389. https://doi.org/10.1890/03-0245

Korndörfer GH, Anderson DL (1997) Use and impact of sugar alcohol residues vinasse and filter on sugarcane production in Brazil. Sugar y azucar 92(3):26–35

Kurihara CH, Venegas VHA, Neves JCL et al (2013) Accumulation of dry matter and nutrients in soybean, as a variable of the productive potential. Rev Ceres 60(5):690–698. https://doi.org/10.1590/S0034-737X2013000500013

Malavolta E (1985) Potassium status of tropical and subtropical region soils. In: Munson R (ed) Potassium in agriculture. ASA CSSA & SSSA, Madison, pp 163–200. https://doi.org/10.2134/1985.potassium.c8

Mallarino AP, Sawyer JE, Barnhart SK (2013) General guide for crop nutrient and limestone recommendations in Iowa. Publ. PM 1688 (Rev.). Iowa State University Extension, Ames. https://store.extension.iastate.edu/product/5232. Accessed 21 May 2020

Marschner H (1995) Mineral nutrition of higher plants, 2nd ed. Academic, San Diego. https://www.elsevier.com/books/mineral-nutrition-of-higher-plants/marschner/978-0-08-057187-4. Accessed 21 May 2020

Matiello JB (2005) Cultura de café no Brasil: novo manual de recomendações. Editora Bom Pastor, Rio de Janeiro

Oliveira MW, Trivelin PCO, Gava GJ et al (1999) Degradação da palhada de cana-de-açúcar. Sci Agric 54(4):803–809. https://doi.org/10.1590/S0103-90161999000400006

Oliveira TK, Carvalho GJ, Moraes RNS (2002) Cover crops and their effects on bean plant in no-tillage system. Pesqui Agropec Brasil 37(8):1079–1087. https://doi.org/10.1590/S0100-204X2002000800005

Oltmans RR, Mallarino AP (2015) Potassium uptake by corn and soybean, recycling to soil and impact on soil test potassium. Soil Sci Soc Am J 79(1):314–327. https://doi.org/10.2136/sssaj2014.07.0272

Pal Y, Wong T, Gilkes R (1999) The forms of potassium and potassium adsorption in some virgin soils from south western Australia. Aust J Soil Res 37(4):695–709. https://doi.org/10.1071/SR98083

Pereira FCBL, Mello LMM, Pariz CM et al (2016) Autumn maize intercropped with tropical forages: crop residues, nutrient cycling, subsequent soybean and soil quality. Rev Brasil de Ciênc do Solo 40:e0150003. https://doi.org/10.1590/18069657rbcs20150003

Prado RM, Caione G, Campos CNS (2013) Filter cake and vinasse as fertilizers contributing to conservation agriculture. Appl Environ Soil Sci, Article ID 581984, 8 p. https://doi.org/10.1155/2013/581984

Raij B van, Cantarella H, Quaggio JA et al (1997) Recomendações de adubação e calagem para o Estado de São Paulo, 2nd ed (Boletim Técnico, 100). Instituto Agronômico, Campinas

Rosolem CA, Nakagawa J (1985) Potassium uptake by soybean as affected by exchangeable potassium in the soil. Commun Soil Sci Plan 16(7):707–726. https://doi.org/10.1080/00103628509367639

Rosolem CA, Steiner F (2017) Effects of soil texture and rates of K input on potassium balance in tropical soil. European Journal of Soil Science, 68: 658–666. 2017. https://doi.org/10.1111/ejss.12460.

Rosolem CA, Calonego JC, Foloni JSS (2003) Potassium leaching from green cover crop residues as affected by rainfall amount. Rev Brasil de Ciênc do Solo 27(2):355–362. https://doi.org/10.1590/S0100-06832003000200015

Rosolem CA, Calonego JC, Foloni JSS (2005) Potassium leaching from millet straw as affected by rainfall and potassium rates. Commun Soil Sci Plan 36(7–8):1063–1074. https://doi.org/10.1081/CSS-200050497

Rosolem CA, Santos FP, Foloni JSS et al (2006) Potássio no solo emconseqüência da adubação sobre a palha de milheto e chuva simulada. Pesqui Agropecu Brasil 41(6):1033–1040. https://doi.org/10.1590/S0100-204X2006000600020

Rosolem CA, Sgariboldi T, Garcia RA et al (2010) Potassium leaching as affected by soil texture and residual fertilization in tropical soils. Commun Soil Sci Plan 41(16):1934–1943. https://doi.org/10.1080/00103624.2010.495804

Rosolem CA, Montans JPTM, Steiner F (2012) Potassium supply as affected by residual potassium fertilization in a Cerrado Oxisol. Rev Brasil de Ciênc do Solo 36:1507–1515. https://doi.org/10.1590/S0100-06832012000500015

Rossato RR (2004) Potential of nitrogen and potassium cycling for the oilseed radish to insert between corn and wheat under no-tillage. Master's thesis, Federal University of Santa Maria (In Portuguese)

Santos DH, Silva MA, Tiritan CS et al (2011) Qualidade tecnológica da cana-de-açúcar sob adubação com torta de filtro enriquecida com fosfato solúvel. Rev Brasil de Engenh Agríc e Ambient 15(5):443–449. https://doi.org/10.1590/S1415-43662011000500002

Santos JD, Lopes da Silva AL, Costa JL et al (2013) Development of a vinasse nutritive solution for hydroponics. J Environ Manag 114(15):8–12. https://doi.org/10.1016/j.jenvman.2012.10.045

Schomberg HH, Steiner JL (1999) Nutrient dynamics of crop residues decomposing on a fallow no-till soil surface. Soil Sci Soc Am J 63:607–613. https://doi.org/10.2136/sssaj1999.03615995006300030025x

Spain JM, Salinas JG (1985) A reciclagem de nutrientes nas pastagens tropicais, Reunião Brasileira de Fertilidade do Solo: proceedings of a conference, ComissãoExecutiva do Plano da LavouraCacaueira, Ilhéus, pp 259–299

Teodoro RB, Oliveira FL, Silva DMN et al (2011) Perennial herbaceous legumes used as permanent cover cropping in the Caatinga Mineira. Rev Ciênc Agron 42(2):292–300. https://doi.org/10.1590/S1806-66902011000200006

Tukey HB (1970) The leaching of substances from plants. Annu Rev Plant Physiol 21:305–324. https://doi.org/10.1146/annurev.pp.21.060170.001513

Vitko LF, Laboski CAM, Andraski TW (2009) Effects of sampling time, soil moisture content, and extractant on soil test potassium levels. In: Thirty-ninth north central extension-industry soil fertility conference, Des Moines, vol 25, 18–19 Nov 2009. International Plant Nutrition Inst, Brookings, pp 124–132

Vitti GC, Luz PHC (2008) Manejo e uso de fertilizantes para cana-de-açúcar. In: Marques MO, Mutton MA, Nogueira TAR, Tasso LC Jr, Nogueira GA, Bernardi JH (eds) Tecnologias na Agroindústria Canavieira. FCAV, Jaboticabal, pp 141–167

Wider RK, Lang GE (1982) Critique of the analytical methods used in examining decomposition data obtained from litter bags. Ecology 63(6):1636–1642. https://doi.org/10.2307/1940104

Zoca SM, Penn CJ, Rosolem CA et al (2014) Coffee processing residues as a soil potassium amendment. Int J Recycl Waste Agric 3:155–165. https://doi.org/10.1007/s40093-014-0078-7

Zörb C, Senbayram M, Peiter E (2014) Potassium in agriculture – status and perspectives. J Plant Physiol 171(9):656–669. https://doi.org/10.1016/j.jplph.2013.08.008

7

Soil Potassium (K) Pools: An Alternative Framework

Michael J. Bell, Michel D. Ransom, Michael L. Thompson, Philippe Hinsinger, Angela M. Florence, Philip W. Moody, and Christopher N. Guppy

Abstract Soil potassium (K) has traditionally been portrayed as residing in four functional pools: solution K, exchangeable K, interlayer (sometimes referred to as "fixed" or "nonexchangeable") K, and structural K in primary minerals. However, this four-pool model and associated terminology have created confusion in understanding the dynamics of K supply to plants and the fate of K returned to the soil in fertilizers, residues, or waste products. This chapter presents an alternative framework to depict soil K pools. The framework distinguishes between micas and feldspars as K-bearing primary minerals, based on the presence of K in interlayer positions or three-dimensional framework structures, respectively; identifies a pool of K in neoformed secondary minerals that can include fertilizer reaction products; and replaces the "exchangeable" K pool with a pool defined as "surface-adsorbed" K, identifying where the K is located and the mechanism by which it is

M. J. Bell (✉)
School of Agriculture and Food Science, The University of Queensland, Brisbane, QLD, Australia
e-mail: m.bell4@uq.edu.au

M. D. Ransom · A. M. Florence
Department of Agronomy, Kansas State University, Manhattan, KS, USA
e-mail: mdransom@ksu.edu

M. L. Thompson
Department of Agronomy, Iowa State University, Ames, IA, USA
e-mail: mlthomps@iastate.edu

P. Hinsinger
Eco&Sols, University of Montpellier, CIRAD, INRAE, Institut Agro, IRD, Montpellier, France
e-mail: philippe.hinsinger@inrae.fr

P. W. Moody
Department of Science, Information Technology and Innovation, Brisbane, QLD, Australia
e-mail: Phil.Moody@nrm.qld.gov.au

C. N. Guppy
School of Environmental and Rural Science, University of New England, Armidale, NSW, Australia
e-mail: cguppy@une.edu.au

held rather than identification based on particular soil testing procedures. In this chapter, we discuss these K pools and their behavior in relation to plant K acquisition and soil K dynamics.

7.1 Introduction

Traditionally, soil potassium (K) has been depicted as occurring in four pools—soil solution K, exchangeable K, interlayer K, and mineral or structural K (Barber 1995). Various publications have estimated the relative sizes of these traditional K pools (for example, Öborn et al. 2005; Hinsinger 2006) with K in primary minerals (90–98% of total soil K) the dominant form of soil K. The K most readily available for plant uptake [exchangeable K (1–2%) and solution K (0.1–0.2%)] represents only a very small fraction of the total soil K, although interlayer K in secondary minerals can also be a significant proportion (1–10%, depending on mineralogy). However, this terminology creates some confusion for a number of reasons. In particular, exchange reactions between the solid and the soil solution phases can result in reversible K movement to or from both surface and interlayer positions of clay minerals, while interlayer K can be found in both primary minerals (e.g., micas) as well as in secondary minerals (e.g., illite). Furthermore, soils with a history of repeated applications of high-K waste materials and/or high rates of band-applied compound fertilizers containing K may also contain neoformed secondary K minerals with poorly understood behavior in terms of K bioavailability.

These considerations have led to the development of an alternative framework to depict soil K pools (Fig. 7.1). There are three key characteristics of this conceptual diagram that differ from the traditional four-pool model. The first is the distinction between micas and feldspars as K-bearing primary minerals, which is based on the presence of K in interlayer positions between mineral layers in phyllosilicate minerals (micas and partially weathered micas) or in the three-dimensional structural frameworks of tectosilicates (feldspars). The flux from structural K in feldspars and micas to soil solution K is unidirectional. However, the flux from interlayer K to soil solution K can be bidirectional for partially weathered micas. The second characteristic is the identification of a pool of K in neoformed secondary minerals that can include fertilizer reaction products. The third characteristic is the replacement of the "exchangeable" K pool with a pool defined as "surface-adsorbed" K. The "surface-adsorbed K" concept clearly identifies where the K is located and the mechanism by which it is held. It avoids ascribing possible locations (surfaces, wedge, and interlayer positions) on the basis of particular soil testing procedures.

Quantification of K in these different K pools, particularly in relation to K available for plant acquisition in time frames relevant to individual crop or pasture seasons (i.e., K bioavailability, Chap. 4), is challenging. This is due to both the limitations of current diagnostic soil testing methods (Chap. 8) and the need to consider crop, season, and soil-specific factors that regulate root dynamics, rhizosphere conditions, and soil moisture dynamics (Chap. 4). In this chapter, we will

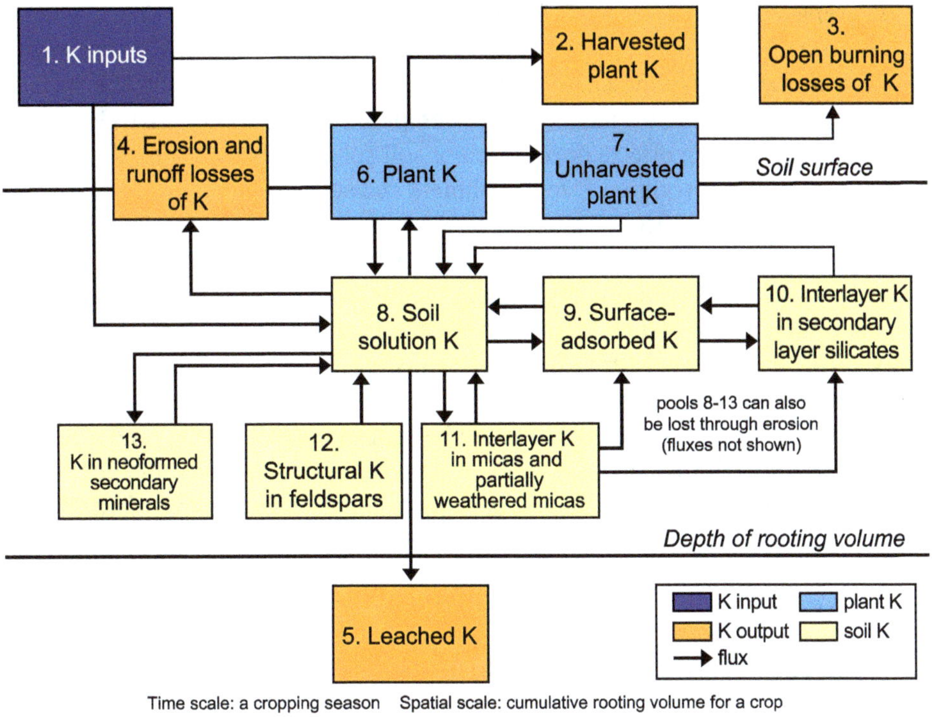

Fig. 7.1 Diagrammatic representation of the K cycle, identifying the soil K pools (rectangles) discussed in this chapter. The arrows denote fluxes of K in soil–plant systems

discuss these K pools and their behavior in relation to plant K acquisition and soil K dynamics.

7.2 Solution Potassium and Potassium Activity

Soil solution K is *the quantity of K dissolved in the aqueous liquid phase of the soil* (Soil Science Glossary Terms Committee 2008). Unless mineral soils have been recently fertilized or amended with manure, soil solution K concentrations are typically low (100–1000 μM) because of selective adsorption of K by some clay minerals (Hinsinger 2006). Leaching losses of K are usually low, but they may occur when K inputs exceed the sum of K holding capacity and plant uptake (Chap. 3) or where preferential flow of water occurs.

As the monovalent cation, K^+ in the soil solution is the form taken up directly by plants (Sparks 1987), with concentrations determined by reactions with solid-phase forms of soil K and with other cations on the exchange complex and in the soil solution, and by soil moisture content (Sparks and Huang 1985). The small amount of K^+ present in the soil solution in all except recently fertilized soils suggests that the rate of K uptake from this pool is likely to be limited as solution K^+ stocks are rapidly depleted (Barrow 1966). Therefore, factors that control solution K concentration ultimately influence uptake rates. These factors include the rates of K diffusion in the soil solution from surrounding, undepleted soil, and the quantity of readily desorbable K on the exchange complex (Evangelou et al. 1994; Barber 1995). The moderating effects of other major soil solution cations (Ca, Mg, and Na) on the kinetics of soil solution K replenishment and subsequent uptake of K by plant roots have been the subject of extensive research, reviewed by Sparks (1987) and Evangelou et al. (1994). This research has shown that the complex interactions among soil solution chemistry, rooting density, and the subsequent competition for K uptake make extrapolation beyond the specific conditions of individual experiments difficult (Barber 1995). There is, however, a general recognition of the importance of cation exchange capacity (CEC) on both the ability of the soil to buffer soil solution K (sometimes measured as buffer capacity, BC_K) and also on K supply to the root (Barber 1981). As CEC increases, there is less K available for plant uptake, even in soils with the same concentrations of exchangeable K (Bell et al. 2009). These relationships have currently not been well defined.

7.3 Surface-Adsorbed Potassium

Surface-adsorbed K is *the quantity of K associated with negatively charged sites on: soil organic matter; planar surfaces of phyllosilicate minerals; and surfaces of iron and aluminum oxides*. However, the location of those charge surfaces and the specificity of those sites for K varies greatly. An example of the contrasting sites was provided by Mengel (1985) for a weathered grain of mica (Fig. 7.2). The **p-position** is *a site on the planar surfaces of phyllosilicate minerals where hydrated K^+ is adsorbed*. In these positions, K ions tend to retain a hydration shell, be weakly and nonspecifically bound, and be in rapid dynamic equilibrium with soil solution K. Exchange reactions in response to altered soil solution concentrations occur very quickly. For example, cation exchange reactions on montmorillonite have been measured to be complete in less than 10^{-1} s (Tang and Sparks 1993), but modelled to be complete much faster, less than 5×10^{-10} s (Bourg and Sposito 2011). As a result, this K pool is considered to be in a form immediately available for movement into the soil solution in response to depletion by plant root uptake (Barber 1995).

Planar surface sorption sites, along with carboxylate and phenolate groups in soil organic matter and negatively charged sites on the surfaces of iron and aluminum oxides, tend to show a greater affinity for divalent cations like Ca^{2+} and Mg^{2+} (Hinsinger 2006). This means that application of soil amendments (lime, limestone, or dolomite), or even frequent rainfall or irrigation in well-drained soils, can result in displacement of surface-adsorbed K from the charged surface by other cations and possible loss from the rooting zone through leaching. Other sorption sites illustrated in Fig. 7.2, the e- and i-positions, show greater affinity for K and a slower interaction with components of the soil solution. These sites are discussed below.

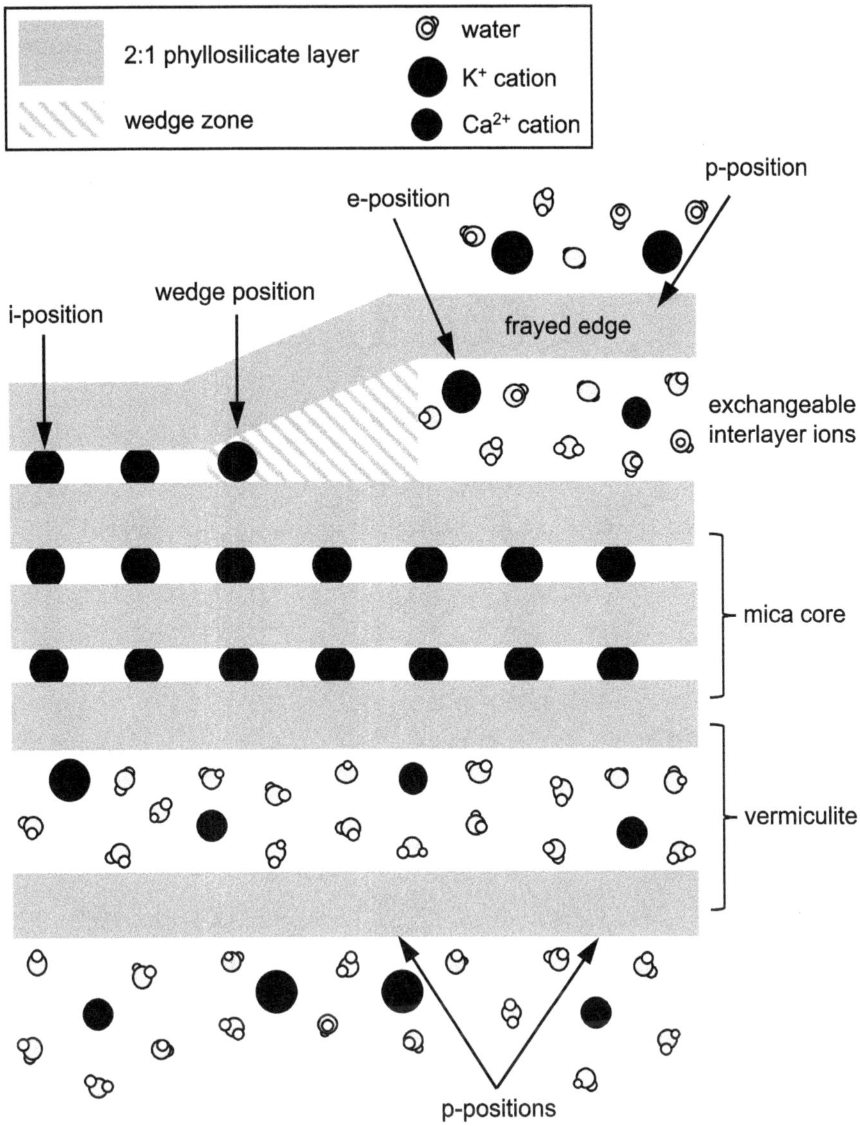

Fig. 7.2 Diagrammatic representation of K adsorption positions for mica, a frayed edge of mica, and an outer layer where mica has transformed to vermiculite. Planar surface (p-position), interlayer (i-position), edge (e-position), and wedge positions are shown. (adapted from Mengel 1985; Mei et al. 2015; Rich 1968)

7.4 Interlayer K in Micas and Partially Weathered Micas

Interlayer K is *the quantity of K bound between layers of phyllosilicate minerals.* Micas are phyllosilicate minerals with each mineral layer composed of two tetrahedral sheets bound on either side of one octahedral sheet (2:1 layer silicates, Fig. 7.3a). Layers carry a net negative charge and are bound together by K^+ ions. Away from the edges of the crystal, these K^+ ions do not have a hydration shell.

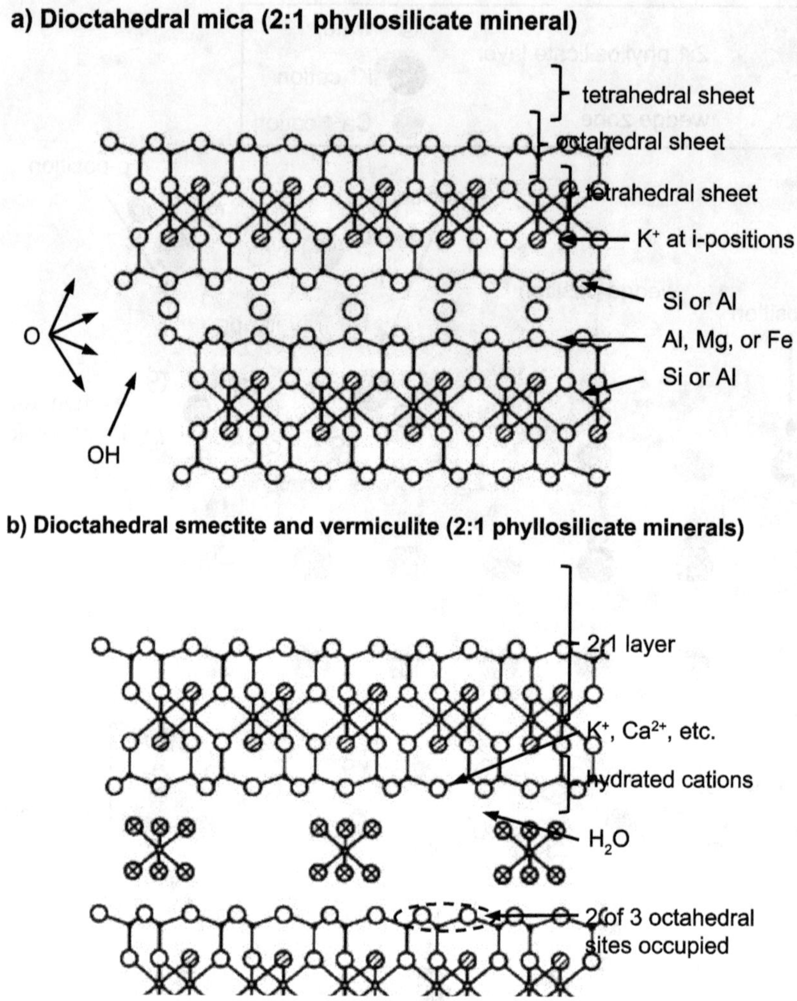

Fig. 7.3 Diagrammatic representation of the structure of (**a**) a dioctahedral mica, and (**b**) vermiculite and smectite. (adapted from Schulze 1989)

They are dehydrated and occupy the ditrigonal cavities in the tetrahedral sheets above and below them in adjacent 2:1 layers (Fig. 7.3a). Muscovite and biotite are two common micas. Pure muscovite is a dioctahedral mica in which two-thirds of the cation positions in the octahedral sheet are filled with trivalent Al^{3+}. Biotite is a trioctahedral mineral in which all the cation positions in the octahedral sheet are filled with divalent cations such as Fe^{2+} and Mg^{2+}. The negative layer charge in these minerals is derived primarily from isomorphic substitution of Al^{3+} for Si^{4+} in the tetrahedral sheets. The layer charge of both muscovite and biotite is close to 1 mol of charge per 10-oxygen formula unit (Thompson and Ukrainczyk 2002).

During mica weathering, the layer charge of these minerals declines and interlayer K leaves from the edges of the particles (Scott and Smith 1967; Barber 1995; Hinsinger 2006). As the net negative charge in the layers decreases, not only is K released from some interlayer positions near the edge of the particles but the space

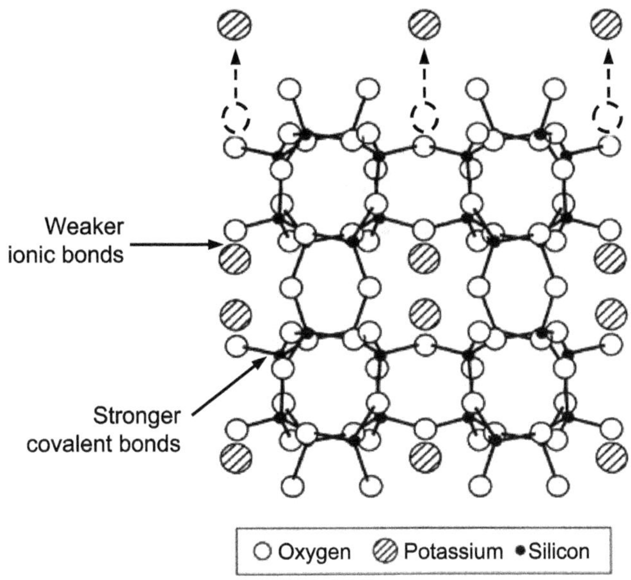

Fig. 7.4 An example structure of a potassium feldspar (orthoclase) and the types of bonds present. Arrows indicate the desorption of K^+ cations from their original positions at the outer surface of the tectosilicate structure, indicated by circles with dotted lines. (adapted from Fenter et al. 2000)

between the layers opens at those locations (Fig. 7.2). These changes allow for freer movement of other cations into that portion of the interlayer region and facilitate exchange reactions. In general, weathering of micas leads to loss of layer charge and transformation to vermiculite, and then to smectite (Wilson 1999; Thompson and Ukrainczyk 2002) (Fig. 7.4).

As micas weather, interlayer adsorption sites become differentiated (Fig. 7.2). The **e-position** is *an interlayer adsorption site where hydrated K^+ is preferentially adsorbed and where K^+ can rapidly diffuse short distances out of the crystal lattice and into the soil solution.* Potassium in this position is rapidly exchangeable with NH_4^+, and NH_4^+ is also preferentially adsorbed at these sites (Bolt et al. 1963; Mengel 1985). The **i-position** is *an interlayer adsorption site where dehydrated K^+ is preferentially adsorbed and where K^+ diffuses more slowly into the soil solution.* A **wedge zone** is *the interlayer volume at the nexus where two adjacent phyllosilicate mineral layers separate farther due to solvation forces.* A **wedge site** is *an interlayer adsorption site at the point where two adjacent layers begin to separate farther due to solvation forces.* Wedge sites on illite are most selective for the dehydrated cesium cation (Cs^+) but are also selective for dehydrated K^+ (Lee et al. 2017).

Hinsinger (2006) noted that while adsorption/desorption of K can occur in the e-positions in frayed edges of weathered clay minerals, this process was considerably slower than that at the planar surfaces discussed in Sect. 7.3. Even slower release can occur with K in the i-positions, because the ionic radius of K^+ and its low hydration energy allow it to dehydrate and fit perfectly into cavities created by the basal plane of oxygen ions in the tetrahedral sheets of phyllosilicates. While adsorption/desorption reactions may still occur under particular circumstances, they are very slow.

Mengel (1985) differentiated K adsorption positions by their Gapon selectivity coefficients. The **Gapon coefficient** is *a factor that quantifies the selectivity for one ion over another at adsorption sites*. The higher the Gapon coefficient is, the greater the selectivity. Using the selectivity for K^+ over Ca^{2+} (Bolt et al. 1963), Mengel (1985) reported that the Gapon coefficients differed by an order of magnitude between adsorption positions. The p-positions had the lowest selectivity for K (Gapon coefficients less than 10), the e-positions had an intermediate selectivity (Gapon coefficients approximately equal to 100), and the i-positions had the highest selectivity for K (Gapon coefficients greater than 1000).

As a mica weathers, transformations occur within very short distances. **Interstratification** is *the occurrence of both high-charge layers of mica and layers with lower charge in the same mineral domain*. In phyllosilicate clays, the high-charge, K-bearing core is typically illite, i.e., a clay mica. Core illite layers are often continuous with near-edge layers where weathering has lowered the charge sufficiently to be classified as vermiculite. The difference in the structural charge of domain cores and of layers near domain edges is usually ignored in the literature, and such particles are lumped into the term “illites.” Potassium that is held in the core interlayers (i-positions) is not surface-adsorbed, is not exchangeable with ammonium, and is not likely to be bioavailable to plants unless subjected to the unique conditions of the rhizosphere (Chap. 4). Indeed, K^+ cations in those positions can be extremely stable. In contrast, the lower-charge interlayer positions near domain edges, e-positions in Fig. 7.2, may retain K from fertilizer amendments and release it back slowly to the soil solution. Because the layer charge there is lower than in unweathered mica layers, K^+ ions are held less tightly and are more susceptible to subsequent release. It should be noted that both primary micas and the high-charge core of illite domains will yield a 1.0-nm *d*–spacing in X-ray diffraction patterns. It is difficult to quantitatively differentiate a primary mica from illite by using routine X-ray diffraction procedures.

7.5 Interlayer Potassium in Secondary Layer Silicates

Secondary layer silicates are *phyllosilicate minerals that are weathering products of primary minerals*. The interlayer compositions of these minerals are variable and can contain a variety of cations and quantities of water. Figure 7.3b shows an example structure that represents two different types of secondary minerals important to K behavior in soils: smectite and vermiculite. A key feature of these minerals is the presence of hydrated cations in the interlayer. The presence of water facilitates the diffusion of cations in and out of the interlayer, making cation exchange possible. Soil tests that rely on cation exchange to measure K will measure some K from interlayers of these minerals.

7.6 Structural Potassium in Feldspar and Feldspathoids

Structural K in feldspars is *the quantity of K in structures of tectosilicate minerals.* Making up nearly 31% of the Earth's crust, feldspars (specifically, orthoclase and microcline) and feldspathoids (e.g., leucine) represent the dominant form of structural K in many soils, with the quantities in soil determined by the intensity of weathering and also erosional and depositional factors (Sparks 1987; Barber 1995). While feldspars are commonly present in the silt and sand fractions of younger and moderately weathered soils (Sparks 1987), alkali feldspars have also been found in the clay fractions of moderately weathered soils (Huang and Lee 1969).

The K in feldspars occurs throughout the mineral structure (Fig. 7.3). In K-feldspars, silicon (Si)-oxygen, and aluminum (Al)-oxygen bonds are stronger, covalent bonds, but K is held in the tectosilicate structure by weaker, ionic bonds (Fogler and Lund 1975). When feldspars weather, the K in the outer layer of the structure is released first, causing structural relaxation of the remaining bonds in the surrounding structure (Fenter et al. 2000). The result is an irreversible dissolution of the mineral's silicate framework during weathering (Sparks 1987; Barber 1995; Hinsinger 2006). While K feldspars can be present in only small quantities or are completely absent in some strongly weathered soils (Sparks 1987), the rates of weathering and subsequent K release may be slowed by the formation of a noncrystalline Si-Al-O skin on the mineral surface (Rich 1972). Rich (1968) found that the rates of K release from feldspars were typically slower than that from micas, although Song and Huang (1988) noted that this order may be altered in the presence of organic acid anions (oxalate and citrate) that can be exuded by roots of some plant species into their rhizospheres.

There is evidence to suggest that feldspars can contribute significant quantities of bioavailable K to plants, with these contributions potentially originating from sand-sized fractions rather than clay-sized fractions. Examples include the work of Rehm and Sorensen (1985), who conducted a 4-year trial that varied fertilizer application rates of K and magnesium (Mg) in a factorial combination on an irrigated Valentine loamy fine sand with aeolian sand parent material. Application rates of fertilizer K ranged from 0–269 kg K ha^{-1}. Based on the levels of ammonium acetate extractable K, Rehm and Sorensen expected maize (*Zea mays* L.) yield to increase when fertilized with K; however, no increase was observed over the study period. They attributed the lack of maize response to the added K to the presence of bioavailable K in feldspars in the fine sand and very fine sand size fractions. Sautter (1964) characterized the Valentine soil as having up to 23% K feldspars in the upper 28 cm (inferred from Fig. 7.1 in that paper). Sadusky et al. (1987) measured K release rates from three US Atlantic Coastal Plains soils that had high quantities of feldspars in the sand fractions, ranging from 6.7–16.0% in the surface horizons and 8.2–24.0% in the subsurface horizons. Potassium release rates were studied for a period of up to 30 days in the presence of both oxalic acid and a cation exchange resin. Because the resin provided a continuous sink for K and kept solution K levels lower than did the oxalic acid, approximately two orders of magnitude more K were released with the resin than with oxalic acid. Most of the K was released in the first 16 days (inferred from Fig. 7.1 in that paper). After 30 days, the total quantities of K released by the resin ranged from 67–92 mg K kg^{-1} soil in the surface horizon samples.

This type of evidence suggests that blanket statements about coarser-textured sandy soils providing little bioavailable K to plants can be inaccurate. The observations of Niebes et al. (1993) that a substantial portion of K had been extracted by rape (*Brassica napus* L.) grown in the coarse (silt and sand) fractions of soils from two long-term fertilizer experiments in Europe support this conclusion. Despite the short duration of this experiment (8 days of plant growth), they found for the coarse silt and sand fractions that 80–100% of the bioavailable K was not originating in the ammonium-exchangeable pool. The mineralogical composition and the K-supplying power of a given particle size fraction must be considered, and future K recommendations need to consider the possible contribution of K feldspars.

7.7 Neoformed Potassium Minerals

Neoformed K minerals are *newly formed minerals created from the reaction of soil solution K with other soil solution ions*. Neoformed K minerals do not include primary or secondary layer silicate minerals, but are considered to be a possible byproduct of fertilizer use in intensive cropping areas, especially where compound fertilizers that include K are applied in bands with high in-band concentrations (e.g., the result of wideband spacings in sugarcane or row crops). Such minerals are probably rare in soils, but they may form under certain conditions.

As an example, we consider the neoformation of potassium taranakite ($K_3Al_5H_6(PO_4)_8 \cdot 18H_2O$) or its noncrystalline analogs. Du et al. (2006) coapplied monocalcium phosphate with KCl on an acid (pH 4.6) soil. They hypothesized that some of the added K displaced Al^{3+} from exchange sites, leading to precipitation of K, P, and Al as a noncrystalline analog of potassium taranakite. Formation of potassium taranakite was also observed by Lindsay (1962) after adding saturated solutions of monocalcium phosphate or monopotassium phosphate to an acidic (pH 4.9) Hartsells fine sandy loam soil. The formation of potassium taranakite with only monocalcium phosphate indicates that to form this precipitate, K may not need to be added but simply needs to be present in solution or in a position (probably the p-position) where it can be readily displaced into solution by fertilizer addition. Potassium taranakite was present at 15 min and 3 h after fertilizer addition, but not after 3 days. When Lindsay added those same saturated fertilizer solutions to a basic (pH 8.3) Webster silty clay loam soil, no taranakite formed. Acid conditions where Fe and Al are present are necessary for potassium taranakite formation.

How the formation of taranakite or other neoformed minerals changes K bioavailability during the growing season of plants has not been well researched and is currently not well known.

7.8 Fixation and Release of Interlayer Potassium

The processes by which K is adsorbed or desorbed from interlayer positions of weathered micas, vermiculite, and high-charge smectite are referred to as fixation and release. In **potassium fixation**, *hydrated K^+ ions move to interlayer positions in phyllosilicate minerals, then dehydrate as the mineral layers contract. In this position, the K^+ is unavailable to plants*. **Potassium release** is when *mineral layers expand, K^+ ions rehydrate, and move to the soil solution, becoming bioavailable to plants*. In addition to environmental factors such as pH, redox potential and temperature, fixation and release are governed by the soil solution concentrations of K and competing cations. The actions of plant root uptake (depletion) or fertilizer application (enrichment) determine the net impact of fixation and release on the dynamics of interlayer K (Schneider et al. 2013). The processes of K interlayer fixation have been studied extensively, due to the focus on exchangeable K as a proxy of the pool of soil K that is bioavailable to plants and the apparent inefficient use of applied K fertilizer in soils where K fixation occurs (e.g., Kovar and Barber (1990)), but there has been less research focus on the release process. Sparks (1987), Barber (1995), Öborn et al. (2005), Hinsinger (2006), and Zörb et al. (2014) provide detailed reviews of much of this work (see also Chap. 4).

In some ways, K fixation can be conceptualized as the reversal of mica weathering (Fig. 7.5). When micas weather, their interlayers expand, and K is released. Even though layer charge declines, there is an increase in CEC because more surface area is available to exchange with cations in extractant solutions used to measure CEC. This greater surface area comes from interlayer positions that have

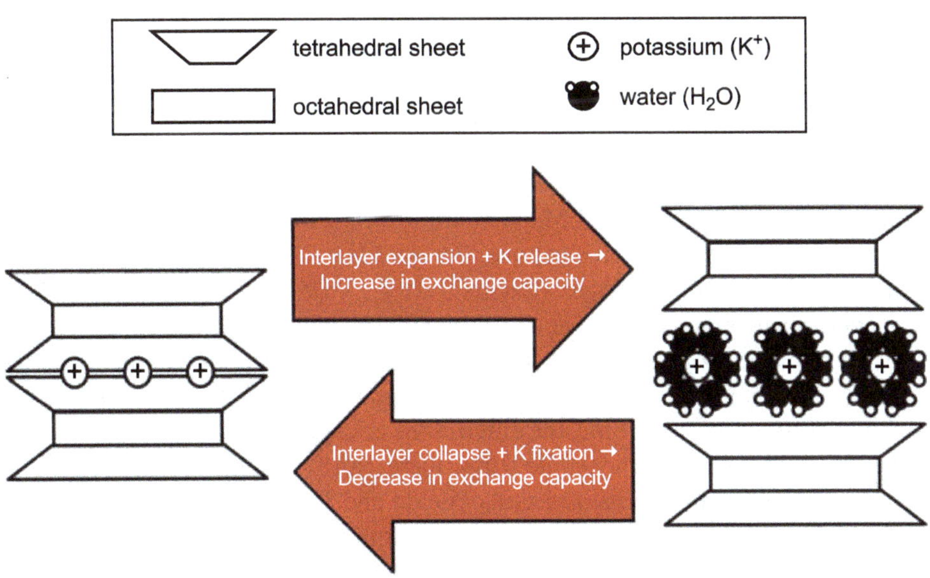

Fig. 7.5 Model of the processes of K fixation and release

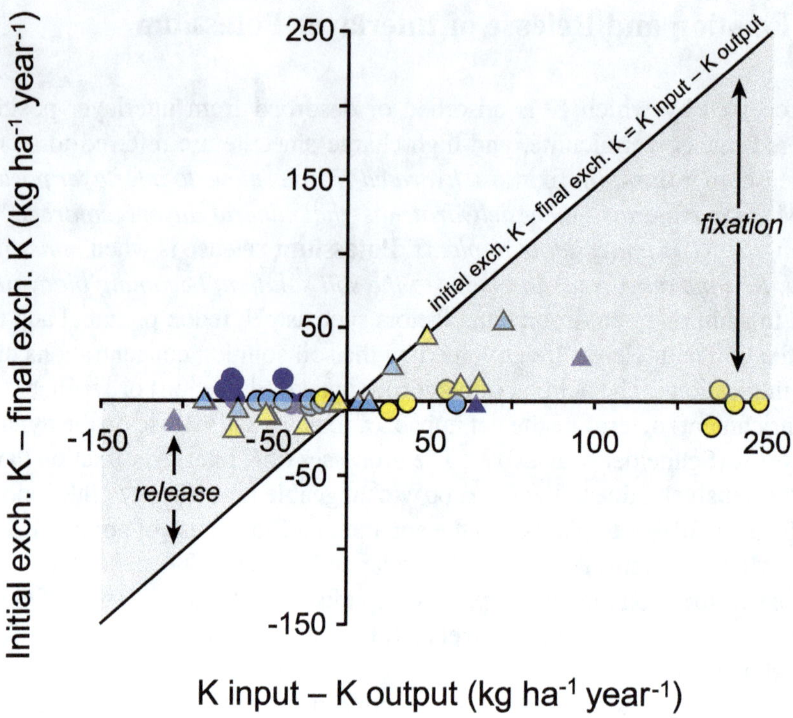

Fig. 7.6 Annual change in exchangeable K (exch. K) in response to annual K budgets in several long-term fertilizer trials in Europe. Potassium inputs are from organic and inorganic K fertilizers, and K outputs occur in harvested produce. The general lack of equivalent changes in exchangeable K in response to surpluses or deficits in K budgets have been interpreted as evidence of K fixation or release, respectively. (Hinsinger 2002)

become available for ion exchange upon interlayer hydration and expansion. When K is fixed between the layers of expanded 2:1 layer silicates, the layers contract, trapping dehydrated K^+ ions in ditrigonal voids in adjacent tetrahedral sheets. The dehydration and contraction of interlayers prevent ion exchange, and an equivalent decrease in CEC is observed. Fixed K becomes nonexchangeable, or at least not exchangeable with NH_4^+ in the 1 M ammonium acetate extractant commonly used to measure CEC and exchangeable cations. For decades, the fixation process has been studied in laboratory settings (Barshad 1948, 1951, 1954; Jackson 1963; Reichenbach and Rich 1975; Wear and White 1951), and its effects can be observed at the field level in some long-term fertilized treatments, for example, at the Rothamsted site discussed by Blake et al. (1999) and other sites in Europe (Hinsinger 2002, Fig. 7.6). At the same time, the unfertilized or negative K balance treatments in long-term studies have demonstrated that there can be significant net release of nonexchangeable K in unfertilized plots (Velde and Peck (2002), Fig. 7.6). However, whether K was released from feldspars, from primary or partially weathered micas, or from secondary phyllosilicate minerals, or simply depletion of unsampled subsoil layers (Kuhlmann and Barraclough 1987; Prasad 2009), was rarely determined.

Potassium ions, in addition to those of other monovalent cations such as NH_4^+, rubidium (Rb^+), and Cs^+, are all fixed in a similar manner (Barshad 1954; Meunier and Velde 2004; Šucha and Širáňová 1991), while ions of monovalent sodium (Na^+) and the divalent cations calcium (Ca^{2+}), magnesium (Mg^{2+}), barium (Ba^{2+}), and strontium (Sr^{2+}) are not fixed to any appreciable extent. This is because monovalent cations have relatively weak energies of hydration compared to the divalent cations and so they are more likely to shed water molecules from the hydration sphere as they enter the charged space of the interlayer region. Furthermore, because of their size, K^+, NH_4^+, Rb^+, and Cs^+ ions can be positioned more stably in the ditrigonal cavities in the tetrahedral sheet than can other monovalent ions like Na^+ or lithium (Li^+) (Reichenbach and Rich 1975). Staunton and Roubaud (1997) reported that the order of increasing affinity of monovalent cations for montmorillonite and illite was generally Na < K < NH_4 < Cs, arguing that the strong retention of Cs may be the result of the more covalent nature of the Cs–clay interaction compared to the electrostatic interactions of the other ions.

7.8.1 Contractive and Expansive Forces

At the molecular scale, most K fixation by 2:1 layer silicates can be thought of most simply as the force of lattice contraction exceeding the forces of lattice expansion with K^+ ions in the interlayer space (Fig. 7.7). Contraction is a result of the attraction of negatively charged sites on the silicate surface for the positively charged cation. Expansion occurs when the layer charge is insufficient to overcome the energies of hydration of the cations in the interlayer (Hurst and Jordine 1964; Kittrick 1966; Kaufhold and Dohrmann 2010). The repulsion of the strong dipoles of the oxygen ions that are fixed in place on the basal planes of the minerals also contribute to repulsion of the layers from one another.

Clay layers may also collapse when there are no interlayer molecules or ions (i.e., water, metal hydroxyl complexes, organic cations, and molecules) to prevent the approach of adjacent layers close enough for van der Waals forces of attraction to exceed the forces of expansion. The strength of van der Waals forces lies in the proximity of clay layers to each other, and the proximity is controlled by the coulombic force of attraction of interlayer cations for interlayer surfaces. The greater the attraction, the closer the cation is held to the interlayer surface. The closer the cation is held, the closer an adjacent clay layer can approach. The closer adjacent clay layers become, the stronger van der Waals attractive forces become. When van der Waals attractive forces exceed the forces of expansion, the clay layers will collapse. Hence, cations with high energies of hydration and large hydrated ionic radii (e.g., Ca^{2+} and Mg^{2+}) are not appreciably fixed because it is difficult to overcome the distance that these hydrated cations place between clay layers and the energy with which they hold onto their surrounding water molecules (Fig. 7.7).

The coulombic force of attraction of interlayer surfaces for interlayer cations is a function of layer charge, charge location, and bond geometries in the layer silicate

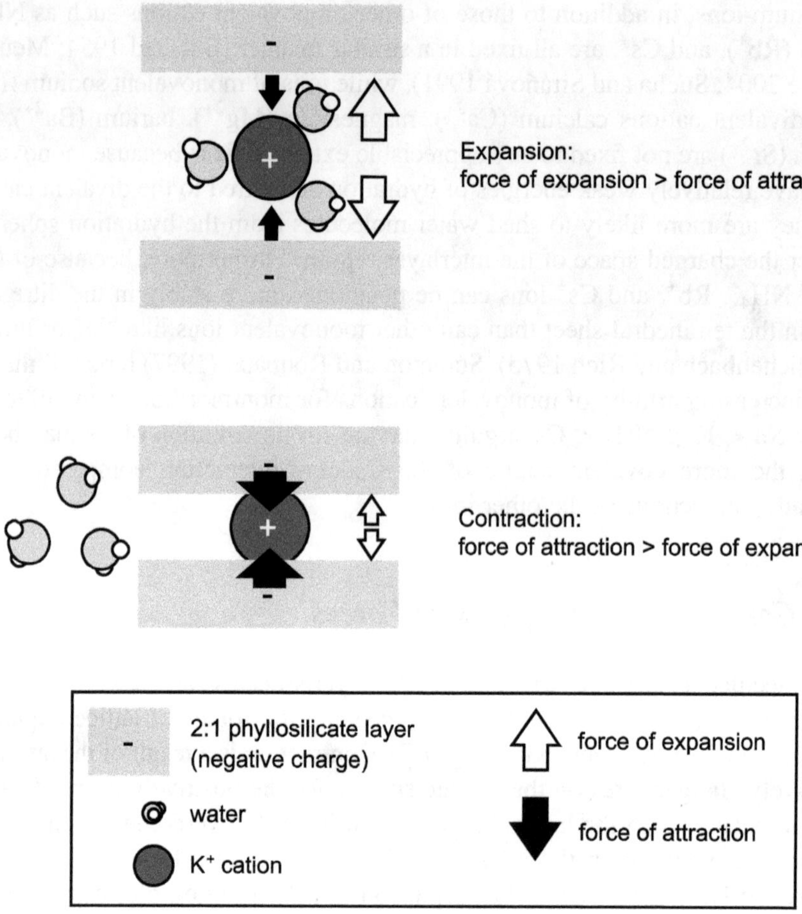

Fig. 7.7 Conceptual model of how the force of attraction and force of expansion contributes to contraction or expansion of 2:1 phyllosilicate layers. (Kaufhold and Dohrmann 2010)

crystal. Increased layer charge equates with increased attraction between interlayer cations and interlayer surfaces. Tetrahedrally derived negative charge is more effective than octahedrally derived negative charge at attracting interlayer cations as a result of its proximity to the basal layer surface. Similarly, bond geometries, particularly the hydroxyl group orientations in the octahedral sheet that are largely a function of cationic occupancies in the octahedral sheet, affect the attraction of interlayer cations for interlayer surfaces by controlling how closely interlayer cations can approach negative layer charge sites.

7.8.2 Factors Affecting Potassium Fixation and Release

Soil mineralogy is the key to understanding K fixation and release because only 2:1 layer silicates fix and release K in interlayer positions. Potassium can be fixed by mica weathering products, i.e., vermiculites and high-charge smectites (Barshad 1951, 1954, Ranjha et al. 1990, Rich 1968, Martin and Sparks 1985). Some authors indicate that K can be fixed by micas per se, but this is not likely. Micas are defined by having a high layer charge (near 1 mol of charge per 10-oxygen formula unit). Once the layer charge declines to a point where K can be reversibly or temporarily "fixed," the mineral is no longer considered a true mica, although it has been derived from mica. These are partially weathered micas that cannot be distinguished from primary micas as previously discussed. Where K-fixing minerals predominate in the clay-sized fraction of the soil, K fixation is often related to the clay content of the soil (Shaviv et al. 1985). However, K fixation has also been documented in silt-sized and even very fine sand-sized vermiculite and hydrobiotite (Harris et al. 1988; Murashkina et al. 2007a, b). The effects of soil type and soil horizon on K fixation occur primarily because the amounts and types of 2:1 layer silicates vary among soil types and soil horizons. Particle size distribution and possible artifacts related to sample grinding are important for assessing the degree of K fixation because they affect the accessibility of interlayer spaces. Smaller particles generally mean easier access to interlayer spaces. In general, however, smaller particles mean more edge area per particle and therefore easier access of K to interlayer spaces where it might be fixed.

The degree to which 2:1 layer silicates fix K is largely a function of layer charge and the distribution of that charge in the octahedral and tetrahedral sheets. Layer charge affects the electrostatic attraction of the layers for K^+ ions (Rich 1968). Other factors being equal, greater layer charge is often correlated with greater potential for fixation (Barshad 1954; Bouabid et al. 1991). Murashkina et al. (2007a) speculated that K fixation in some soils dominated by smectite may be due to high-charge smectites that are transitional to vermiculite. Charge location, however, also influences K fixation. Potassium-fixation capacity has been found to be well correlated with isomorphic substitution in tetrahedral sheets, but poorly correlated with isomorphic substitution in octahedral sheets (Bouabid et al. 1991). This is likely because negative charge originating in the tetrahedral sheet is closer to K^+ ions in the interlayer space (Reid-Soukup and Ulery 2002).

In acidic soils, the presence of hydroxy Al interlayers in vermiculite and smectite may affect K fixation (Saha and Inoue 1998). The hydroxy Al interlayers act as obstructions between 2:1 layers that restrict the collapse of the interlayer space around the K^+ ions. Blockage of the surface results in a decrease in cation exchange capacity and causes K to become more exchangeable and less likely to be fixed. In addition, hydroxy interlayers may slow the entry and exit of exchanging cations.

Due to the impact of layer charge on K fixation, the redox state of structural Fe in 2:1 layer silicates can also influence K fixation. The chemical reduction of structural Fe in both smectites and vermiculites has been shown to lead to increases in negative layer charge and K fixation. Furthermore, reduced Fe (i.e., Fe^{2+}) in the tetrahedral sheet appears to have a greater impact on K fixation than reduced Fe in the octahedral sheet (Chen et al. 1987; Dong et al. 2003; Favre et al. 2006; Florence et al. 2017; Stucki et al. 1984, 2000). This is likely due to increased coulombic attraction

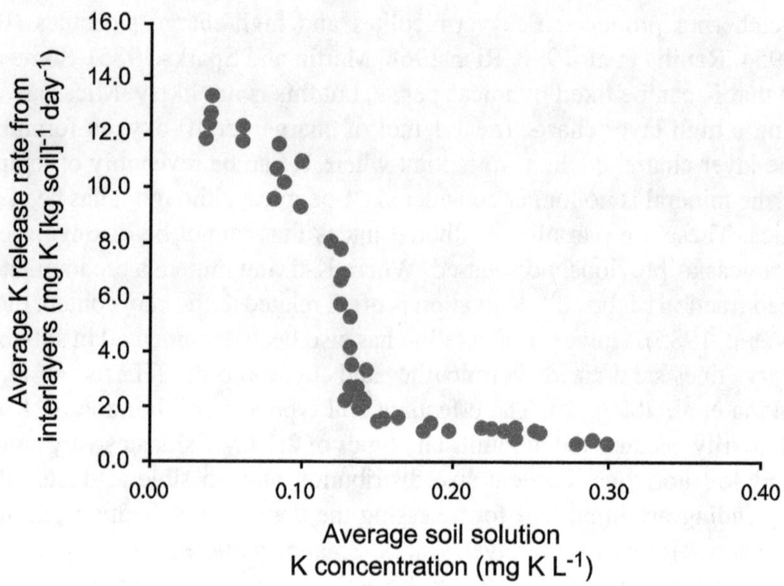

Fig. 7.8 The effect of solution K concentration on the rate of release of nonexchangeable K from a moderately weathered Luvisol derived from loess; reproduced from Hinsinger (2002) and adapted from Springob and Richter (1998)

between interlayer surfaces and K^+ ions, facilitating interlayer dehydration and collapse.

The oxidation of structural Fe^{2+} in the octahedral sheets of micas may lead to interlayer expansion and K release. Conversely, the reduction of structural Fe^{3+} in expanded 2:1 layer silicates can lead to interlayer collapse and K fixation (Scott and Amonette 1988). Note that the oxidation of structural Fe in micas can either stabilize or destabilize interlayer K, depending on both the pH of the environment and on whether oxidation leads to the ejection of the Fe atoms from the octahedral sheet (Thompson and Ukrainczyk 2002). When the oxidation of octahedral Fe in biotite, for example, leads to Fe ejection, nearby hydroxyl ions tend to orient toward the empty octahedral site, allowing interlayer K ions to nest closer into the ditrigonal cavity and subsequently to be held more tenaciously (Barshad and Khishk 1970). Although redox-driven changes in K fixation and release might explain some temporal fluctuations in soil exchangeable K, more work needs to be done to further test the role of Fe redox processes on K fixation and release.

Release of interlayer K is facilitated by the expansion of interlayer spaces when K is replaced by cations with larger hydrated radii (e.g., Ca^{2+} and Mg^{2+}), as illustrated in Fig. 7.2. Adequate soil moisture and low solution K concentrations are prerequisites for the release of interlayer K during the weathering process, with the key role played by solution K concentration, as demonstrated in Fig. 7.8 (Springob and Richter 1998; Hinsinger 2002). Key considerations in assessing the potential contribution of released interlayer K for plant uptake are the following: (1) release of K potentially bioavailable to plants is limited to the effective diffusion distance around roots and root hairs—i.e., the zones of greatest solution K depletion correspond to the rhizosphere, which typically comprises less than about 3% of the soil volume (Hinsinger et al. 2005); and (2) release of K is also favored by a high concentration of H^+ ions (i.e., low pH). This is explained in greater detail in Chap. 4.

7.9 Interpreting "Exchangeable Potassium"

While the strengths and weaknesses of specific soil test procedures are discussed in detail in Chap. 8, the diagnostic tests most commonly associated with the measurement of surface-adsorbed K is termed "exchangeable K." **Exchangeable K** is the *K extracted from a sample of soil* via *cation exchange using a solution of a specified composition under a specific set of conditions*. The quantity of exchangeable K extracted from a given soil therefore varies according to the extracting solution and conditions used. Conditions typically involve displacement of K^+ by NH_4^+ ions. Exchangeable K is generally considered to be bioavailable to plants.

The continuum of sorption positions in soil minerals and organic matter is typically distinguished by the affinity for K and the speed of the chemical reactions with simulated soil solutions in soil testing procedures. Accurate quantification of the K in the surface-adsorbed pool can be difficult, especially in soils where there are significant amounts of weathered mica grains or secondary phyllosilicate minerals. In soils where the charge is predominantly associated with organic matter or kaolinite, exchangeable K determinations are a quantification of surface-adsorbed K. In such contexts, this measure can be a reliable index of the capacity of the soil to supply bioavailable K to plants over an extended period (Obreza and Rhoads 1988).

A common misconception is that K exchangeable with NH_4^+ captures only surface-adsorbed K. When soils contain significant amounts of phyllosilicate minerals with hydrated interlayers (e.g., smectite or vermiculite), NH_4^+ will exchange not only with surface-adsorbed K but also with a variable amount of interlayer K. The presence of water between individual layers allows cations to diffuse in and out of interlayer positions, making cation exchange possible. Ammonium is especially able to replace interlayer K in hydrated layer silicates because its size and hydration energy are similar to those of K.

The rates of exchange with NH_4^+ are variable and difficult to predict, depending on the minerals involved, the proportions of planar, edge, wedge and interlayer adsorption positions, and thus the selectivity of adsorption sites for K (Mengel 1985; Sparks 1987; Lin 2010; Römheld and Kirkby 2010). For example, the rate of K desorption from kaolinite and smectite is usually rapid (Sparks and Jardine 1984), while from vermiculite and micaceous minerals (i.e., illite) it tends to be much slower (Sparks 1987). In these situations, Carey and Metherell (2003), among others, hypothesized that the two-stage extraction of K by tetraphenyl boron could be related to the e-position, including wedge zones (rapid) and i-positions (slow).

In most situations, the volume of soil immediately adjacent to roots is small compared to the entire soil volume, so there is unlikely to be a simple, quantitative relationship between K that is exchangeable with ammonium in a soil test procedure and that acquired by plants during a growing season. However, even in experiments in which soil volumes are small and root density is very high, as is typically the case in nutrient depletion assays, not all ammonium-exchangeable K is bioavailable to plants. Various authors have proposed the concept of a lower accessible limit of the exchangeable K pool, below which plants may not be able to extract K (e.g., Tabatabai and Hanway 1969; Schneider et al. 2016). Termed the "plant minimal exchangeable K" this term could perhaps be more simply described as the "minimal plant-accessible K." The size of this fraction increases with increasing clay and CEC (measured at the pH of the soil (Schneider et al. 2016)), suggesting that as clay content increases the number of K-specific adsorption sites also increases. It may be hypothesized that this fraction of the measured exchangeable K could be a useful predictor of a "plant minimal solution K concentration" below which plants are unable to take up K.

7.10 Mineral Transformations

7.10.1 Reversible Changes in Interlayer Potassium

Long-term K removal in cropping systems can result in accelerated degradation of micas and an increase in secondary phyllosilicate minerals such as vermiculite and smectite (Barré et al. 2008). Conversely, long-term fertilization with K has been shown to increase the apparent concentration of mica-like minerals in soils, even though minerals with layer charges high enough to classify as micas were not being formed (Scheffer et al. 1960; Ross et al. 1989; Meunier and Velde 2004; Barré et al. 2008). Ross et al. (1989) reported such observations after only 4 years of high K fertilization. Indeed, changes in the proportion of collapsed interlayers in clay minerals of rhizosphere soil materials over the course of a single growing season have been reported by Barré et al. (2007b) and by Adamo et al. (2016).

Changes consistent with these reports have been reported from both cropped and pasture systems. For example, a 15-year study of alfalfa (*Medicago Sativa* L.) cropping on a loess-derived soil (De-Cheng et al. 2011) showed topsoil exchangeable K decreased due to alfalfa forage uptake and removal, but total K in the topsoil increased, likely due to K uptake and translocation by the alfalfa roots from deep in the soil. A similar phenomenon of enrichment of illite-like clay in temperate region soils has been hypothesized to result from the redistribution of K from deeper soil horizons to the topsoil horizon through root uptake and plant residue deposition (Barré et al. 2007a, 2009).

7.10.2 *Implications for Building and Depleting Soil Fertility*

The fixation and release of K from interlayer positions in phyllosilicate minerals represent an important buffering system for plant-available reserves of soil K where those minerals occur. Release of K from interlayers can be particularly important under continuous cropping when rates of added K have been inadequate to replace K removed in harvested crop biomass. There is also evidence that fixation of K in these minerals stores K when it is added to soils at rates that exceed removal or when it is redistributed to surface horizons by crop residue deposition, suggesting a degree of resilience that can be exploited to manage soil fertility in the longer term.

The immediate challenge to the sustainable exploitation of soil K is to develop simple diagnostic indicators of the likely K status of these buffered systems. Candidates for this approach would include assays such as those developed by Cassman et al. (1990) and refined by Murashkina et al. (2007a), in which the extent of fixation of added K fertilizer was ascertained. These assays would not only provide information needed to develop an effective K fertilizer application strategy (discussed in Chap. 12), but could also provide an indicator of the extent of K depletion of the interlayer buffer system. Such knowledge could allow the K buffer system to be exploited when K status was moderate or high, but replenished when K status was low to avoid irreversible degradation of these minerals. Another useful approach could be to develop region- or soil-specific predictions of the minimal plant-accessible K that are based on an easily determined parameter like clay content or CEC.

7.11 Short-Term Transformations in the Rhizosphere

The low concentration of K in soil solutions is thought to be an important driver for the release of K through mineral dissolution and interlayer K release. Plant roots play a key role in depleting soil solution K concentrations in the rhizosphere. Hinsinger (1998) reported that solution K concentrations can decrease by 2–3 orders of magnitude to as little as 2–3 μM in the vicinity of plant roots, and at these concentrations the release of structural and interlayer K can occur at high rates (Springob and Richter 1998, Fig. 7.8). However, it is also clear that this dissolution process is accentuated by root exudates and other rhizosphere characteristics (e.g., low pH). For example, plants can release conjugate bases of a variety of organic acids and those anions complex Al, contributing to dissolution of aluminosilicate framework minerals like feldspars (Chap. 4). Roots of maize (*Zea mays* L.) and rape (*Brassica napus* L.) can release citrate and malate (Hoffland et al. 1989; Pellet et al. 1995), while roots of bak choi (*Brassica rapa* ssp.) (Wang et al. 2000) and radish (*Raphanus raphanistrum* ssp.) (Zhang et al. 1997) have been shown to release tartrate. These ions accelerate the dissolution process at mineral surfaces by complexing and solubilizing Al^{3+} from minerals. Zörb et al. (2014) suggested that the generation of these exudates is driven by the soil solution K concentration and could be initiated when K concentrations fall to less than 10–20 μM. Plants also release hydronium ions, which lower the pH in the zone next to the root and contribute to weathering by creating local charge imbalances when they form new bonds with oxygen ions at mineral surfaces (Brantley 2003). Much more detailed coverage of these and other rhizosphere transformations is presented in Chap. 4.

7.12 Nonexchangeable Potassium as a Functional Pool

Nonexchangeable K is *soil K that is not measured by soil tests that rely on exchange or displacement of K by another cation.* The current lack of analytical techniques that can successfully differentiate between K that resides as structural K in feldspars; interlayer pools in micas, partially weathered micas and secondary clay minerals; or in neoformed K minerals in fertilized soils (pools 9–13 in Fig. 7.1) presents real problems for predicting the size and behavior of sources of K that are potentially bioavailable to plants. Collectively, these pools can make a major contribution to the K uptake by crops. Hinsinger (2006) estimated that this aggregation of K pools could release up to 100 kg K ha^{-1} $year^{-1}$, which is a significant proportion of plant K demand in many agricultural systems. This is substantiated by the K budgets in the treatments without K fertilization of long-term fertilizer trials in Europe (Fig. 7.6). The contributions of interlayer and structural K to crop uptake will obviously vary with soil type and mineralogy. For example, Moody and Bell (2006) demonstrated a significant contribution of nonexchangeable K (measured as the difference between exchangeable K and K extracted using sodium tetraphenyl borate) to plant uptake in some Vertisols, but effectively no contribution in others. Subsequent work indicated that the majority of the nonexchangeable K taken up by different plant species in the 15–30 soils studied was from dissolution of structural K rather than K release from near-edge interlayer positions. The apparent lack of significant release of K from near-edge interlayer positions in this study may reflect the highly weathered nature of the Australian soils. Nonexchangeable, but near-edge, interlayer K could make a significant contribution to plant uptake from less weathered soils in temperate regions where partially weathered micas or vermiculite are present (Barré et al., 2007b, 2008).

The relative strengths and weaknesses of the main analytical methods used to quantify nonexchangeable K (extraction with boiling HNO_3 or sodium tetraphenyl borate) are discussed in detail in Chap. 8. However, given the inability of current analytical soil test methods to differentiate between these nonexchangeable K sources, that is, structural and near-edge interlayer K, Hinsinger (2006) has suggested they could be referred to collectively as "nonexchangeable K." This term has merit, as it recognizes the current limitations and uncertainty surrounding diagnostic testing and the variability in interpretation of soil test results across soil types with differing mineralogy.

7.13 Classifying Soils According to Their Potassium Behavior

Soil Taxonomy, the system of soil classification used in the United States (Soil Survey Staff 2014) groups soils into classes that have similar behavior, use and management, and productivity. The lowest level of classification is the family level, which emphasizes soil properties relevant to the potential use and management of the soil. Differentiating criteria at the family level include particle size classes, mineralogy classes, and cation activity classes. In the United States, *Soil Taxonomy* is the basis for soil surveys, which are made by the National Cooperative Soil Survey. These surveys are typically intended for general agriculture and land use planning and are made at scales ranging from 1:12,000 to 1:31,680. Although soil surveys are not intended to be used for making fertilizer recommendations, they could be used to group and identify soils with potential for K fixation or release. This would help researchers identify sites for future K fertility trials or help producers and consultants identify soils that need K fertility recommendations designed to compensate for K fixation or release. Currently, *Soil Taxonomy* does not include specific classification criteria for recognizing soils with potential for K fixation or release. However, the system could be modified by extrapolating mineralogical and K-fixation data for horizons at well-studied sites to other land areas.

Effective recognition of K-fixing or K-releasing soils in *Soil Taxonomy* would require knowledge of the particle size distribution, mineralogy, and cation exchange capacity of multiple soil horizons, including the surface horizons that are not currently used in the system's mineralogy classes at the family level. An assessment of the soil's cation exchange capacity (e.g., before and after heating a K-saturated sample of clay fractionated from the soil, as described by Ransom et al. (1988)) could be combined with the properties listed above to identify soils with a significant potential for buffering plant-available K. Additional work would be required to see if a similar procedure could be used to recognize soils with a potential for K-fixation using other soil classification systems such as the World Reference Base for Soil Resources and the Australian Soil Classification System. However, neither soil classification system includes clay mineralogy and cation exchange capacity as classification criteria used to predict soil responses for use and management.

Even without formal revisions to soil classification systems, informal, local models to predict the K-supply characteristics and locations of K-fixing or K-releasing soils could be useful. For example, O'Green et al. (2008) developed a method to classify the K-supply characteristics of soils in vineyards of the Lodi Winegrape District of California in the United States. Their approach correlated readily available soil survey information about parent material age, mineralogy, and weathering intensity with levels of exchangeable K and the K-fixation potential for soils in the region. While their model was specifically designed for this region or for wine-grape cropping systems, it is innovative and the concepts could be applied elsewhere.

7.14 Lessons Learned from Long-Term Experiments

Exhaustive cropping currently remains the most effective way of quantifying the bioavailable fraction of a nonexchangeable K pool, although these results may vary considerably with crop species (and perhaps even genotypes) and experimental conditions. However, long-term experiments offer an opportunity to explore the importance of nonexchangeable soil K reserves, i.e., pools 9–12 in Fig. 7.1. An example that quantifies the long-term impact of different K inputs/balances on soil K content in different K pools at the end of a monitoring period is the study of Blake et al. (1999), with this extended to cover other long-term experiments in France, Germany, and Poland by Hinsinger (2002, 2006, Fig. 7.6). Others have measured the change in soil K stocks in different pools over a defined period between treatment sequences or soil sampling events, e.g., Kaminski et al. (2010) and Madaras et al. (2014). Still others have reported qualitative changes in secondary phyllosilicate minerals in response to either K fertilizer use or unbalanced K removal (Barré et al. 2007b, 2008). Collectively, such studies have been able to demonstrate the dynamic nature of soil K, the importance of mineralogy on K fixation and release (the latter by either desorption or dissolution) and the impact of net K balance on the direction of K fluxes.

Many of these studies cannot conclusively identify the pools of soil K that contribute to the overall K balance of the systems investigated. Some are limited by their soil sampling strategy, which typically focusses on the cultivated layer (0–25 cm) or even the upper 10 cm in minimum and zero-tillage systems. The sources and sinks of K may well be from deeper soil layers, especially in tropical and subtropical environments where soil temperatures do not limit root exploration and access to subsoil moisture. Nutrient stores in deeper layers are keys to productive agricultural systems (Bell et al. 2009), but changes in soil K in these layers are typically not measured (Prasad 2009). Pradier et al. (2017) have measured significant changes of K pools at considerable depth, down to 4 m when sampling the rhizosphere of eucalypt trees in Brazil. They observed an increase of exchangeable K in the rhizosphere, possibly related to weathering of K-bearing minerals, as speculated by these authors. Even those that have endeavored to account for changes in the amount of available K in different profile layers have found it difficult to precisely identify the K pool acting as the predominant source or sink for K. This relates to the limitations of existing soil test methods, discussed in Chap. 8, which are effectively limited to identifying changes in exchangeable and nonexchangeable K. Both measures will access different K pools in soils with different mineralogy, making extrapolation to other soil types and cropping systems challenging.

7.15 Prognosis

A better understanding of the allocation of K bioavailability to plants among different functional pools in agricultural soils will allow the development of more defensible and economically justifiable fertilizer recommendations. This will particularly apply to soils where there are significant amounts of nonexchangeable K in either (or both of) three-dimensional framework or layer silicate minerals. The development of such an understanding will require the use of a combination of diagnostic soil K tests, including those that will provide an assessment of the likely availability of applied K fertilizer. While this will increase the costs of soil testing, the greater certainty provided around understanding the K status of the particular field under management will add considerable confidence to fertilizer decisions. Further research is clearly required on this topic, as is research on soil sampling protocols that allow quantification of K status in the soil layers from which the crop acquires significant quantities of K.

References

Adamo P, Barré P, Cozzolino V, Di Meo V, Velde B (2016) Short term clay mineral release and re-capture of potassium in a *Zea mays* field experiment. Geoderma 264:54–60. https://doi.org/10.1016/j.geoderma.2015.10.005

Barber SA (1981) Soil chemistry and the availability of plant nutrients. In: Stelly M (ed) Chemistry in the soil environment, Special Publication No. 40. American Society of Agronomy, Madison, pp 1–12. https://doi.org/10.2134/asaspecpub40.c1

Barber SA (1995) Soil nutrient bioavailability: a mechanistic approach, 2nd edn. Wiley, New York

Barré P, Velde B, Abbadie L (2007a) Dynamic role of "illite-like" clay minerals in temperate soils: facts and hypotheses. Biogeochemistry 82(1):77–88. https://doi.org/10.1007/s10533-006-9054-2

Barré P, Velde B, Catel N, Abbadie L (2007b) Soil-plant potassium transfer: impact of plant ac-tivity on clay minerals as seen from X-ray diffraction. Plant Soil 292:137–146

Barré P, Montagnier C, Chenu C, Abbadie L, Velde B (2008) Clay minerals as a soil potassium reservoir: observation and quantification through X-ray diffraction. Plant Soil 302:213–220

Barré P, Berger G, Velde B (2009) How element translocation by plants may stabilize illitic clays in the surface of temperate soils. Geoderma 151(1–2):22–30. https://doi.org/10.1016/j.geoderma.2009.03.004

Barrow NJ (1966) Nutrient potential and capacity. II. Relationship between potassium potential and buffering capacity and the supply of potassium to plants. Aust J Agric Res 17:849–861

Barshad I (1948) Vermiculite and its relation to biotite as revealed by base exchange reactions, x-ray analyses, differential thermal curves, and water content. Am Mineral 33(11–12):655–678. http://www.minsocam.org/ammin/AM33/AM33_655.pdf. Accessed 11 May 2020

Barshad I (1951) Cation exchange in soils: I. Ammonium fixation and its relation to potassium fixation and to determination of ammonium exchange capacity. Soil Sci 72(5):361–371

Barshad I (1954) Cation exchange in micaceous minerals: II. Replaceability of ammonium and potassium from vermiculite, biotite, and montmorillonite. Soil Sci 78(1):57–76

Barshad I, Khishk FM (1970) Factors affecting potassium fixation and cation exchange capacities of soil vermiculite clays. Clay Clay Miner 18:127–113

Bell MJ, Moody PW, Harch GR, Compton B, Want PS (2009) Fate of potassium fertilisers applied to clay soils under rainfed grain cropping in south-east Queensland, Australia. Aust J Soil Res 47(1):60–73. https://doi.org/10.1071/SR08088

Blake L, Mercik S, Koerschens M, Goulding KWT, Stempen S, Weigel A, Poulton PR, Powlson DS (1999) Potassium content in soil, uptake in plants and the potassium balance in three European long-term field experiments. Plant Soil 216(1–2):1–14. https://doi.org/10.1023/A:1004730023746

Bolt GH, Sumner ME, Kamphorst A (1963) A study of the equilibria between three categories of potassium in an illitic soil. Soil Sci Soc Am J 27:294–299. https://doi.org/10.2136/sssaj1963.03615995002700030024x

Bouabid R, Badraoui M, Bloom PR (1991) Potassium fixation and charge characteristics of soil clays. Soil Sci Soc Am J 55(5):1493–1498. https://doi.org/10.2136/sssaj1991.03615995005500050049x

Bourg IC, Sposito G (2011) Molecular dynamics simulations of the electrical double layer on smectite surfaces contacting concentrated mixed electrolyte (NaCl-$CaCl_2$) solutions. J Colloid Interf Sci 360(2):701–715. https://doi.org/10.1016/j.jcis.2011.04.063

Brantley SL (2003) Reaction kinetics of primary rock-forming minerals under ambient conditions. In: Holland HD, Turekian KK (eds) Treatise on geochemistry, vol 5. Elsevier, Amsterdam, pp 73–117. https://doi.org/10.1016/B0-08-043751-6/05075-1

Carey PL, Metherell AK (2003) Rates of release of non-exchangeable potassium in New Zealand soils measured by a modified sodium tetraphenyl-boron method. New Zeal J Agric Res 46 (3):185–197. https://doi.org/10.1080/00288233.2003.9513546

Cassman KG, Bryant DC, Roberts BA (1990) Comparison of soil test methods for predicting cotton response to soil and fertilizer potassium on potassium fixing soils. Commun Soil Sci Plan 21 (13–16):1727–1743. https://doi.org/10.1080/00103629009368336

Chen SZ, Low PF, Roth CB (1987) Relation between potassium fixation and the oxidation state of octahedral iron. Soil Sci Soc Am J 51(1):82–86. https://doi.org/10.2136/sssaj1987.03615995005100010017x

De-Cheng LI, Velde B, Feng-Min LI, Zhang G-L, Zhao M-S, Huang L-M (2011) Impact of long-term alfalfa cropping on soil potassium content and clay minerals in a semi-arid loess soil in China. Pedosphere 21(4):522–531. https://doi.org/10.1016/S1002-0160(11)60154-9

Dong H, Kostka JE, Kim J (2003) Microscopic evidence for microbial dissolution of smectite. Clay Clay Miner 51(5):502–512. https://doi.org/10.1346/CCMN.2003.0510504

Du Z, Zhou J, Wang H, Du C, Chen X (2006) Potassium movement and transformation in an acid soil as affected by phosphorus. Soil Sci Soc Am J 70:2057–2064. https://doi.org/10.2136/sssaj2005.0409

Evangelou VP, Wang J, Phillips RE (1994) New developments and perspectives on soil potassium quantity/intensity relationships. Adv Agron 52:173–227. https://doi.org/10.1016/s0065-2113(08)60624-0

Favre F, Stucki JW, Boivin P (2006) Redox properties of structural Fe in ferruginous smectite. A discussion of the standard potential and its environmental implications. Clay Clay Miner 54 (4):466–472. https://doi.org/10.1346/CCMN.2006.0540407

Fenter P, Teng H, Geissbühler P, Hanchar JM, Nagy KL, Sturchio NC (2000) Atomic-scale structure of the orthoclase (001)-water interface measured with high-resolution x-ray reflectivity. Geochim Cosmochim Ac 64(21):3663–3673. https://doi.org/10.1016/S0016-7037(00)00455-5

Florence AM, Ransom MD, Mengel DB (2017) Potassium fixation by oxidized and reduced forms of phyllosilicates. Soil Sci Soc Am J 81(5):1247–1255. https://doi.org/10.2136/sssaj2016.12.0420

Fogler HS, Lund K (1975) Acidization III – the kinetics of the dissolution of sodium and potassium feldspar in HF/HCl acid mixtures. Chem Eng Sci 30(11):1325–1332. https://doi.org/10.1016/0009-2509(75)85061-5

Harris WG, Hollien KA, Yuan TL, Bates SR, Acree WA (1988) Nonexchangeable potassium associated with hydroxy-interlayered vermiculite from coastal plain soils. Soil Sci Soc Am J 52 (5):1486–1492. https://doi.org/10.2136/sssaj1988.03615995005200050053x

Hinsinger P (1998) How do plant roots acquire mineral nutrients? Chemical processes involved in the rhizosphere. Adv Agron 64:225–265. https://doi.org/10.1016/S0065-2113(08)60506-4

Hinsinger P (2002) Potassium. In: Lal R (ed) Encyclopedia of soil science. Marcel Dekker, New York, pp 1354–1358

Hinsinger P (2006) Potassium. In: Lal R (ed) Encyclopedia of soil science, vol 2, 2nd edn. Taylor & Francis, New York, pp 1354–1358

Hinsinger P, Gobran GR, Gregory PJ, Wenzel WW (2005) Rhizosphere geometry and heterogeneity arising from root-mediated physical and chemical processes. New Phytol 168:293–303

Hoffland E, Findenegg GR, Nelemans JA (1989) Solubilization of rock phosphate by rape. II Local root exudation of organic acids as a response to P-starvation. Plant Soil 113(2):161–165. https://doi.org/10.1007/BF02280176

Huang PM, Lee SY (1969) Effect of drainage on weathering transformations of mineral colloids of some Canadian prairie soils. In: Proceedings of the international clay conference, Tokyo, 5–10 September 1969

Hurst CA, Jordine ESA (1964) Role of electrostatic energy barriers in expansion of lamellar crystals. J Chem Phys 41(9):2735–2745. https://doi.org/10.1063/1.1726345

Jackson ML (1963) Interlayering of expansible layer silicates in soils by chemical weathering. Clay Clay Miner 11:29–46

Kaminski J, Moterle DF, Rheinheimer DS, Gatiboni LC, Brunetto G (2010) Potassium availability in a Hapludalf soil under long term fertilization. Rev Bras de Ciênc Solo 34(3):783–791. https://doi.org/10.1590/S0100-06832010000300020

Kaufhold S, Dohrmann R (2010) Stability of bentonites in salt solutions II. Potassium chloride solution – initial step of illitization? Appl Clay Sci 49:98–107. https://doi.org/10.1016/j.clay.2010.04.009

Kittrick JA (1966) Forces involved in ion fixation by vermiculite. Soil Sci Soc Am Pro 30 (6):801–803. https://doi.org/10.2136/sssaj1966.03615995003000060040x

Kovar JL, Barber SA (1990) Potassium supply characteristics of thirty-three soils as influenced by seven rates of potassium. Soil Sci Soc Am J 54(5):1356–1361. https://doi.org/10.2136/sssaj1990.03615995005400050026x

Kuhlmann H, Barraclough PB (1987) Comparison between the seminal and nodal root systems of winter wheat in their activity for N and K uptake. J Plant Nutr Soil Sci 150(1):24–30. https://doi.org/10.1002/jpln.19871500106

Lee J, Park S-M, Jeon E-K, Baek K (2017) Selective and irreversible adsorption mechanism of cesium on illite. Appl Geochem 85:188–193. https://doi.org/10.1016/j.apgeochem.2017.05.019

Lin Y-H (2010) Effects of potassium behaviour in soils on crop absorption. Afr J Biotechnol 9 (30):4638–4643 https://www.ajol.info/index.php/ajb/article/view/82964. Accessed 11 May 2020

Lindsay WL, Frazer AW, Stephenson HF (1962) Identification of reaction products from phosphate fertilizers in soils. Soil Sci Soc Am Proc 26:446–452. https://doi.org/10.2136/sssaj1962.03615995002600050013x

Madaras M, Koubová M, Smatanová M (2014) Long-term effect of low potassium fertilization on its soil fractions. Plant Soil Environ 60(8):358–363. https://doi.org/10.17221/290/2014-PSE

Martin HW, Sparks DL (1985) On the behavior of nonexchangeable potassium in soils. Commun Soil Sci Plan 16(2):133–162. https://doi.org/10.1080/00103628509367593

Mei L, Tao H, He C, Xin X, Liao L, Wu L, Lv G (2015) Cd^{2+} exchange for Na^+ and K^+ in the interlayer of montmorillonite: experiment and molecular simulation. J Nanomater 2015:925268. https://doi.org/10.1155/2015/925268

Mengel K (1985) Dynamics and availability of major nutrients in soils. In: Stewart BS (ed) Advances in soil science, vol 2. Springer, New York, pp 66–131. https://www.springer.com/us/book/9781461295587. Accessed 11 May 2020

Meunier A, Velde B (2004) Illite: origins, evolution, and metamorphism. Springer, Berlin

Moody PW, Bell MJ (2006) Availability of soil potassium and diagnostic soil tests. Aust J Soil Res 44(3):265–275. https://doi.org/10.1071/SR05154

Murashkina MA, Southard RJ, Pettygrove GS (2007a) Potassium fixation in San Joaquin Valley soils derived from granitic and nongranitic alluvium. Soil Sci Soc Am J 71(1):125–132. https://doi.org/10.2136/sssaj2006.0060

Murashkina MA, Southard RJ, Pettygrove GS (2007b) Silt and fine sand fractions dominate K fixation in soils derived from granitic alluvium of the San Joaquin Valley, California. Geoderma 141:283–293. https://doi.org/10.1016/j.geoderma.2007.06.011

Niebes JF, Hinsinger P, Jaillard B, Dufey JE (1993) Release of non exchangeable potassium from different size fractions of two highly K-fertilized soils in the rhizosphere of rape (*Brassica napus* cv Drakkar). Plant Soil 155(156):403–406

Öborn I, Andrist-Rangel Y, Askekaard M, Grant CA, Watson CA, Edwards AC (2005) Critical aspects of potassium management in agricultural systems. Soil Use Manag 21(s1):102–112. https://doi.org/10.1111/j.1475-2743.2005.tb00414.x

Obreza TA, Rhoads FM (1988) Irrigated corn response to soil-test indices and fertilizer nitrogen, phosphorus, potassium, and magnesium. Soil Sci Soc Am J 52:701–706. https://doi.org/10.2136/sssaj1988.03615995005200030020x

O'Geen AT, Pettygrove S, Southard RJ, Minoshima H, Verdegaal PS (2008) Soil-landscape model helps predict potassium supply in vineyards. Calif Agric 62(4):195–201. https://doi.org/10.3733/ca.v062n04p195

Pellet DM, Grunes DL, Kochian LV (1995) Organic acid exudation as an aluminum-tolerance mechanism in maize (*Zea mays* L.). Planta 196(4):788–795. https://doi.org/10.1007/BF01106775

Pradier C, Hinsinger P, Laclau JP, Pouillet JP, Guerrini IA, Goncalves JLM, Asensio V, Abreu-Junior CH, Jourdan C (2017) Rainfall reduction impacts rhizosphere biogeochemistry in eucalypts grown in a deep Ferralsol in Brazil. Plant Soil 414:339–354

Prasad R (2009) Potassium fertilization recommendations for crops need rethinking. Indian J Fert 5 (8):31–33

Ranjha AM, Jabbar A, Qureshi RH (1990) Effect of amount and type of clay minerals on potassium fixation in some alluvial soils of Pakistan. Pak J Agric Sci 27(2):187–192. https://www.pakjas.com.pk/papers/1260.pdf. Accessed 11 May 2020

Ransom MD, Bigham JM, Smeck NE, Jaynes WF (1988) Transitional vermiculite-smectite phases in Aqualfs of southwestern Ohio. Soil Sci Soc Am J 52(3):873–880. https://doi.org/10.2136/sssaj1988.03615995005200030049x

Rehm GW, Sorensen RC (1985) Effects of potassium and magnesium applied for corn grown on an irrigated sandy soil. Soil Sci Soc Am J 49:1446–1450. https://doi.org/10.2136/sssaj1985.03615995004900060023x

Reichenbach HGv, Rich CI (1975) Fine-grained micas in soils. In: Gieseking JE (ed) Soil components, vol 2: inorganic components. Springer, New York, pp 59–95. https://www.springer.com/us/book/9783642659195. Accessed 11 May 2020

Reid-Soukup DA, Ulery AL (2002) Smectites. In: Dixon JB, Schulze DG (eds) Soil mineralogy with environmental applications. SSSA Book Series, no 7. Soil Science Society of America, Madison, pp. 467–499. https://doi.org/10.2136/sssabookser7.c15

Rich CI (1968) Mineralogy of soil potassium. In: Kilmer VJ, Younts SE, Brady NC (eds) The role of potassium in agriculture. American Society of Agronomy, Madison, pp 79–108. https://doi.org/10.2134/1968.roleofpotassium.c5

Rich CI (1972) Potassium in soil minerals. In: potassium in soil: proceedings of the 9th colloquium of the international Potash Institute, Landshut, pp 15–31. https://www.ipipotash.org/udocs/potassium_in_soil.pdf. Accessed 11 May 2020

Römheld V, Kirkby EA (2010) Research on potassium in agriculture: needs and prospects. Plant Soil 335(1–2):155–180. https://doi.org/10.1007/s11104-010-0520-1

Ross GJ, Cline RA, Gamble DS (1989) Potassium exchange and fixation in some southern Ontario soils. Can J Soil Sci 69(3):649–661. https://doi.org/10.4141/cjss89-064

Sadusky MC, Sparks DL, Noll MR, Hendricks GJ (1987) Kinetics and mechanisms of potassium release from sandy Middle Atlantic coastal plain soils. Soil Sci Soc Am J 51:1460–1465. https://doi.org/10.2136/sssaj1987.03615995005100060011x

Saha UK, Inoue K (1998) Hydroxy-interlayers in expansible layer silicates and their relation to potassium fixation. Clay Clay Miner 46(5):556–566. https://doi.org/10.1346/CCMN.1998.0460509

Sautter EH (1964) Potassium-bearing feldspars in some soils of the sandhills of Nebraska. Soil Sci Soc Am J 28:709–710. https://doi.org/10.2136/sssaj1964.03615995002800050037x
Scheffer VF, Welte E, Reichenbach HGV (1960) Über den Kaliumhaushalt und Mineralbestand des Göttinger E-Feldes. Z Pflanz Bodenkunde 88(2):115–128. https://doi.org/10.1002/jpln.19600880203
Schneider A, Tesileanu R, Charles R, Sinaj S (2013) Kinetics of soil potassium sorption–desorption and fixation. Commun Soil Sci Plan 44(1–4):837–849. https://doi.org/10.1080/00103624.2013.749442
Schneider A, Augusto L, Mollier A (2016) Assessing the plant minimal exchangeable potassium of a soil. J Plant Nutr Soil Sci 179(4):584–590. https://doi.org/10.1002/jpln.201600095
Schulze DG (1989) An introduction to soil mineralogy. In: Dixon JB, Weed SB (eds) Minerals in soil environments. Soil Science Society of America, Madison, pp 1–34. https://doi.org/10.2136/sssabookser1.2ed
Scott AD, Amonette J (1988) Role of iron in mica weathering. In: Stucki JW, Goodman BA, Schwertmann U (eds) Iron in soils and clay minerals, series C: mathematical and physical sciences, vol 217. NATO ASI Series. Springer, New York, pp 537–624. https://www.springer.com/us/book/9789027726131. Accessed 11 May 2020
Scott AD, Smith SJ (1967) Visible changes in macro mica particles that occur with potassium depletion. Clay Clay Miner 15(1):357–373. https://doi.org/10.1346/CCMN.1967.0150138
Shaviv A, Mattigod SV, Pratt PF, Joseph H (1985) Potassium exchange in five southern California soils with high potassium fixation capacity. Soil Sci Soc Am J 49(5):1128–1133. https://doi.org/10.2136/sssaj1985.03615995004900050011x
Soil Survey Staff (2014) Keys to soil taxonomy, 12th edn. United States Department of Agriculture Natural Resources Conservation Service, Washington, DC. https://www.nrcs.usda.gov/wps/PA_NRCSConsumption/download?cid=stelprdb1252094&ext=pdf. Accessed 11 May 2020
Song SK, Huang PM (1988) Dynamics of potassium release from potassium-bearing minerals as influenced by oxalic and citric acids. Soil Sci Soc Am J 52(2):383–390. https://doi.org/10.2136/sssaj1988.03615995005200020015x
Sparks DL (1987) Potassium dynamics in soils. In: Stewart BS (ed) Advances in soil science, vol 6. Springer, New York, pp 1–63. https://www.springer.com/us/book/9781461291121. Accessed 11 May 2020
Sparks DL, Huang PM (1985) Physical chemistry of soil potassium. In: Munson RD (ed) Potassium in agriculture. American Society of Agronomy, Madison, pp 201–276. https://doi.org/10.2134/1985.potassium.c9
Sparks DL, Jardine PM (1984) Comparison of kinetic equations to describe potassium-calcium exchange in pure and in mixed systems. Soil Sci 138(2):115–122
Springob G, Richter J (1998) Measuring interlayer potassium release rates from soil materials. II. A percolation procedure to study the influence of the variable 'solute K' in the <1...10 uM range. J Plant Nutr Soil Sci 161(3):323–329. https://doi.org/10.1002/jpln.1998.3581610321
Staunton S, Roubaud M (1997) Adsorption of 137Cs on montmorillonite and illite: effect of charge compensating cation, ionic strength, concentration of Cs, K, and fulvic acid. Clay Clay Miner 45:251–260. https://doi.org/10.1346/CCMN.1997.0450213
Stucki JW, Golden DC, Roth CB (1984) Preparation and handling of dithionite-reduced smectite suspensions. Clay Clay Miner 32(3):191–197. https://doi.org/10.1346/CCMN.1984.0320306
Stucki JW, Wu J, Gan H, Komadel P, Banin A (2000) Effects of iron oxidation state and organic cations on dioctahedral smectite hydration. Clay Clay Miner 48(2):290–298. https://doi.org/10.1346/CCMN.2000.0480216
Šucha V, Širáňová V (1991) Ammonium and potassium fixation in smectite by wetting and drying. Clay Clay Miner 39(5):556–559. https://doi.org/10.1346/CCMN.1991.0390511
Tabatabai MA, Hanway JJ (1969) Potassium supplying power of Iowa soils at their "minimal" levels of exchangeable potassium. Soil Sci Soc Am Pro 33(1):105–109. https://doi.org/10.2136/sssaj1969.03615995003300010029x

Tang L, Sparks DL (1993) Cation-exchange kinetics on montmorillonite using pressure-jump relaxation. Soil Sci Soc Am J 57(1):42–46. https://doi.org/10.2136/sssaj1993.03615995005700010009x

Thompson ML, Ukrainczyk L (2002) Micas. In: Dixon JB, Schulze DG (eds) Soil mineralogy with environmental applications. Soil Science Society of America Book Series no. 7. Soil Science Society of America, Madison, pp 431–466. https://doi.org/10.2136/sssabookser7.c14

Velde B, Peck T (2002) Clay mineral changes in the morrow experimental plots, University of Illinois. Clay Clay Miner 50(3):364–370

Wang JG, Zhang FS, Cao YP, Zhang XL (2000) Effect of plant types on release of mineral potassium from gneiss. Nutr Cycl Agroecosyst 56(1):37–44. https://doi.org/10.1023/A:1009826111929

Wear JI, White JL (1951) Potassium fixation in clay minerals as related to crystal structure. Soil Sci 71(1):1–14

Wilson MJ (1999) The origin and formation of clay minerals in soils: past, present and future perspectives. Clay Miner 34(1):7–25

Zhang FS, Ma J, Cao YP (1997) Phosphorus deficiency enhances root exudation of low molecular weight organic acids and utilization of sparingly soluble inorganic phosphates by radish (*Raghanus sativus* L.) and rape (*Brassica napus* L.) plants. Plant Soil 196:261–264. https://doi.org/10.1023/A:1004214410785

Zörb C, Senbayram M, Peiter E (2014) Potassium in agriculture – status and perspectives. J Plant Physiol 171(9):656–669. https://doi.org/10.1016/j.jplph.2013.08.008

8

Soil Potassium Test: An Integrated Study

Michael J. Bell, Michael L. Thompson, and Philip W. Moody

Abstract The purpose of this chapter is to describe how bioavailable soil K is assessed or predicted by soil tests. Soil testing commonly refers to the collection of a sample of soil representative of a field or agronomic management unit and, by way of extraction using chemical reagents, determination of the quantity of a nutrient that can be related to plant uptake or yield. Normally only a small fraction of the total quantity of the nutrient present in the soil is extracted during the procedure, but if that amount can be correlated with actual crop uptake or overall crop productivity, then the soil test is deemed to have useful predictive power.

Soil tests are routinely used to guide applications of fertilizer to soil so that crop demand for nutrients can be met effectively and economically. Here, we summarize the procedures involved in collecting a representative soil sample for K analysis, outline how that sample should be prepared for laboratory analysis, highlight the principles and mode of action of routine soil tests, and explore some common issues that may confound the correlation between a soil K test result and plant K acquisition or crop yield. Soil testing methods are discussed in the context of their relationship to the different forms of soil K and the in-soil chemical processes that may change these forms into K that can be taken up by roots.

M. J. Bell (✉)
School of Agriculture and Food Sciences, The University of Queensland, Brisbane, QLD, Australia
e-mail: m.bell4@uq.edu.au

M. L. Thompson
Department of Agronomy, Iowa State University, Ames, IA, USA
e-mail: mlthomps@iastate.edu

P. W. Moody
Department of Science, Information Technology and Innovation, Brisbane, QLD, Australia
e-mail: Phil.Moody@nrm.qld.gov.au

8.1 Sample Collection and Preparation

Regardless of the soil analytical method used, one of the greatest challenges in deriving a prediction of the fertilizer requirement for a field from a soil analysis is the accuracy with which the soil samples reflect the fertility status of the field, and specifically the parts of the soil profile that are exploited by crop roots. Therefore, a soil sample that purports to represent the K status of the crop root zone should be collected from the soil layers with most intense root activity during growth stages when K uptake is critical. The following aspects are of particular importance for K, given its relative immobility in all except coarse-textured soils, and given the relatively low proportion of plant K that is removed at harvest in many grain and horticultural crops.

8.1.1 Vertical Stratification

Most fertilizer K, as well as that from animal dung and urine, is typically applied to the soil surface, or only into shallow profile layers. In addition, a substantial proportion of the crop K content from all except forage and sugarcane (*Saccharum* spp.) crops is returned to the soil surface in crop residues. Tillage will redistribute residues and fertilizer K within the plow layer, but the increasing proportion of fields under minimum or zero tillage management, combined with less inversion tillage in conventional tillage systems, is increasing the importance of surface layer enrichment (e.g., Barré et al. 2009). Apparent stratification of surface soil K can be accentuated (typically) by plant K uptake from the subsoil that is not replaced in fertilizer programs (Kuhlmann 1990; Chap. 12), although this can be moderated in lighter textured soils by K leaching into the deeper profile layers (Williams et al. 2018). Collectively, these effects typically result in net K depletion from soil layers immediately below the cultivated zone in tilled systems or below the depth of fertilizer band application in no-till systems. The most substantial K depletion occurs in soil profile layers with high root densities and in drier environments (especially in clayey soils) where those soil layers retain sufficient moisture to support an extended period of root activity.

A soil sampling program should therefore ideally determine K status in multiple soil profile layers, with the temporal frequency of analysis of each layer determined by the predicted or assumed rate of net K depletion/enrichment. Few, if any, commercial testing programs currently implement such a structured approach, although the value of testing deeper soil profile layers has been demonstrated for both mobile and immobile nutrients (Bell et al. 2013a, b).

8.1.2 Spatial Heterogeneity in Response to Agronomic Management

Fertilizer K is often applied in bands, especially in row-crop systems, either as a single nutrient or as part of a fertilizer blend (Chap. 12). Without adequate soil mixing through tillage, the presence of residual K in old fertilizer bands will potentially distort soil sample K content. This spatial heterogeneity represents a particular challenge for soil sampling in minimum and no till systems, where soil mixing during seed bed preparation is limited.

Non-uniform distribution of crop residues across a field can also introduce spatial heterogeneity in soil nutrient status (Brennan et al. 2004), with effects on K more pronounced due to the relatively high proportion of crop K returned in residues—especially in grain and oilseed crops (Pluskie et al. 2018; Chap. 5). This heterogeneity is accentuated by consistent placement of crop residue windrows in precision-controlled traffic systems and where residues are deliberately concentrated to facilitate windrow burning as a means of reducing the seed bank of herbicide-resistant weeds. The increasing width of broadacre grain harvesters also makes uniform residue distribution more difficult to achieve. The result is increasing heterogeneity of soil K that needs to be recognized in devising an appropriate soil sampling strategy.

8.1.3 Sample Drying and Handling

Once collected, the soil sample should be prepared for analysis in a way that does not depart from the sample preparation method used in the soil test—crop yield (or crop K uptake) calibration studies on which the interpretation guidelines for the soil test are based. Soil samples are typically air- or oven-dried (generally at 40 °C) and then ground/crushed (typically <2 mm) to create a homogenous sample from which a portion is selected for analysis. However, the drying process can itself influence the results obtained from laboratory analyses, particularly estimates of bioavailable K (Martins et al. 2015). These effects are most pronounced in soils with mineralogy that supports K fixation and release (e.g., with significant amounts of illite, vermiculite, or smectite). In soils with known K fixation and release characteristics, soil drying can either increase, decrease, or have no appreciable impact on extractable K concentrations, depending on the soil's K status at sampling.

The likelihood of an increase in extractable K upon air-drying depends primarily on two factors: the amount of initially extractable and soluble K in the sample and the degree to which the sample is dried (Scott et al. 1957; Scott and Smith 1968; Haby et al. 1988). In general, when soil K concentrations are low, K is released upon drying of the sample, probably in response to multiple mechanisms, including exchange of cations like Ca^{2+} and H_3O^+ that increase in solution concentration as water evaporates, as well as "scrolling" of the weathered edges of clay sheets

(McLean and Watson 1985). However, when the initial concentrations of surface-adsorbed and soluble K are high, those concentrations will increase even further as water evaporates, and in response, K ions are more likely to move into wedge and interlayer positions of 2:1 minerals, leading to contraction of some interlayer spaces. Both release and fixation of K ions are therefore likely to occur simultaneously during drying of such samples, but the mechanisms that favor release will dominate in low-K samples and vice versa in high K samples. The "crossover point," that is, the initial K status at which the ammonium-extractable K in moist samples is similar to that in dry samples, varies for each soil, most likely dependent on clay concentration, organic matter concentration, and soil pH.

Potassium fertilizer recommendations are most often developed using correlations between crop response and K extracted from air-dried samples with a uniform and low moisture content. However, some studies of soils with abundant 2:1 layer silicates have shown better correlations between soil K extractions from field-moist soil samples and crop responses to K fertilizer (Hanway et al. 1961, 1962; Barbagelata and Mallarino 2013). Adoption of an approach that uses field-moist samples may increase temporal variability in soil test results due to variation in moisture content, as well as requiring new approaches to homogenizing samples collected from soils with poor soil structure or high clay contents. However, if the improvement in prediction of bioavailable K status and crop fertilizer responsiveness is sufficient, such an approach will be warranted.

In soils where the mineralogy does not promote K fixation and release, effects of soil drying on extractable K are less important, and normal sample drying can be conducted without affecting the quantum of extractable K. While there are fewer studies of drying effects on extractable K in such soils, a recent investigation (Williams et al. 2017) found that soil drying method had no impact on the ability of Mehlich-3 soil extractions to predict responsiveness of soybeans to K in coarse-textured, sandy soils. Therefore, a knowledge of soil mineralogy and/or the presence of K fixation and release properties is necessary to develop soil drying protocols that do not interfere with the determination of bioavailable soil K status and that might confound the development of fertilizer recommendations.

8.2 What Are the Forms of Potassium in Soil?

Here we summarize the forms of K that are identified in the soil K cycle diagram (Fig. 1.1 in Chap. 1; Fig. 7.1 in Chap. 7). Potassium occurs in several pools, which are indicated by *boxes* in the diagram. The key constituents of each soil K pool, and the process by which K^+ ions move from one pool to another, are discussed in detail in Chaps. 1 and 7. Importantly, from the perspective of plant K uptake, it is the K ions in the soil solution that are most critical, since only that K can move to and into a root. While the application of fertilizer or contributions from plant residues can directly replenish soil solution K, several different solid-phase pools can also supply the ion to the liquid phase by means of a variety of physical, chemical, or biological processes.

Overall, in considering the development of soil tests to determine the plant-available K status of a soil, K ions associated with minerals may be classified in four possible locations that are referenced in the K cycle diagram (Fig. 1.1 in Chap. 1). These are adsorbed on the exterior surfaces of negatively charged clay particles (*Pool 9*), in interlayer positions of clay-size illite, vermiculite, or smectite (*Pool 10*), deeply embedded in interlayer positions of mica particles (*Pool 11*), or embedded in the structure of feldspar crystals (*Pool 12*).

The main source of solution K replenishment from solid-phase pools is desorption of the ion from mineral surfaces and some clay interlayers. *Pools 9* and *10* can supply K to the soil solution and therefore to plants, although the mechanisms and rate of movement from solid phase to solution phase vary. The degree to which K ions enter the soil solution from these sources depends on several factors: the concentration of K^+ in the solution, competition from other cations in the solution, the amount and location of negative charge in the mineral crystals, the activity of hydronium ions at the crystal surfaces, the water content of the soil, the redox potential of the soil, the abundance of hydroxy-Al polymers in interlayer positions, and the activity of low-molecular-mass organic anions that can complex Al and therefore degrade mineral surfaces. Electrostatically adsorbed K ions on exterior surfaces (*Pool* 9) are readily susceptible to exchange with other cations in the soil solution or to displacement by high concentrations of other cations in soil tests.

Potassium ions that occur in the interior of silt- and sand-size particles as structural components of primary micas and feldspars (*Pools 11* and *12*), on the other hand, are not very accessible to the soil solution and therefore are assumed to contribute to plant-available K supplies very slowly. The rate of contribution may be increased when *Pools 9* and *10* are locally depleted by plant uptake or leaching or when hydrolysis and complexation reactions accelerate weathering of the primary minerals, but the rates of release generally remain insufficient to support crop production without fertilizer K amendments.

8.3 How Is Potassium Released from Different Solid-Phase Forms?

8.3.1 Potassium in Fertilizer and Crop Residues

Potassium ions must be in the soil solution before they can be taken up by plant roots in the transpiration stream. In an agricultural context, most inorganic K fertilizers are water soluble, so fertilizer K does not usually persist in solid granules in the soil for long after its application. As long as the soil is moist (i.e., soil moisture content is greater than that at permanent wilting point), there will be a rapid increase in soil solution K in response to fertilizer application.

In plant cells, K occurs in both the cytoplasm and in the liquid in the vacuole. As invertebrates and microbial enzymes attack the cells of crop residues, the cells' primary membranes and tonoplasts are broken, and cell fluids merge with the soil solution. Cell walls and membranes of crop residues are also disrupted by drying or burning, allowing K ions to rapidly enter the soil solution.

8.3.2 *Surface-Adsorbed (Exchangeable) Potassium*

Electrostatically adsorbed K ions may be released from mineral surfaces or organic components of a soil in response to low K concentrations in the soil solution. The degree to which this happens depends on the concentrations of K and other ions in the soil solution as well as the quantity of K that is adsorbed on mineral and organic matter surfaces. Ions that are electrostatically adsorbed on exchangers like minerals and organic matter are not bound tightly to specific sites, but they are in equilibrium with ions in the liquid phase. This can be illustrated in the symbolic exchange reaction below, where X = one mole of negative charge associated with the exchanger (mineral surface or organic matter). Ca^{2+} is chosen as a model divalent cation because it is usually more abundant in the soil than other cations, but here it stands in for other "exchangeable" cations like Mg^{2+} or Na^{+}.

$$\mathrm{K}X + 0.5\mathrm{Ca}^{2+} \rightleftarrows \mathrm{Ca}^{2+}_{0.5}X + \mathrm{K}^{+}$$

The equilibrium constant for this reaction is a selectivity coefficient, K_s, that expresses the preference for the exchanger to host Ca^{2+} over K^{+} ions. The brackets indicate molar concentrations of ions in the solution, and the exchange-phase ion concentrations are moles of the ions per unit mass of the solid-phase.

$$K_s = \frac{\mathrm{Ca}^{2+}_{0.5}X}{\mathrm{K}X} \cdot \frac{[\mathrm{K}^{+}]}{[\mathrm{Ca}^{2+}]^{0.5}} \tag{8.1}$$

The first term on the right side of Eq. (8.1) is a ratio of charges associated with Ca^{2+} to the charges associated with K at the surface. The second term is a ratio of the concentration of K^{+} in solution to the square root of the concentration of Ca^{2+} ions in the solution. In general, the larger the value of K_s is for a given soil or particular mineral, the greater is that material's preference for Ca over K. Selectivity is a function of the positive charge and the radius of each of the cations as well as the amount of negative charge and its location in the mineral.

When the exchange reaction is at equilibrium, there is still movement of ions between the exchanger and the solution, but the rate of the forward reaction is equal to the rate of the backward reaction. The value of K_s is relatively constant (at constant temperature and ionic strength), so the product of the ratios of charges in solution and charges near the solid surface, as shown in the equation, is also constant. Therefore, if the concentration ratio of K^+ to other exchangeable ions (like Ca^{2+}) in the solution changes, the ratio of those charges on the exchanger's surface must also change.

This means that when the concentration of K^+ in the liquid phase drops below the concentration at equilibrium because of root uptake or leaching (i.e., the numerator of the solution term in Eq. (8.1) declines), release of surface-adsorbed K^+ to the solution (the left -to- right reaction) will begin and continue until equilibrium is re-established. Also, if the free Ca^{2+} concentration in the solution near a mineral surface increases, K^+ will be displaced as the reaction shifts to the right. Such displacement by Ca^{2+} is likely to occur, for example, as transpiration-induced mass flow of solution from the soil matrix moves into the rhizosphere and past the mineral edges and surfaces where K is adsorbed. Similarly, if the free Ca^{2+} concentration in the solution local to a mineral surface decreases—say, by complexation of Ca^{2+} with organic anions (e.g., oxalate or citrate), by leaching, by precipitation of insoluble apatites (calcium phosphates) as a result of ammonium-based phosphate fertilizer addition, or by preferential uptake of Ca^{2+} by a root—then the backward reaction will be favored, and K^+ will be more likely to move back from the solution phase to the exchange phase until equilibrium is again reached.

8.3.3 *Chemical Weathering*

Potassium-bearing minerals are transformed by both physical and chemical weathering. Here we focus on chemical weathering by describing some molecular-scale chemical reactions that are responsible for the release of K from structural positions in feldspar and mica crystals. Hydronium ions are released from roots into the rhizosphere to maintain electrochemical balance of charges when roots absorb cations (e.g., Ca^{2+}, Mg^{2+}, and NH_4^+) via the transpiration stream. Another source of hydronium ions is carbonic acid that forms when root-respired CO_2 enters the soil solution [$CO_2(g) + H_2O \rightarrow H_2CO_3 + H_2O \rightarrow H_3O^+ + HCO_3^-$]. Hydronium ions ($H_3O^+$) in the soil solution may play two kinds of roles in the weathering of mineral surfaces. First, they can be attracted to oxygen anions in $\equiv$Al-O-Si $\equiv$ bonds near crystal edges (where $\equiv$ represents bonds to adjacent O^{2-} ions). When the bridging O accepts a proton from H_3O^+, the Al-O bond at that location is weakened to the point of breaking, leaving $\equiv$Si-OH. The remaining H_2O molecule bonds with $\equiv$Al to form $\equiv$Al-OH_2^+. As Al-O-Si bonds are broken in such hydrolysis reactions, gaps in the crystal open, and there is greater opportunity for structural K^+ ions near crystal edges to escape to the solution. Second, as crystal edges begin to break up and deteriorate, H_3O^+ may also exchange for K^+ ions in the structure, further accelerating the release of K^+.

Dissolution of aluminosilicate minerals is also promoted by the complexation of Al^{3+} ions in solution by soluble organic anions such as oxalate and citrate. Low-molecular-mass organic anions are commonly exuded from roots or released during decomposition of crop residues and soil organic matter. At mildly acid to neutral soil solution pH values (pH 5–7), each organic anion often has two or more negative charges, and the product of the reaction with Al may be a very stable, soluble complex (e.g., the complex of Al with oxalate: $[C_2O_2^{2-}(Al^{3+})^+]$). This reaction effectively limits re-precipitation of Al^{3+} ions at the mineral surface and thus prevents the degraded surface from being sealed with a poorly crystalline gel, thus allowing more opportunities for K^+ ions to escape to the solution phase.

A third potential mechanism of chemical weathering is the oxidation of Fe atoms in crystals of K-bearing minerals that contain Fe, such as biotite. Oxidation from Fe^{2+} to Fe^{3+}occurs when the redox potential of the solution around the crystal is high, such as in well-drained soils where O_2 from the atmosphere or H_2O_2 derived from rainfall are dissolved in the solution. Oxidation of Fe means that the positive charge on the Fe atom increases, so the *net* negative charge near that location in the crystal decreases, making nearby K^+ ions more susceptible to exchange reactions with other cations. Eventually, the charge imbalance caused by oxidation can result in expulsion of Fe ions from the crystal, further weakening the structure and allowing even more K to move to the solution phase.

8.4 How Do Soil Tests Assess Plant-Available Potassium?

8.4.1 Soil-Test Development

The goal of any soil fertility test is to provide crop producers with a rapid, inexpensive, reproducible value that can be used to predict the need for, or outcome of, soil amendments or other management actions. Soil K test values are intended to provide guidance for whether application of K fertilizer would be beneficial in a typical growing season and, if so, by how much. For example, most soil tests for K assume that readily plant-available K ions in a soil sample are adsorbed to minerals and organic matter by electrostatic forces, and they can therefore be readily displaced by a high solution concentration of another cation, such as ammonium. Therefore, several soil-test procedures involve mixing or leaching a soil sample with a solution with a high concentration of an ammonium salt and measuring the amount of K that is moved to the liquid phase as a result. The displacement process is fast, the salt is inexpensive, and when all experimental parameters are standardized, the procedure itself is reproducible. Such displacement reactions are extreme examples of chemical exchange reactions. Several decades of international research efforts have focused on methods to relate concentrations of this "exchangeable" soil K to crop demand, crop uptake, and crop yield. Sometimes this form of K is called "plant-available," although the correlation between, say, ammonium-displaceable K and K taken up by plants is an indirect inference.

However, exchangeable soil K tests are not always highly correlated with plant response. In some soils, soil-test K values may suggest the need for fertilizer K, but when it is not applied, the crop is still able to remove enough K from the soil to produce a respectable (but possibly sub-optimal) yield. The sources of K in such

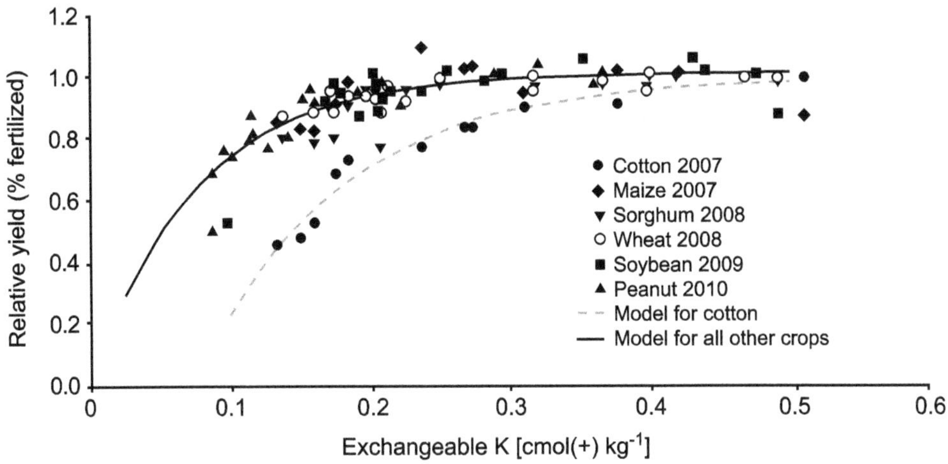

Fig. 8.1 Relationship between exchangeable K (by ammonium displacement) [cmol(+) kg^{-1}] in the top 10 cm of the soil profile and relative grain or fiber yield for different crops grown over a sequence of crop seasons on an Oxisol soil in northeastern Australia

soils may include hydroxy-interlayered vermiculite, biotite, or feldspars. Furthermore, most soil K test extractions do not provide information about the *rate* of K release from the soil during the growing season, and it is often the rate of release that determines whether a crop response to fertilizer will occur.

Over the time frame of a single season, correlations between exchangeable K and crop K uptake or grain yield can be used to identify a "critical" soil-test K value (or range) that indicates K sufficiency for different crops; for example, cotton has a higher critical value for exchangeable K (by ammonium displacement) than other species (Fig. 8.1).

However, when adequate K needs to be supplied over longer time periods (such as occurs with repeated biomass removal in forage cropping, or perennials, or high biomass-high K demand crops such as sugarcane), there are many soils from which plants are able to extract more than the initially exchangeable K pool. For example, Fig. 8.2 illustrates that K removal by crops may average almost 160% of the measured change in exchangeable K (by ammonium displacement). In this situation, *pools 9, 10*, and/or *11* (depending on soil mineralogy) are buffering soil solution K (and thereby, exchangeable K). The change in K extracted by tetraphenyl borate (TPB) (TB-K1h—described later) was a better indicator of the cumulative K removal in biomass than the change in exchangeable K (by ammonium displacement) (Fig. 8.2).

Plant K uptake over an even longer period can be derived from multiple K pools in the soil. For example, a study of empirically defined K pools of an Oxisol in northeastern Australia shows that change in the soil profile exchangeable K (by ammonium displacement) to a depth of 90 cm only accounted for 41% of the K removed from the total K pool over four decades (Fig. 8.3). It is apparent that over this extended uptake period, soil solution K was buffered not only by exchangeable K but also by K pools that are reflected by the TB-K 1 h. extraction, the boiling 1 M

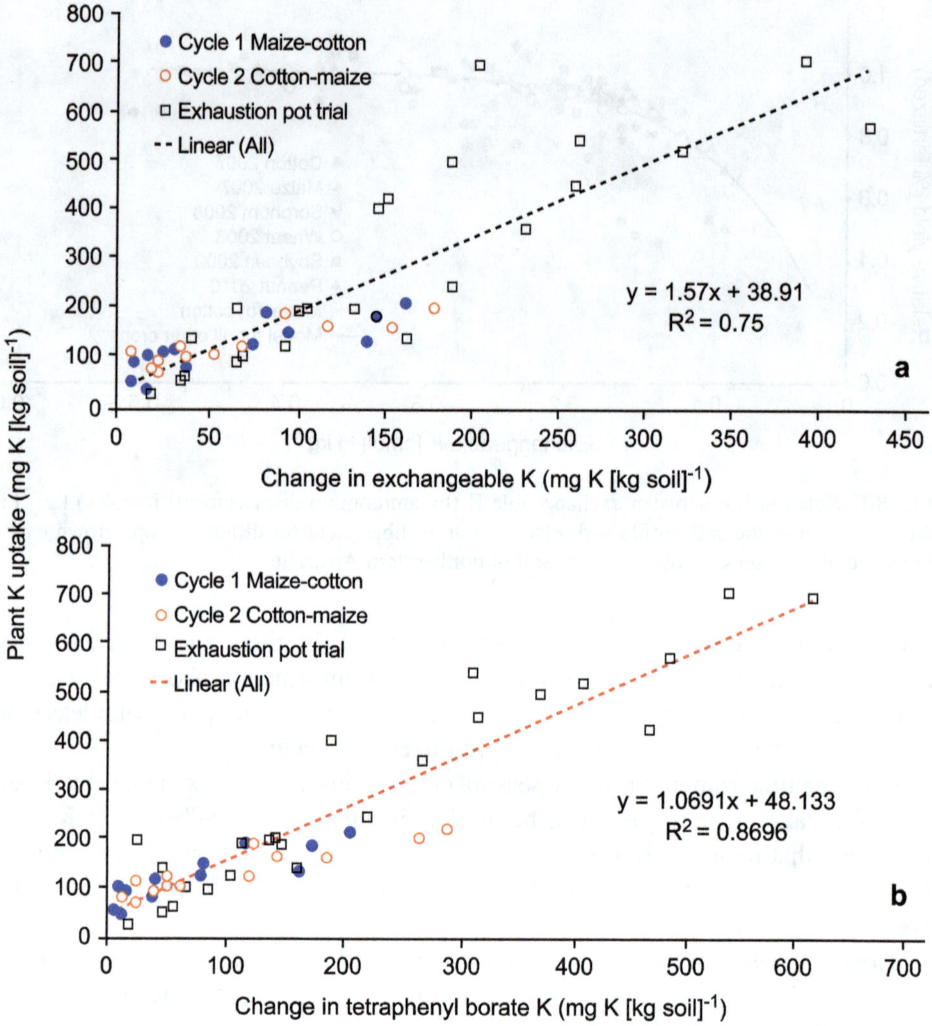

Fig. 8.2 The change in (**a**) exchangeable K (by ammonium displacement) and (**b**) tetraphenyl borate-extractable K (1 h) in response to plant K removal resulting from growing sequential maize (*Zea mays*) and cotton (*Gossypium* spp.) (or vice versa) crops, or a sequence of forage harvests (exhaustion pot trial) on neutral to alkaline medium-heavy clay soils in Australia

nitric acid extraction (described later), and the residual K fraction not extracted by any of the other extractants (Fig. 8.3).

The availability of K to plants is therefore determined by: (1) the quantities of K present in the different soil pools depicted in Fig. 1.1 (Chap. 1); (2) the rate of replenishment of soil solution K from those pools as K is removed by plant uptake or other processes; and (3) the period over which crop K uptake is measured. It is apparent that soil K tests reflecting one or more of these K supply factors will be required to predict the bioavailable K status in different cropping systems. The most

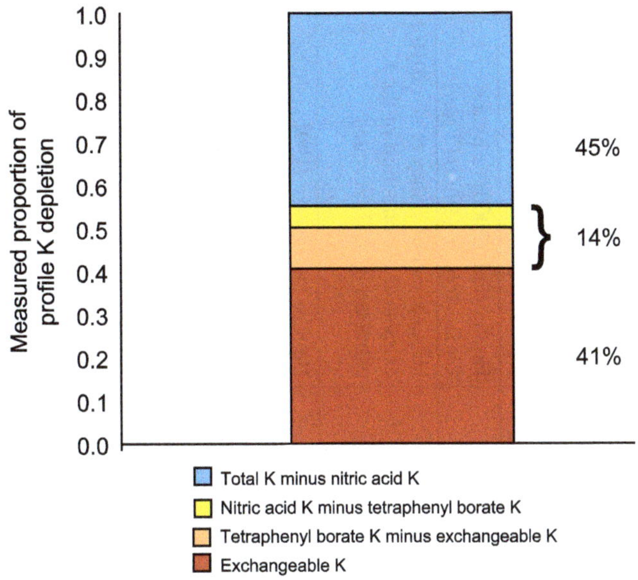

Fig. 8.3 The proportional change in soil K content to a depth of 90 cm resulting from 40 years of annual grain or forage cropping that can be accounted for by different soil tests. Data are from paired cropped and uncropped sites on an Oxisol in northeastern Australia

useful predictive test(s) will be the one(s) best correlated with the factor(s) governing K availability in any particular soil and growing season.

Accordingly, commonly used soil K tests have been grouped together in Table 8.1 on the basis of the processes that occur during the extraction: (1) equilibrated solution; (2) cation exchange; (3) acid dissolution; (4) multiple mechanisms, including complexation; and (5) rate of solution K replenishment. In addition, for a chemical extractant to be attractive for routine commercial application, it is advantageous if other nutrients (e.g., Ca, Mg, or P) can also be extracted by the same test and similarly correlated with plant uptake. Depending on the processes occurring during the extraction, the soil-extractant ratio, the extraction time and the extraction pH, inferences can be drawn about which K pools are likely to be contributing to the quantum of extracted K.

To organize the following discussion of soil tests to assess plant-available K, we have listed the extractions by numbers in Table 8.1, and we refer to those numbers in the narrative.

8.4.2 *Soil Tests for Assessing Soil Solution Potassium*

One soil K test intended to reflect K concentrations similar to those that roots might encounter in the soil solution is an extraction with 0.01 M $CaCl_2$ (Houba et al. 2000; extraction [1] in Table 8.1). The concentration of the salt simulates a typical ionic strength of soil solution, although the ionic strength of 0.06 M is higher than that often observed in highly weathered soils (≈0.005 M, Gillman and Bell 1978). The Ca^{2+} ions in the extractant are expected to partially displace electrostatically bound K^+, NH_4^+, Mg^{2+}, and Na^+ ions, and the Cl^- ions are expected to partially displace

Table 8.1 Soil tests to assess bioavailable K

Extraction process	Method and reagents	Concentration of reagents	Initial pH	Solid-solution ratio and other parameters	References and comments
Equilibrated solution	1. $CaCl_2$	$CaCl_2$ 0.01 M	Buffered by soil	1 g:10 mL, 2 h or 7 day	Houba et al. (2000) (2 h) and Cassman et al. (1990) (7 day)
	2. H3A multi-nutrient extractant	Li citrate, 0.010 M; citric acid, 0.003 M; malic acid, 0.003 M; oxalic acid, 0.004 M	4.4	1 g:10 mL, 5 min	H3A extract of Haney et al. (2010); extracts about one-third of the K extracted by NH_4 acetate
Ammonium exchange or displacement	3. Ammonium acetate at pH 7.0	NH_4 acetate, ~1 M	7.0	1 g:10 mL 2.5 g:25 mL 2.5 g:50 mL	Helmke and Sparks (1996) (shake 5 min), Gavlak et al. (2005) (shake 30 min), and Burt and Soil Survey Staff (2014) (leach 12 h on vacuum extractor)
	4. Ammonium acetate at pH 4.8 (Modified Morgan)	NH_4 acetate, 1.25 M; NH_4OH, 0.62 M	4.8	4 cm^3:20 mL; 5 min	McIntosh (1969) and Wolf and Beegle (2011)
	5. Ammonium chloride	NH_4Cl, 1 M	Buffered by soil	2.5 g:50 mL; 12 h on mechanical vacuum extractor	Burt and Soil Survey Staff (2014)
	6. Ammonium bicarbonate (AB-DTPA) at pH 8	NH_4HCO_3, 1 M; diethylene triamine pentaacetic acid (DTPA), 0.005 M	7.6	3 g:20 mL; 15 min	Soltanpour and Schwab (1977); designed for alkaline soils; NH_4^+ used to displace base cations; bicarbonate used to displace phosphate; DTPA added to complex trace metals
	7. Ammonium lactate	NH_4 lactate, 0.1 M; CH_3COOH, 0.4 M	3.75	5 g:100 mL; 90 min	Egnér et al. (1960) and Salomon (1998); used in Sweden for P and K; NH_4^+ exchange with K and other base cations; pH buffered by acetic acid

Exchange by other cations	8. Colwell	Na bicarbonate, 0.5 M	8.5	1 g:100 mL, 16 h	Colwell (1963)
	9. Na exchange—Morgan	Na acetate, 0.73 M; acetic acid, 0.52 M	4.8	10 cm^3:50 mL; 5 min	Morgan (1941) and Wolf and Beegle (2011)
	10. CALS (Ca acetate and lactate)	Ca acetate, 0.05 M; Ca lactate, 0.05 M;	4.1	1 g:20 mL, 90 min	Schüller (1969) and Zbíral and Němec (2005); developed primarily as an extractant for P; used in Germany for P and K
	11. $BaCl_2$ (with NH_4^+) exchange	$BaCl_2$, 0.1 M; NH_4Cl, 0.1 M	Buffered by soil	2 g:20 mL, 2 h shake in water 1 h before adding the $BaCl_2$ and NH_4Cl	Gillman (1979) and Gillman and Sumpter (1986); developed for variable charge soils, but suitable for all soils
Acid dissolution	12. Mehlich 1—Double acid	HCl, 0.05 M; H_2SO_4, 0.025 M	1.2	4 cm^3:20 mL, 5 min	Mylavarapu and Miller (2014); developed for acid, sandy soils
	13. Nonexchangeable K: Strong acid dissolution	HNO_3, 1 M	0.1	2.5 g:25 mL, 386 K, 25 min	McLean and Watson (1985); nonexchangeable K computed by difference between the total K removed and that exchangeable with 1 M NH_4^+ acetate, pH 7
Multiple mechanisms	14. Mehlich 3	CH_3COOH, 0.2 M; NH_4NO_3, 0.25 M; NH_4F, 0.015 M; HNO_3, 0.013 M; EDTA, 0.001 M	2.5 ± 0.1	1 g:10 mL, 5 min	Mehlich (1984); not recommended for K in calcareous soils; pH buffer; NH_4^+ exchange with K and other base cations; F^- complexes Al and Si; hydrolytic attack at mineral surfaces; complexes trace metal nutrients
	15. Kelowna with EDTA or DTPA	CH_3COOH, 0.25 M; NH_4F, 0.015 M; EDTA or DTPA, (0.001 or 0.005 M)	2.7–3.0	1:10 by volume, 5 min	Van Lierop and Gough (1989); pH buffer; F complexes Al, Si; NH_4^+ exchange with base cations

(continued)

Table 8.1 (continued)

Extraction process	Method and reagents	Concentration of reagents	Initial pH	Solid-solution ratio and other parameters	References and comments
	16. Modified Kelowna	CH_3COOH, 0.025 M; NH_4 acetate, 0.25 M; NH_4F, 0.015 M	4.9	1:10 by volume, 5 min	Qian et al. (1994); designed to extract more K than the original Kelowna extraction; pH buffer; NH_4^+ exchange with K and other base cations; F^- complexes Al, Si; NH_4^+ exchanges with base cations
	17. Lancaster	Solution A: HCl, 0.05 M; Solution B: glacial acetic acid, 1.57 N, plus compounds listed below: malonic acid, 0.063 M; malic acid, 0.089 M; NH_4F, 0.032 M; aluminum chloride hexahydrate 0.12 M	1.4 4.0	1st step—5 g:5 mL HCl for 5 min, no shaking Second step—add 20 mL of solution B to first step, 10 min shaking	Cox (2001) and Oldham (2014); pH buffer; Intended to complex Al; F^- sufficient to complex Al, but concentration is thought to be low enough to prevent precipitation with Ca and Mg; Al is also added to prevent F^- precipitation with Ca and Mg
Sinks	18. Cation exchange resin: Na exchange for K	Na-saturated cation exchange resin (sulfonated polystyrene)	6.8	1:1 volume ratio; 16 h shaking with a previously disaggregated sample	van Raij et al. (1986). Na-saturated cation exchange resin (sulfonated polystyrene) coupled with bicarbonate-saturated anion exchange resin (quaternary ammonium functionality) for anions. 1 M $NaHCO_3$ is used to saturate the resins
	19. Nonexchangeable K: Na tetraphenylboron	Na tetraphenylboron, 0.2 M; NaCl, 1.7 M; EDTA, 0.01 M		0.5 g:3 mL; 1, 5, 15, or 60 min incubations	Cox et al. (1996, 1999)

Water-soluble K is extracted with all of these extractants, but its concentration is assumed to be negligible in non-saline soils and is ignored weakly held phosphate species, SO_4^{2-}, and NO_3^- ions. Other nutrients and contaminants may also be determined in the extract. Because the soil-solution ratio and the physical disruption caused by shaking the soil sample with the extractant do not simulate normal soil conditions, the extracted nutrient concentrations must be understood as indices of nutrient concentrations one might expect in a soil solution, not absolute values.

A number of studies have shown that the $CaCl_2$ extraction solubilizes less K than the ammonium acetate, Mehlich 3, or ammonium lactate extractions (e.g., Simonis and Setatou 1996; Baier and Baierova 1998; Zbíral and Němec 2005; Woods et al. 2005; Salomon 1998). The differences in the extraction results are probably due to the relatively low concentration of Ca in the $CaCl_2$ extractant as well as to the lower hydration energy of ammonium, which allows it to penetrate clay interlayer spaces more effectively than Ca. While there is typically a strong correlation between $CaCl_2$-extracted and ammonium-extracted K, generally, 0.01 M $CaCl_2$ extracts only 30–80% of the K removed by the more aggressive extractions. The amount of K solubilized increases at lower clay concentrations, suggesting that the efficiency of Ca^{2+} to exchange with K^+ declines when more clay is present (i.e., when the cation exchange capacity is higher or when clay microaggregates are less likely to be rapidly dispersed).

Haney et al. (2010) proposed a multi-element extraction, called H3A ([2] in Table 8.1), intended to simulate the solution in the rhizosphere by a suite of dilute organic bases like those that might be exuded by roots, including citrate, oxalate, and malate. The only cation present in this extractant is Li^+, and it is present at a low concentration. Comparisons of nutrient concentrations released by the H3A extractant with those of other, more aggressive extractions have been few, and we are not aware of calibration studies relating solubilized nutrient concentrations with crop growth or yield. In a comparison of K extracted by the H3A extractant with K extracted by ammonium acetate, H3A solubilized only about one-third of the ammonium-extractable K in 60 soil samples from the continental United States (Haney et al. 2010). The H3A extraction has recently been modified by removing Li citrate from the reagents in the extraction solution (Haney et al. 2017).

The assumption behind both the $CaCl_2$ and H3A extractions is that extractants similar to the soil solution will provide more accurate knowledge of the instantaneous bioavailability of soil nutrients. While the extracted concentrations of K may be correlated with those of other extractants, significant advantages of the $CaCl_2$ and H3A extractants over ammonium-based procedures for predicting seasonal K needs for a crop have yet to be demonstrated.

8.4.3 *Soil Tests for Assessing Surface-Adsorbed Potassium*

The ammonium ion, NH_4^+, has similar size, charge, and hydration energy to the K^+ ion. For this reason, ammonium has been preferred as the cation most likely to replace K that is surface-adsorbed or located in readily accessible interlayer positions of soil minerals. Several soil-test extractions have been developed for this purpose. These include extractants with high concentrations ($\geq$1 M) of ammonium acetate (e.g., [3], [4] in Table 8.1), ammonium chloride ([5] in Table 8.1), and ammonium bicarbonate ([6] in Table 8.1) (Burt and Soil Survey Staff 2014; Wolf and Beegle 2011; Soltanpour and Schwab 1977). The 1 M ammonium acetate and ammonium chloride extractions also promote displacement of other exchangeable cations such as Ca^{2+}, Mg^{2+}, and Na^+. The ammonium lactate extraction [7] employs ammonium at a lower concentration (0.1 M) than the previously mentioned extractions.

In high pH, calcareous soils, exchangeable Ca and Mg do not limit crop nutrition, and the ammonium bicarbonate extraction (AB-DTPA) focuses not only on extraction of bioavailable K^+, but also on phosphate ions (presumably by exchange of HCO_3^- with HPO_4^{2-}). DTPA is added to the AB-DTPA extractant to complex micronutrient trace metals (Zn, Fe, Mn, and Cu). The modified Morgan extraction ([4] in Table 8.1) has an even higher concentration of ammonium (1.87 M) and is used to extract K, Ca, Mg, P, Cu, Mn, and Zn.

Cations other than ammonium may be used to displace K^+ from exchange sites; these include Na^+ (e.g., Colwell extraction, [8], in Table 8.1 and the original Morgan extraction, [9], in Table 8.1), Ca^{2+} (CALS extraction: [10] in Table 8.1), and Ba^{2+} (the $BaCl_2$ extraction, [11], in Table 8.1). In soils where all exchange sites are equally accessible, these ions are expected to displace surface-adsorbed K^+ effectively, especially when employed at high concentrations compared to the concentration of extractable K in the soil sample. However, the larger hydrated radius and hydration energy of Na make its entry into interlayer spaces difficult, so it is much less likely to displace K^+ ions in those locations. Similarly, the much higher energies of hydration of Ca^{2+} and Ba^{2+} ions mean that the hydration spheres of these divalent cations are also stable enough to prevent entry into interlayer spaces, even though the ions are competitive with K at external mineral surfaces. Primarily for this reason, ammonium-based extractants are preferred to assess bioavailable K in soils where 2:1 layer silicates are present and interlayer K is likely to occur.

Extractants that exchange ammonium or other cations for K are also distinguished from one another by the pH of the extraction. Maintaining a constant, buffered pH during the extraction period (7.0 for 1 M ammonium acetate, 7.6 for AB-DTPA, 4.8 for the modified Morgan extractant, 8.5 for the Colwell extractant), while not expected to affect significantly the amount of K extracted, may affect the extractability of other nutrients. The pH of the modified Morgan extraction (pH 4.8) was chosen to simulate the pH of the soil solution in equilibrium with high concentrations of CO_2 (g) and organic base anions, as the solution adjacent to a root hair may be during active root respiration. For acidic soils that are not highly weathered, pH ~7 is a common target for optimizing plant growth; thus 1 M ammonium acetate at pH 7 may better represent K ions that would be bioavailable if the field soil pH were adjusted by adding lime. In the AB-DTPA extraction, pH is maintained high enough that calcite (calcium carbonate) will not be dissolved during the extraction. The pH of the ammonium acetate extraction may also be adjusted (e.g., to 8.5) to minimize dissolution of calcite in calcareous soils.

8.4.4 *Soil Tests for Dissolving Interlayer/Structural Potassium*

Some soil K tests have been developed to simulate the chemical processes that lead to K release from pools other than exchange sites. For example, the Mehlich-1 extraction ([12] in Table 8.1) creates a low-pH environment so that nutrients (base cations, P, and micronutrients) will be solubilized by hydronium exchange or by hydrolysis reactions. The Mehlich-1 extract relies on 0.05 M HCl and 0.025 M H_2SO_4 to provide an excess of hydronium ions to partially dissolve nutrient-bearing minerals as well as to displace electrostatically adsorbed base cations and to compete with complexed micronutrient metals at variable-charge sites.

The strong acid extraction described by McLean and Watson (1985) [13] comprises boiling a soil sample in 1 M nitric acid for 25 min, amplifying the intensity of the hydrolysis reaction to dissolve K-bearing silicates. "Non-exchangeable" K is estimated by the difference between the acid-soluble K and ammonium-displaceable K (1 M NH_4^+ acetate, pH 7). While the results of this extraction have been correlated with plant uptake of K in some studies, at least a portion of the K released is likely to derive from interlayers of primary micas and the interiors of feldspar crystals—i.e., K that would not be very available to plants in a single growing season.

8.4.5 *Soil Tests that Combine Multiple Mechanisms of Potassium Dissolution*

Several multi-element soil extractants have been developed that employ ammonium salts in concentrations lower than those of the preceding extractions, but they also include reagents that promote mechanisms other than cation exchange for solubilizing K. For example, the concentration of NH_4^+ ions in the Mehlich-3 extractant ([14] in Table 8.1) is 0.265 M. In the Kelowna extractant with EDTA ([15] in Table 8.1) and the modified Kelowna extractant at pH 4.9 ([16] in Table 8.1), it is 0.015 M and 0.265 M, respectively. In the Lancaster extractant ([17] in Table 8.1), the concentration of NH_4^+ ions is 0.032 M. Typically, K extracted by these methods is compared with the 1 M ammonium acetate extraction for effectiveness. By using the Kelowna extraction (with EDTA or DTPA), van Lierop and Gough (1989) reported that ~20% less K was solubilized, on average, in 100 Canadian soils than by using the 1 M ammonium acetate extraction. However, in a subsequent modification of the Kelowna extractant, increasing the ammonium ion concentration to 0.265 M increased extractable K by ~30% (i.e., similar to the ammonium acetate extraction for soils with extractable K < 450 mg kg^{-1}) and significantly improved the relationship between extractable K and K uptake by canola (Qian et al. 1994).

In the Mehlich 3, Kelowna (both versions), and Lancaster extractions, all or a portion of the ammonium added to the extractant is in the form of NH_4F. The fluoride ion is added primarily because it forms strong complexes with Al^{3+}, helping to dissolve Al oxyhydroxides and release adsorbed orthophosphate ions. Mehlich (1984) reported that 6–8% more K was extracted by Mehlich 3 than by 1 M ammonium acetate in a suite of 105 soils from the southern and eastern United States. It may be speculated that the added F^- can also contribute to dissolution of Al from hydroxy-interlayered vermiculite, thereby facilitating $NH_4^+ - K^+$ exchange. This complexation reaction may be compared with weathering of micas in the presence of organic compounds. Low-molecular-mass organic anions (e.g., oxalate, malate, and citrate) may complex Al^{3+}, limit re-precipitation of Al^{3+} ions at mineral surfaces, and therefore promote K exchangeability.

8.4.6 Soil Tests for Assessing the Rate of Solution Potassium Replenishment

By varying the extraction time, soil K tests that provide a sink for K released into the soil solution can be used to assess the rate of solution K replenishment in response to depletion by plant uptake or other processes. The mixed bed cation-anion exchange resin method ([18] in Table 8.1) was designed to simulate plant uptake of nutrients by providing a strong sink for both cations and anions. The sink creates a strong disequilibrium that favors nutrient release from the soil sample. In the case of K, one cation exchange resin that has been used is a Na-saturated sulfonated polystyrene that strongly adsorbs cations. The K extracted by this method from Oxisols, Ultisols, and Alfisols with low cation exchange capacity (CEC) in Brazil was very similar to K extracted by 1 M ammonium acetate and 0.025 M H_2SO_4 (van Raij et al. 1986). The method has been successfully adapted for routine processing of large numbers of soil samples in a single, overnight extraction period.

The tetraphenyl borate extraction method ([19] in Table 8.1) was developed by Cox et al. (1996, 1999) on the basis of work by Scott and colleagues (e.g., Scott and Reed 1962) and exploits the strong complexing power of tetraphenyl borate for alkali metals. Short reaction times have been correlated with release of K from vermiculite interlayers, but the longer the reaction is allowed to proceed, the more K can be pulled from primary minerals, too.

Variations of both the resin method and the tetraphenyl borate method could be used to characterize the rate of release of K into solution from the various soil pools. In a glasshouse experiment, Moody and Bell (2006) found that the absolute changes in TB-K (1 h) in 37 soils of diverse chemistry were more highly correlated with exhaustive cumulative crop K uptake than changes in TB-K (15 min), with regression slopes of 0.99 (±0.04) and 1.12 (±0.07), respectively. These results demonstrate that the rate of replenishment of solution K is important to K bioavailability, but this effect is unlikely to be captured by a single extraction. Multiple temporal assays to measure the rate of K release are probably not feasible in a commercial laboratory setting, and the resin or TPB approaches will likely remain as research tools to better understand the interactions among the various K pools.

8.5 Difficulties Relating Soil Test Potassium to Crop Acquisition

A number of the processes already discussed in this chapter will clearly affect the usefulness with which a soil test can predict crop K acquisition. These include the applicability of the soil K test extraction method to estimate the K pools that contribute to plant uptake, the period for which K availability is being predicted (single crop, multiple harvests, multiple growing seasons), and the intensity of K demand during the growth period.

However, as illustrated in Fig. 8.4, even the same crop species and cultivar growing on two contrasting soil types within the same crop region in northeast Australia exhibited very different responses to increasing soil test K, despite similar maximum yields and crop K demand. The soils in Fig. 8.4 (an Oxisol with a CEC of 10 cmol(+)/kg and a Vertisol with a CEC of 60 cmol(+)/kg) both supplied K from the exchangeable K pool. However, as exchangeable K increased in response to K fertilizer addition, grain sorghum growing in the Oxisol was able to accumulate K much more efficiently, and to a greater extent, than grain sorghum growing in the Vertisol. Despite the choice of soil test methods appropriate for assessing soil K pools that can meet crop K demand and sampling depths that reflect crop root activity under those environmental conditions, these different patterns of crop K acquisition demonstrate the impact of other factors that should be considered when interpreting a soil test K result. These are discussed briefly below.

Fig. 8.4 The relationship between exchangeable K (by ammonium displacement) and accumulation of K in aboveground biomass of grain sorghum grown in each of three different growing seasons on an acidic Oxisol (CEC 10 cmol (+)/kg) and an alkaline Vertisol (CEC 60 cmol(+)/kg)

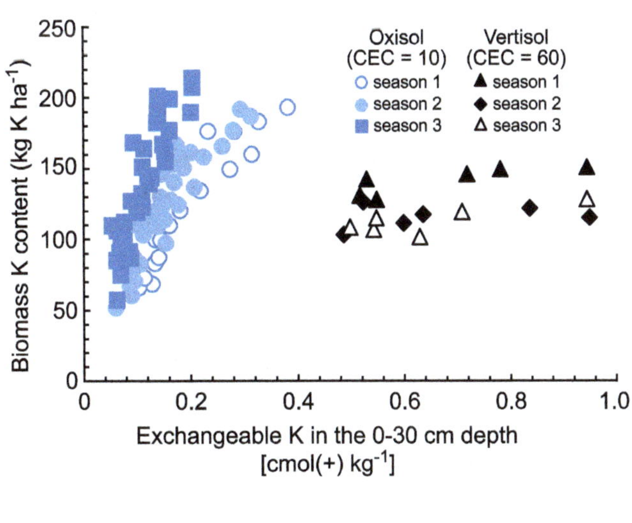

8.5.1 Rates of Resupply to Potassium-Depleted Zones Around Active Roots

The replenishment of soil solution K in response to depletion by plant root uptake is dependent on: (1) the solution K concentration in undepleted soil; (2) mass flow of that soil solution to the root in response to crop water extraction; and (3) diffusion of K along a concentration gradient between the depleted rhizosphere soil solution and that of the surrounding undepleted soil (see Chap. 7).The contrasting relationships between exchangeable K and K acquisition by sorghum (Fig. 8.4) illustrate the effect of soil properties on aspects relating to both (1) and (3). The much higher CEC in the Vertisol results in lower soil solution K concentrations in undepleted soil (e.g., Bell et al. 2009), which by itself can limit K supply to the root (Barber 1995). However, this soil characteristic will also limit the K concentration gradient that can develop between the depleted rhizosphere and the undepleted soil solution. In addition, physical impediments related to poor soil structure and porosity affect the tortuosity of the diffusion path that K ions must traverse to reach the depleted rhizosphere, and they can therefore influence the rate of rhizosphere replenishment (e.g. Barber 1995; Dodd et al. 2013). Solution K in the Oxisol, which is porous and strongly structured, has a much less tortuous diffusion path than in the Vertisol and therefore allows more rapid rates of K diffusion into the rhizosphere. Collectively, these soil characteristics contribute to the more rapid plant uptake of K in the Oxisol in response to incremental increases in exchangeable K that are presented in Fig. 8.4. From the perspective of soil test interpretation and K fertilizer recommendations, lighter textured soils that have low K buffer capacities (i.e., limited ability to hold K in pools other than the soil solution) and can support high K diffusion rates are likely to have lower critical soil test K concentrations (e.g., Brennan and Bell 2013) and will respond to lower rates of applied K (e.g., Bell et al. 2009).

8.5.2 Root System Architectures and Their Interaction with Soil Moisture

This topic is covered in depth in Chaps. 4 and 12, respectively, from the perspective of the relative advantage of different root morphologies in systems with contrasting seasonal moisture availability and profile K distributions. These characteristics interact with the efficacy of different K fertilizer application strategies. They are especially relevant in the consideration of the depths from which soil samples are to be collected, as these need to reflect the root characteristics of the target species for which K availability or fertilizer requirement is being predicted. The data presented in Fig. 8.1 provide a good example of the relative disadvantage of the coarse root system of cotton in being able to efficiently acquire K from a soil profile in which K is strongly stratified, in comparison to root systems of other crop species grown in the same soil and seasonal conditions. Other examples of the impact of root morphology on K acquisition include the report by Witter and Johansson (2001) that illustrated the advantage of deeply rooted forage species such as lucerne/alfalfa (*Medicago sativa*) and chicory (*Cichorium intybus*) in acquiring K from subsoil layers.

The interaction of root morphological characteristics with seasonal moisture availability is a further complication in choosing the depth of soil sampling, the critical soil test concentration used to determine adequacy of soil K status, and any subsequent fertilizer placement strategy. Soil layers that are periodically dry during the growing season are less likely to contribute a substantial proportion of plant K uptake. Therefore, cropping systems where seasonal variation in rainfall amount and distribution alter the reliance on topsoil and subsoil K reserves provide additional challenges to predicting K availability from soil test results.

8.5.3 *Variation in Root System Attributes that Allow Plants to Exploit Different Potassium Pools*

Chapter 4 focuses on the foraging strategies that plants use to increase the volume of the rhizosphere from which they acquire K, and provides a more detailed coverage of this topic. Briefly, there are many examples where more aggressive depletion of the rhizosphere K by plant root systems [e.g., by ryegrass (*Lolium*) (Barré et al. 2007, 2008)], an ability to more extensively lower rhizosphere pH (Hinsinger et al. 1993), or an ability to release exudates that promote the dissolution of K-bearing silicates (Song and Huang 1988) can provide plant species or genotypes with greater access to less readily available K pools. These characteristics could potentially affect the choice of diagnostic soil test if the differences are large enough. However, it is more likely that they would simply reduce the precision with which a particular soil test could identify a critical soil concentration above which fertilizer responses would be less probable. Given the move toward diversity in crop rotations and the speed with which new cultivars are introduced into agricultural systems, the ability to precisely define soil test-crop response relationships for a single crop species is likely to remain challenging.

8.5.4 *Specificity of Soil Test Potassium-Crop Response Relationships and the Role of Trial Databases*

There are clearly many challenges that will constrain our ability to develop robust soil test-crop response relationships that can cope with spatial variability in soil types, seasonal variability in access to different soil profile layers, and different genotypes and species in a crop rotation. These challenges have contributed to the commonly reported site-specific nature of soil test-crop response relationships for K (e.g., Brouder et al. 2015). While greater process-level understanding may help explain the reasons for variation in soil test-crop response relationships, the rate of knowledge gain is unlikely to be able to keep up with the rate of management-induced changes in profile K status and distribution, or the changes in cultivars being delivered from breeding programs.

One approach to increasing the rate of accumulation of soil test-crop response data is the development of searchable databases that allow aggregation of available data at an appropriate scale (e.g., regionally or on the basis of soil type). An example of this has been the Better Fertilizer Decisions for Cropping database and database interrogator developed in Australia (Whatmuff et al. 2013), which currently houses in excess of 5500 historic data sets from trials that have been used to develop soil test-crop response relationships for N, P, K, and S. New experimental data are also being added to this database, allowing not only the greatest density of trials to build soil test calibrations but also an opportunity to explore the impact of time and management changes on critical soil test values (e.g., the change from conventional to zero tillage for less mobile nutrients like P and K). A relevant example of the use of that database for comparing critical exchangeable K concentrations for wheat (*Triticum aestivum*) grown on contrasting soil types is the paper by Brennan and Bell (2013). The use of such databases as repositories for data from national or international research programs would allow opportunities for collaborative approaches to the development of new soil test-crop response relationships.

8.6 Lessons Learned from Long-Term Experiments

The chapter in this book that discusses the relationship between changes in soil test values in response to K mass balance (Chap. 10), and a recent review of the lessons that long-term experiments can provide with respect to K management (Goulding et al. 2017), came to similar conclusions. While providing clear insights into the dynamics of K in soil and plant systems, long-term experiments clearly demonstrated that current soil tests targeting bioavailable K do not provide a reliable benchmark of the impact of practices on the size of the bioavailable K pool or of the long-term sustainability of K management practices. An example has also been presented in Chap. 7 (Fig. 7.6—reproduced from Hinsinger 2002), showing that despite wide variation in K balance between management strategies, there was no consistent pattern of change in exchangeable K in the soil layers monitored.

Similarly, the example provided in Fig. 8.3 in this chapter shows that even when the K status of subsoil layers (to 90 cm) was considered, and a variety of commercially available soil tests were used (exchangeable K, tetra phenyl borate-K, and nitric acid-extractable K), only a little over 50% of the soil K depletion resulting from long-term cropping could be accounted for. Such results have significant implications for the use of commercial soil testing procedures to monitor long-term changes in K fertility. A more detailed analysis of soil and crop removal data from the cropped fields depicted in Fig. 8.3 is provided by Fig. 8.5. This shows

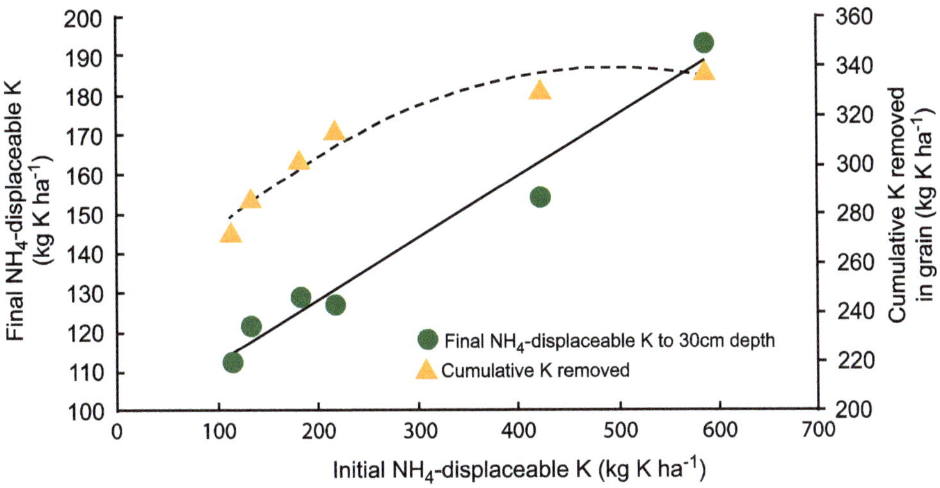

Fig. 8.5 The relationship between the initial quantum of exchangeable K (by ammonium displacement) to 30-cm depth (kg K ha^{-1}) in an Oxisol and the cumulative K removal in harvested produce over a 13-year cropping period (right) and the impact of that cumulative crop removal on the final exchangeable K remaining in the same profile layers (left). Each point represents the mean of nine replicate plots in which differing soil K concentrations were initially established by fertilizer addition

that despite a fivefold variation in initial exchangeable K (achieved through fertilizer addition), the cumulative removal of K in harvested produce only varied by a maximum of 25% over a sequence of 13 cropping seasons. However, the impact of that crop K removal on soil exchangeable K content was related strongly to the initial stocks of exchangeable K, rather than to the amount of K removed. Soils with the lowest initial exchangeable K exhibited little or no net change in exchangeable K over the cropping period, despite cumulative K removal of 270–290 kg K/ha. Conversely, soils with the highest reserves of exchangeable K recorded net changes in exchangeable K that were similar to the quantum of K removed.

These considerations are a reminder that soil testing methods targeting K pools that are bioavailable in single seasons are not always the best method of monitoring the K status of agricultural fields. The longer-term K dynamics that can occur in response to K removal, as well as the action of roots in different parts of the soil profile, also need to be considered.

8.7 Concluding Remarks

A primary consideration in the choice of analytical method and interpretation of the resulting soil test K values would seem to be the kind of cropping system for which recommendations are being made. We can define at least three scenarios from a crop demand perspective. These are: (1) very short- and short-duration crops (vegetables, cereals, and grain legumes—all with a single harvest); (2) longer season crops like sugarcane; and (3) perennial or forage crops, where the K demand is extended over multiple years and/or multiple forage harvests per season.

If a "quick" soil test is used to predict bioavailable K for the next crop season of a short duration crop, then exchangeable K is probably the best indicator of K that the crop is likely to access, with different crops likely having different "critical ranges" that may also be dependent on the soil's cation exchange capacity. There is not enough time for the crop rhizosphere conditions to extensively degrade minerals, and there will probably not be enough wetting and drying cycles for fixation and release to have a major impact on K availability. The major variable here is to adequately characterize readily available K in the soil layers where there is a high root density (and plant-available moisture).

For longer season crops like sugarcane, the rate of K release or replenishment over the longer time frame is likely to be more significant, as there is more time for the crop root system to access K from both surface-adsorbed and interlayer or structural K pools. In this situation, the difference between exchangeable K and non-exchangeable K (perhaps assessed by the TPB or nitric acid extractions) may correlate with crop K uptake and provide a better indication of whether or not the soil is capable of supplying K in the medium term (over several months).

The final cropping systems of interest are the multiple forage harvest or multiple crop season systems (e.g., lucerne/alfalfa), where demand is high and persistent over an extended period of time (years). Here, the more slowly available K (interlayer and structural pools) becomes critically important for the sustained removal and replenishment of plant-available K. Exhaustive cut-and-remove pot trials are extreme examples of this, because inevitably there is a restricted soil volume as well as a very high root density, in addition to the high and prolonged K demand. In these situations, the rate of release/replenishment of slowly available K pools quickly becomes the dominant factor in determining fertilizer K responsiveness.

References

Baier J, Baierova V (1998) Hundredth molar calcium chloride extraction procedure. Part IV: calibration with conventional soil testing methods for potassium. Commun Soil Sci Plant Anal 29:1641–1648. https://doi.org/10.1080/00103629809370056

Barbagelata PA, Mallarino AP (2013) Field correlation of potassium soil test methods based on dried and field-moist soil samples for corn and soybean. Soil Sci Soc Am J 77:318–327. https://doi.org/10.2136/sssaj2012.0253

Barber SA (1995) Soil nutrient bioavailability: a mechanistic approach, 2nd edn. Wiley, New York

Barré P, Montagnier C, Chenu C et al (2007) Clay minerals as a soil potassium reservoir: observation and quantification through X-ray diffraction. Plant Soil 302(1–2):213–220. https://doi.org/10.1007/s11104-007-9471-6

Barré P, Velde B, Fontaine C et al (2008) Which 2: 1 clay minerals are involved in the soil potassium reservoir? Insights from potassium addition or removal experiments on three temperate grassland soil clay assemblages. Geoderma 146(1–2):216–223. https://doi.org/10.1016/j.geoderma.2008.05.022

Barré P, Berger G, Velde B (2009) How element translocation by plants may stabilize illitic clays in the surface of temperate soils. Geoderma 151(1–2):22–30. https://doi.org/10.1016/j.geoderma.2009.03.004

Bell MJ, Moody PW, Harch GR et al (2009) Fate of potassium fertilisers applied to clay soils under rainfed grain cropping in south-East Queensland, Australia. Aust J Soil Res 47:60–73. https://doi.org/10.1071/sr08088

Bell MJ, Moody PW, Anderson GC et al (2013a) Soil phosphorus—crop response calibration relationships and criteria for oilseeds grain legumes and summer cereal crops grown in Australia. Crop Pasture Sci 64(5):499–513. https://doi.org/10.1071/CP12428

Bell MJ, Strong W, Elliott D et al (2013b) Soil nitrogen—crop response calibration relationships and criteria for winter cereal crops grown in Australia. Crop Pasture Sci 64(5):442–460. https://doi.org/10.1071/CP12428

Brennan RF, Bell MJ (2013) Soil potassium-crop response calibration relationships and criteria for field crops grown in Australia. Crop Pasture Sci 64(5):514–522. https://doi.org/10.1071/cp13006

Brennan RF, Bolland MDA, Bowden JW (2004) Potassium deficiency, and molybdenum deficiency and aluminium toxicity due to soil acidification, have become problems for cropping sandy soils in South-Western Australia. Aust J Exp Agric 44(10):1031–1039. https://doi.org/10.1071/EA03138

Brouder S, Bell MJ, Moody PW (2015) How site-specific are soil test calibration relationships for potassium? In: Murrell TS, Mikkelsen RL (eds) Frontiers in potassium science: developing a roadmap to advance the science of potassium soil fertility evaluation, Kailua-Kona, Hawaii. International Plant Nutrition Institute, Peachtree Corners, GA, pp 18–21

Burt R, Soil Survey Staff (ed) (2014) Kellogg soil survey laboratory methods manual. Soil survey investigations report no 42. Version 5.0. Kellogg Soil Survey Laboratory, National Soil Survey Center, Natural Resources Conservation Service, U.S. Department of Agriculture. Lincoln, NE. https://www.nrcs.usda.gov/Internet/FSE_DOCUMENTS/stelprdb1253872.pdf. Accessed 11 May 2020

Cassman KG, Bryant DC, Roberts BA (1990) Comparison of soil test methods for predicting cotton response to soil and fertilizer potassium on potassium fixing soils. Commun Soil Sci Plant Anal 21(13–16):1727–1743. https://doi.org/10.1080/00103629009368336

Colwell JD (1963) The estimation of the phosphorus fertilizer requirements of wheat in southern New South Wales by soil analysis. Aust J Exp Agric Anim Hus 3:190–198. https://doi.org/10.1071/EA9630190

Cox MS (2001) The Lancaster soil test method as an alternative to the Mehlich-3 soil test method. Soil Sci 166:484–489

Cox AE, Joern BC, Roth CB (1996) Nonexchangeable ammonium and potassium in soils with a modified sodium tetraphenylboron method. Soil Sci Soc Am J 60:114–120. https://doi.org/10.2136/sssaj1996.03615995006000010019x

Cox AE, Joern BC, Brouder SM et al (1999) Plant-available potassium assessment with a modified sodium tetraphenylboron method. Soil Sci Soc Am J 63:902–911. https://doi.org/10.2136/sssaj1999.634902x

Dodd K, Guppy CN, Lockwood PV et al (2013) The effect of sodicity on cotton: does soil chemistry or soil physical condition have a greater role? Crop Pasture Sci 64:806–815. https://doi.org/10.1007/s11104-009-0196-6

Egnér H, Riehm H, Domingo WR (1960) Chemical extraction methods for phosphorus and potassium. Chemical analyses of soil as a basis for determining soil fertility. Lantbr Högsk Annlr 26:199–215. (in German)

Gavlak RG, Horneck DA, Miller RO (2005) Plant, soil and water reference methods for the Western region, 3rd ed. WREP 125. https://www.naptprogram.org/files/napt/western-states-method-manual-2005.pdf. Accessed 11 May 2020

Gillman GP (1979) A proposed method for the measurement of exchange properties of highly weathered soils. Aust J Soil Res 17:129–139

Gillman GP, Bell LC (1978) Soil solution studies on weathered soils from North Queensland. Aust J Soil Res 16:67–77. https://doi.org/10.1071/SR9790129

Gillman GP, Sumpter EA (1986) Modification to the compulsive exchange method for measuring exchange characteristics of soils. Soil Res 24:61–66. https://doi.org/10.1071/SR9860061

Goulding KWT, Johnston AE, Mallarino AP (2017) What can long-term experiments teach us about potassium management? In: Murrell TS, Mikkelsen RL (eds) Frontiers of potassium science, Rome, Italy. International Plant Nutrition Institute, Peachtree Corners, GA, pp O11–O18. https://www.apni.net/k-frontiers/. Accessed 29 May 2020

Haby VA, Sims JR, Skogley EO, Lund RE (1988) Effect of sample pretreatment on extractable soil potassium. Commun Soil Sci Plant Anal 19:91–106. https://doi.org/10.1080/00103628809367922

Haney RL, Haney EB, Hossner LR et al (2010) Modifications to the new soil extractant H3A-1: a multi-nutrient extractant. Commun Soil Sci Plant Anal 41(12):1513–1523. https://doi.org/10.1080/00103624.2010.482173

Haney RL, Haney EB, Smith DR et al (2017) Removal of lithium citrate from H3A for determination of plant available P. Open J Soil Sci 7:301–314. https://doi.org/10.4236/ojss.2017.711022

Hanway JJ, Barber SA, Bray RH et al (1961) North central regional potassium studies: I. Field studies with alfalfa. North central regional publication no.124. Iowa Agric Home Econ Exp Stn Res Bull 34(494):Article 1

Hanway JJ, Barber SA, Bray RH et al (1962) North central regional potassium studies: III. Field studies with corn. North central regional publication no. 135. Iowa Agric Home Econ Exp Stn Res Bull 503

Helmke PA, Sparks DL (1996) Lithium, sodium, potassium, rubidium, and cesium. In: Sparks DL et al (ed) Methods of soil analysis part 3—chemical methods. SSSA book series 5.3. SSSA, ASA, Madison, WI, pp 551–574. https://doi.org/10.2136/sssabookser5.3.c19

Hinsinger P (2002) Potassium. In: Lal R (ed) Encyclopedia of soil science, 1st edn. Marcel Dekker, New York, pp 1035–1039

Hinsinger P, Elsass F, Jaillard B et al (1993) Root-induced irreversible transformation of a trioctahedral mica in the rhizosphere of rape. J Soil Sci 44:535–545. https://doi.org/10.1111/j.1365-2389.1993.tb00475.x

Houba VJG, Temminghoff EJM, Gaikhorst GA et al (2000) Soil analysis procedures using 0.01 M calcium chloride as extraction reagent. Commun Soil Sci Plant Anal 31(9–10):1299–1396. https://doi.org/10.1080/00103620009370514

Kuhlmann H (1990) Importance of the subsoil for the K nutrition of crops. Plant Soil 127 (1):129–136. https://doi.org/10.1007/bf00010845

Martins PO, Slaton NA, Roberts TL et al (2015) Comparison of field-moist and oven dry soil on Mehlich-3 and ammonium acetate extractable soil nutrient concentrations. Soil Sci Soc Am J 79:1792–1803. https://doi.org/10.2136/sssaj2015.03.0094

McLean EO, Watson ME (1985) Soil measurements of plant-available potassium. In: Munson RD (ed) Potassium in agriculture. American Society of Agronomy, Madison, WI, pp 277–308

Mclntosh JL (1969) Bray and Morgan soil extractants modified for testing acid soils from different parent materials. Agron J 61:259–265. https://doi.org/10.2134/agronj1969.00021962006100020025x

Mehlich A (1984) Mehlich 3 soil test extractant: a modification of Mehlich 2 extractant. Commun Soil Sci Plant Anal 15(12):1409–1416. https://doi.org/10.1080/00103628409367568

Moody PW, Bell MJ (2006) Availability of soil potassium and diagnostic soil tests. Aust J Soil Res 44(3):265–275. https://doi.org/10.1071/SR05154

Morgan MF (1941) Chemical soil diagnosis by the universal soil testing system. Connecticut Agric Exp Stn Bull 450. Univ of Connecticut, Storrs.

Mylavarapu R, Miller R (2014) Mehlich-1. In: Sikora FJ, Moore KP (eds) Soil test methods from the Southeastern United States. Southern Cooperative Series Bulletin No. 419, pp 95–100. isbn#1581614195

Oldham JL (2014) Lancaster. In Sikora FJ, Moore KP (eds) Soil test methods from the Southeastern United States. Southern Cooperative Series Bulletin No. 419, pp 111–117. isbn#1581614195

Pluskie W, Walker R, Young J (2018) Just how much are nutrients redistributed unevenly across the paddock when canola and wheat is windrowed. Grains Research and Development Corporation. https://grdc.com.au/resources-and-publications/grdc-update-papers/tab-content/grdc-update-papers/2018/02/just-how-much-are-nutrients-redistributed-unevenly-across-the-paddock-when-canola-and-wheat-is-windrowed. Accessed 11 May 2020

Qian P, Schoenau JJ, Karamanos RE (1994) Simultaneous extraction of available phosphorus and potassium with a new soil test: a modification of Kelowna extraction. Commun Soil Sci Plant Anal 25(5–6):627–635. https://doi.org/10.1080/00103629409369068

Salomon E (1998) Extraction of soil potassium with 0.01 M calcium chloride compared to official Swedish methods. Commun Soil Sci Plant Anal 29:2841–2854. https://doi.org/10.1080/00103629809370159

Schüller H (1969) Die CAL-Methode, eine neue Methode zur Bestimmung des pflanzenverfügbaren Phosphates in Bäden. Z Pflanzenernähr Däng Bodenkd 123:48–63

Scott AD, Reed MG (1962) Chemical extraction of potassium from soils and micaceous minerals with solutions containing sodium tetraphenylboron: III. Illite. Proc Soil Sci Soc Am 26:45–48. https://doi.org/10.2136/sssaj1962.03615995002600010012x

Scott AD, Smith SJ (1968) Mechanism for soil potassium release by drying. Proc Soil Sci Soc Am 32:443–444. https://doi.org/10.2136/sssaj1968.03615995003200030049x

Scott AD, Hanway JJ, Stickney EM (1957) Soil potassium-moisture relations: I. potassium release observed on drying Iowa soils with added salts or HCl. Proc Soil Sci Soc Am 21:498–501. https://doi.org/10.2136/sssaj1957.03615995002100050010x

Simonis AD, Setatou HB (1996) Assessment of available phosphorus and potassium in soils by the calcium chloride extraction method. Commun Soil Sci Plant Anal 27:685–694

Soltanpour PN, Schwab AP (1977) A new soil test for simultaneous extraction of macro- and micro-nutrients in alkaline soils. Commun Soil Sci Plant Anal 8:195–207. https://doi.org/10.1080/00103627709366714

Song SK, Huang PM (1988) Dynamics of potassium release from potassium-bearing minerals as influenced by oxalic and citric acids. Soil Sci Soc Am J 52:383–390. https://doi.org/10.2136/sssaj1988.03615995005200020015x

Van Lierop W, Gough NA (1989) Extraction of potassium and sodium from acid and calcareous soils with the Kelowna multiple element extractant. Can J Soil Sci 69:235–242. https://doi.org/10.4141/cjss89-024

van Raij B, Quaggio JA, Silva NM (1986) Extraction of phosphorus, potassium, calcium, and magnesium from soils by an ion-exchange resin procedure. Commun Soil Sci Plant Anal 17:547–566. https://doi.org/10.1080/00103628609367733

Whatmuff G, Reuter DJ, Speirs SD (2013) Methodologies for assembling and interrogating N, P, K and S soil test calibrations for Australian cereal, oilseed and pulse crops. Crop Pasture Sci 64:424–434. https://doi.org/10.1071/CP12424

Williams AS, Parvej MR, Holshouser DL et al (2017) Correlation of field-moist, oven-dry and air-dry soil potassium for mid-Atlantic USA soybean. Soil Sci Soc Am J 81:1586–1594. https://doi.org/10.2136/sssaj2016.10.0324

Williams AS, Parvej MR, Holshouser DL et al (2018) Correlation and calibration of soil-test potassium from different soil depths for full-season soybean on coarse-textured soils. Agron J 110:369–379. https://doi.org/10.2134/agronj2017.06.0344

Witter E, Johansson G (2001) Potassium uptake from the subsoil by green manure crops. Biol Agric Hortic 19:127–141. https://doi.org/10.1080/01448765.2001.9754917

Wolf AM, Beegle DB (2011) Recommended soil tests for macronutrients. In: Sims JT, Wolf A (eds) Recommended soil testing procedures for the Northeastern United States. Northeast regional bulletin no. 493, 3rd edn. Agricultural Experiment Station, University of Delaware, Newark, DE, pp 39–47

Woods MS, Ketterings QM, Rossi FS (2005) Effectiveness of standard soil tests for assessing potassium availability in sand rootzones. Soil Sci 170:110–119

Zbíral J, Němec P (2005) Comparison of Mehlich 2, Mehlich 3, CAL, Schachtschabel, 0.01 M $CaCl_2$ and *aqua regia* extractants for determination of potassium in soils. Commun Soil Sci Plant Anal 36:795–803. https://doi.org/10.1081/CSS-200043404

9

Determination and Analysis of Potassium in Plant

T. Scott Murrell and Dharma Pitchay

Abstract Several methods exist for evaluating plant nutritional status. Looking for visual deficiency symptoms is perhaps the simplest approach, but once symptoms appear, crop performance has already been compromised. Several other techniques have been developed. All of them require correlation studies to provide plant performance interpretations. Reflectance is a remote sensing technique that detects changes in light energy reflected by plant tissue. It has proven successful in detecting nutrient deficiencies but does not yet have the ability to discriminate among more than one deficiency. Chemical assays of leaf tissue, known as tissue tests, require destructive sampling but are the standard against which other assessments are compared. Sufficiency ranges provide concentrations of each nutrient that are considered adequate for crop growth and development. They consider nutrients in isolation. Other approaches have been developed to consider how the concentration of one nutrient in tissue impacts the concentrations of other nutrients. These approaches strive to develop guidelines for maintaining nutrient balance within the plant. All approaches require large data sets for interpretation.

Ideally, a diagnostic test, whether of the soil or plant, meets four requirements: (1) it is easy and inexpensive to perform; (2) it definitively identifies a potassium (K) deficiency; (3) it allows farmers time to respond; and (4) it leads to interventions with high probabilities of success. Soil tests have been widely used and have been useful for assessing plant-available K, but they do have diagnostic limitations. As Hall (1905) stated over 100 years ago:

T. S. Murrell (✉)
African Plant Nutrition Institute and Mohammed VI Polytechnic University, Ben Guerir, Morocco

Department of Agronomy, Purdue University, West Lafayette, IN, USA
e-mail: s.murrell@apni.net

D. Pitchay
Department of Agricultural and Environmental Sciences, Tennessee State University, Nashville, TN, USA
e-mail: dpitchay@tnstate.edu

> One of the main problems placed before the agricultural chemist is the estimation of the requirements of a given soil for specific manures.... For various reasons the obvious method of determining the proportions of nitrogen, phosphoric acid, and potash in the soil fails in many cases to give the required information.... Hence from time to time attempts have been made to attack the problem from another side and to use the living plant as an analytical agent. The scheme is to take a particular plant grown upon the soil in question, and determine in its ash the proportions of constituents like phosphoric acid and potash. Any deviations from the normal in these proportions may then be taken as indicating deficiency or excess of the same constituent in the soil and therefore the need or otherwise for specific manuring in that direction.

Hall's statement elucidates the shortcomings of a soil test as the sole guide for K management and captures the rationale for adding plant measurements to the suite of nutritional assessments. This chapter provides an overview of the many "attempts. . .to use the living plant as an analytical agent."

9.1 Visual Symptoms of Potassium Deficiency

When K is deficient, several processes are impaired (Marschner 2002). Low K inhibits enzyme activation, making plants more susceptible to fungal attack. Impaired stomatal activity results in poor control over gas exchange, impairing photosynthesis and water control, making plants more susceptible to stresses from drought, frost, water uptake, and soil salinity. Low K also impairs proton (H^+) exchange across the thylakoid membranes in chloroplasts, resulting in worsening symptoms under higher light intensity (Marschner and Cakmak 1989). Transport of photosynthates can also be impaired, resulting in a buildup of sugars and a reduction in protein and starch synthesis, lowering the plant's dry weight. Impaired lignification of vascular bundles may result in weaker stalks and increased lodging. Potassium is an abundant cation (K^+) found in the cytoplasm, providing cell turgidity and rigidity by maintaining the osmotic potential. The K^+ concentration can be anywhere between 10 and 200 mM and in some cases as high as 500 mM in guard cells and pulvini of Fabaceae family member species. Lack of K results in reduced cell size and number, causing reduced growth and affecting nyctinasty in the Fabaceae family. Because K is mobile in the plant, it can be remobilized from older tissue to younger tissue when uptake from the soil is insufficient; therefore, visual symptoms generally occur first on older tissue, often the most recently matured.

Learning to recognize visual symptoms of deficiency requires training to become familiar with symptoms that can be crop specific (Table 9.1). Potassium deficiency may first appear as deep green plants with shorter and fewer internodes and smaller leaves, followed by the rapid development of necrotic spots along the margins and across leaf blades of recently matured leaves. In most cases, necrotic lesions begin without prior chlorotic lesions. In some cases, chlorosis develops in the tissue surrounding necrotic spots as the necrosis enlarges in advanced stages, or as necrosis is followed by chlorosis on recently matured and maturing leaves.

Table 9.1 Descriptions of visual symptoms of K deficiency for several horticultural crops

Crop	Description
Brassica oleracea L. (broccoli)	Internodes of maturing and recently matured leaves develop purplish concentric circles with green, normal-looking tissues in the center. Gradually, the undersides of these leaves' petioles shrivel, followed by the development of irregular sunken light green concentric circles with normal islands on tissues (Fig. 9.1a). The lesions progress further (Fig. 9.1b–d) with severe shriveling of petioles and the spreading of purplish pigmentation on the internodes. Eventually, necrosis develops across the entire lamina.
Cucumis sativus L. (cucumber)	In the early stages of appearance, the leaf area is reduced (Fig. 9.2a, b). As symptoms worsen, leaves appear deeper green (Fig. 9.2c), and lesions of grayish necrosis develop on recently matured leaves and on matured leaves below them (Fig. 9.2d).
Spinacia oleracea L. (spinach)	Lesions of light greenish, sunken, irregular necrotic spots develop on recently matured leaves (Fig. 9.3a). The lesions rapidly coalesce and progress to distinct whitish necrotic spots on interveinal tissue on the acropetal leaf area (Fig. 9.3b, c).
Cucurbita pepo L. (zucchini squash)	In the early stages, leaves may appear darker green (Fig. 9.4a). Necrotic spots of light whitish pin-head sized lesions develop on interveinal tissue across the leaf blades closer to the primary veins (Fig. 9.4b), which then enlarge to irregular sunken necrotic spots and eventually into large necrotic spots of 0.5–1.0 cm (Fig. 9.4c).
Lactuca sativa L. (romaine lettuce)	Lesions of dark grayish irregular sunken necrotic spots begin at the acropetal area of recently matured leaves (Fig. 9.5a). These necrotic lesions rapidly coalesce into large irregular necrotic patches randomly across the lamina (Fig. 9.5b, c).
Carica papaya L. (papaya)	Necrosis begins as greenish-gray patches just inside the leaflet margins of recently matured leaves and rapidly progresses to large necrotic areas and the tissue collapses along the terminal margins (Fig. 9.6a, b). As patches increase in size and number toward the leaf base, areas between the necrotic patches turn chlorotic and the leaves eventually abscise (Fig. 9.6c).
Solanum tuberosum L. (potato)	Plants initially appear deeper green and compact due to fewer nodes and shorter internodes (Fig. 9.7a). Further K deprivation results in minor puckering and crinkling of leaves, with slight cupping of recently matured leaves (Fig. 9.7b). Plant growth is reduced. These symptoms progress rapidly to severe puckering and reduced leaf size (Fig. 9.7c). Randomly spaced necrotic areas first appear on the abaxial leaflet surface and along the secondary and tertiary veins of outermost leaflets of recently matured leaves. The symptoms gradually progress to the adjacent leaflets and then to the remaining leaflets. As the deficiency becomes severe, dark necrotic spots develop on the adaxial surface of maturing leaves and necrotic areas enlarge.
Arachis hypogaea L. (Peanut)	Recently matured/matured leaves initially appear slightly darker green (Fig. 9.8a). As the K deprivation extends, lesions of light grayish irregular sunken necrotic spots begin to develop across the lamina (Fig. 9.8b). The lesions rapidly progress to distinct brownish necrotic spots surrounded by chlorotic tissues (Fig. 9.8c).

Fig. 9.1 Potassium deficiency symptoms exhibited on *Brassica oleracea* L. (broccoli). For a description, see Table 9.1

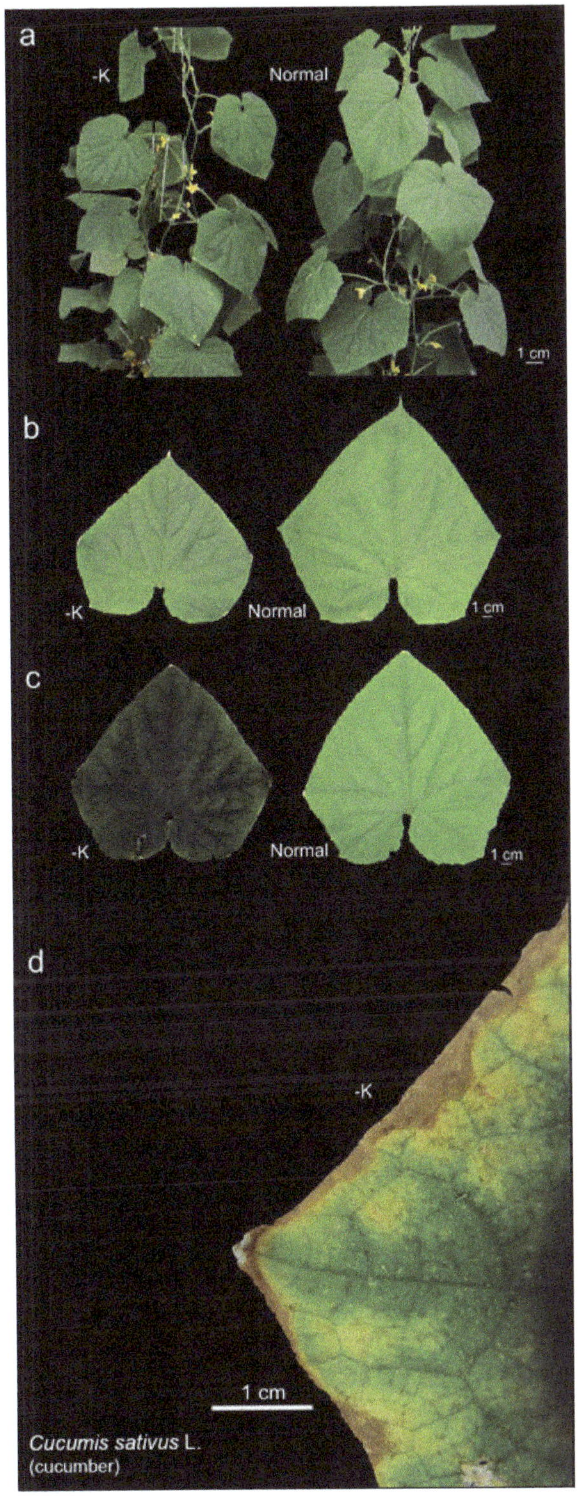

Fig. 9.2 Potassium deficiency symptoms exhibited on *Cucumis sativus* L. (cucumber). For a description, see Table 9.1

Fig. 9.3 Potassium deficiency symptoms exhibited on *Spinacia oleracea* L. (spinach). For a description, see Table 9.1

Fig. 9.4 Potassium deficiency symptoms exhibited on *Cucurbita pepo* L. (zucchini squash). For a description, see Table 9.1

Fig. 9.5 Potassium deficiency symptoms exhibited on *Lactuca sativa* L. (romaine lettuce). For a description, see Table 9.1

Fig. 9.6 Potassium deficiency symptoms exhibited on *Carica papaya* L. (papaya). For a description, see Table 9.1

Fig. 9.7 Potassium deficiency symptoms exhibited on *Solanum tuberosum* L. (potato). For a description, see Table 9.1

Fig. 9.8 Potassium deficiency symptoms exhibited on *Arachis hypogaea* L. (peanut). For a description, see Table 9.1

Nutrient deficiency symptoms can be used as a rapid diagnostic tool for identifying factors that might limit crop yield and quality. Deficiency symptoms are not the ideal approach for dealing with nutritional shortages, because by the time symptoms are visible, plant productivity has already been impaired by the period of malnutrition preceding the appearance of symptoms, a condition called "hidden hunger."

9.2 Light Reflectance

Potassium deficiency can cause changes in both individual plant organs and, collectively, the crop canopy. When leaf tissue becomes chlorotic or necrotic, it no longer reflects light the same way it did when it was healthy. Visibly, leaves change from green to yellow or brown, but changes also occur outside the visible spectrum, particularly at the nexus of visible and infrared wavelengths.

When incident energy hits the surface of a leaf or other plant organ, it is either absorbed, reflected, or transmitted through the tissue. Several types of detectors can measure, over a range of wavelengths, the energy not absorbed by the plant. Under controlled conditions, positioning both the light source and the energy detector on the same side of the plant tissue measures reflected energy. Positioning the light source behind the tissue and the detector in front of the tissue measures transmitted energy. In spectral analysis, the reflectance is the reflected energy expressed as a percentage of the incoming energy. To calculate this percentage, the energy of the light source must be known. Some instruments provide their own light source of known energy. Instruments without a light source require reference materials with a known reflectance, such as a white panel (Albayrak et al. 2011). A higher reflectance means a lower energy absorbance. Because energy measurements do not require contact with plant tissue, they are considered a type of "remote sensing." For agricultural uses, energy detectors have been mounted on tractors, airplanes, unmanned aerial vehicles, and satellites.

Figure 9.9 is an example of a reflectance spectrum. Reflectance (%) is plotted over a range of wavelengths (nm). Reflectance in the visible spectrum, 400–700 nm, is less than reflectance in the infrared spectrum (700 nm–1 mm). Chlorophylls, carotenoids, and anthocyanins absorb light in the visible spectrum (Knipling 1970), and combined they absorb more energy in the blue and red wavelengths and less in the green wavelengths (Shull 1929). The longer wavelength infrared light is not absorbed by the aforementioned phytochemicals, resulting in a rapid increase in reflectance at the end of the visible and the beginning of the infrared spectrum, termed the "red edge" (Horler et al. 1983). Farther into the longer wavelength range of the infrared spectrum (not shown in Fig. 9.9) several compounds in the plant absorb energy. Among these are protein, oil, starch, lignin, cellulose, water, and sugar (Curran 1989). Additionally, water absorbs energy across both visible and infrared spectra.

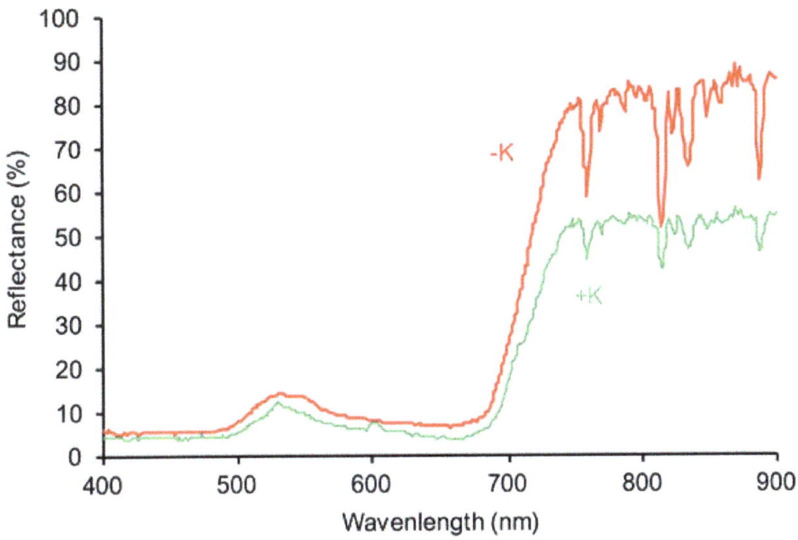

Fig. 9.9 Reflectance spectra for wheat (*Triticum aestivum* L.) without K (−K) and with K (+K), and all other nutrients at sufficient levels. (adapted from Ayala-Silva and Beyl 2005)

A red edge at lower wavelength is associated with K deficiency. Figure 9.9 shows two spectra for wheat (*Triticum aestivum* L.), one where K was added to the nutrient solution (+K) and the other where K was omitted (−K) (Ayala-Silva and Beyl 2005). The K deficiency resulted in a red edge at shorter wavelengths and increased reflectance in the visible range (400–700 nm). When tissue yellows or becomes necrotic due to nutritional deficiencies or other damage, less chlorophyll is available to absorb energy in the visible spectrum, and reflectance increases (Shull 1929). Also, when tissue dries, it typically reflects more energy (Woolley 1971). A confounding factor in analyzing shifts in the red edge is that as plant tissue ages, the red edge shifts to longer wavelengths (Horler et al. 1983; Milton et al. 1991). Therefore, a red edge occurring at shorter wavelengths under K deficiency could be interpreted as an "...inhibition of the normal shift to longer wavelengths" (Milton et al. 1991).

Spectral analysis usually involves creating indexes that are diagnostic of plant nutritional status. Commonly used metrics of nutritional status are K concentration in dry tissue [g K (kg dry matter)$^{-1}$] and total K accumulation in plant biomass (kg K ha^{-1}). Indexes are created from the combinations of wavelength reflectance measurements that correlate most strongly with K concentration or K accumulation. In remote sensing applications, spectra are the result not only of the absorption of light by the crop canopy but also absorption by water and other compounds in the atmosphere and in the soil not covered by the canopy.

Vegetative indexes are calculated by a normalization technique that uses ratios of wavelength combinations. The normalized difference vegetation index (NDVI) is a commonly used ratio. It is a combination of reflectance in the near-infrared spectrum (R_{NIR}) and reflectance in the red region (R_{red}). Specifically, NDVI = (R_{NIR} − R_{red})/

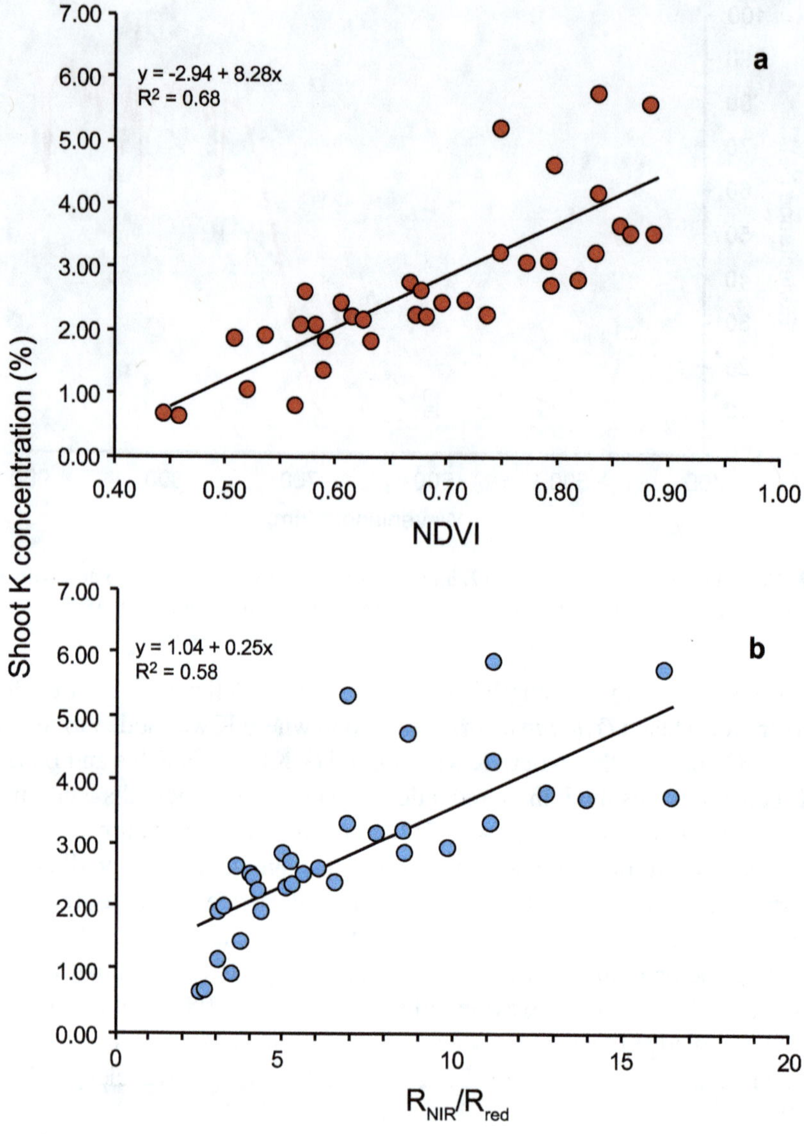

Fig. 9.10 Correlation of *Vicia villosa* Roth (hairy vetch or winter vetch) shoot tissue K concentration with (**a**) the normalized difference vegetative index (NDVI) and (**b**) the ratio of near-infrared reflectance to reflectance in the red region (R_{NIR}/R_{red}). (adapted from Albayrak et al. 2011)

($R_{NIR} + R_{red}$) (Tarr et al. 2005). Another often used index is the simple ratio R_{NIR}/R_{red} (Birth and McVey 1968).

For example, Albayrak et al. (2011) correlated NDVI and the simple ratio to leaf K concentrations of shoot tissues of three vetch (*Vicia*) species. Figure 9.10 shows the correlations with both indexes for *Vicia villosa* Roth, which was representative of the correlations of the other two *Vicia* species. The NDVI correlation was somewhat better than that of the simple ratio. Many other vegetative indices have been tested using tissue K concentration: green-to-red and near-infrared-to-green ratios (Gómez-

Casero et al. 2007); green normalized difference vegetation index (GNDVI), soil adjusted vegetation index (SAVI), optimized soil adjusted vegetation index (OSAVI), N_870_1450 and N_1645_1715 (Mahajan et al. 2014; Pimstein et al. 2011); P_1080_1460, S_660_1260, and S_660_1080 (Mahajan et al. 2014); vegetation vigor index (VVI) (Noori and Panda 2016); the GM Index, the Vogelmann Index (VOG), the Green Leaf Index (GLI), the normalized difference index (NDI), and a simple ratio RI (Anderson et al. 2016); and the re-normalized difference vegetation index (RDVI) (Guo et al. 2017). Other normalization procedures are continuum removal and band depth (BD) normalization (Kokaly and Clark 1999); continuum-removed derivative reflectance (CRDR), the normalized band depth ratio (BDR), and the normalized band depth index (NBDI) (Mutanga et al. 2004).

Statistical techniques are commonly used to create multivariate models of wavelengths that account for the most variation in either tissue K concentration or total K accumulation. The approach is to develop a statistical model with one set of data, termed a calibration or training data set, then test how well that model predicts either K concentration or total K accumulation in another data set, termed the validation, prediction, or test data set (Geladi and Kowalski 1986). The most widely used statistical technique has been stepwise regression; however, uninformed use of this approach can lead to overfitting, making a model "...unlikely to be replicated if the experiment is repeated" (Mutanga et al. 2004). Another criticism of stepwise regression has been that the wavelengths most highly correlated with plant chemical content may not relate to known absorption features of those chemicals (Curran et al. 1992). To address concerns of overfitting, researchers are turning to other techniques to build models, such as partial least squares (PLS) regression (Geladi and Kowalski 1986), kernel partial least squares regression (KPLSR), and support vector regression (SVR) (Pullanagari et al. 2016).

The major difficulty that has yet to be overcome is the inability of reflectance to discriminate among more than one deficient element. Predictive models have been built for individual nutrients, but not for combinations of nutrients. Fridgen and Varco (2004) conducted a study with a factorial combination of nitrogen (N) and K fertilizer application rates applied to cotton (*Gossypium hirsutum* L.). They built a predictive model for leaf N content using partial least squares regression. When K was limiting, the model performed poorly, as measured by the correlation between predicted and observed N content ($r^2 = 0.06$); however, when K was adequate, the model performed markedly better ($r^2 = 0.70$). From their experiment examining several isolated nutrient deficiencies, Pacumbaba and Beyl (2011) concluded that spectral analysis could detect nutrient stress but was not able yet to identify which nutrient was causing the stress.

An additional consideration is when to take reflectance measurements. Early detection of K deficiency is the goal. As discussed in the previous section, K deficiency can progress rapidly from "hidden hunger" to complete tissue necrosis. Creating diagnostic interpretations of reflectance spectra needs to consider the dynamic nature of K deficiency during the plant's lifecycle.

9.3 Plant Tissue Chemical Content

The most widely adopted determinations of plant tissue K content are those performed in the laboratory on tissue samples collected from the field. Farmers or consultants gather samples of specific tissue at specific growth stages (Jones 1998). After the laboratory receives the samples, technicians dry them, typically in a forced-air oven, then grind them to reduce particle size. A small subsample of the ground material is weighed out and digested in an acid solution (Campbell and Plank 1998; Hanlon 1998). The quantity of K in an aliquot of the digestate is determined using either an atomic absorption spectrophotometer (Hanlon 1998) or an inductively coupled plasma atomic emission spectrometer (Isaac and Johnson 1998). Although laboratory determination of K content is the scientific standard, it is not available everywhere, since it requires capital to set up the laboratory, an appropriate business plan to sustain laboratory operations, trained personnel, well-developed infrastructure to transport samples and maintain equipment, and quality assurance and quality control procedures to ensure results are accurate and precise.

For meaningful interpretation, tissue test results must be correlated to crop performance. The stronger the correlation, the more useful tissue analyses become as a diagnostic tool. Yield has traditionally been the performance metric of universal interest; however, crop quality and plant health can be just as important, depending on the crop and the requirements for marketable yield. The objective of correlations is to identify test levels associated with K sufficiency and balanced nutrition.

Balance considers concentrations of other nutrients in plant tissues. The tissue concentration of K is influenced by the supply of other cations such as ammonium, magnesium (Mg), calcium (Ca), iron, manganese, copper, and zinc. The growing environment also plays an important role in the determination of the critical level of K, since tissue K concentrations are influenced by factors such as incident radiation, light intensity, air and soil temperature, soil moisture, etc. The demand for K also changes during the season, as discussed in Chap. 1, with increases in many species during the reproductive stage, especially during flowering and fruiting. The following sections discuss approaches that have been developed to create diagnostic criteria.

9.3.1 Sufficiency Ranges (SR)

Macy (1936) formulated many of the concepts used today for interpreting tissue test correlations. He assembled data from studies that added or subtracted a given nutrient from the plant-available supply and measured changes in biomass production and nutrient concentration in plant tissue. He defined yield response as the decrease in any given yield from the maximum yield. He plotted those yield decreases against the associated tissue nutrient concentrations to create correlations like the one in Fig. 9.11. His approach considered one nutrient at a time and assumed levels of other nutrients were adequate. He defined "poverty adjustment" as the

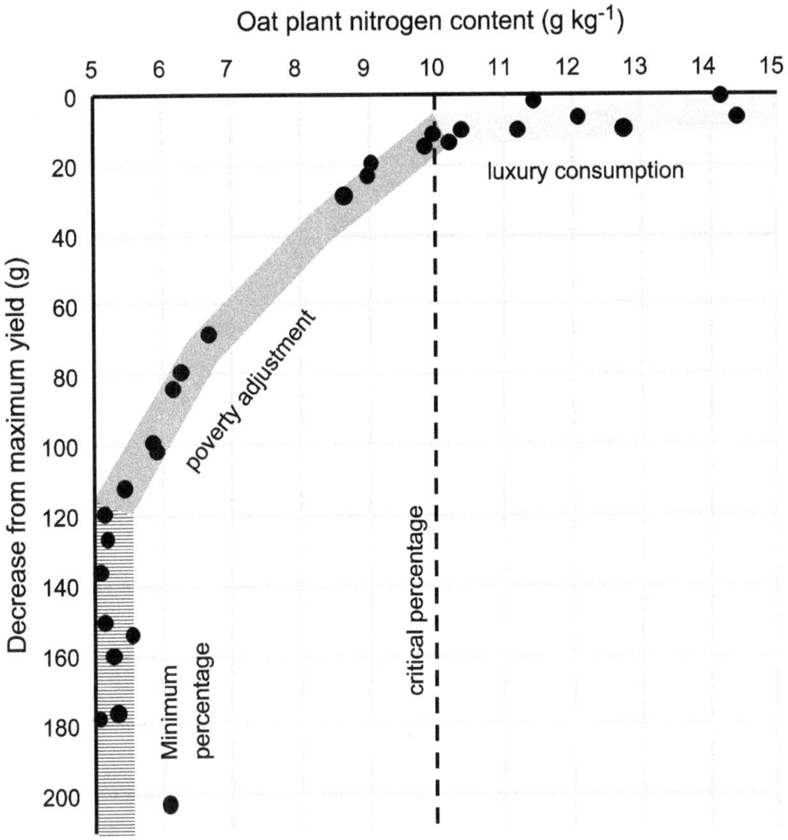

Fig. 9.11 Interpretation of oat (*Avena sativa* L.) tissue test correlation data into three categories: minimum percentage, poverty adjustment, and luxury consumption, the latter two divided by the critical percentage. (Macy 1936)

range where biomass yield climbed toward maximum yield as tissue nutrient concentration increased. At higher nutrient concentrations, biomass no longer increased, a range he termed "luxury consumption." He then defined the "critical percentage" as the tissue concentration dividing poverty adjustment from luxury consumption. He also observed that a range of the largest yield decreases occurred within a relatively narrow range of low nutrient concentrations. He termed this range "minimum percentage." As Macy's concept was adopted by others, relative yield or actual yield was used as the dependent variable rather than the decrease from maximum yield. Relative yield is the ratio of observed yield to maximum yield. Research has been going on for decades on numerous crops to develop correlations between tissue concentrations and crop yields, and studies are periodically summarized in reviews such as Hardy et al. (1967) and Westerman (1990). While correlations of yield or relative yield with tissue concentrations provide the basis for diagnosing a K deficiency, they do not indicate how much K needs to be applied to rectify the deficiency. Calibration experiments, like those discussed in Chap. 1, provide that kind of information.

The selection of models fit to correlation data influences the determination of critical level or range. As an example, we compare models fit to data correlating maize (*Zea mays* L.) yield to K concentration in the leaf blade opposite and below the ear, sometimes called the "sixth leaf" or the "ear leaf." Tyner (1947) modeled his ear leaf correlation data set using actual yields and a linear model. His data did not have a clear "flex point" where yield no longer increased with K concentration (no luxury consumption). He set the critical concentration at the concentration where only two higher yielding observations at lower concentrations fell outside the standard error interval. Stammer and Mallarino (2018), correlating leaf concentrations to maize and soybean (*Glycine max* (L.) Merr.) relative yields, defined a critical range, rather than a single level, from two different least squares statistical models fit to the same data. The inflection of the linear-plateau model set the lower limit of the range and the plateau of the quadratic-plateau model set the upper limit. In addition to those already mentioned, other statistical models commonly used in nutrient response or correlation studies are: linear segmented models (Anderson and Nelson 1975); quadratic, exponential, and square root functions (Cerrato and Blackmer 1990; Mombiela et al. 1981); and the "Cate-Nelson" procedure (Cate and Nelson 1971).

Empirical data comprising tissue test correlations are gathered under specific conditions. When conditions change, interpretations may also change. For instance, Kovács and Vyn (2017) correlated yields and ear leaf K concentrations of modern maize hybrids and found that the lower limit of the sufficiency range in existing interpretations was too low. Their more recently collected data revealed that a higher range in tissue K concentrations needed to be recommended to farmers. This example demonstrates that tissue test correlations must be reevaluated when current practices or growing conditions differ from the historical ones upon which the recommendations were based. Unfortunately, funds for conducting such research are usually sparse and only sporadically available. Unmet funding needs have limited scientists' ability to update data sets in a timely manner, resulting in interpretations and recommendations that may reflect older cropping practices, genetics, and climatic conditions.

Tissue K concentration is affected by factors other than the quantity of plant-available K in soil. Over 100 years ago, Hall (1905) noted that environmental factors can alter nutrient concentrations as much as variations in the K content of the soil. To quiet the noise from other factors, Mallarino and Hagashi (2008) used relative rather than absolute tissue concentrations. In their field experiments, they calculated relative concentrations for a given season at a given site (site-year) by dividing tissue K concentrations from unfertilized plots by the tissue K concentration of the highest treatment mean. They then aggregated data across all site-years and examined the relationship between soil test level and either absolute or relative tissue concentration. Relative tissue concentrations were better related to changes in soil test levels than absolute concentrations; however, they did not examine the relationship between relative concentration and relative yield, likely because only small yield responses to K fertilization occurred in the study.

Table 9.2 Interpretations of leaf K concentrations of mature Valencia and navel orange (*Citrus sinensis* (L.) Osbeck) leaves. (Embleton et al. 1978)

Interpretation	Leaf K concentration g K (kg dry matter)$^{-1}$
Deficient	<4.0
Low	4.0–6.9
Optimum	7.0–10.9
High	11.0–20.0
Excess	>23.0

Although correlation studies are essential, they do not completely define the process used to develop diagnostic criteria for tissue tests. To quote Melsted et al. (1969):

> Critical plant composition values can seldom be derived through a single carefully designed experiment. Rather they evolve through hundreds of fertility trials and the resulting thousands of plant analyses. Slight variations in plant composition may result through differences in individual plants, in plant varieties, in seasonal changes, in nutrient levels, in nutrient ratios, and in various other factors. There is no good way to evaluate or balance some of these differences except through personal judgement.

To elucidate the process of deriving tissue test interpretations for production settings, we explore an example from the citrus industry in California. We begin with the current recommendations. Embleton et al. (1978) published the interpretations for leaf tissue K concentration reproduced in Table 9.2 which are still used today. Farmers and consultants compare their own test results with those in the table to assess the K status of their orange trees. The interpretations are valid for specific plant tissue: 5–7-month-old, spring cycle, terminal leaves from nonfruiting, and nonflushing shoots that are 0.9–1.5 m above the ground. These interpretations apply to orange (*Citrus sinensis* (L.) Osbeck), grapefruit (*Citrus paradisi* Macfad.), and lemon (*Citrus limon* (L.) Osbeck). To reach the desired number of fruit per tree, farmers manage fertilizer K to keep tissue tests in the optimum range (7.0–10.9 g kg^{-1}). To attain the desired fruit size and quality, a higher concentration in the range of 10.0–12 g kg^{-1} is recommended. Although not shown in Table 9.2, ranges are provided for other nutrients too. The presentation of interpretation ranges for individual nutrients with associated descriptions of tissue at a given range of positions on the plant and an associated age range is representative of guidelines in use for many crops and nutrients (Mills and Jones 1996; Shear and Faust 2011; Uchida 2000). Commonly, the optimum range is called the "sufficiency range." Baldock and Schulte (1996) classified sufficiency ranges as an independent nutrient index (INI), because they represent the sufficiency of one nutrient independently of other nutrients.

The tissue K interpretations in Table 9.2 are a result of a long period of work. We explore the process followed to create the entries in that table. Doing so reveals the many steps required to create tissue test interpretations.

Southern California farmers' desire for evidence-based management practices for oranges sparked the long process of creating tissue test interpretations. The Citrus Experiment Station was established at Riverside, California in 1907. Field trials were also started that year. Research summaries published in 1922 (Vaile 1922) and 1942 (Parker and Batchelor 1942) concluded that N was the limiting nutrient for Valencia and navel oranges, not K; however, Chapman and Brown (1943) stated that despite the lack of experimental evidence, many farmers continued to believe that K was necessary to sustain fruit yield and quality.

Chapman and his colleagues began investigating K nutrition more closely. Their initial work focused on characterizing the effects of K deficiency, sufficiency, and excess on the growth and fruit production of Valencia and navel orange trees (Chapman and Brown 1943; Chapman et al. 1947). To do this, they initially conducted greenhouse studies using nutrient solutions with and without K to compare differences in leaf tissue concentrations as well as visual deficiency symptoms. They also conducted "controlled-culture experiments" on 4-year old trees growing in large (700 L) pots outdoors under ambient conditions and fed with nutrient solutions with varying levels of K (3–7 ppm K, 30–40 ppm K, and 300–400 ppm K) (Chapman et al. 1947). They noted that in the early stages of K deficiency, trees showed only faint symptoms and set fruit normally, but at harvest, their fruit was much smaller. Although the K-deficient trees did not exhibit clear visual symptoms, the K concentrations in their leaves were much lower than those where K was sufficient. Driven by their concern that "incipient" K deficiencies likely went unnoticed in commercial orchards (termed "hidden hunger" earlier in this chapter), Chapman and Brown (1950) set up a series of investigations to determine if the K concentration in leaves could serve as a diagnostic test.

Although they had been analyzing leaves in their work, Chapman and Brown (1950) were not sure leaves were the best tissue for diagnosing tree nutrient status. They took samples from a long-term experiment where navel orange trees had been growing for over 13 years on differentially fertilized plots. Given the long time they had to respond to differences in fertility, trees on these plots were the most likely to exhibit clear differences in tissue K concentration. Unfortunately, the effects of K could not be isolated, since S was applied with K. They sampled: leaves; small, "pencil-sized" terminal shoots (twigs); blossoms; immature fruit; and mature fruit. An excerpt from their results is in Table 9.3. Comparing the pair of treatments without K (N and N + P) to treatment with K (N + P + K + S and manure) showed that leaves were the plant organs most sensitive to changes in K nutritional status. In a separate study, Chapman and Brown (1950) also tested petioles and found they offered no advantage over leaves. Foundational work of others on a variety of crops, summarized by Ulrich (1952), showed that nutrient concentrations in vegetative organs of plants were often more sensitive to changes in soil nutrient supply than those in fruits, seeds, or tubers.

Chapman and Brown (1950) investigated the best time to sample leaf tissue. Under southern California conditions, orange trees, which are evergreen, have three cycles of new leaf growth (flushes) during the following approximate times of the year: April (first or spring flush), June (second flush), and August through September (third flush). Experiments examining the time of sampling revealed that K concentrations generally decreased in leaves as they aged, a phenomenon noted by many other researchers working on many other crops, for example, Thomas (1937)

Table 9.3 Tissue K concentrations in various organs of navel orange trees (*Citrus sinensis* (L.) Osbeck) growing for over 13 years on differentially fertilized plots in a long-term experiment. (Chapman and Brown 1950)

	Twig bark	Twig wood	Leaves	Blossoms	Immature fruit	Mature fruit
Treatment[a]	g K (kg dry matter)$^{-1}$				g K (kg fresh weight)$^{-1}$	
N	3.9	1.2	6.5	12.9	2.5	1.4
N + P	3.3	1.5	6.8	12.9	2.4	1.4
N + P + K + S	2.1	1.1	10.7	12.6	2.6	1.5
Manure	3.8	2.0	13.2	15.5	3.1	1.4

[a]N = nitrogen applied as urea; P = phosphorus applied as triple superphosphate; K + S = potassium and sulfur co-applied as potassium sulfate; manure = dairy manure

studying potato leaves, Bell et al. (1987) examining soybean leaves, and Xue et al. (2016) analyzing rice (*Oryza sativa* L.) leaves. Using data from the "controlled-culture experiments," Chapman and Brown (1950) determined that K concentrations in 12-month-old leaves taken from shoots that had been spring flush, fruit-bearing terminal shoots the previous year had the lowest K concentration (Fig. 9.12). They posited that by 12 months, some of the K in these older leaves had translocated to newly developing tissue. The K concentration in the older leaves had the clearest separation between deficient, slightly deficient, and ample K nutrition; however, accurately identifying these leaves in commercial orchards was questionable. Also, sampling at this leaf age was limited to a short period in the spring. Both of these practical limitations led to recommendations for sampling younger leaves, 4–7 months old, located immediately behind the fruit on spring flush, terminal fruiting shoots (Chapman 1949, 1960; Chapman and Brown 1950; Embleton et al. 1962). These leaves were more readily identifiable and could be sampled over a several-month period, giving them practical value as a diagnostic test.

Chapman and Brown (1950) conducted additional studies to provide further details on sampling leaves. They examined how high to sample on the tree and found that leaf K concentrations at the top (3.0–4.9 m) were slightly lower than other positions. Samples taken at 1.5–3.0 m were no different than those at 0.6–1.5 m. The lowest height was easiest to reach and that height has carried through to the current recommendation of 0.9–1.5 m. Other investigations by Chapman and Brown (1950) showed that leaf K concentrations were relatively constant on various sides of the tree. Large and small leaves were also similar in concentration. Chapman and Brown (1950) summarized their own as well as other scientists' data, including international data sets, and published the first set of interpretations.

In later work, Jones et al. (1955) and Embleton et al. (1962) found that sampling the youngest fully expanded leaf from nonfruiting, rather than fruiting, terminal shoots resulted in more reliable interpretations. The current recommendation in California is to sample 5–7-month-old, spring cycle, terminal leaves from nonfruiting, and nonflushing shoots (Embleton et al. 1978). That approach has proven so reliable that it is used by the citrus industry worldwide.

In developing sufficiency ranges, both Chapman and Embleton gathered data from farmers' orchards. They surveyed leaf concentrations in orchards of

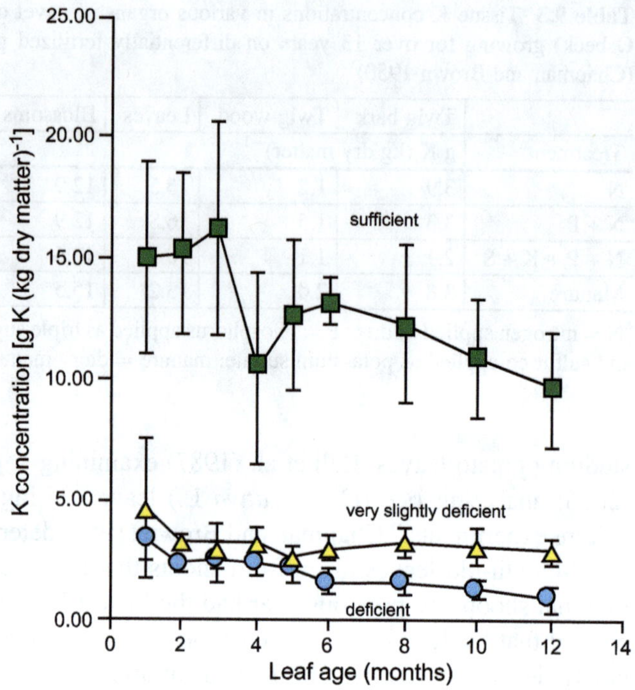

Fig. 9.12 Temporal changes in Valencia and navel orange (*Citrus sinensis* (L.) Osbeck) leaves taken from spring flush, fruit-bearing terminals, based on data in Table 6, Chapman and Brown (1950), and experiment descriptions in Chapman et al. (1947). Error bars are ±2 standard deviations

top-producing farmers. Additionally, Embleton worked with growers whose orchards were deficient in one or more nutrients and established fertilizer rate trials there. Stemming from such practical research, Embleton et al. (1974) stated that one of the reasons for setting 7.0 g kg^{-1} as the threshold between "low" and "optimum" was their experience that it was difficult to increase leaf K concentration once it fell below this level. Also, Embleton et al. (1978) looked beyond the number of fruit per tree and examined other quality parameters, such as fruit size and quality. On-farm research, on-farm surveys, on-farm experience, and goals for not only yield but also crop quality helped define the interpretations used today.

An important observation by Chapman and Brown (1950) was that K concentrations in leaves were influenced by the concentrations of other nutrients and vice versa. Table 9.4 shows that when K was deficient, the leaf concentrations of N, Ca, and Mg increased. Additionally, when the other listed nutrients were deficient, the concentration of K in leaves increased.

The implication of the findings in Table 9.4 is that accurate interpretations of adequate and higher K concentrations depend upon knowing the concentrations of other nutrients. Higher K concentrations considered in isolation could be misinterpreted as sufficient when other nutrients are deficient. Chapman and Brown (1950) concluded that low leaf K concentrations were diagnostic and would not produce a false positive for K deficiency. However, when other nutrients were limiting, higher K concentrations could produce a false negative for K deficiency. The possibilities for misdiagnosis have led to ongoing research on how to consider more than one nutrient at a time when interpreting tissue analyses.

Table 9.4 Directional changes in nutrient concentrations of orange (*Citrus sinensis* (L.) Osbeck) leaves when other nutrients are deficient. (adapted from Chapman and Brown 1950)

Nutrient	Directional change in K leaf concentration due to a deficiency in the nutrient in the first column	Directional change in leaf concentration of the element in the first column due to a deficiency in K[a]
Nitrogen	Increase	Increase
Phosphorus	Increase	?
Calcium	Increase	Increase
Magnesium	Increase	Increase
Sulfur	Increase	?
Boron	Increase	?
Iron	Increase	?
Zinc	Increase	?

[a]? indicates the direction is not known with certainty

9.3.2 *Diagnosis and Recommendation Integrated System (DRIS)*

The importance of nutrient interactions on tissue concentrations was an early observation. Hall (1905) noted that the total quantity of sodium (Na) and K, summed together, was relatively constant but the cation that contributed more to the sum was the one in greater supply in the soil. He observed that,

> Any abundance of soda acts as a diluent and reduces the proportion of potash in the mangel ash, even though the plant may have an excess of potash available. In consequence the normal proportion of potash in the ash of the mangel will vary with factors other than the potash content of the soil...

The most well-known approach for considering multiple nutrients is the Diagnosis and Recommendation Integrated System (DRIS), developed by Beaufils (1973) in South Africa from his work on maize and rubber trees. The objective of DRIS is to rank nutrients in their order of sufficiency to one another. The rank reveals which nutrients are most limiting and which are least, relative to each other. The rank does not, however, indicate whether a crop will respond favorably to fertilization if the lowest-ranked nutrient or nutrients are added (Sumner 1990). Baldock and Schulte (1996) classified DRIS as a dependent nutrient index (DNI), because it considers two or more nutrients in its interpretations.

DRIS begins by assembling all available data, typically hundreds to thousands of records, into a database. The minimum data required for each record are yield and the tissue nutrient concentrations associated with that yield. This large data set is parsed into at least two subpopulations based on yield: low and high. Often, the yield delineating the subpopulations is chosen from experience. For example, Sumner (1977a) set the delineating soybean grain yield at 2.6 Mg ha^{-1}.

The next step is to create parameters, which are ratios of two nutrient concentrations. Continuing with the soybean example, Sumner (1977a) created the following parameters for the nutrients N, phosphorus (P), and K: N/P, N/K, and K/P

Table 9.5 An example of DRIS parameters and norms for N, P, K concentrations in soybean (*Glycine max* (L.) Merr.) leaves derived from 879 observations in the low-yielding subpopulation and 366 observations in the high-yielding subpopulation. (excerpted from Sumner 1977a)

	High-yielding subpopulation					
Parameter	Mean or "norm"[a]	Standard deviation (sd)	Coefficient of variation (CV) (%)	Variance	Variance in the low-yielding subpopulation	Ratio of variances (low/high)[b]
N/P	13.77	2.72	20	7.40	19.89	2.69**
N/K	2.43	0.50	21	0.25	1.88	7.52**
K/P	5.97	1.47	25	2.16	6.81	3.15**

[a]Because units of concentrations ratios are unity, units of all descriptive statistics except the coefficient of variation are also unity. The means of each parameter from the high-yielding subpopulation are the DRIS "norms"

[b]** Indicates the variance ratio is highly statistically significant

(Table 9.5). He calculated descriptive statistics for each one: mean, variance, standard deviation (sd), and coefficient of variation (CV). Beaufils (1973) advised ensuring the distribution of each parameter in each yield subpopulation be normally distributed. None of the statistics for the parameters have units, since nutrient concentrations are expressed as ratios of one another. In the soybean example, Sumner calculated statistics for N/P, N/K, and K/P within each yield subpopulation. Table 9.5 shows all of these statistics for the high yield subpopulation but displays only the variance for the low-yielding one. To determine if a parameter should be included, Sumner (1977a) calculated the ratio of the low-yielding variance to the high-yielding variance. The assumption was that higher yields were less variable than lower yields. If the variance ratio was statistically significant, that parameter was included. Beaufils and Sumner (1976) interpreted significance to mean that, "...a relationship between yield and...plant composition is possible and that this relationship...can be exploited to suit diagnostic purposes." Table 9.5 shows that the variance ratios of all three parameters were highly statistically significant, so he included all of them.

In DRIS, the means of each parameter from the high-yielding subpopulation are "norms." When the yield is normally distributed across levels of a parameter, the parameter mean is associated with the highest yields in the high yield subpopulation. Walworth et al. (1986) noted that high-yielding subpopulations were less likely to be skewed and more likely to follow the normal distribution, providing an additional reason for using them when creating norms. Norms are the standards used for interpreting tissue concentrations from individual data sets and represent the target levels of each parameter. The DRIS norms are the basis for ordering nutrients according to their sufficiency in the plant. There are two ways of creating this ranking: (1) graphically using a DRIS chart or (2) numerically using DRIS indexes.

9.3.2.1 DRIS Chart

The DRIS chart (Fig. 9.13) ranks three nutrients qualitatively. When more than three nutrients are analyzed, the user interprets multiple DRIS charts simultaneously, as outlined in Beaufils and Sumner (1976). In the soybean example, the chart is composed of three axes, one for each parameter. The arrowheads on each axis indicate the direction of the increase in each parameter. The intersection of the three axes is the norm for each parameter (the "norms" in Table 9.5): 13.77 for N/P, 2.43 for N/K, and 5.97 for K/P. The inner circle has a radius of (2/3) × sd for each parameter (Sumner 1977a). For example, the radius of the circle for the N/P line is (2/3) × 2.72 = 1.813. Adding and subtracting that value from the N/P norm (13.77) calculates the two intersection points of the inner circle on the N/P line: 13.77 + 1.813 = 15.6 and 13.77–1.813 = 12.0. The outer circle has a radius of (4/3) × sd. Calculations are repeated for the other two lines to find their intersection points with both circles. The arrows next to the element symbols denote the balance of the nutrients on either side of a given axis. Keeping with the N/P axis, the inner circle represents nutrient balance between N and P, denoted by the horizontal arrows next to each nutrient. Within this circle, tissue concentrations are close to norms and are therefore associated with the highest yields in the high yield subpopulation. Moving upward on the N/P axis to the zone between the two concentric circles, the level of N begins to become too high, denoted by the upward slanted arrow, and the level of P starts to become too low, denoted by the downward slanted arrow. This is a transition zone. Moving still farther upward outside the second circle is nutrient imbalance. N is too high (arrow straight up) and P is too low (arrow straight down). An N/P ratio outside the second circle is associated with the lowest yields in the high

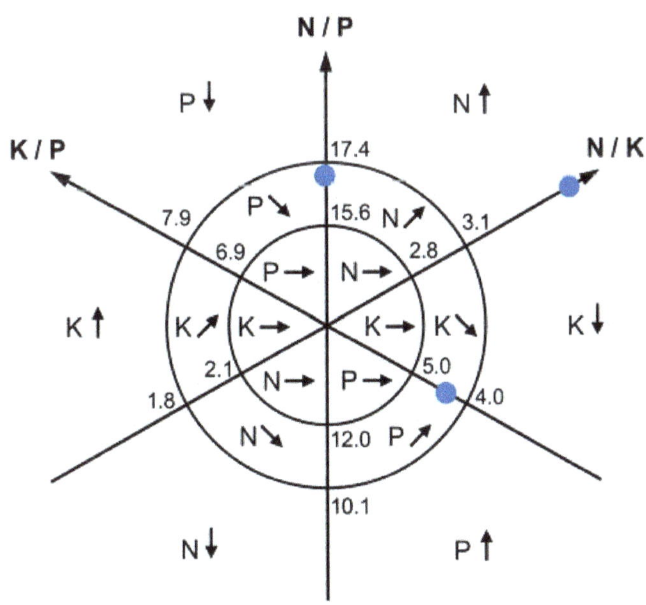

Fig. 9.13 DRIS chart for the parameters N/P, N/K, and K/P. The radius of the inner circle is (2/3) × sd and that of the outer circle is (4/3) × sd. The blue points are the values of the parameters for the second entry (92 days after emergence) in Table 9.6: N/P = 17.00; N/K = 3.86; K/P = 4.40

yield subpopulation. Imbalances between N and P move in the opposite direction downward from the center on the N/P axis, with P being too high and N being too low outside the second circle on the bottom of the N/P axis. Similar interpretations exist for the other nutrients on the other two axes (N/K and K/P).

DRIS charts like that in Fig. 9.13 rank three nutrients from most deficient to least deficient. For example, Sumner (1977a) applied the soybean DRIS norms in Table 9.5 to soybean tissue concentrations he interpolated from a study by Hanway and Weber (1971) (Table 9.6). The general method for using a DRIS chart was best

Table 9.6 Soybean (*Glycine max* (L.) Merr.) leaf tissue nutrient concentrations from Hanway and Weber (1971) as interpolated by Sumner (1977a), concentration ratios, and rank in order of most yield-limiting to least yield-limiting

	N[a]	P	K	Parameters			
Days after emergence	g (kg dry matter)$^{-1}$			N/P	N/K	K/P	Rank[b]
73	64.5	4.2	17.5	15.36	3.69	4.17	$K > P \approx N$
92	42.5	2.5	11.0	17.00	3.86	4.40	$K > P > N$
102	29.0	2.0	9.5	14.50	3.05	4.75	$K > P \approx N$

[a]Tissue data are from the top leaves, above node 14

[b]Rank differs slightly from that published by Sumner (1977a) who used computational methods rather than the DRIS chart to rank nutrients

described by Sumner (1977b). We apply that method to the soybean tissue concentrations listed in the second entry in Table 9.6 (92 days after emergence): 42.5 g N kg^{-1}, 2.5 g P kg^{-1}, and 11.0 g K kg^{-1}.

First, we calculate tissue nutrient concentration ratios to create parameters that match the axes in Fig. 9.13, shown in columns 5–7 in Table 9.6: N/P = 17.00, N/K = 3.86, and K/P = 4.40.

Next, we locate 17.00 along the N/P axis (top leftmost point in Fig. 9.13) and see that it falls between the inner and outer circles. We then find the nutrients and arrows to the left and right of the axis where the point is located. To the left of the N/P axis is a P\ and to the right is an N/. In the typeset used in this chapter, "\" denotes an arrow slanted downward and "/" denotes an arrow slanted upward. For each nutrient, we tally only the horizontal or downward-facing arrows. We record a "\" next to P, as shown in the first step of the tally in Table 9.7.

Next, we locate N/K = 3.86 along the N/K axis, which falls outside the outer circle (point in the upper right of Fig. 9.13). To the right of the N/K axis outside the outer circle is a K↓ and to the left of the N/K axis is an N↑. We record only the downward-facing arrow and add it to K in the second step of the tally in Table 9.7. The \ arrow next to P from the first step is copied down into the second step.

Next, we locate K/P = 4.40 along the K/P axis, which falls between the inner and outer circles (the point in the lower right in Fig. 9.13). To the right of this axis next to the point is a K\ and to the left of this axis is a P/. We record only the downward slanted arrow next to K in step 3 in Table 9.7, again copying all arrows from the previous step.

Table 9.7 Ranking of the N, P, and K tissue concentrations listed in Table 9.6 for the entry associated with samples taken 92 days after soybean (*Glycine max* (L.) Merr.) emergence

Progressive steps in the tally	Parameter interpreted	Parameter value	Nutrient with a horizontal or downward arrow[a]	Tally of horizontal and downward arrows[b]
1	N/P	17.00	P\	N P\ K
2	N/K	3.86	K↓	N P\ K↓
3	K/P	4.40	K\	N P\ K↓\
Nutrient not assigned an arrow	N	→		N→ P\ K↓\
Final ranking (from most limiting to least yield-limiting)				K > P > N

[a]An arrow slanting downward is denoted by "\" in the typeset of this chapter. These arrows are read from the DRIS chart in Fig. 9.13

[b]Downward slanting arrows are given a lower ranking (less limiting) than straight downward arrows (more limiting), so in order of most to least yield-limiting, the combination "↓\" denotes a more limiting nutrient than "\" alone

By convention, after all, three parameters have been read from the DRIS chart, the nutrient not yet assigned an arrow receives a horizontal one (→), which we have added in the fourth entry in Table 9.7.

The last step is to order the nutrients from most limiting to the least limiting. Straight downward arrows are more yield-limiting than slanted downward arrows. In the example in Table 9.7, K with both a ↓ and a "\" is more limiting than P with only a "\." Finally, N is the least limiting since it has no downward arrows. The final ranking, from most to least yield-limiting is therefore: K > P > N. Repeating this process for the other two entries in Table 9.6 (73 and 102 days after emergence) fills out the remaining rankings for the earlier sampled and later sampled soybean tissue. We see that in each entry in Table 9.6, K is identified as the most limiting nutrient, with P possibly the next limiting, and N the least limiting.

9.3.2.2 DRIS Indexes

DRIS indexes rank multiple nutrients quantitatively. Nutrient indexes combine functions of each parameter. For the soybean example, three functions exist: *f* (N/P), *f*(N/K), *f*(K/P) (Sumner 1977a). Each function (not displayed here) is composed of (1) the distance a particular parameter is from its norm, (2) a weighting factor, and (3) a sensitivity coefficient that is simply a multiple of 10. The distance is negative when the parameter is less than its norm and positive when the parameter is greater than its norm. The distance is zero when the parameter equals its norm. Distances are on a continuous scale. The weight in the function is the inverse of the coefficient of variation (CV) associated with the parameter norm. The DRIS index for a particular nutrient is an average of the functions containing that nutrient. Keeping to the three nutrients N, P, and K, their respective indexes are:

$$\begin{aligned} \text{N Index} &= I_N = [f(\text{N/P}) + f(\text{N/K})]/2 \\ \text{P Index} &= I_P = [-f(\text{N/P}) - f(\text{K/P})]/2 \\ \text{K Index} &= I_K = [f(\text{K/P}) - f(\text{N/K})]/2 \end{aligned} \tag{9.1}$$

Because weights are in the functions, the average that calculates an index is a weighted average. The index calculation assigns a negative sign to a function that has the nutrient of interest in the denominator and a positive sign to a function with the nutrient of interest in the numerator. Therefore, for the K index, *f*(N/K) is negative and *f*(K/P) is positive. Indexes provide a continuous scale for ranking nutrients as opposed to the DRIS chart which provides only qualitative rankings.

When Sumner (1977a) used indexes rather than a chart to rank the data in Table 9.6, his ranking was K > P > N, and this ranking was the same across all plant ages. This is notable, since the actual concentrations of N, P, and K in tissue decreased over time in Table 9.6. Sumner considered the stability of rankings with plant age one of the major contributions of DRIS to tissue test interpretations. Interpretations for sufficiency ranges, discussed previously, are for specific growth stages. Sumner (1990) argued that the ranking stability arose from the use of tissue concentration ratios. Nutrient concentrations are normally reported on a dry matter basis, such as g K (kg dry matter)$^{-1}$. In ratios of two concentrations, such as K/N, the dry matter units cancel each other, making the ratios less sensitive to changes in dry matter content as tissue ages.

When setting up nutrient ratios, Sumner (1990) advised accounting for how nutrient concentrations of each element changed as tissue aged. The most consistent ranking of nutrients over time occurred when all nutrient concentrations in the analysis changed in the same direction, for instance becoming less concentrated, as is the case for N, P, and K. However, some nutrients increase in concentration at later growth stages, like Ca in sugarcane (*Saccharum officinarum* L.) leaf tissue (Sumner 1990). Taking the inverse (1/Ca) of concentrations that increase with tissue age shifts their direction to match those that decrease. So instead of using a ratio of N/Ca, one would instead use a ratio of N/(1/Ca), which results in the product N × Ca. Creating such consistency in directional change produced much more consistent rankings across tissue ages (Sumner 1990), and increased the accuracy of predicting yield responses (Hallmark et al. 1988). In addition, Hallmark et al. (1988) observed that not taking the inverse of Ca concentration resulted in Ca often appearing in rankings as the most limiting nutrient.

Jones (1981) suggested two modifications to the DRIS procedure. DRIS used one function when a parameter was greater than its norm and another function when a parameter was less than its norm. Jones (1981) pointed out that the functions did not weight parameter values equivalently. The difference in weighting biased the indexes. In his first modification, Jones advocated for a single functional form that weighted variances equivalently regardless of how a parameter value compared to its norm (Eq. 9.2). In that equation, the parameter function $f(p_{ij})$ is equal to the value of parameter j in a given tissue sample i (p_{ij}) minus the norm of that parameter (M_j) divided by the standard deviation of the norm of that parameter (sd_j). Both M_j and sd_j are calculated using the high yield subpopulation. The weight in Eq. (9.2) is $1/\text{sd}_j$.

The equation calculates negative values when the parameter is less than its norm, positive values when it is greater, and zero when the two are equal. In later work, DRIS adopted Jones' single equation as the parameter function (Hallmark et al. 1987).

$$f(p_{ij}) = (p_{ij} - M_j)/\mathrm{sd}_j \tag{9.2}$$

Jones (1981) also questioned the use of only variance ratios, like those in Table 9.5, to select which parameters to include. He observed that parameter means between high and low-yielding subpopulations could be significantly different even though their variances were not. In his second modification, he argued for testing means in addition to variances for selecting parameters. He also noted that more parameters would be statistically selected as the number of observations in the yield subpopulations increased.

One of the assumptions of the DRIS system was that yield distributions across levels of any ratio were normally distributed. Beverly (1987a, b) demonstrated that in some cases, those ratios were positively skewed. Lack of normality created different norms for the same ratio when the numerator and denominator in that ratio were switched. Taking the natural log of the concentration ratio corrected this problem. The log transformation was the major modification. That change also led to a simplification of the index calculation.

Elwali and Gascho (1984) created the nutrient balance index (NBI) for DRIS, also termed by some the nutrient imbalance index (NII). It is the sum of the absolute values of all of the nutrient indexes. For N, P, and K the NBI is:

$$\mathrm{NBI} = |I_\mathrm{N}| + |I_\mathrm{P}| + |I_\mathrm{K}| \tag{9.3}$$

The closer the nutrient balance index is to 0, the more balanced are the nutrient concentrations. The NBI does not indicate which nutrients are out of balance. It indicates only the overall magnitude of imbalance.

9.3.3 The Modified DRIS System (M-DRIS)

DRIS, as originally created, did not have a method for predicting the probability of crop response to applications of nutrients that it ranked as most yield-limiting. Researchers have tested several methods targeted at fulfilling this need. Jones (1981) observed in his sugarcane data set that negative, rather than positive, index values more accurately identified when crops would respond. Walworth et al. (1986), working with alfalfa (*Medicago sativa* L.), proposed including an index for dry matter (DM) when ranking nutrients. The DM index served as an internal standard. Nutrients with indexes less than the DM index were likely yield-limiting and alfalfa was more likely to respond to additions of those nutrients. Alfalfa performed better when applying all nutrients with indexes less than the DM index compared to applying only the nutrient with the most negative index. Walworth

(1986) did provide the caveat that using the DM index as a delineation level for crop response might work best for crops where total DM production is desired, like forage crops. Hallmark et al. (1987) included DM as a factor when diagnosing nutrient balance in soybean leaves. They confirmed that including DM improved predictions of crop response for grain crops too and that such improvements were not limited to only forage crops.

Including DM as a factor is what differentiates the modified DRIS (M-DRIS) from DRIS. Additionally, M-DRIS uses the single-parameter function (Eq. 9.2) proposed by Jones (1981). As Walworth et al. (1986) noted, including DM as a factor made M-DRIS more susceptible to changes in plant age than DRIS.

9.3.4 Plant Analysis with Standardized Scores (PASS)

Baldock and Schulte (1996) developed the Plant Analysis with Standardized Scores (PASS) system to combine the respective strengths of sufficiency range and DRIS interpretations. They did this to better relate indexes to probabilities of crop response. The PASS system consists of two sections. The PASS Dependent Nutrient Index (PASS DNI) is an interpretation based on DRIS. The PASS Independent Nutrient Index (PASS INI) is an interpretation based on sufficiency ranges.

PASS DNI creates indexes for nutrient ratios. It uses the Jones (1981) parameter function in Eq. (9.2). Like DRIS, the PASS DNI for a given nutrient is the average of all parameter functions using nutrient ratios containing that nutrient. Unlike DRIS, PASS DNI includes only "common response nutrients." To be classified as a common response nutrient: (a) the crop must have a high requirement of that nutrient, and (b) the correlation of the concentrations of that nutrient to the yield responses of that crop must have a well-defined critical concentration. Failure to meet both of these criteria categorize a nutrient as a "rare response nutrient."

PASS INI is an index for nutrient concentrations, not nutrient ratios. It uses the critical concentration (C_j) as the reference for its values. The function that transforms nutrient concentration to the PASS INI is based on Eq. (9.2). Instead of using the parameter mean (M_j), Baldock and Schulte substituted the critical concentration plus one standard deviation (C_j + sd$_j$) based on the assumption that $M_j \approx (C_j + \mathrm{sd}_j)$. They also introduced a sensitivity coefficient by multiplying the numerator in Eq. (9.2) by 10:

$$\text{PASS INI}_{ij} = 10\left[p_{ij} - \left(C_j + \mathrm{sd}_j\right)\right]/\mathrm{sd}_j$$

which simplifies to

$$\text{PASS INI}_{ij} = \left[10\left(p_{ij} - C_j\right)/\mathrm{sd}_j\right] - 10 \tag{9.4}$$

According to Eq. (9.4), when a sample nutrient concentration is equal to the critical concentration, PASS INI equals −10. Values between or equal to −10 and 10 are sufficient, which is equivalent to the range $M_j \pm \mathrm{sd}_j$. Index values less than −10 are deficient. The more negative an index becomes, relative to −10, the greater the chances become that a crop will respond to an application of that nutrient. Index values greater than 10 indicate that the nutrient concentration is high and that crops are not likely to respond to additions of that nutrient. PASS INI uses the same division of nutrients as PASS DNI (common response and rare response).

PASS uses three interpretation categories: (1) probable nutrient deficiency, (2) slightly possible nutrient deficiency, and (3) nutrient sufficiency. PASS places common response nutrients with PASS INIs below −10 in the "probably nutrient deficient" category. It places rare response nutrients with PASS INIs below −10 in the "slightly possible nutrient deficiency" category. Also into this category go nutrients that have both PASS INIs less than zero and PASS INI and PASS DNI sums that are less than −10. Into the third category, "nutrient sufficiency," go all nutrients not already in the first two categories.

When making fertilizer decisions, a combination of PASS INI and PASS DNI is helpful. PASS INI identifies nutrients that are yield-limiting. If more than one nutrient is yield-limiting, PASS DNIs order them in a rank from most yield-limiting to least. This ranking helps prioritize which nutrients to apply.

To simplify interpretations and give greater weight to PASS INI values that were more negative, Baldock and Schulte (1996) created the PASS yield index (PASS YI). The PASS YI reassigns all values greater than −10 to zero, since no crop response is expected at those index levels. All values less than −10 are squared, giving exponentially greater weight to diminishing tissue concentrations. The PASS YI subtotals all of the squared values across all nutrients within each group: both common and rare response nutrients. The subtotal of the sums of squares in the common response group is multiplied by 2 to give it more weight than the subtotal of the sums of squares of the rare response group. After the multiplication, the two subtotals are added to calculate the PASS YI across all nutrients. Because PASS YI uses squared values, its minimum is zero. That minimum value means all nutrients considered are at or above their respective critical concentrations and no crop response to any of them is likely. With no restrictions from nutrient limitations, yields are expected to be higher. Conversely, as PASS YI values become greater, more nutrient restrictions exist and yields will likely be lower unless nutrients are applied.

The PASS approach has not been tested by researchers to the extent DRIS has been. We were able to locate only two relatively recent studies that compared PASS to other methods of interpretation (Simón et al. 2013; Urricariet et al. 2004). Both showed PASS to be a promising diagnostic approach.

9.3.5 *Compositional Nutrient Diagnosis (CND)*

Compositional nutrient diagnosis (CND) arose from analysis techniques developed for geological sediment compositions (Aitchison 1982, 1983). Compositions are made up of individual compounds. Each compound comprises a given percent of the total composition. The percentages of all compounds add up to 100%. Traditional statistical approaches assume that levels of factors are unbounded; however, in compositions, there is an upper bound (100%) to the level of any one factor. In addition, if the percentage of one compound increases, the percentage of at least one other compound must decrease. Therefore, compositions are not independent. Finally, compositional data are not normally distributed, as Beverly (1987a, b) pointed out when working with DRIS. Consequently, compositions require their own statistical approaches.

Parent and Dafir (1992) adapted the work of Aitchison (1982, 1983) to plant tissue nutrient compositions. They termed their approach "compositional nutrient diagnosis" or CND. Parent created two approaches to CND: (1) one that uses centered log ratios (CND-*clr*) and (2) one that uses isomeric log ratios (CND-*ilr*). Both approaches examine the interactions of nutrients in plant tissue. Compositional nutrient diagnosis falls under Baldock and Schulte's (1996) classification as a DNI. CND-*clr* was developed first and shared many characteristics with DRIS (Parent and Dafir 1992). Both CND-*clr* and DRIS analyze all possible combinations of nutrients first, leaving the user to interpret the results afterward. CND-*ilr* was developed to provide the user the ability to incorporate knowledge of nutrient interactions into the analysis ahead of time (Parent 2011). Therefore CND-*ilr* allows users to test for certain interactions in a given sample.

9.3.5.1 CND-*clr*

CND-*clr* examines all possible interactions of one nutrient concentration with all other measured concentrations (Parent and Dafir 1992). In this regard, it is fundamentally different than the DRIS approach which considers the interaction of only two nutrient concentrations at a time. Parameters in CND-*clr* differ from those in DRIS. Interaction parameters formed from N, P, and K in DRIS are nutrient concentrations ratios like N/P, N/K, and K/P. In CND-*clr*, interaction parameters for N, P, and K are logarithms of ratios, like $\log[N/(N \times P \times K \times R)^{1/4}]$, $\log[P/(N \times P \times K \times R)^{1/4}]$, and $\log[K/(N \times P \times K \times R)^{1/4}]$. The denominator in all three CND-*clr* parameters is the same and is the geometric mean of all measured nutrient concentrations and the "filling-up value" R. The filling-up value is the percent remaining after summing all of the nutrient concentrations: $R = 100 - (N + P + K)$. The filling-up value is always an additional term in the geometric mean. The general formula for the interaction of any one nutrient with all other elements is:

$$V_{ij} = \log\left[x_{ij}/G_i\right] \tag{9.5}$$

where V_{ij} is the parameter for nutrient j in sample i, x_{ij} is the concentration of nutrient j in sample i, and G_i is the geometric mean of the following concentrations: nutrient j in sample i (x_{ij}); the filling-up parameter R; and the concentrations of all other nutrients in sample i.

The filling-up parameter and the logarithmic function come from considering leaf tissue to be a closed compositional system (Parent and Dafir 1992). The percentages of all compounds and elements in tissue must add up to 100%. Increasing the percentage of one element decreases the percentage of at least one other element or compound. Adjustments in composition do not have to occur with other elements that are measured. They may occur with any number of compounds not measured but included in the filling-up value, resulting in a lower value for R.

Just as in DRIS, CND-*clr* computes indexes for each nutrient (Parent and Dafir 1992). The equation CND-*clr* uses to calculate indexes is similar to the equation DRIS uses (Eq. 9.2) to calculate parameter functions:

$$I_{ij} = \left(V_{ij} - V_j^*\right)/\mathrm{sd}_j^* \tag{9.6}$$

where I_{ij} is the CND-clr index for nutrient j in a given tissue sample i, V_{ij} is the CND-*clr* parameter for nutrient j in sample i (Eq. 9.5), V_j^* is the CND-*clr* norm for nutrient j (the average V_j of the high-yielding subpopulation), and sd_j^* is the standard deviation of V_j in the high yield subpopulation. In DRIS, the yield level separating high and low-yielding subpopulations is usually chosen from experience. For CND-*clr*, Khiari et al. (2001) developed a statistical method to separate those populations. In DRIS, Eq. (9.2) calculates parameter values for each ratio of two nutrients. Since a given nutrient may be in more than two ratios, DRIS requires Eq. (9.1) to combine all ratios containing that nutrient. In CND-*clr*, an equation like Eq. (9.1) is not needed, since each nutrient has only one log ratio (Eq. 9.5). Just like in DRIS, CND-*clr* indexes classify nutrients in order of their limitation (Parent et al. 1994a), with more negative values indicating greater deficiency and more positive values indicating greater excess.

Akin to the DRIS NBI (Eq. 9.3), CND-*clr* provides quantitative evaluation of overall nutrient balance. Two metrics have been developed for this purpose. The first one developed was a Euclidian distance d (Parent et al. 1994a, b). The greater the value of d, the more overall nutrient imbalance exists. The second metric developed was the CND r^2 value, defined as the sum of the squares of all CND-*clr* indexes for a particular sample (Khiari et al. 2001). The farther CND r^2 is from zero, the greater the nutrient imbalance.

For practical use in production settings, a tool has been developed to implement CND-*clr* analyses for guava (Rozane et al. 2012). Users enter nutrient concentrations as received from an analytical laboratory, and the tool returns CND-*clr* indexes for each nutrient, presented in three ways: numerically, displayed in a radar chart, and displayed in a bar chart. The tool also provides the associated CND r^2 value.

9.3.5.2 CND-*ilr*

CND-*ilr* compares two user-specified subsets of nutrients. The user selects these groups ahead of analysis. Table 9.8 provides an example of how a user can create a subset (Parent 2011). The first row (Sample) is the concentration of nutrients in a leaf tissue sample. The second through sixth rows are sets that define the groups to compare. Each set examines two groups. In a given set, all nutrients with a 1 are in one group and all nutrients with a −1 are in the other group. By assigning 1s and −1s, users can group nutrients in meaningful ways. The number of sets is equal to one less than the number of components of the composition, including the filling-up value *R*. Because the composition is made up of six components (N, P, K, Ca, Mg, and *R*), 5 sets can be compared.

After creating subsets, the next step is to calculate isometric log ratio (ilr) coordinates for each set *i*. The general formula is:

$$\mathrm{ilr}_i = [(r \times s)/(r+s)]^{1/2} \times \ln[\mathrm{g}(x+)/\mathrm{g}(x-)] \tag{9.7}$$

where ilr_i is the ilr coordinate of set *i*, *r* is the number of components assigned a 1 in set *i*, *s* is the number of components assigned a −1 in set *i*, g(*x*+) is the geometric mean of the percentages of components assigned a 1 in set *i*, and g(*x*−) is the geometric mean of the percentages of components assigned a −1 in set *i*. For example, for set 4 in Table 9.8, K versus Ca + Mg, the ilr coordinate for the sample (ilr_4) is $[(1 \times 2)/(1 + 2)]^{1/2} \times \ln[(2.64)/(1.15 \times 0.11)^{1/2}] = 1.637$. According to Parent (2011), the average ilr coordinate for this set in the high-yielding subpopulation is $\mathrm{ilr}_4^* = 1.154$. The difference $\mathrm{ilr}_4 - \mathrm{ilr}_4^*$ determines how far the ilr of sample set 4 is from that of the same set in the high-yielding subpopulation: 1.637–1.154 = 0.483. This difference is the second highest of the sets and indicates that K is out of balance with Ca + Mg (set 4, Table 9.8). The difference that was greater belonged to set 5, which showed that Ca was out of balance with Mg. Because it had a greater difference, Ca and Mg were more out of balance than

Table 9.8 Orthogonal partitions of N, P, K, Ca, Mg concentrations, and *R* (the filling-up value) nutrient concentrations subgroups for an apple (*Malus domestica* Borkh.) leaf analysis. (Parent 2011)

	Components						
	N	P	K	Ca	Mg	*R*	Interpretation
	Concentration (%)						
Sample:	2.50	0.22	2.64	1.15	0.11	93.38	
	Orthogonal partitions						
Set 1:	1	1	1	1	1	−1	Nutrients versus filling-up value
Set 2:	1	1	−1	−1	−1	0	Anions versus cations
Set 3:	1	−1	0	0	0	0	N versus P
Set 4:	0	0	1	−1	−1	0	K versus Ca + Mg
Set 5:	0	0	0	1	−1	0	Ca versus Mg

were K and Ca + Mg. Parent (2011) concluded that the Mg supply needed to be increased and K fertilization needed to be either decreased or halted. The power of CND-*ilr* is the ability to test specific nutrient combinations for their relative balance.

9.3.6 Multiple Regression Approaches

Statistical approaches, like multiple regression, are another avenue for considering more than one nutrient at a time when evaluating nutrient concentrations in plant tissue. Lissbrant et al. (2010) used a combination of cluster analysis, logistic regression, and concentration ratios like that in DRIS to predict alfalfa yield in an experiment examining various rates of applied P and K. Cluster analysis classified crop yield into groups, then those groups were further classified as "acceptable" (high and medium-high-yielding groups) or "unacceptable" (all lower-yielding groups). To best predict acceptable or unacceptable performance for May cuttings, binary logistic regression identified a combination of tissue P, K, and the K/P ratio as model factors. For the later June cutting, the regression model included only tissue K and the K/P ratio. Other statistical approaches have also been developed, like multiple regression using combinations of concentrations of multiple nutrients (Martinez et al. 2003; Srivastava et al. 2001).

9.3.7 Metabolite Profiles

Potassium-deficient plants have different metabolite profiles than plants with sufficient K. For instance, shoot tissue of K-deficient *Arabidopsis* exhibited higher concentrations of the soluble sugars sucrose, glucose, and fructose; higher concentrations of several basic or neutral amino acids; and lower concentrations of the acid amino acids glutamic acid and aspartic acid (Armengaud et al. 2009). In their review, Amtmann et al. (2008) concluded that while increased concentrations of soluble sugars, organic acids, and amino acids were often observed in tissue from K-deficient plants, there was great variability and not all crops showed increases. Some crops, in fact, showed decreases in some of these same metabolites. A particularly insightful observation by Armengaud et al. (2009) was that the observed changes in metabolite concentrations, "...were not always related to tissue K content." This demonstrates the shortcomings of using tissue K concentration as the sole metric of K deficiency. While metabolite profiles have the potential to be diagnostic of K deficiency, there is much yet to be learned before they can be incorporated into fertilizer recommendations and on-farm decision-making.

9.3.8 *Potassium Content in Plant Sap*

Another approach to diagnosing K status is to analyze the K content of plant sap. Sap is extracted by pressing plant tissue with tools such as a pliers (Syltie et al. 1972), a garlic press (Gangaiah et al. 2016), or a press attached to a syringe (Burns and Hutsby 1984). Freshly sampled petioles and leaf midribs are the typical samples. Hand-held analytical equipment makes in-field testing possible. Hand-held meters with ion-selective electrodes provide quantitative evaluations of the K concentration in the sap. Also available are test strips or spot tests that react the sap with reagents to create a color. The user compares that color with standardized colors representing different K concentrations. Unless the color matches perfectly with that of a standard, the user must interpolate concentrations between two adjacent color standards (Burns and Hutsby 1984). With test strips or spot tests, there is a limited range of detection, and the user cannot infer values beyond either end of that range (Syltie et al. 1972).

When sampling sap, it is important to recognize that its chemical composition changes throughout the day. Meitern et al. (2017) sampled branches of hybrid aspen (*Populus tremula* L. x *Populus tremuloides* Michx.) saplings. They found that the K concentration in xylem sap increased quickly after 6:00 am, plateaued at 12:00 pm to 3:00 pm, then decreased again. These changes started after dawn as photon flux density increased, air temperature increased, and relative humidity decreased. For this reason, diagnostic interpretations need to specify the times of day for collecting samples.

Like tissue K concentration, sap K concentration also changes with organ and age. Age is usually specified by position on the plant. For instance, the "uppermost" leaf specifies the youngest leaf. Burns (1992) observed that when K supply was cut off from actively growing lettuce (*Lactuca sativa* L.), sap K concentration dropped more quickly in the youngest leaf than in the older ones. This meant the youngest leaf was more sensitive to changes in K supply, making it a good choice for diagnosis. Further support for sampling young tissue was provided by Vruegdenhil and Koot-Gronsveld (1989) when they observed that the K concentration in the sap was highest in the uppermost fully developed leaf of castor bean (*Ricinus communis* L.).

Creating diagnostic interpretations for plant sap K concentrations requires correlations to yield or other metrics of crop performance, like those previously discussed for tissue K concentration. As an example, Hochmuth et al. (1993) grew eggplant (*Solanum melongena* L.) on a K deficient soil and applied incremental rates of K fertilizer. Several times during the season they used a hand-held ion-specific electrode to measure the K concentration in plant sap extracted from petioles of the youngest fully expanded leaves. At the same time, they also collected whole leaf samples, including petioles, and analyzed them for K concentration with traditional laboratory techniques. Third, they measured total marketable eggplant yield, summed over the yields of various market grades throughout the season. They determined critical plant sap K concentrations and confirmed sap analysis as a viable

Table 9.9 Sufficiency ranges for sap K concentration in petioles of the most recently matured leaves and K concentration in the most recently matured, whole leaves of various vegetable crops sampled during the day from 9:00 to 16:00. (excerpted from Hochmuth 1994a)

		Petiole sap K concentration	Whole leaf K concentration
Crop[a]	Growth stage	mg K L^{-1}	g kg^{-1}
Eggplant	First fruit (5 cm long)	4500–5000	35–50
Pepper	First open flowers	3000–3200	45–50
Potato	First open flowers	4500–5000	30–50
Tomato (field)	First open flowers	3500–4000	35–40
Watermelon	Vines 15 cm long	4000–5000	35–40

[a]Eggplant (*Solanum melongena* L.); pepper (*Capsicum* spp.), potato (*Solanum tuberosum* L.), tomato (*Solanum lycopersicum* L.), watermelon (*Citrullus lanatus* (Thunb.) Matsum. & Nakai)

diagnostic tool. In a later review of 10 years of his and his colleagues' research, Hochmuth (1994a) published interpretations of sap K concentration for other vegetable crops, along with tissue K concentration (Table 9.9). Those interpretations are used in nutrient management guidance for growers (Hochmuth 1994b).

Some studies focus on correlating sap K concentration to leaf K concentration and do not include correlations to yield. This is often done in exploratory studies examining new analytical methods (Gangaiah et al. 2016; He et al. 1998; Iseki et al. 2017). Eventually, however, each new method must be correlated to yield or some metric of crop performance.

9.4 Conclusions

We have reviewed a number of ways scientists have determined the nutritional status of plants. Visual symptoms detect moderate to severe deficiencies where plant metabolism has already been irreversibly and negatively impacted. Measuring light reflectance is non-invasive and non-destructive, but to date, methods have not uncovered spectral combinations specific to K nutritional status. Tissue sampling is destructive but has been the most diagnostic approach to date. Sufficiency ranges consider nutrients in isolation and do not account for nutrient interactions. Interpretation methods that do account for interactions are DRIS, M-DRIS, PASS, CND-*clr*, CND-*ilr*, and multifactor statistical models. Metabolite profiles show promise, but more research is needed to determine if the level of certain metabolites or their interactions with other components are diagnostic of K deficiency. Sap analysis provides rapid results while in the field. Interpretation of results has been so far limited to sufficiency ranges. Sap analysis results can be highly variable because of diurnal fluctuations in K content.

Moving forward, an important theme throughout all of the tissue sampling interpretations is the value of large, crop-specific data sets composed of nutrient contents and associated yield and quality levels. At the least, such a data set must centralize data from as many high-yielding production settings as possible. Large data sets representing high-yielding and/or high-quality crops have been used for creating sufficiency ranges as well as norms for DRIS, M-DRIS, PASS, CND-*clr,* and CND-*ilr*. Indeed, this was Beufils (1973) original vision. He saw the need for large, multinational databases that contained large amounts of meta-data for each yield observation. He divided these meta-data into two categories: (1) "external characters" comprised of soil properties, climatic conditions, and farming practices, and (2) "internal characters" comprised of data on the chemical and physical characteristics of various plant organs, including nutrient concentration. In his vision, data could come from farmers' fields or controlled experiments. Data from both sources would be merged and used to create norms. Further, querying large databases rich in meta-data could potentially guide a user to enough relevant studies to develop quantifiable recommendations, such as rates of specific K sources to apply, to rectify any given nutrient deficiency. While many isolated databases have been developed, there is a lot more to be done, both in centralization as well as in completeness of meta-data, to realize a vision Beaufils had decades ago but which is just as relevant today.

References

Aitchison J (1982) The statistical analysis of compositional data. J R Stat Soc B Methodol 44 (2):139–177. https://doi.org/10.1111/j.2517-6161.1982.tb01195.x

Aitchison J (1983) Principle component analysis of compositional data. Biometrika 70(1):57–65. https://doi.org/10.1093/biomet/70.1.57

Albayrak S, Basayigit L, Türk M (2011) Use of canopy- and leaf-reflectance indices for the detection of quality variables of *Vicia* species. Int J Remote Sens 32(4):1199–1211. https://doi.org/10.1080/01431161003762389

Amtmann A, Troufflard S, Armengaud P (2008) The effect of potassium nutrition on pest and disease resistance in plants. Physiol Plant 133(4):682–691. https://doi.org/10.1111/j.1399-3054.2008.01075.x

Anderson RL, Nelson LA (1975) A family of models involving intersecting straight lines and concomitant experimental designs useful in evaluating response to fertilizer nutrients. Biometrics 31(2):303–318. https://doi.org/10.2307/2529422

Anderson G, van Aardt J, Bajorski P, Vanden Heuvel J (2016) Detection of wine grape nutrient levels using visible and near infrared 1nm spectral resolution remote sensing. Proc SPIE 9866, Autonomous air and ground sensing systems for agricultural optimization and phenotyping, 98660H. https://doi.org/10.1117/12.2227720

Armengaud P, Sulpice R, Miller AJ, Stitt M, Amtmann A, Gibon Y (2009) Multilevel analysis of primary metabolism provides new insights into the role of potassium nutrition for glycolysis and nitrogen assimilation in Arabidopsis roots. Plant Physiol 150(2):772–785. https://doi.org/10.1104/pp.108.133629

Ayala-Silva T, Beyl CA (2005) Changes in spectral reflectance of wheat leaves in response to specific macronutrient deficiency. Adv Space Res 35(2):305–317. https://doi.org/10.1016/j.asr.2004.09.008

Baldock JO, Schulte EE (1996) Plant analysis with standardized scores combines DRIS and sufficiency range approaches for corn. Agron J 88(3):448–456. https://doi.org/10.2134/agronj1996.00021962008800030015x

Beaufils EA (1973) Diagnosis and recommendation integrated system (DRIS). Soil Science Bulletin No. 1. University of Natal, Pietermaritzburg

Beaufils EA, Sumner ME (1976) Application of the DRIS approach for calibrating soil and plant factors in their effect on yield of sugarcane. Proc South African Sugar Technologists' Association (SASTA), June 1976, pp 118–124

Bell RW, Brady D, Plaskett D, Loneragan JF (1987) Diagnosis of potassium deficiency in soybean. J Plant Nutr 10(9–16):1947–1953. https://doi.org/10.1080/01904168709363740

Beverly RB (1987a) Comparison of DRIS and alternative nutrient diagnostic methods for soybean. J Plant Nutr 10(8):901–920. https://doi.org/10.1080/01904168709363619

Beverly RB (1987b) Modified DRIS method for simplified nutrient diagnosis of 'Valencia' oranges. J Plant Nutr 10(9–16):1401–1408. https://doi.org/10.1080/01904168709363672

Birth GS, McVey GR (1968) Measuring the color of growing turf with a reflectance spectrophotometer. Agron J 60:640–643. https://doi.org/10.2134/agronj1968.00021962006000060016x

Burns IG (1992) Influence of plant nutrient concentration on growth rate: use of a nutrient interruption technique to determine critical concentrations of N, P, and K in young plants. Plant Soil 142(2):221–233. https://doi.org/10.1007/BF00010968

Burns IG, Hutsby W (1984) Development and evaluation of rapid tests for the estimation of phosphate and potassium in plant sap. Commun Soil Sci Plant Anal 15(12):1463–1480. https://doi.org/10.1080/00103628409367573

Campbell CR, Plank CO (1998) Preparation of plant tissue for laboratory analysis, chapter 3. In: Kalra YP (ed) Handbook of reference methods for plant analysis. CRC Press, Boca Raton, FL, pp 37–49

Cate RB Jr, Nelson LA (1971) A simple statistical procedure for partitioning soil test correlation data into two classes. Soil Sci Soc Am Proc 35(4):658–660. https://doi.org/10.2136/sssaj1971.03615995003500040048x

Cerrato ME, Blackmer AM (1990) Comparison of models for describing corn yield response to nitrogen fertilizer. Agron J 82(1):138–143. https://doi.org/10.2134/agronj1990.00021962008200010030x

Chapman HD (1949) Citrus leaf analysis: nutrient deficiencies, excesses and fertilizer requirements of soil indicated by a diagnostic aid. Calif Agric 3(11):10–14. http://calag.ucanr.edu/Archive. Accessed 19 May 2020

Chapman HD (1960) Leaf and soil analyses as guides for citrus fertilizer practices in southern California orchards. Calif Agric 14(10):13–14. http://calag.ucanr.edu/Archive. Accessed 19 May 2020

Chapman HD, Brown SM (1943) Potash in relation to citrus nutrition. Soil Sci 55(1):87–100

Chapman HD, Brown SM (1950) Analysis of orange leaves for diagnosing nutrient status with reference to potassium. Hilgardia 19:501–540. https://doi.org/10.3733/hilg.v19n17p501. Accessed 19 May 2020

Chapman HD, Brown SM, Rayner DS (1947) Effects of potash deficiency and excess on orange trees. Hilgardia 17:619–650. https://doi.org/10.3733/hilg.v17n19p619

Curran PJ (1989) Remote sensing of foliar chemistry. Remote Sens Environ 30(3):271–278. https://doi.org/10.1016/0034-4257(89)90069-2

Curran PJ, Dungan JL, Macler BA, Plummer SE, Peterson DL (1992) Reflectance spectroscopy of fresh whole leaves for the estimation of chemical concentration. Remote Sens Environ 39 (2):153–166. https://doi.org/10.1016/0034-4257(92)90133-5

Elwali AMO, Gascho GJ (1984) Soil testing, foliar analysis, and DRIS as guides for sugarcane fertilization. Agron J 76(3):466–470. https://doi.org/10.2134/agronj1984.00021962007600030024x

Embleton TW, Jones WW, Labanauskas CK (1962) Sampling orange leaves—leaf position important. Calif Citrograph 47:382–396

Embleton TW, Jones WW, Platt RG, Burns RM (1974) Potassium nutrition and deficiency in citrus. Calif Agric 28(8):6–8. http://calag.ucanr.edu/Archive. Accessed 19 May 2020

Embleton TW, Jones WW, Platt RG (1978) Leaf analysis as a guide to citrus fertilization. In:

Reisenauer HM (ed) Soil and plant-tissue testing in California. Bulletin 1879. Division of Agricultural Sciences, University of California, Berkeley, pp 4–16. https://books.google.com/books?id=BXssCAAAQBAJ&source=gbs_book_other_versions. Accessed 19 May 2020

Fridgen JL, Varco JJ (2004) Dependency of cotton leaf nitrogen, chlorophyll, and reflectance on nitrogen and potassium availability. Agron J 96:63–69. https://doi.org/10.2134/agronj2004.0063

Gangaiah C, Ahmad AA, Nguyen HV, Radovich TJK (2016) A correlation of rapid Cardy meter sap test and ICP spectrometry of dry tissue for measuring potassium (K^+) concentrations in pak choi (*Brassica Rapa* Chinensis group). Commun Soil Sci Plant Anal 47(17):2046–2052. https://doi.org/10.1080/00103624.2016.1208752

Geladi P, Kowalski BR (1986) Partial least-squares regression: a tutorial. Anal Chim Acta 185:1–17. https://doi.org/10.1016/0003-2670(86)80028-9

Gómez-Casero MT, López-Granados F, Peña-Barragán JM, Jurado-Expósito M, García-Torres L, Fernández-Escobar R (2007) Assessing nitrogen and potassium deficiencies in olive orchards through discriminant analysis of hyperspectral data. J Am Soc Hortic Sci 132(5):611–618. https://doi.org/10.21273/JASHS.132.5.611

Guo R, Zhao MZ, Yang ZX, Wang GJ, Yin H, Li JD (2017) Simulation of soybean canopy nutrient contents by hyperspectral remote sensing. Appl Ecol Environ Res 15(4):1185–1198. https://doi.org/10.15666/aeer/1504_11851198

Hall AD (1905) The analysis of the soil by means of the plant. J Agric Sci Camb 1(1):65–68. https://doi.org/10.1017/S0021859600000162

Hallmark WB, Walworth JL, Sumner ME, de Mooy CJ, Pesek J, Shao KP (1987) Separating limiting from non-limiting nutrients. J Plant Nutr 10(9–16):1381–1390. https://doi.org/10.1080/01904168709363670

Hallmark WB, de Mooy CJ, Morris HF, Pesek J, Shao KP, Fontenot JD (1988) Soybean phosphorus and potassium deficiency detection as influenced by plant growth stage. Agron J 80 (4):586–591. https://doi.org/10.2134/agronj1988.00021962008000040009x

Hanlon EA (1998) Elemental determination by atomic absorption spectrophotometry, chapter 20. In: Kalra YP (ed) Handbook of reference methods for plant analysis. CRC Press, Boca Raton, FL, pp 157–164

Hanway JJ, Weber CR (1971) N, P, and K percentages in soybean (*Glycine max* (L.) Merrill) plant parts. Agron J 63(2):286–290. https://doi.org/10.2134/agronj1971.00021962006300020027x

Hardy GW, Halvorson AR, Jones JB, Munson RD, Rouse RD, Scott TW, Wolf B (eds) (1967) Soil testing and plant analysis. Part II: plant analysis. Soil Science Society of America, Madison, WI. https://doi.org/10.2136/sssaspecpub2.frontmatter

He Y, Terabayashi S, Namiki T (1998) Comparison between analytical results of plant sap analysis and the dry ashing method for tomato plants cultured hydroponically. J Plant Nutr 21 (6):1179–1188. https://doi.org/10.1080/01904169809365476

Hochmuth GJ (1994a) Efficiency ranges for nitrate-nitrogen and potassium for vegetable petiole sap quick tests. HortTechnology 4(3):218–222. https://doi.org/10.21273/HORTTECH.4.3.218

Hochmuth GJ (1994b) Plant petiole sap-testing for vegetable crops. CIR1144. Horticultural Sciences Department, University of Florida. http://edis.ifas.ufl.edu/cv004. Accessed 19 May 2020

Hochmuth GJ, Hochmuth RC, Donley ME, Hanlon EA (1993) Eggplant yield in response to potassium fertilization on a sandy soil. HortScience 28(10):1002–1005

Horler DNH, Dockray M, Barber J (1983) The red edge of plant leaf reflectance. Int J Remote Sens 4(2):273–288. https://doi.org/10.1080/01431168308948546

Isaac RA, Johnson WC Jr (1998) Elemental determination by inductively coupled plasma atomic emission spectrometry, chapter 21. In: Kalra YP (ed) Handbook of reference methods for plant analysis. CRC Press, Boca Raton, FL, pp 165–170

Iseki K, Marubodee R, Ehara H, Tomooka N (2017) A rapid quantification method for tissue Na^+ and K^+ concentrations in salt-tolerant and susceptible accessions in *Vigna vexillata* (L.) A. Rich. Plant Prod Sci 20(1):144–148. https://doi.org/10.1080/1343943X.2016.1251826

Jones CA (1981) Proposed modifications of the diagnosis and recommendation integrated system (DRIS) for interpreting plant analyses. Commun Soil Sci Plant Anal 12(8):785–794. https://doi.org/10.1080/00103628109367194

Jones JB Jr (1998) Field sampling procedures for conducting a plant analysis, chapter 2. In: Kalra YP (ed) Handbook of reference methods for plant analysis. CRC Press, Boca Raton, FL, pp 25–35

Jones WW, Embleton TW, Johnston JC (1955) Leaf analysis a guide to fertilizer need. Calif Citrograph 40:332–347

Khiari L, Parent LE, Tremblay N (2001) Selecting the high-yield subpopulation for diagnosing nutrient imbalance in crops. Agron J 93(4):802–808. https://doi.org/10.2134/agronj2001.934802x

Knipling EB (1970) Physical and physiological basis for the reflectance of visible and near-infrared radiation from vegetation. Remote Sens Environ 1(3):155–159. https://doi.org/10.1016/S0034-4257(70)80021-9

Kokaly RF, Clark RN (1999) Spectroscopic determination of leaf biochemistry using band-depth analysis of absorption features and stepwise multiple linear regression. Remote Sens Environ 67 (3):267–287. https://doi.org/10.1016/S0034-4257(98)00084-4

Kovács P, Vyn TJ (2017) Relationships between ear-leaf nutrient concentrations at silking and corn biomass and grain yields at maturity. Agron J 109(6):2898–2906. https://doi.org/10.2134/agronj2017.02.0119

Lissbrant S, Brouder SM, Cunningham SM, Volenec JJ (2010) Identification of fertility regimes that enhance long-term productivity of alfalfa using cluster analysis. Agron J 102:580–591. https://doi.org/10.2134/agronj2009.0300

Macy P (1936) The quantitative mineral nutrient requirements of plants. Plant Physiol 11 (4):749–764. https://doi.org/10.1104/pp.11.4.749

Mahajan GR, Sahoo RN, Pandey RN, Gupta VK, Kumar D (2014) Using hyperspectral remote sensing techniques to monitor nitrogen, phosphorus, sulfur and potassium in wheat (*Triticum aestivum* L.). Precis Agric 15:499–522. https://doi.org/10.1007/s11119-014-9348-7

Mallarino AP, Higashi SL (2008) Assessment of potassium supply for corn by analysis of plant parts. Soil Sci Soc Am J 73(6):2177–2183. https://doi.org/10.2136/sssaj2008.0370

Marschner H (2002) Mineral nutrition of higher plants, 2nd edn. Academic Press, New York

Marschner H, Cakmak I (1989) High light intensity enhances chlorosis and necrosis in leaves of zinc, potassium, and magnesium deficient bean (*Phaseolus vulgaris*) plants. J Plant Physiol 134 (3):308–315. https://doi.org/10.1016/S0176-1617(89)80248-2

Martinez HEP, Souza RB, Bayona JA, Venegas VHA, Sanz M (2003) Coffee-tree floral analysis as a mean of nutritional diagnosis. J Plant Nutr 26:1467–1482. https://doi.org/10.1081/PLN-120021055

Meitern A, Õunapuu-Pikas E, Sellin A (2017) Circadian patterns of xylem sap properties and their covariation with plant hydraulic traits in hybrid aspen. J Plant Physiol 213:148–156. https://doi.org/10.1016/j.jplph.2017.03.012

Melsted SW, Motto HL, Peck TR (1969) Critical plant nutrient composition values useful in interpreting plant analysis data. Agron J 61(1):17–20. https://doi.org/10.2134/agronj1969.00021962006100010006x

Mills HA, Jones JB (1996) Plant analysis handbook II: a practical sampling, preparation, analysis, and interpretation guide. Micro-Macro, Athens, GA

Milton NM, Eiswerth BA, Ager CM (1991) Effect of phosphorus deficiency on spectral reflectance and morphology of soybean plants. Remote Sens Environ 36(2):121–127. https://doi.org/10.1016/0034-4257(91)90034-4

Mombiela F, Nicholaides JJ III, Nelson LA (1981) A method to determine the appropriate mathematical form for incorporating soil test levels in fertilizer response models for recommendation purposes. Agron J 73(6):937–941. https://doi.org/10.2134/agronj1981.00021962007300060007x

Mutanga O, Skidmore AK, Prins HHT (2004) Predicting in situ pasture quality in the Kruger

National Park, South Africa, using continuum-removed absorption features. Remote Sens Environ 89(3):393–408. https://doi.org/10.1016/j.rse.2003.11.001

Noori O, Panda SS (2016) Site-specific management of common olive: remote sensing, geospatial, and advanced image processing applications. Comput Electron Agric 127:680–689. https://doi.org/10.1016/j.compag.2016.07.031

Pacumbaba RO Jr, Beyl CA (2011) Changes in hyperspectral reflectance signatures of lettuce leaves in response to macronutrient deficiencies. Adv Space Res 48(1):32–42. https://doi.org/10.1016/j.asr.2011.02.020

Parent LE (2011) Diagnosis of the nutrient compositional space of fruit crops. Rev Bras Frutic 33 (1):321–334. https://doi.org/10.1590/S0100-29452011000100041

Parent LE, Dafir M (1992) A theoretical concept of compositional nutrient diagnosis. J Am Soc Hortic Sci 117(2):239–242. https://doi.org/10.21273/JASHS.117.2.239

Parent LE, Cambouris AN, Muhawenimana A (1994a) Multivariate diagnosis of nutrient imbalance in potato crops. Soil Sci Soc Am J 58(5):1432–1438. https://doi.org/10.2136/sssaj1994.03615995005800050022x

Parent LE, Isfan D, Tremblay N, Karam A (1994b) Multivariate nutrient diagnosis of the carrot crop. J Am Soc Hortic Sci 119(3):420–426. https://doi.org/10.21273/JASHS.119.3.420

Parker ER, Batchelor LD (1942) Effect of fertilizers on orange yields. Agricultural Experiment Station Bulletin 673. University of California Press, Berkeley. https://archive.org/details/effectoffertiliz673park. Accessed 19 May 2020

Pimstein A, Karnieli A, Bansal SK, Bonfil DJ (2011) Exploring remotely sensed technologies for monitoring wheat potassium and phosphorus using field spectroscopy. Field Crops Res 121 (1):125–135. https://doi.org/10.1016/j.fcr.2010.12.001

Pullanagari RR, Kereszturi G, Yule IJ (2016) Mapping of macro and micro nutrients of mixed pastures using airborne AisaFENIX hyperspectral imagery. ISPRS J Photogramm Remote Sens 117:1–10. https://doi.org/10.1016/j.isprsjprs.2016.03.010

Rozane DE, Natale W, Parent LE, dos Santos EMH (2012) The CND-Goiaba 1.0 software for nutritional diagnosis of guava (*Psidium guajava* L.) 'Paluma', in Brazil. Acta Hortic 959:161–166. https://doi.org/10.17660/ActaHortic.2012.959.19

Shear CB, Faust M (2011) Nutritional ranges in deciduous tree fruits and nuts In: Janick J (ed) Hortic Rev, 2 p142–163. doi:https://doi.org/10.1002/9781118060759.ch3

Shull CA (1929) A spectrophotometric study of reflection of light from leaf surfaces. Bot Gaz 87 (5):583–607. https://jstor.org/stable/2471112. Accessed 19 May 2020

Simón MD, Nieves-Cordones M, Sánchez-Blanco MJ, Ferrández T, Nieves M (2013) Nutrient requirements of *Chamaerops humilis* L. and *Washingtonia robusta* H. Wendl palm trees and their long-term nutritional responses to salinity. J Plant Nutr 36(9):1466–1478. https://doi.org/10.1080/01904167.2013.794240

Srivastava AK, Singh S, Huchche AD, Ram L (2001) Yield-based leaf and soil-test interpretations for Nagpur mandarin in Central India. Commun Soil Sci Plant Anal 32(3–4):585–599. https://doi.org/10.1081/CSS-100103030

Stammer AJ, Mallarino AP (2018) Plant tissue analysis to assess phosphorus and potassium nutritional status of corn and soybean. Soil Sci Soc Am J 82(1):260–270. https://doi.org/10.2136/sssaj2017.06.0179

Sumner ME (1977a) Preliminary N, P, and K foliar diagnostic norms for soybeans. Agron J 69 (2):226–230. https://doi.org/10.2134/agronj1977.00021962006900020008x

Sumner ME (1977b) Preliminary NPK foliar diagnostic norms for wheat. Commun Soil Sci Plant Anal 8(2):149–167. https://doi.org/10.1080/00103627709366709

Sumner ME (1990) Advances in the use and application of plant analysis. Commun Soil Sci Plant Anal 21(13–16):1409–1430. https://doi.org/10.1080/00103629009368313

Syltie PW, Melsted SW, Walker WM (1972) Rapid tissue tests as indicators of yield, plant composition, and soil fertility for corn and soybeans. Commun Soil Sci Plant Anal 3 (1):37–49. https://doi.org/10.1080/00103627209366348

Tarr AB, Moore KJ, Dixon PM (2005) Spectral reflectance as a covariate for estimating pasture productivity and composition. Crop Sci 45(3):996–1003. https://doi.org/10.2135/cropsci2004.0004

Thomas W (1937) Foliar diagnosis: principles and practice. Plant Physiol 12(3):571–599. https://doi.org/10.1104/pp.12.3.571
Tyner EH (1947) The relation of corn yields to leaf nitrogen, phosphorus, and potassium content. Soil Sci Soc Am J 11:317–323. https://doi.org/10.2136/sssaj1947.036159950011000C0059x
Uchida RS (2000) Recommended plant tissue nutrient levels for some vegetable, fruit, and ornamental foliage and flowering plants in Hawaii. In: Silva JA, Uchida RS (eds) Plant nutrient management in Hawaii's soils: approaches for tropical and subtropical agriculture, University of Hawaii, Manoa, pp 57–65. https://www.ctahr.hawaii.edu/oc/freepubs/pdf/pnm0.pdf. Accessed 19 May 2020
Ulrich A (1952) Physiological bases for assessing the nutritional requirements of plants. Annu Rev Plant Physiol 3:207–228. https://doi.org/10.1146/annurev.pp.03.060152.001231
Urricariet S, Lavado RS, Martin L (2004) Corn response to fertilization and SR, DRIS, and PASS interpretation of leaf and grain analysis. Commun Soil Sci Plan 35(3–4):413–425. https://doi.org/10.1081/CSS-120029722
Vaile RS (1922) Fertilizer experiments with citrus trees. Agricultural Experiment Station Bulletin 345. University of California, Berkeley Press. https://archive.org/details/fertilizerexperi345vail. Accessed 19 May 2020
Vreugdenhil D, Koot-Gronsveld AM (1989) Measurements of pH, sucrose and potassium ions in the phloem sap of castor bean (*Ricinus communis*) plants. Physiol Plant 77(3):385–388. https://doi.org/10.1111/j.1399-3054.1989.tb05657.x
Walworth JL, Sumner ME, Isaac RA, Plank CO (1986) Preliminary DRIS norms for alfalfa in the southeastern United States and a comparison with Midwestern norms. Agron J 78 (6):1046–1052. https://doi.org/10.2134/agronj1986.00021962007800060022x
Westerman RL (ed) (1990) Soil testing and plant analysis, 3rd edn. Soil Science Society of America, Madison, WI. https://doi.org/10.2136/sssabookser3.3ed.frontmatter
Woolley JT (1971) Reflectance and transmittance of light by leaves. Plant Physiol 47(5):656–662. https://doi.org/10.1104/pp.47.5.656
Xue X, Lu J, Ren T, Li L, Yousaf M, Cong R, Li X (2016) Positional difference in potassium concentration as diagnostic index relating to plant K status and yield level in rice (*Oryza sativa* L.). Soil Sci Plant Nutr 62(1):31–38. https://doi.org/10.1080/00380768.2015.1121115

10

Soil Test Changes and the Availability of Potassium (K)

David W. Franzen, Keith Goulding, Antonio P. Mallarino, and Michael J. Bell

Abstract The exchangeable fraction of soil potassium (K) has been viewed as the most important source of plant-available K, with other sources playing smaller roles that do not influence the predictive value of a soil test. Thus, as K mass balance changes, the soil test should change correspondingly to be associated with greater or reduced plant availability. However, soil test changes and the availability of K to plants are influenced by many other factors. This chapter reviews research on soil test K changes and the relation to crop uptake and yield. A mass-balance relationship is rarely achieved from the measurement of exchangeable K because of the potential for buffering of K removal from structural K in feldspars and from interlayer K in primary and secondary layer silicates. Similarly, surplus K additions can be fixed in interlayer positions in secondary layer silicates, or potentially sequestered in sparingly soluble neoformed secondary minerals, neither of which is measured as exchangeable K. In addition, soil moisture, temporal differences in exchangeable K with K uptake by crops, K leaching from residues, clay type, organic matter contribution to the soil CEC, and type of K amendment confound attempts to relate K additions and losses with an exchangeable K soil test. Research is needed to create regionally specific K soil test procedures that can predict crop response for a subset of clays and K-bearing minerals within specific cropping systems.

D. W. Franzen (✉)
North Dakota State University, Fargo, ND, USA
e-mail: david.franzen@ndsu.edu

K. Goulding
Rothamsted Research, Harpenden, Hertfordshire, UK
e-mail: keith.goulding@rothamsted.ac.uk

A. P. Mallarino
Department of Agronomy, Iowa State University, Ames, IA, USA
e-mail: apmallar@iastate.edu

M. J. Bell
School of Agriculture and Food Sciences, The University of Queensland, Brisbane, QLD, Australia
e-mail: m.bell4@uq.edu.au

10.1 Introduction

If K nutrition of crops were a simple system, with additions of K held by the soil and released as needed, then the K supply and relative availability could be accurately predicted by a simple soil extraction. However, the reality is that the prediction of K availability can be as difficult to understand as that of N and P. While N availability prediction is confounded by temperature, rainfall, soil moisture condition, biological transformation through oxidation and reduction products, and nutrient cycling in plants and soil organisms, K nutrition is affected by most of those factors, with the exception of biological redox products, but with the addition of the physical and chemical properties of the soil and its mineralogy, mainly of the K-bearing minerals. Although students are often taught that the exchangeable K "pool" in the soil is the major source of plant-available K and the reservoir for most fertilizer K applied, the reality is that the equilibrium reactions between the soil solution K, adsorbed K, interlayer K in primary and secondary layer silicates and in structural pools in minerals such as potassium feldspar can be rapid and have a significant influence on the soil test and crop production.

The dynamics of sources of K and interactions between sources have been summarized and described in many diagrams, mostly in the flowchart in Fig. 7.1 developed after the 2017 Frontiers in Potassium Science Workshop in Rome (Bell et al. 2017a, b), and discussed in detail in Chap. 7 of this book. Discussion of the dynamics depicted in Fig. 10.1 suggests that equilibria between exchangeable K, solution K, and K additions as fertilizer are always rapid, with each being measured in hours or days, not weeks, months, and years (Krauss and Johnston 2002). We will attempt to capture the difficulty of using a mass-balance approach to crop K nutrition from the biological, temporal, clay mineral, and primary mineral components depicted in Fig. 10.1.

10.2 The Mass-Balance Approach

The basic premise of a mass-balance approach is that when K is added to the soil, most often as soluble K fertilizers, such as potassium chloride or potassium sulfate, the K ions are retained at the soil cation exchange sites with the potential to be released later into the soil solution and taken up by the crop. The most commonly used measurement techniques for quantifying K on the exchange sites is displacement with NH_4^+ in a variety of extraction techniques, such as 1M-ammonium acetate. In the USA and the UK, the mass-balance approach is a common strategy for K fertilization, particularly where a buildup and maintenance approach to crop fertilization is recommended, but also to maintain desirable soil-test values when response-based information is used to decide fertilization rates for low-testing soils. In the strict buildup–maintenance approach, the fertilizers are applied at rates intended to replace crop grain or forage removal and, if necessary, the soil test is

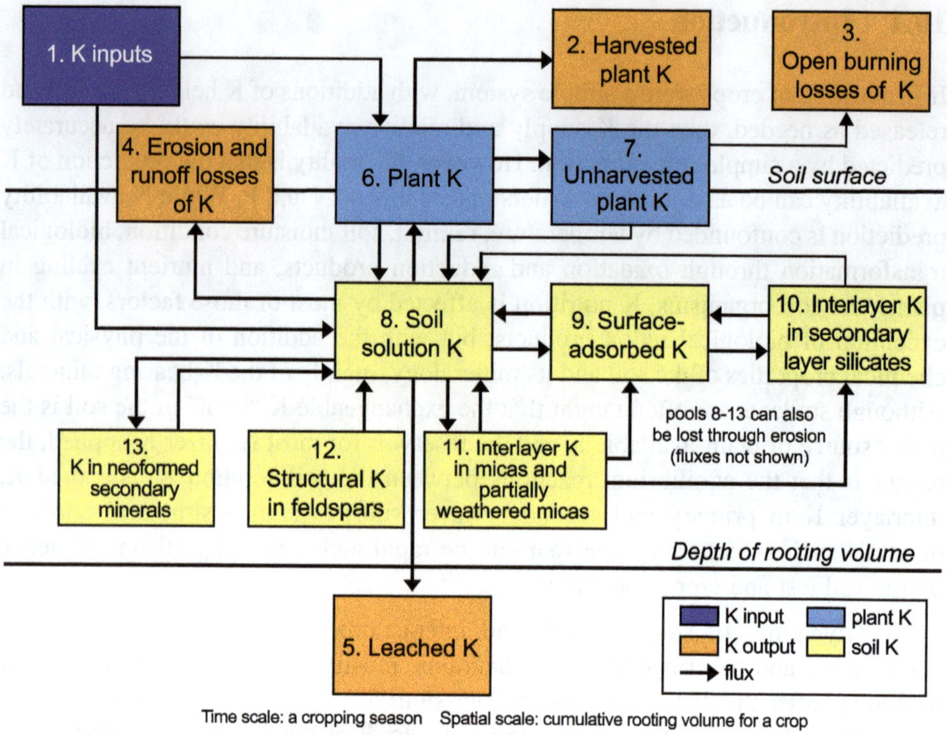

Fig. 10.1 Sources of soil K and interactions between sources. (Chap. 7)

increased through the addition of extra fertilizer to attain a certain soil test concentration regarded as critical or optimum for K supply. In Illinois, USA for example, it is assumed that the amount of fertilizer K recommended to increase the K soil test by 1 mg kg^{-1} in the medium to higher CEC (cation exchange capacity) soils of central and northern Illinois is about 10 kg K ha^{-1} (Fernandez and Hoeft 2015), but this is a broad average. In contrast, Missouri USA recommendations indicate that 63 kg K ha^{-1} will increase K soil test by 1 mg kg^{-1} (Buchholz 2004). In the UK, a rough estimate of soil test increase with K addition is that 3–5 kg K ha^{-1} results in a 1 mg kg^{-1} increase in soil test K (Potash Development Association 2011). Since these recommendations are empirically based within the region affected, one can assume that there are soil characteristics that contribute to the different responses of the resulting soil test K with K fertilizer addition.

In Illinois, the mean fertilizer K application rates and mean yields in two 12.5 ha fields were documented for 40 years starting in 1960 (Franzen 1993). The rates of applied K were greater than crop removal until 1982, which resulted in increased soil test K concentrations. From 1982 until 1992 no additional fertilizer K was added and K removal by corn (*Zea mays* L.) and soybean (*Glycine max* L. Merr.) harvests was estimated for all years. The results are shown in Fig. 10.2. The rate at which the soil K test increased with K fertilizer addition was different from the rate of decline with crop removal. The results show a "hysteresis" effect. At a site near Thomasboro, IL,

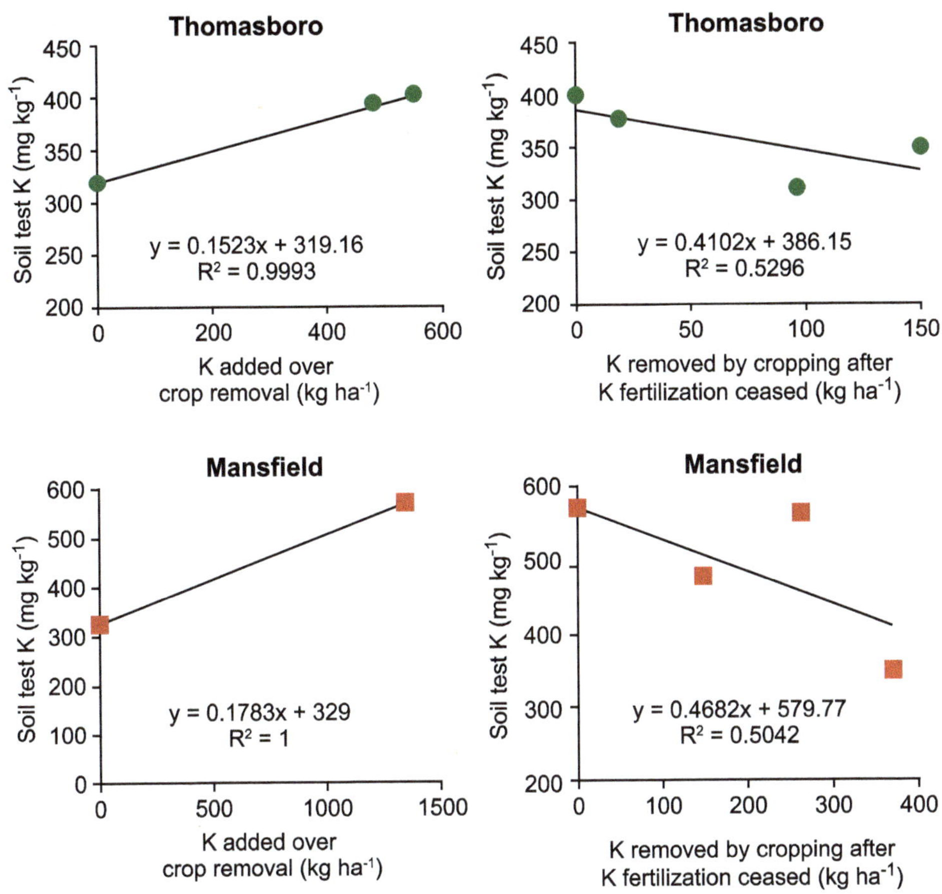

Fig. 10.2 Cumulative net K added per hectare to Thomasboro (upper left) and Mansfield (lower left) Illinois 12.5 ha fields over a period of years, compared to K removed by crop removal at Thomasboro (upper right) and Mansfield (lower right) after K fertilization ceased in 1982 at both sites

USA, the rate of soil test K increase was 1 mg kg^{-1} for 6.6 kg K ha^{-1} added as fertilizer over crop removal. After K fertilization ceased, the rate of soil test K decline was 1 mg kg^{-1} for 2.4 kg K ha^{-1} in crop removal. At a site near Mansfield, IL, USA, the rate of soil test K increase was 1 mg kg^{-1} for 5.8 kg ha^{-1} added as fertilizer over crop removal. After K fertilization ceased, the rate of soil test K decline at Mansfield was 1 mg kg^{-1} for 2.2 kg K ha^{-1} in crop removal.

In the Parana state of Brazil, a K application experiment on 12 tropical soils with basalt, shale, sandstone, and alluvial parent materials resulted in reductions in exchangeable K after the initial 2-year soybean/pearl millet cropping sequence with comparably small reductions in exchangeable K during the following 4 years (Steiner et al. 2015). In K-fertilized treatments in the same soils, exchangeable K increased in all soils and reached a plateau of values after the first two crops. The subsequent exchangeable K values were similar through the following four crop years despite continued K fertilization. Thus, the mass-balance approach appeared to be operating during the first 2 years, but not in the subsequent 4 years. What was not considered by the researchers, but has been observed in Australia (Bell et al. 2009) is that K can be extracted by roots from deeper soil layers not measured in traditional soil testing approaches (e.g., the cultivated layer), especially when surface soil K availability is depleted (unfertilized treatments), or leached to deeper soil depths and retained at these depths when surplus K applications are made (Bell et al. 2017a, b). In both instances, the surface layer soil K tests may remain relatively unchanged. In unfertilized treatments, the K uptake from deeper soil layers, residue return to surface layers and the relatively small proportion of biomass K actually removed in harvested grain (especially in cereals) will help to stabilize topsoil K concentrations. Conversely, in fertilized K treatments, the K-specific adsorption sites may become saturated with K, with subsequent K fertilization resulting in leaching of surplus K into subsoils. Bell et al. (2009) support K determination in their soils at deeper depths than those usually considered to account for these aspects.

In Alabama, USA, Cope (1981) summarized 50 years of K fertilization in four sandy low CEC (<5 cmol(+) kg^{-1}) soils and one silt loam soil with CEC about 10 cmol(+) kg^{-1}. He found that, in the unfertilized plots, the soil test K concentration initially decreased, but reached an equilibrium after 28 years at most; the second sampling being in 1957, 28 years following study initiation. A K fertilizer application rate of 18 kg K ha^{-1} resulted in no soil test change, while K application rates greater than 18 kg K ha^{-1} resulted in soil test K increases in all soils. The soil test K during and at the termination of the experiment was related to the CEC of the soils. So, in this experiment with kaolinitic and quartz-based soil parent material soils, the mass-balance approach to K fertilization was effective.

In a summary of a series of five 16-year K fertilizer rate experiments in Iowa (Villavicencio 2011), researchers found that when K was not applied, the rate of decrease in soil test K (15-cm depth) did not match the K removal rate from the grain in the short term (Fig. 10.3). There was very high temporal soil test K variability from year to year that the yearly K removal in the grain harvest seldom explained. The soil test and K removal trends approximately matched over several years, however, except when decreasing soil-test K reached a minimum plateau at around 120 mg kg^{-1}. At this point, in spite of continuous K removal, the soil test decrease may have been buffered by soil release from other pools in the topsoil or subsoil, or supplemented by deeper profile soil K. The soil clays at the experimental sites were dominantly smectitic. In addition, as the soil test K concentration decreased, corn and soybean yield was not always increased by K fertilization. Yield increases were frequent and common only at three sites where soil test K was initially <170 mg kg^{-1}, which placed them in Iowa State University recommendation categories of "optimal" and lower.

Results from the then 130-year-old Garden Clover experiment at Rothamsted, UK showed that the change in exchangeable K as a percentage of a positive K balance was on average only 39% of the K applied, while it was only 38% of the K removed when the K balance was negative (McEwen et al. 1984).

Thirty-year experiments in the UK, Germany, and Poland compared K balances with or without K fertilizer or farmyard manure (Blake et al. 1999). The recovery

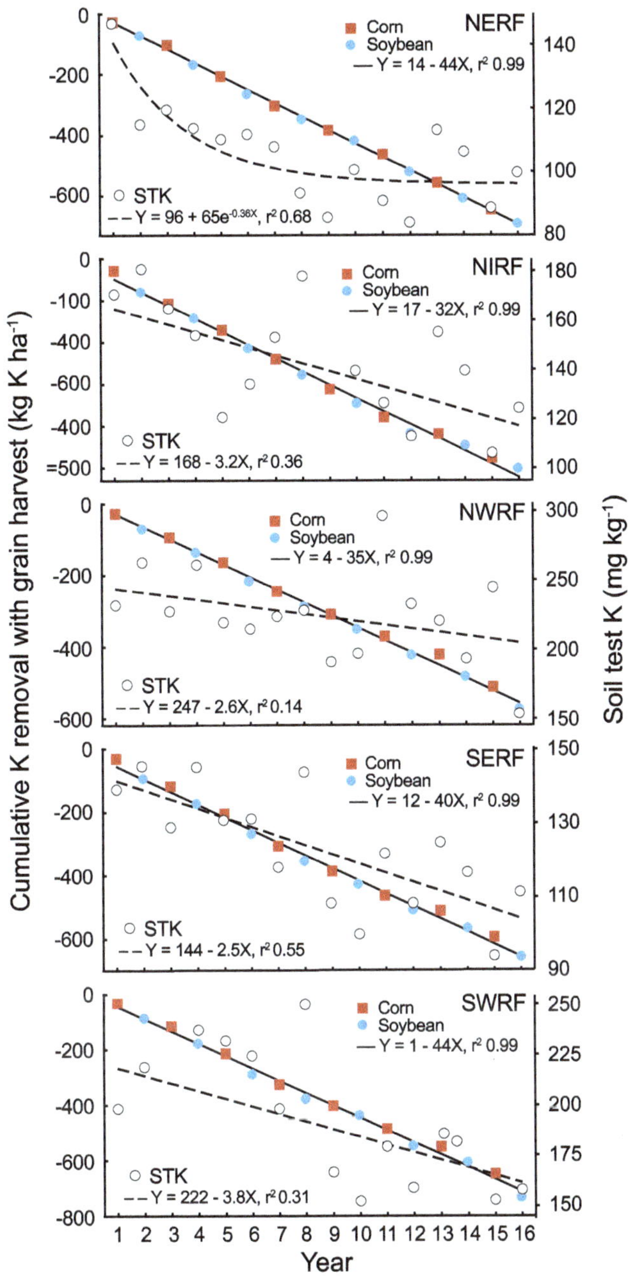

Fig. 10.3 Trends over time for cumulative K removal with grain harvest and soil-test K for samples collected each year from the non-fertilized plots in long-term Iowa K fertilization experiments. NERF, NIRF, NWRF, SERF, and SWRF are different experiment sites throughout Iowa. (from Villavicencio 2011)

rate of K from mineral fertilizers was less than 62% and decreased with increasing CEC. At Rothamsted, the fixation capacity of the clays was also a significant factor. In Germany and Poland, there was lower utilization of fertilizer K when manures were applied because K from manures were taken up in preference. A Canadian study of the kinetics of K adsorption by organic matter showed that organic matter adsorbed-K is much more accessible by plants and adsorption of K onto organic matter exchange sites is more rapid (Wang and Huang 2001). These experiments again showed that a mass-balance approach is not entirely useful because the source of K, the characteristics of the CEC sites, and the soils interact to confound any "check-book" balance that may be attempted. An attempt to understand this was suggested by Addiscott and Johnston (1971). They proposed that by increasing the soil organic matter content through application of farmyard manure, along with regulating the soil moisture supply to minimize K leaching, more K would be retained on the organic matter exchange sites and less would be moved into non-exchangeable forms.

Despite all these problems, the mass-balance approach is still the basis for many national and regional recommendation systems as noted above. For example, Dobermann et al. (1996) recommended that efficient K management for rice should be based on the K balance, but modified by the achievable yield target and the effective K-supplying power of the soil.

A variation on the mass-balance approach is the "Ideal Soil" CEC balance ratios championed by Bear in New Jersey (1951) and later by Albrecht at the University of Missouri (Albrecht 1975). The premise of this approach is that the *amount* of K in the soil is not as important as the *ratio* of K to that of the other major cations, Ca and Mg. A commonly used ratio of K in an "ideal" soil is somewhere between 2 and 5% of the base exchange capacity (Graham 1959). Although this approach is still used in parts of the USA by fertilizer and amendment sales organizations, its scientific validity has been refuted. A review of the use of this Base Cation Saturation Ratio approach was authored by Kopittke and Menzies (2007). The review concludes that the data do not support the existence of an ideal ratio and that its use would result in inefficient use of resources. The Albrecht approach has also been strongly challenged in the popular agricultural press by a number of agronomists (Miles et al. 2013).

10.3 Temporal Nature of K Soil Test Values

The previous studies have demonstrated that a purely mass-balance approach does not always account for soil test differences in exchangeable K values and that the use of some ideal ratio of cations in directing K application is not effective. The deficiency of a mass-balance approach is also reinforced by the temporal nature of exchangeable K soil test values that have been recorded in soils with mineralogy that supports K fixation and release. A 20-year study using twice-monthly soil sampling at the 0–15 cm depth for exchangeable K (1M-ammonium acetate) was conducted at Urbana, IL and Brownstown, IL, USA. Most of this data was lost and never published, but 9 years of the Urbana work was published in 2005 (Peck and Sullivan 1995). Using tabular values, Franzen (2011) imposed a seasonal repeated analysis using the statistical package SAS (PROC UCM time series analysis using 24 data points per cycle) to the data and found that the relative K values were related to soil moisture at the time of sampling. Starting on April 1, 1986, in each year (Fig. 10.4), the extractable K is highest in winter when soil moisture is greatest and lowest in late

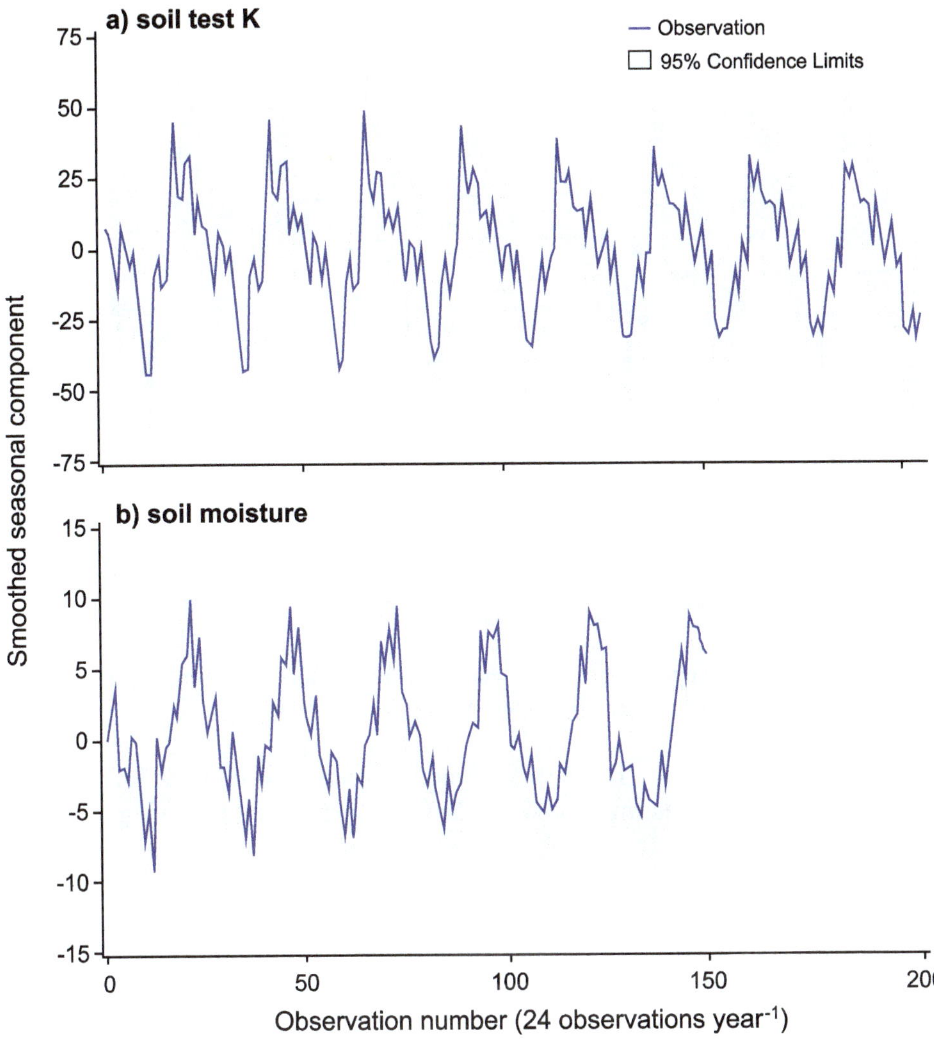

Fig. 10.4 (**a**) Soil test K. (**b**) Soil moisture. Seasonality of soil K (top image) over 9 years, compared with soil moisture (bottom image over 6 years. Soil samples 0–15 cm obtained twice monthly from April 1, 1986, through 1994. Soil moisture was available for the first 6 years only. (from Franzen 2011)

summer when the soil is driest and K supply was decreased by corn uptake. The soils in this study were dominated by smectite clay.

Recent K studies in North Dakota, USA using illite-dominant and smectite-dominant clays have indicated that the seasonal variation of the K soil tests is minor on soils with a smectite/illite ratio <3.5, but relatively high on soils with a smectite/illite ratio >3.5 (Fig. 10.5). The North Dakota climate is not favorable for winter soil sampling; however, the North Dakota data shows the greatest K extraction by 1M-ammonium acetate in early spring, with values decreasing as the season

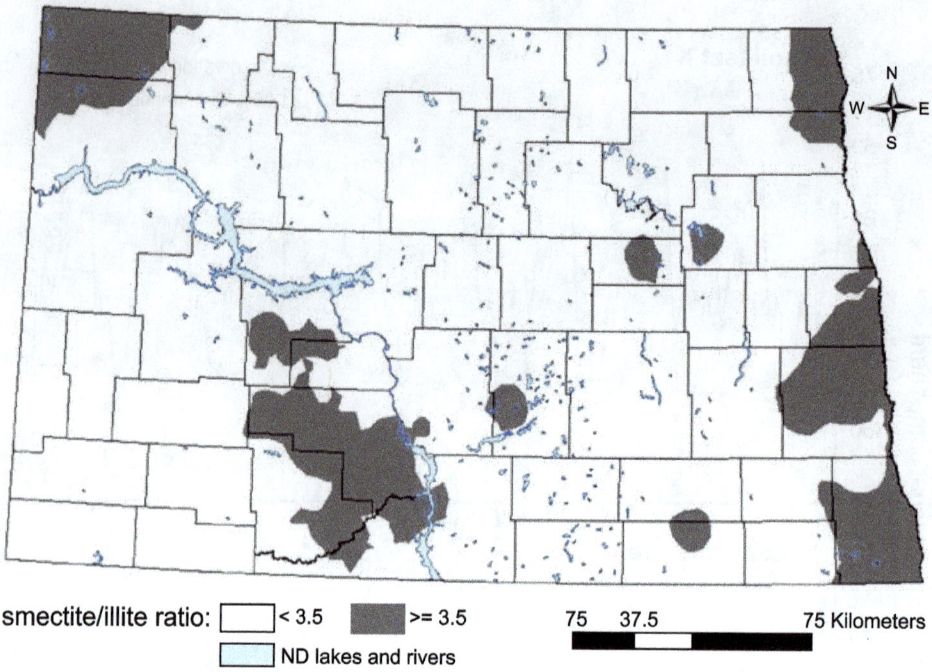

Fig. 10.5 Smectite/illite ratio in the clay fraction of North Dakota soils from a state-wide soil sampling survey conducted spring, 2017. (Franzen, unpublished data)

progresses to drier months. These data are not as clearly related to soil moisture as those from Illinois. To determine whether the seasonality was related to crop uptake, both fallow and cropped (corn) check plots were sampled twice each month from planting to harvest. The seasonality of the K test and its decrease through the growing season was present at both the smectitic and illitic sites when the soils were cropped (Fig. 10.6). When the soils had a smectite/illite ratio <3.5, the K test tended to remain relatively constant, while in a soil with a smectite/illite ratio >3.5, the K test levels tended to decrease in the drier part of the summer (between 12 and 17 weeks after planting) and increase when soils were moist (week 10, 18, 19). This is consistent with fixation during dry conditions, and release when the soils are moist. The fixation in these soils is temporary, and release and fixation of K are relatively rapid and reversible based on these data.

10.4 Crop Residue Recycling in K Mass-Balance Considerations

Part of the reason for the apparent seasonality in the K test is uptake of K by the crop and its release back into the soil after physiological maturity. Mean corn vegetative K content at physiological maturity in a series of Iowa K rate experiments (Oltmans and Mallarino 2015) was 93 kg K ha^{-1} for sites with a K yield response and 101 kg K

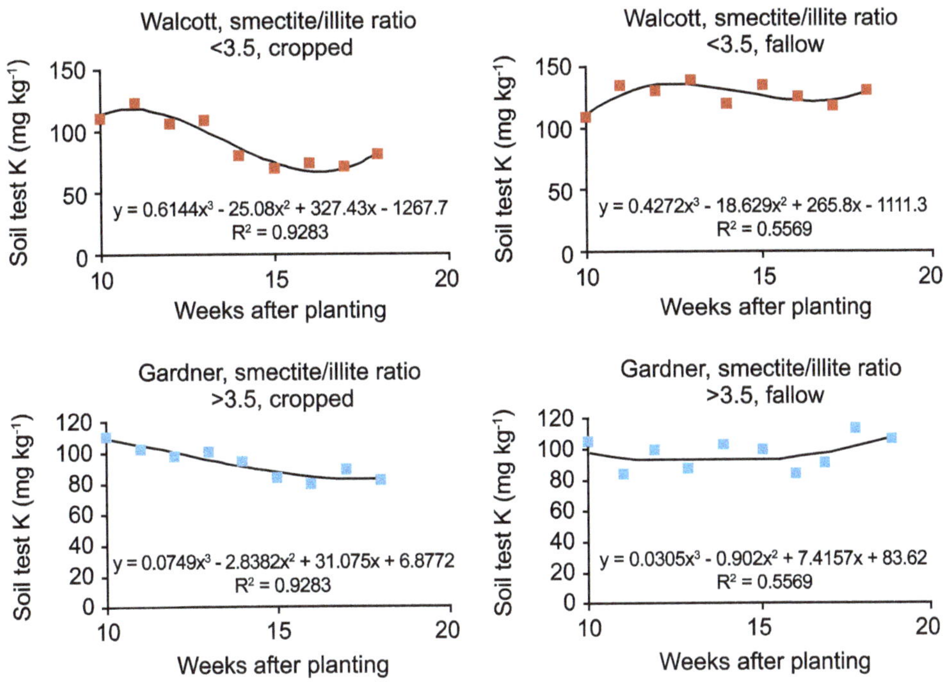

Fig. 10.6 Change in dry soil test K concentrations from 10 weeks after planting to 20 weeks after planting in a soil with a smectite/illite ratio in the soil clay fraction < 3.5 and a soil with a smectite/illite ratio > 3.5 under corn and fallow. (Breker and Franzen, unpublished; North Dakota, USA)

ha^{-1} for sites with no K response. For soybean, the mean K accumulated in vegetative parts at physiological maturity was 95 kg ha^{-1} for all sites. From physiological maturity, K is lost from the vegetative portion of the plants, with mean K in vegetative tissues in corn at harvest at sites with a K yield response of 66 kg ha^{-1}, and 67 kg ha^{-1} at sites with no K response. In soybean, the mean residue content of K declined to 41 kg ha^{-1} at harvest. These data indicate that soil sampling at physiological maturity, which is a practice utilized by some crop consultants, may result in lower soil test K values than if the samples were taken at harvest or later in the fall. The residue K content of corn and soybean was about 25 kg ha^{-1} and 54 kg ha^{-1} less, respectively, at harvest compared to physiological maturity. Losses from crop residues also contribute to the soil K pool.

The soybean residue in the Oltmans and Mallarino (2015) study continued to lose K throughout the fall until late January, when the K content reached about 10 kg ha^{-1}. Loss of K from corn residue steadily declined to about 30 kg ha^{-1} in early December, then decreased again during the spring thaw. Precipitation following physiological maturity explained much of the residue K decline. Across all sites, the soil test K concentration (15-cm depth) was higher in spring than in the fall, and the magnitude of difference was linearly related with the K loss from residue, although the relationship was much better in soybean residue (r^2 0.56) than in corn residue (r^2 0.16). These data indicate that soil sampling at physiological maturity, which is a practice utilized by some crop consultants, may result in lower soil test K values than if the samples were taken at harvest or later in the fall due to K leaching into the soil from plant residues.

The cycling of K from crop residue back to soil is a noted component of many plant ecosystems (Jobbagy and Jackson 2004). Plants mobilize K from deeper in the soil in natural systems, and their leaves and other vegetative parts are returned to the soil, resulting in an accumulation of K near the surface. Corn and soybean and other crops also return K to the soil surface; corn returning a much larger percentage than soybean (Oltmans and Mallarino 2015). Most grain crops at harvest, including wheat and corn, do not contain high K concentrations in the seed that is removed, with soybean being an exception.

Although there is often a relationship between exchangeable K and crop yield response to K addition, the relationship is less frequent with absolute crop yield. The relationship is also seldom mass-balance based, as the previous discussion would indicate. In soils with a sandy texture (Mendes et al. 2016; Alfaro et al. 2004) and in clay soils with deep cracks followed by high rainfall (Alfaro et al. 2004), K losses from the soil system by leaching are common, resulting in much apparently available exchangeable K being unavailable to the crop early in the growing season; this can result in a yield penalty. In high-clay soils with significant content of smectitic clays, deep cracks, sometimes over 1 m in depth, may form. During rainfall, some of the topsoil washes into the cracks, resulting in loss of K to deeper soil depths (Alfaro et al. 2004). In most soils, the K leached from the 0–15 cm typical K-sampling depth would be retained at deeper soil depths; therefore, a multi-depth approach in soils with surface layer K leaching potential may be important to explaining mass-balance and response to K at lower soil test values.

10.5 Clay Chemistry and K Response

According to research by Sharpley (1989) in his study of 102 soils from the continental US and Puerto Rico, water-soluble K was closely related to exchangeable K within soils of similar clay-type, but not between clay-types. The release of K from exchangeable sources into solution increased from smectitic to "mixed" clays to kaolinitic clays. Sharpley recommended that an analysis of exchangeable K and K reserves (as nitric acid-extractable K) would provide a better indication of K supply for crops. This had been suggested by Goulding (1984) and Goulding and Loveland (1986) and is discussed in detail in Chaps. 7 and 8 in this book.

An Australian study reported K fertilization practices in Red Ferrosol, low CEC soils, and Black and Grey Vertisols, which are moderate to high clay-content soils. Analyses of exchangeable K, soil solution K, and the activity of K in soil solution varied 6–7-fold between soil types (Bell et al. 2009). The management of K for optimal crop production varied with soil. The Vertisols had a high K buffer capacity (BC_K) and applications of K were not as effective at increasing crop K accumulation as those in the low CEC Red Ferrosols, where BC_K was significantly lower. A better K fertilization approach for the Vertisols was to apply a K fertilizer in concentrated bands rather than a broadcast application that resulted in lower soil solution K concentrations within the soil volume. In the high clay Vertisols with high CEC (50–60 cmol(+) kg^{-1}), a deep band of K with N and P was more effective at overcoming K deficiency due to root proliferation around the K band (Chap. 12).

In a study of 23 K rate experiments in North Dakota in 2014 and 2015, the response of corn at sites below the current exchangeable K critical level of 150 mg K kg^{-1} was not predicted at 10 sites (Table 10.1). The relative corn yield of the unfertilized check in relation to soil test K at these sites is shown in Fig. 10.7, with the relationship between exchangeable K and relative yield very weak. A multiple regression analysis of relative yield with the potassium feldspar concentration of the mineral fraction of the 23 sites and the relative clay mineral percentage of the clay fraction is shown in Table 10.2. This analysis suggested that the best prediction of yield response to K fertilization in these North Dakota soils would need to consider both clay mineralogy and the soil test K concentration. Unfortunately, many agricultural regions do not have access to clay mineralogy data, and so these refinements can be difficult to implement.

Using a Ward minimum variance clustering technique, the sites clustered into those with smectite/illite ratios < 3.5 and those > 3.5 (Fig. 10.5). Using K response data from the sites falling into those categories, the resulting responses indicated that a critical level of 150 mg K kg^{-1} was appropriate for site with a smectite/illite ratio $<$ 3.5 and a critical level of 200 mg K kg^{-1} was more appropriate for sites with a smectite/illite ratio > 3.5 (Fig. 10.8).

Different articles in this book focus directly on the importance of different soil K pools (Bell et al. 2017a, Chap. 7) in relation to bioavailability, and on the relationship between various soil tests and plant-availability of soil K (Bell et al. 2017b, Chap. 8), but it is relevant to mention here some concepts and findings. Potassium application and its interaction with clays, particularly a change in clay type with K addition or loss, has been documented and could confound expected K test results. Barre et al. (2007) hypothesized that the illite-dominance of prairie soils in temperate regions was due to the production and stabilization of illite from K redistributed from deeper soil strata to surface strata through root uptake and residue deposition to the surface. Barre et al. (2009) provided a brief review of studies supporting the hypothesis of illite construction and deconstruction under prairie vegetation and offered a model that might predict clay stability under K addition or loss through intensive cropping without adequate K replenishment. A study from China in a loess-derived soil containing illite and chlorite focused on topsoil total K compared to a reduction in exchangeable K during 15 years of continuous alfalfa cropping. The topsoil exchangeable K decreased due to alfalfa forage uptake and removal; however, the total K in the topsoil increased due to K additions from deep soil K taken up by the alfalfa roots and deposited in the upper strata of the soil, producing a highly crystalline illitic-like clay (Li et al. 2011).

Singh and Goulding (1997), using X-ray diffraction, looked for changes in micaceous minerals that might accompany 150 years of continuous cropping by wheat, with and without K fertilizer, on the Broadbalk experiment at Rothamsted, UK, but found none, although deep plowing had changed the mineralogy and K

Table 10.1 A series of 23 K fertilizer rate studies in corn in North Dakota, with clay mineralogy of the clay fraction and potassium feldspar content of the mineral portion of the soil. Initial pre-plant soil test K concentrations (1M-ammonium acetate on dry soil) are also indicated along with expected yield increase compared to the yield increase experienced

Site, year	K test (mg kg^{-1})	Expected yield increase	Actual yield increase	Potassium feldspar (%)	Smectite-illite (%)
Buffalo, 2014	100	**Y**	**N**	7.1	85–11
Walcott E, 2014	100	Y	Y	5.8	84–13
Wyndmere, 2014	100	**Y**	**N**	6.1	72–22
Milnor, 2014	100	**Y**	**N**	11.7	35–57
Gardner, 2014	115	Y	Y	5.3	76–20
Fairmount, 2014	140	**Y**	**N**	8.0	80–14
Walcott W, 2014	80	**Y**	**N**	7.3	52–40
Arthur, 2014	170	**N**	**Y**	1.7	85–11
Valley City, 2014	485	N	N	9.0	70–23
Page, 2014	200	N	N	5.7	74–20
Absaraka, 2015	113	**Y**	**N**	9.9	84–14
Arthur, 2015	125	Y	Y	9.5	85–12
Barney, 2015	170	N	N	6.3	79–16
Casino, 2015	120	Y	Y	6.4	85–12
Dwight, 2015	110	Y	N	6	82–15
Fairmount1, 2015	188	**N**	**Y**	5.6	87–10
Fairmount2, 2015	118	**N**	**Y**	7.4	79–12
Leonard N, 2015	380	N	N	6.9	70–25
Leonard S, 2015	190	N	N	5.5	52–41
Milnor, 2015	118	Y	Y	8.6	74–20
Prosper, 2015	205	N	N	9.2	83–14
Valley City, 2015	200	N	N	5.6	65–30
Walcott, 2015	109	Y	Y	6.2	47–48

Bold font denotes site where expected yield response or nonresponse was not recorded

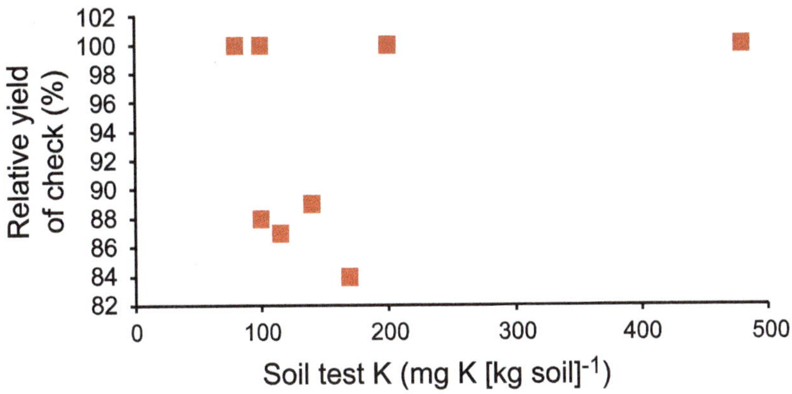

Fig. 10.7 Relative yield of check (with no added K) in North Dakota K fertilizer rate studies in 2014 and the relationship to soil test K

Table 10.2 Multiple linear regression of possible factors relating to relative corn grain yield of 2014 and 2015 K fertilizer rate trials in North Dakota from Table 10.1

Factor	K test	K-feldspar	Illite	Smectite
K test	1.0			
K-feldspar	0.17	1.0		
Illite	–0.03	–0.32	1.0	
Smectite	0.05	0.33	–0.99	1.0
Relative yield	0.29	–0.0002	0.32	–0.25

content of the surface (0–23 cm) soil. Evidence for changes in illite with K additions and losses through cropping was found from examination of long-term treatments in the Morrow Plots at the University of Illinois Experiment Station at Urbana, Illinois, USA (Velde and Peck 2002). Soil samples from the plots were archived from 1913. The differences in clay mineralogy between 1913 and 1996 indicate changes in clay mineralogy with cropping. These changes were small in the corn–oats–hay rotations, but there was a significant loss of illitic clay from the continuous corn rotation with no amendments. The authors considered that the stability of illite/smectite under the corn–oats–hay rotations was the result of K inputs from non-clay K-containing minerals restoring the K lost from the clay minerals under this less intense cropping system. The use of NPK fertilizer was adopted in 1955, and this restored the illite content of the clay mineralogy in the continuous corn plots to what was measured in 1913. The authors speculated that the addition of K replenished the clay mineral K, stabilizing and restoring illite/smectite integrity. The illites/smectites serve as a reservoir of K in prairie soils, but the clays can be degraded under continuous cropping.

Previous (Clover and Mallarino 2009) and ongoing unpublished work in Iowa has shown that in smectite-dominant soils (but with smaller contents of illite and vermiculite), the measurement of both soil-test K and non-exchangeable K explains the effects of K additions and removal by corn and soybean on post-harvest soil K. Also, more rapid than expected relative changes between exchangeable K and a combination of interaction with K in neoformed secondary minerals, structural K in

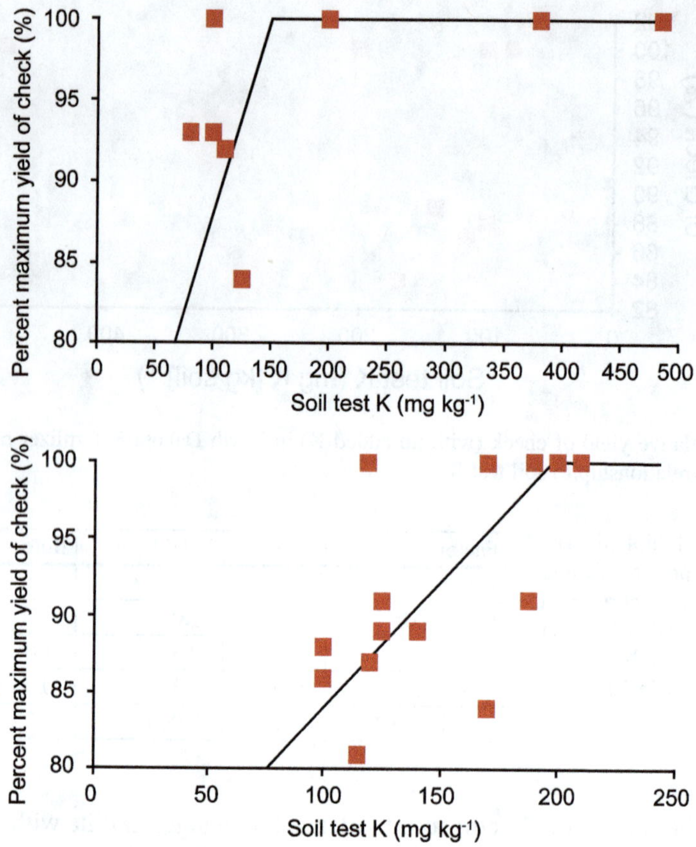

Fig. 10.8 Percent maximum corn check yields in North Dakota K fertilizer rate studies at sites with smectite/illite ratio < 3.5 (top), and smectite/illite ratio >3.5 (bottom) (Franzen, unpublished data)

feldspars, interlayer K in micas and partially weathered micas, and interlayer K in secondary layered silicates explained part of the soil K variation between one crop and the next.

An important review of K chemistry was presented by Sparks and Huang (1985). One of the main points of their review was to highlight experiments in which the replenishment of K in the soil solution from non-exchangeable K in clays and potassium feldspars was measured in hours and days, i.e., was fast enough to meet crop needs (see also Krauss and Johnston 2002). They also noted that wetting and drying affects the movement of K in and out of clay interlayers. Soil wetness increases the content of reduced iron within the clay interlayers, which has a large hydrated radius. The larger radius facilitates the movement of K^+ out of the clay interlayers. Dry soil increases the concentration of oxidized iron within the clay interlayers, blocking the outflow of K^+ from the clays.

Some soils retain K^+ on drying, including smectitic soils in temperate regions (e.g., Barbagelata and Mallarino 2012) due to increased soil solution K concentration pushing the equilibrium between the soil solution and the soil clay in the direction of interlayer K^+ (Dowdy and Hutcheson 1963).

10.6 Relative Unresponsiveness in K Removal in Harvested Grain, Despite Wide Variability in Crop K Status and Responsiveness to K Fertilizer Application

Variation in grain, seed, or kernel K concentration of eight crops was analyzed as a function of crop yield response to increasing soil exchangeable K in a long-term field experiment on an Oxisol soil in Queensland, Australia (Bell et al. 2009). Despite large yield responses to increasing soil K availability in most crops (3–10-fold range in yields of cottonseed and peanut kernels, and nearly twofold ranges in yields of wheat and sorghum), there was effectively no variation in K concentration in the harvested product. The lack of change in corn grain K concentration despite variation in soil K availability was also observed by Oltmans and Mallarino (2015). There is, however, large species variation in the rate of K removal (e.g., removal of K in soybean grain is about 5 times the rate of removal in sorghum grain—Table 10.3). This means that growers can budget to replace K removal in a crop rotation reasonably accurately—something that is harder to do in other nutrients where both yield and grain concentration vary in response to differences in soil supply.

Table 10.3 Response of grain/seed/kernel yield of different crops to soil K supply grown in a strongly K-limiting soil or a soil with excess K supply, and the resulting variation in K concentration in grain/seed

		Strongly K-limited soil		Excess soil K supply	
Crop	Harvested portion	Yield (kg ha^{-1})	Seed/grain/ kernel K (g kg^{-1})	Yield (kg ha^{-1})	Seed/grain/ kernel K (g kg^{-1})
Corn (*Zea mays* L.)	Grain	4200	2.7	6240	2.9
Sorghum (*Sorghum bicolor* L.)	Grain	5210	3.4	8690	3.2
Wheat (*Triticum aestivum* L.)	Grain	3700	3.9	5710	3.7
Cotton (*Gossypium hirsutum* L.)	Seed	380	8.9	3840	8.8
Peanut (*Arachis hypogaea* L.)	Kernel	810	6.9	2720	7.0
Chickpea (*Cicer arietinum* L.)	Grain	2220	8.1	3010	8.5
Mungbean (*Vigna radiata* L.)	Grain	1150	13.1	1706	11.7
Soybean (*Glycine max* L. Merr.)	Grain	2170	16.3	3270	16.5

From a long-term experiment on an Oxisol near Queensland, Australia, described in Bell et al. (2009)

10.7 Potassium Losses Due to Erosion from Wind and Water

A seldom considered source of K loss that may contribute to regional K balance is the loss of topsoil due to wind and water erosion. An analysis of nutrient losses in North Dakota from the time of first plowing (1880 through 1920, depending on the location within the state) is underway, and complete for phosphorus (P). Total P loss from wind erosion in North Dakota since the time of plowing is estimated to be 17 M tons of P (Franzen 2016). This is equivalent to over 200 years of P application by North Dakota farmers to the 10 M ha of state cropland at fertilizer rates commonly applied in 2016. Total K loss from wind erosion since plowing is estimated to be much higher. Analysis of dust originating from North Dakota topsoil collected in eastern US cities in the 1930s contained 19 times more P and 45 times more K than samples obtained in regions of origin after the storms (Hansen and Libecap 2004). Topsoil loss in large portions of the state not under no-till or modified no-till systems is still on-going. A site northwest of Grand Forks, ND characterized in 1958 and re-characterized in 2014 revealed 48 cm of topsoil loss due to wind erosion over the 56 years (Montgomery 2015).

10.8 Summary

The mass balance of K in the surface soil is a function of: (1) K added in fertilizers and manures; (2) plant redistribution of subsoil K to the surface; (3) K losses due to leaching in low CEC soils, and sampling to deeper depths than the 0–15 cm depth to determine subsoil contribution; (4) K removal with grain, forage, and crop residues; and (5) K lost in soil erosion from wind/water. A mass-balance relationship based on the measurement of exchangeable K is only rarely achieved because of the rapid equilibrium between soil solution K and its relatively rapid exchange with K in K-bearing primary minerals and clay interlayers. In addition, soil moisture, temporal differences in exchangeable K with K uptake by crops, K leaching from residues, clay type, organic matter contributions to the soil CEC and type of K amendment have confounded many attempts to relate K additions and losses with K soil test. Research is needed to create regionally specific K soil test procedures that can predict crop response within a subset of clays and K-bearing minerals within specific cropping systems.

References

Addiscott TM, Johnston AE (1971) Potassium in soils under different cropping systems 2. The effects of cropping systems on the retention by the soils of added K not used by crops. J Agr Sci Camb 76:553–561. https://doi.org/10.1017/S0021859600069550

Albrecht WA (1975) The Albrecht papers. Vol 1: Foundation concepts. Acres USA, Kansas City

Alfaro MA, Jarvis SC, Gregory PJ (2004) Factors affecting potassium leaching in different soils. Soil Use Manage 20:182–189. https://doi.org/10.1111/j.1475-2743.2004.tb00355.x

Barbagelata PA, Mallarino AP (2012) Field correlation of potassium soil test methods based on dried and field-moist soil samples for corn and soybean. Soil Sci Soc Am J 77:318–327. https://doi.org/10.2136/sssaj2012.0253

Barre P, Velde B, Abbadie L (2007) Dynamic role of 'illite-like' clay minerals in temperate soils: facts and hypotheses. Biogeochemistry 82:77–88. https://doi.org/10.1007/s10533-006-9054-2

Barre P, Berger G, Velde B (2009) How element translocation by plants may stabilize illitic clays in the surface of temperate soils. Geoderma 151:22–30. https://doi.org/10.1016/j.geoderma.2009.03.004

Bell MJ, Moody PW, Harch GR, Compton B, Want PS (2009) Fate of potassium fertilisers applied to clay soils under rainfed grain cropping in south-east Queensland, Australia. Aust J Soil Res 47:60–73. https://doi.org/10.1071/SR08088

Bell MJ, Mallarino AP, Moody P, Thompson M, Murrell S (2017a) Soil characteristics and cultural practices that influence potassium recovery efficiency and placement decisions. In: Proceedings of the frontiers of potassium workshop, Rome, Italy, 25–27 Jan 2017. https://www.apni.net/k-frontiers/ . Accessed 29 May 2020

Bell MJ, Ransom MD, Thompson ML, Florence AM, Moody PW, Guppy CN (2017b) Soil potassium pools with dissimilar bioavailability. In: Proceedings of the frontiers of potassium workshop, Rome, Italy, 25–27 Jan 2017. https://www.apni.net/k-frontiers/. Accessed 29 May 2020

Blake L, Mercik S, Koerschens M, Goulding KWT, Stempen S, Weigel A, Poulton PR, Powlson DS (1999) Potassium content in soil, uptake in plants and the potassium balance in three European long-term field experiments. Plant Soil 216:1–14. https://doi.org/10.1023/A:1004730023746

Buchholz DD (2004) Soil test interpretations and recommendations handbook. Univ. of Missouri College of Agriculture, Division of Plant Sciences, Columbia, MO. http://aes.missouri.edu/pfcs/soiltest.pdf. Accessed 12 May 2020

Clover MW, Mallarino AP (2009) Potassium removal and post-harvest soil potassium fractions in corn-soybean rotations as affected by fertilization. In: Agronomy abstracts. CD-ROM. ASA-CSSA-SSSA, Madison, WI

Cope JT Jr (1981) Effects of 50 years of fertilization with phosphorus and potassium on soil test levels and yields at six locations. Soil Sci Soc Am J 45:342–347. https://doi.org/10.2136/sssaj1981.03615995004500020023x

Dobermann A, Cassman KG, Sta CPC, Adviento MAA, Pampolino MF (1996) Fertilizer inputs, nutrient balance, and soil nutrient-supplying power in intensive, irrigated rice systems. II: Effective soil K-supplying capacity. Nutr Cycl Agroecosyst 46:11–21. https://doi.org/10.1007/BF00210220

Dowdy RH, Hutcheson TB (1963) Effect of exchangeable potassium level and drying on release and fixation of potassium by soils as related to clay mineralogy. Soil Sci Soc Am J 27:31–34. https://doi.org/10.2136/sssaj1963.03615995002700010014x

Fernandez F, Hoeft RG (2015) Illinois agronomy handbook. Chapter 8 Managing soil pH and crop nutrients. http://extension.cropsciences.illinois.edu/handbook/pdfs/chapter08.pdf. Accessed 12 May 2020

Franzen DW (1993) Spatial variability of plant nutrients in two Illinois fields. PhD Dissertation, University of Illinois, Urbana, IL. http://hdl.handle.net/2142/72615 . Accessed 12 May 2020

Franzen DW (2011) Variability of soil test potassium in space and time. In: Proceedings of the North Central extension industry soil fertility conference, 16–17 Nov 2011. Des Moines, IA. IPNI, Brookings, SD, pp 74–82

Franzen DW (2016) A history of phosphate export from North Dakota and the region. https://www.youtube.com/watch?v=vQF0hy2crH0&feature=youtu.be. Accessed 12 May 2020

Goulding KWT (1984) The availability of potassium in soils to crops as measured by its release to a calcium-saturated cation exchange resin. J Agric Sci Camb 103:265–275. https://doi.org/10.1017/S0021859600047213

Goulding KWT, Loveland PJ (1986) The classification and mapping of potassium reserves in soils of England and Wales. J Soil Sci 37:555–565. https://doi.org/10.1111/j.1365-2389.1986.tb00387.x

Graham ER (1959) An explanation of theory and methods of soil testing. Bull. 734. Missouri Agr Exper Station, Columbia

Hansen ZK, Libecap GD (2004) Small farms, externalities and the Dust Bowl of the 1930's. J Polit Econ 112:665–694. https://doi.org/10.1086/383102

Jobbagy EG, Jackson RB (2004) The uplift of soil nutrients by plants: Biogeochemical consequences across scales. Ecology 85:2380–2389. https://doi.org/10.1890/03-0245

Kopittke PM, Menzies NW (2007) A review of the use of the basic cation saturation ratio and the 'ideal' soil. Soil Sci Soc Am J 71:259–265. https://doi.org/10.2136/sssaj2006.0186

Krauss A Johnston AE (2002) Assessing soil potassium, can we do better? In: 9th International congress of soil science, Faisalabad, Pakistan, 18–20 Mar 2002. https://doi.org/10.3923/ijss.2016.36.48

Li DC, Velde B, Li FM, Zhang GL, Zhao MS, Huang LM (2011) Impact of long-term alfalfa cropping on soil potassium content and clay minerals in a semi-arid loess soil in China. Pedosphere 21:522–531. https://doi.org/10.1016/S1002-0160(11)60154-9

McEwen J, Johnston AE, Poulton PR, Yeoman DP (1984) Rothamsted garden clover - red clover grown continuously since 1854. Yields, crop and soil analyses 1956-1982. Rothamsted Report for 1983, pp 225–237. https://doi.org/10.23637/ERADOC-1-34111

Mendes W da C, Junior JA, da Cunha PCR, da Silva AR, Evangelista AWS, Casaroli D (2016) Potassium leaching in different soils as a function of irrigation depths. Rev Bras Eng Agr Amb 20:972–977. https://doi.org/10.1590/1807-1929/agriambi.v20n11p972-977

Miles N, van Antwerpen, R, Fey M, Edmeades D, Farina M, Lambrechts JJN, Hoffman JE, Hardie-Pieters AG, Clarke CE, Rozanov DA, Ellis F, du Preez CC, Manson A., Thibaud G, Roberts V (2013) The Albrecht System: uneconomical and outdated! Farmers Weekly, 24 May 2013, pp 8–9. http://www.farmersweekly.co.za/opinion/by-invitation/the-albrecht-system-uneconomical-outdated/ . Accessed 12 May 2020

Montgomery BL (2015) Evaluating dynamic soil change in the Barnes soil series across eastern North Dakota. MS Thesis. 521 pages. North Dakota State University, Fargo, ND. http://search.proquest.com/docview/1686112990. Accessed 12 May 2020

Oltmans RR, Mallarino AP (2015) Potassium uptake by corn and soybean, recycling to soil, and impact on soil test potassium. Soil Sci Soc Am J 79:314–327. https://doi.org/10.2136/sssaj2014.07.0272

Peck TR, Sullivan ME (1995) Twice monthly field soil sampling for soil testing to evaluate reproducibility of soil test levels. In: Illinois fertilizer conference proceedings, 22–25 Jan 1995

Potash Development Association (2011) Soil analysis. Key to nutrient management planning. The Potash Development Association, Leaflet 24. Middlesbrough, UK. http://www.pda.org.uk/pda_leaflets/24-soil-analysis-key-to-nutrient-management-planning/. Accessed 12 May 2020

Sharpley AN (1989) Relationship between soil potassium forms and mineralogy. Soil Sci Soc Am J 53:1023–1028. https://doi.org/10.2136/sssaj1989.03615995005300040006x

Singh B, Goulding KWT (1997) Changes with time in the potassium content and phyllosilicates in the soil of the Broadbalk continuous wheat experiment at Rothamsted. Eur J Soil Sci 48:651–659. https://doi.org/10.1111/j.1365-2389.1997.tb00565.x

Sparks DL, Huang PM (1985) Physical chemistry of soil potassium. In: Munson RD (ed) Potassium in agriculture. ASA, CSSA, SSSA, Madison, WI, pp 201–276. https://doi.org/10.2134/1985.potassium

Steiner FM, Lana M do C, Zoz T, Frandoloso JF (2015) Changes in potassium pools in Paraná soils under successive cropping and potassium fertilization. Semin Cienc Agrar 36:4083–4098. https://doi.org/10.5433/1679-0359.2015v36n6Supl2p4083

Velde B, Peck T (2002) Clay mineral changes in the Morrow Experimental Plots, University of Illinois. Clay Miner 50:364–370. https://doi.org/10.1346/000986002760833738

Villavicencio CX (2011) Relationships between potassium fertilization, removal with harvest, and soil-test potassium in corn-soybean rotations. Iowa State University Graduate Theses and Dissertations. http://lib.dr.iastate.edu/cgi/viewcontent.cgi?article=1150&context=etd. Accessed 12 May 2020

Wang FL, Huang PM (2001) Effects of organic matter on the rate of potassium adsorption by soils. Can J Soil Sci 81:325–330. https://doi.org/10.4141/S00-069

11

Potassium (K) Input–Output Balance and their Impacts

Kaushik Majumdar, Robert M. Norton, T. Scott Murrell, Fernando García, Shamie Zingore, Luís Ignácio Prochnow, Mirasol Pampolino, Hakim Boulal, Sudarshan Dutta, Eros Francisco, Mei Shih Tan, Ping He, V. K. Singh, and Thomas Oberthür

K. Majumdar (✉) · S. Zingore · S. Dutta · T. Oberthür
African Plant Nutrition Institute and Mohammed VI Polytechnic University, Ben Guerir, Morocco
e-mail: k.majumdar@apni.net; s.zingore@apni.net; s.dutta@apni.net; t.oberthur@apni.net

R. M. Norton
Faculty of Veterinary & Agricultural Sciences, The University of Melbourne, Parkville, VIC, Australia
e-mail: rnorton@unimelb.edu.au

T. S. Murrell
African Plant Nutrition Institute and Mohammed VI Polytechnic University, Ben Guerir, Morocco

Department of Agronomy, Purdue University, West Lafayette, IN, USA
e-mail: s.murrell@apni.net

F. García
Balcarce, Argentina

L. I. Prochnow · E. Francisco
Plant Nutrition Science and Technology, Piracicaba, SP, Brazil
e-mail: LProchnow@npct.com.br

M. Pampolino
Calabarzon, Philippines

H. Boulal
African Plant Nutrition Institute, Settat, Morocco
e-mail: h.boulal@apni.net

M. S. Tan
Dell Technologies, Pulau Pinang, Malaysia

P. He
Institute of Agricultural Resources and Regional Planning, Chinese Academy of Agricultural Sciences, Beijing, China

V. K. Singh
Division of Agronomy, ICAR-Indian Agricultural Research Institute, New Delhi, India

Abstract Estimating nutrient mass balances using information on nutrient additions and removals generates useful, practical information on the nutrient status of a soil or area. A negative input–output balance of nutrients in the soil results when the crop nutrient removal and nutrient losses to other sinks become higher than the nutrient inputs into the system. Potassium (K) input–output balance varies among regions that have different climates, soil types, cropping systems, and cropping intensity. This chapter illustrates the farm-gate K balances in major production areas of the world and their impacts on native K fertility and crop yields. On-farm and on-station research examples show significant negative K balances in South Asia and Sub-Saharan Africa, while China, the USA, Brazil, and countries of the Latin America Southern Cone highlighted continued requirement of location-specific K application to maintain crop yields and soil K fertility status at optimum levels.

11.1 Concepts of Soil Nutrient Balance

Soil nutrient balance is *an account of the total inputs and outputs of a particular nutrient in an agroecosystem* (NAL 2020). Soil nutrient balance is the principle of mass balance applied to crop nutrients. Mass balance accounts for the matter entering, present in, and leaving a system. Öborn et al. (2003) separated soil nutrient balances into three categories: farm-gate, field, and farm-system budgets. These types of balances compare nutrient imports to nutrient exports. Farm-gate balances are not limited to farms but can be calculated at a variety of scales, depending on the data available. In this chapter, we focus on K farm-gate balances at the state/province and national levels across major production areas of the world.

11.1.1 Potassium Removal and Use for Different Cropping Systems and Geopolitical Boundaries

Because of its economic and environmental importance, there is increasing interest in developing ways to evaluate the efficiency and effectiveness of fertilizer use on farms, as well as at regional and national scales. **Partial nutrient balance** (PNB) is, *for a given nutrient, the sum of outputs divided by the sum of inputs* (Table 11.1). **Partial factor productivity** (PFP) is, *for a given nutrient, biomass yield divided by the sum of inputs*. Both can provide some guidance on system-level efficiency relative to nutrient use. These two metrics have been used to describe system performance in relation to nutrient use at continental (Ladha et al. 2003), national (Lassaletta et al. 2014), regional (Edis et al. 2012), and at farm-gate (Gourley et al. 2012) scales, and by industry (McLaughlin et al. 1992). Although trends in N and P use have been presented (Zhang et al. 2015; Lassaletta et al. 2014), there are few reports of these trends for K other than Fixen et al. (2015). This chapter seeks to

Table 11.1 Commonly used K use efficiency metrics and typical ranges for cereal crops

Potassium use efficiency abbreviation, calculations (calc), and units[a]	Equation[b]	Typical ranges for cereals[c]
PFP_K: partial factor productivity calc: biomass yield/sum of K inputs units: kg biomass (kg K)$^{-1}$	Y/I_K	75–200
PNB_K: partial nutrient balance calc: sum of K outputs/sum of K inputs units: unitless	O_K/I_K	0.7–0.9
$PNBI_K$: partial nutrient balance intensity calc: sum of K inputs – sum of K outputs units: kg K ha^{-1} or kg K_2O ha^{-1}	$I_K–O_K$	–
AE_K: agronomic efficiency calc: increase in biomass yield/sum of K inputs units: kg biomass (kg K input)$^{-1}$	$(Y_{+K} - Y_{-K})/I_K$	8–20
RE_K: recovery efficiency calc: increase in K uptake/sum of K inputs units: unitless	$(U_{+K} - U_{-K})/I_K$	0.3–0.5

[a]Generalized from Dobermann (2007) to apply to all inputs and outputs in the K cycle
[b]Y, biomass yield; I_K, sum of K inputs; O_K, sum of K outputs; Y_{+K}, biomass yield where K was added; Y_{-K}, biomass yield where K was not added; U_{+K}, plant K uptake where K was added; U_{-K}, plant K uptake where K was not added
[c]Fixen et al. (2015)

provide a selection of case studies from different regions and at different scales on the removal and use of K within farming systems.

11.1.2 Metrics for Nutrient Use Efficiency

Nutrient use efficiency (NUE) is *an evaluation of crop performance based on the quantity of a given nutrient input*. Nutrient use efficiency is a broad term and is quantified in various ways by various metrics. Table 11.1 shows a selection of nutrient use efficiency terms and their definitions. Partial nutrient balance and PFP have already been discussed. **Partial nutrient balance intensity** (PNBI) is, *for a given nutrient, the sum of inputs minus the sum of outputs on an area basis*. Where data or estimates exist, outputs and inputs can include all of those in the K cycle (Fig. 1.2, Chap. 1). **Agronomic efficiency** (AE) is, *for a given nutrient, the increase in biomass yield divided by the associated sum of inputs*. Positive returns to fertilizer investments are indicated when AE is greater than the ratio of fertilizer price to crop price. **Recovery efficiency** (RE) is, *for a given nutrient, the increase in uptake divided by the associated sum of inputs*. Recovery efficiency ranges from 0 to 1, with 1 interpreted as complete uptake of all of the nutrients applied. Of these metrics, PNB and PNBI are most often used to evaluate soil nutrient balances.

At the field, farm, or region scale, PNB_K and $PNBI_K$ are often calculated using only harvested plant K and fertilizer K (inorganic and organic). Calculating PNB_K

and $PNBI_K$ with just these data is widely performed, but only indicates the fate of harvest nutrients and does not consider other transfer or retention fates. Partial nutrient balance has been recommended by the International Fertilizer Association (IFA 2020), the EU Nitrogen Expert Panel (2015), and the Global Partnership on Nutrient Management (Norton et al. 2015) as the most appropriate measure of nutrient use efficiency because data are generally available at farm and national levels from which it can be calculated.

In the context of K, when the sum of K outputs equals the sum of K inputs, $PNB_K = 1$ and $PNBI_K = 0$. When more K is added than removed, $PNB_K < 1$ and $PNBI_K > 0$ (positive). The nutrient not removed can either be stored in the soil and/or flow through to the environment. When more K is removed than supplied, $PNB_K > 1$ and $PNBI_K < 0$ (negative), indicating that the soil is being depleted of K, lowering soil fertility. The extent to which this depletion can continue without affecting yield depends on the level of soil reserves as well as the rate at which K becomes plant available.

Partial nutrient balance does not describe pathways of internal K transformation within a system (e.g., K dissolution or fixation in soils). It is not necessarily a direct quantitative estimate of K loss from the system, because K not removed in the harvest might remain on site in the soil. Over the long term, however, changes in soil K stocks are usually small relative to inputs and outputs, and therefore, low PNB_K values over multiple years are reasonably reliable indirect indicators of K depletion.

The selection of NUE indicators should be considered in the light of the purpose of the undertaking. An indicator may be used by growers at field scale or as a statement of accountability at a regional and/or industry scale. The two reasons—while not mutually exclusive—do require clarity of purpose and transparency of data used to derive them. None of the indicators reference soil health or soil nutrient concentrations, so they are incomplete in their description of sustainability impacts. Because marginal nutrient use declines as the nutrient is supplied, the highest values of many indicators occur at the lowest level of application which is also likely the lowest yield. More discussion on selecting appropriate nutrient performance indicators can be found in Fixen et al. (2015) and Norton et al. (2015).

11.1.3 Uncertainties in Estimating Nutrient Balances

Nutrient balances provide perspective on the extent of nutrient sources relative to crop demand and may be helpful in identifying opportunities to improve nutrient use efficiency. However, unlike a financial balance sheet, nutrient balances involve considerable uncertainty, particularly at regional and national scales. These uncertainties derive from regional variations in crop K concentration, inadequate information on nutrient removal by some crops, lack of information on the contribution of manure, inability to account for nutrient loss by runoff and erosion, poor fertilizer use data, and other factors. Because of these sources of error, at best, nutrient balance is only a partial balance (Roberts and Majumdar 2017).

Deriving these ratios relies on reliable data on crop production (e.g., FAOSTAT), fertilizer use (e.g., IFA industry statistics), and crop product nutrient density (e.g., FAOSTAT). While of interest at a general level, the data do not provide information that can be used for system improvement and is just a reporting method. The metrics are more an assessment of the inherent K fertility of the system, the type of crop produced, and the farming system employed.

While production quantities are reasonably well known, the amount of grain retained on-farm for seed, animal feed, or domestic use is not often included. Also, the area of crop production could be the areas of the country, of agriculture, of arable farming, the area sown, fertilized, or harvested. The nutrient concentration of manures and organic supplements included in the budget approach is quite variable. For example, the sugar industry in Australia has an apparent high PNB_K because calculations do not often include K-rich by-products from sugar mills that are recycled back onto cane fields.

A second aspect of the uncertainties is the concentration of the nutrient in the product removed as well as in nutrient inputs. For example, Norton (2012) reported that K concentrations in wheat grain varied by ±14% of the mean value for dry grain of 4606 mg K kg^{-1}. As a consequence, any PNB_K or PFP_K is likely to have a 10–15% error embedded in the data used to derive the metric.

Thirdly, there are few reliable data sets on the use of fertilizers on different crops, and the best current data at the national level was reported by Heffer (2013), although regional agricultural and resource management groups may also hold similar data from farm surveys, various agricultural agencies, or the fertilizer industry. Even so, not many sources disaggregate the data to fertilizer use by production region and crop, which is really the detail required for growers and advisors to make system-level improvements.

Finally, a single fertilizer application may carry through to a second and often different crop, such as in a maize–soybean rotation. The residual nutrient carryover and then removal by the second crop is not accounted for, similar to not taking account of K released from soil minerals in the balance calculations.

The critical aspects of developing these metrics are to ensure that the data are transparent, auditable, referenced, consider all nutrient sources, and are regionally relevant and appropriate to the intention as to how the metrics are to be interpreted.

11.1.4 Interpreting Nutrient Balance Information

The first and most significant issue to consider when interpreting a nutrient metric is the degree of limitation that the particular nutrient imposes on the system studied. If the nutrient is not limiting crop growth due to high soil reserves or other biotic or abiotic influences, then the value of PNB_K and PFP_K will be high, as little K fertilizer is applied relative to yield and nutrient removal. This can give an unrealistic impression of the potential returns on K investment from the PFP_K value and an over-assessment of the degree of soil depletion occurring from PNB_K. In such cases,

$PNBI_K$ may be a more accurate indicator of soil depletion since it is an extensive or area-based rather than intensive metric.

Where K is the most limiting nutrient, over the long term and within the bounds of errors associated with the data, it is desirable for PNB_K to approach unity, so that input and output are balanced. When $PNB_K < 0.5$, there is probably an opportunity for using evidence-based nutrient management principles to improve efficiency. At the other extreme, when $PNB_K > 1.0$, it is likely that efficiency cannot be improved further without risking the depletion of soil supply. However, this should not imply that PNB_K values between 0.5 and 1.0 are necessarily acceptable, because, as already noted, a PNB_K value of, say 0.7, maybe good for some systems in some places and not so good for other systems in other places.

While mean values are useful, downscaled nutrient performance indicators from dairy farms (Gourley et al. 2012) and grain farms (Norton 2017) are highly variable and generally skewed. While the distribution of these values can be informative for growers as benchmarks for nutrient performance in a participatory research setting, valid comparisons can only be made among similar systems.

Trends in efficiency metrics can be viewed in a broader background against economic development in general. An economic Kuznets curve (Kuznets 1955) identifies that as an economy develops, resource-use metrics like PNB initially indicate unsustainability (such as PNB >> 1) as resources are exploited, but then resource use becomes more sustainable as inputs become economic. So nutrient PNBs are likely to reflect the stage of economic development and agricultural industrialization as well as production systems management.

11.2 Australia

In Australia, grazing land accounts for 87% of agricultural land use, with 16% of land under improved pastures. Around 50 M ha is used for crop production, with less than 5% irrigated. Farmers produce around 40 Mt of grains annually, with wheat (*Triticum aestivum* L.) (24 Mt) and barley (*Hordeum vulgare* L.) (7 Mt) as the main grains. Sugarcane (*Saccharum giganteum* (Walter) Pers.), cotton (*Gossypium hirsutum* L.), and viticulture are worth a total of around $4.2 billion annually, while other horticultural crops add another $8 billion. The Australian beef, sheep, and dairy industries are largely pasture-based, and the gross value of slaughtering is over $7.3 billion, while dairy products ($4.7 billion) and wool ($2.6 billion) are also significant industries. Grain and red meat production are highly variable due to seasonal conditions, and growers are careful with the allocation of production resources (ABARES 2016).

All the K fertilizer used in Australia is imported, and the annual peak quantity of imported KCl was ~480 kt 2004/2007, equivalent to ~239 kt K, assuming KCl averaged 60% K_2O. The annual peak of K_2SO_4 imports was 60 kt in 2012, or 25 kt K, calculated using a 50% K_2O concentration for K_2SO_4. Long-term K use has been ~170 kt of K, but during the "Millennium Drought" total K fertilizer use

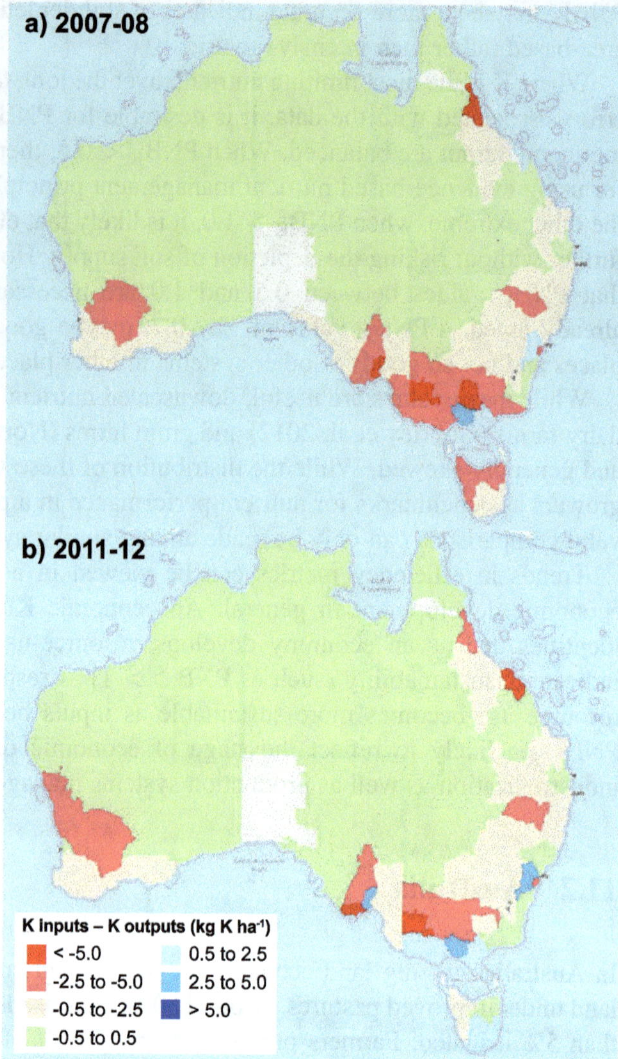

Fig. 11.1 Partial nutrient balance intensity ($PNBI_K$, kg K ha^{-1}) across different natural resource management regions across Australia for (**a**) 2007–2008 and (**b**) 2011–2012. Values reported are the means for each 2-year period. In general, the red regions indicate where nutrient removal is more than nutrient supply. (OzDSM 2020)

declined to a low of 134 kt K in 2009, but has since recovered to 227 kt K according to the 2016–2017 report from Fertilizer Australia (Drew 2018). This amount makes up less than 1% of the global K used.

Using the agricultural production and fertilizer use data, PNB_K can be calculated at national and regional scales, with regional fertilizer use data derived from farm surveys (ABS 2016). In aggregate, the national PNB_K was 2.9 for the audited period, and $PNBI_K$ was –0.6 kg K ha^{-1}, with the denominator used as the area of land used for agricultural production. These values are consistent with the National Land and Water Resources Audit (2001) which reported that K use was around one-third of the amount of K removed.

The patterns across different Australian resource management regions for two audit periods are shown in Fig. 11.1. The distribution of the balances largely reflects the balance of enterprises within each region, as well as the inherent K fertility of the

Table 11.2 Partial K balance (PNB_K) and the nutrient balance intensity ($PNBI_K$) averaged for 2008 and 2010: mean rates are derived from the reported fertilizer use and the areas fertilized; the proportion of the total K fertilizer used by each industry was derived from the survey data

Industry	Partial nutrient balance (PNB_K) Unitless	Partial nutrient balance intensity ($PNBI_K$) kg K ha^{-1}	Mean application rate of K kg K ha^{-1}	Industry proportion of K use %
Grain and livestock	3.1	3.7	2	14
Other grains	5.5	4.1	1	19
Rice	6.9	14.8	2	0
Cotton	0.5	−10.6	23	7
Sugarcane	7.6	84.9	13	5
Vegetables (outdoors)	1.2	4.8	21	6
Tree fruits and vines	1.3	3.7	14	10
Sheep farming, specialized	2.9	3.1	2	4
Beef cattle farming, specialized	1.5	3.1	6	16
Sheep-beef cattle farming	2.9	1.7	2	4
Dairy cattle farming	1.6	5.5	10	17

ABS (2016)

soils. In essence, the areas where K was in the largest deficit were in the sugar-growing areas in Queensland and the lower rainfall grain-growing regions of Western Australia, South Australia, and Central Queensland. It should be noted that the data used to generate these maps did not include any recycled materials such as mill wastes from sugar processing or manures used as inputs into crop production.

Edis et al. (2012) used farm survey data that included fertilizer inputs estimated for each industry to disaggregate PNB_K and $PNBI_K$ by commodity. These data are summarized in Table 11.2, which shows that all industries except cotton have more K removed than applied as fertilizer. Cotton is usually grown in rotation with other annual crops so the true K balance is confounded by fertilizer practices on those other crops. There is much less K applied than is removed and the largest apparent deficits are in the grains, sugar, and dairy industries. Recycled K-rich mill wastes are not included for the sugar industry, and PNB_K and $PNBI_K$ for the dairy industry do not include K supplied to pastures that is ultimately derived from feeds purchased from outside the farm-gate. Gourley et al. (2012) reported that K from cattle feed averaged 25 kg K ha^{-1} compared to a fertilizer input of 32 kg K ha^{-1} from data collected on 44 dairy farms across Australia, so that total K input in this industry may be underestimated.

While it may be of interest to compare industries or regions, there are important limitations in the data presented in both Fig. 11.1 and Table 11.2. The survey sample sizes are small which leads to up-scaling errors, and in these data, the up-scaled national K use is about 65% of the fertilizer industry-reported consumption. Under-reporting of K use in these farm surveys is likely a consequence of imprecise survey questions that, for example, do not discriminate fertilizer product application rate from nutrient rate. Regional differences in product nutrient density, rotations used, and the extent of the use of recycled matter all make the actual values imprecise and of limited value in drawing conclusions about the efficiency of different production systems.

However, the major significant deficiency in these types of regional or industry-based data is that averages provide little or no intelligence to growers on their farm-level balances. The data collected by Gourley et al. (2012) for the dairy industry and Norton and vanderMark (2016) for the grains industry gives error terms around the inputs, as well as the derived metrics concerning nutrient use efficiency.

11.2.1 Southern Australian Grain Farms

While regional performance indicators are of interest in a policy sense, the collection and collation of nutrient removal and use at farm or field levels are more important to growers, as these data inform them about how their specific management practices have built up or depleted nutrients over time. Such an understanding will help them make decisions about appropriate interventions to address any imbalances.

Norton (2017) reported nutrient performance indicators from a survey of 474 fields (34,900 ha) between 2010 and 2014 in south-eastern Australia. Nutrient balances (nitrogen [N], phosphorus [P], K, and sulfur [S]) for each field over the audit period were estimated from fertilizer use, stubble management (burned, removed, grazed), and crop yield. Grain and hay yields were recorded in the farm records, and regional wheat grain nutrient concentrations for wheat (Norton 2012) and canola (*Brassica napus* L.) (Norton 2014a) were used to estimate removal in grains. Other values were derived from the values used in the NLWA (2001). The summary presented here is for the K balances alone.

Even though 20% of surveyed fields received K fertilizer, with an average K application rate of <10 kg K ha^{-1} $year^{-1}$, where K fertilizer was used, the application rates were about 90 kg K ha^{-1} on canola and 66 kg K ha^{-1} on cereals. On the fields where K was applied, PNB_K and PFP_K were calculated with the median PNB_K of 3.0, and only seven of the fields surveyed showed more K use than removal over the audit period. Even where K was used, 12 fields had $PNB_K > 5$ (Fig. 11.2a). The PFP_K values where K was used had a median of 350 kg grain kg^{-1} K (Fig. 11.2b).

The low use of K in eastern Australia in particular can be explained by the generally high soil test K values, indicating that supplementary K was not required. Christy et al. (2015) re-analyzed soil test data collected in the NLWA (2001) to assess the proportion of areas where a response to K was likely based on the soil test

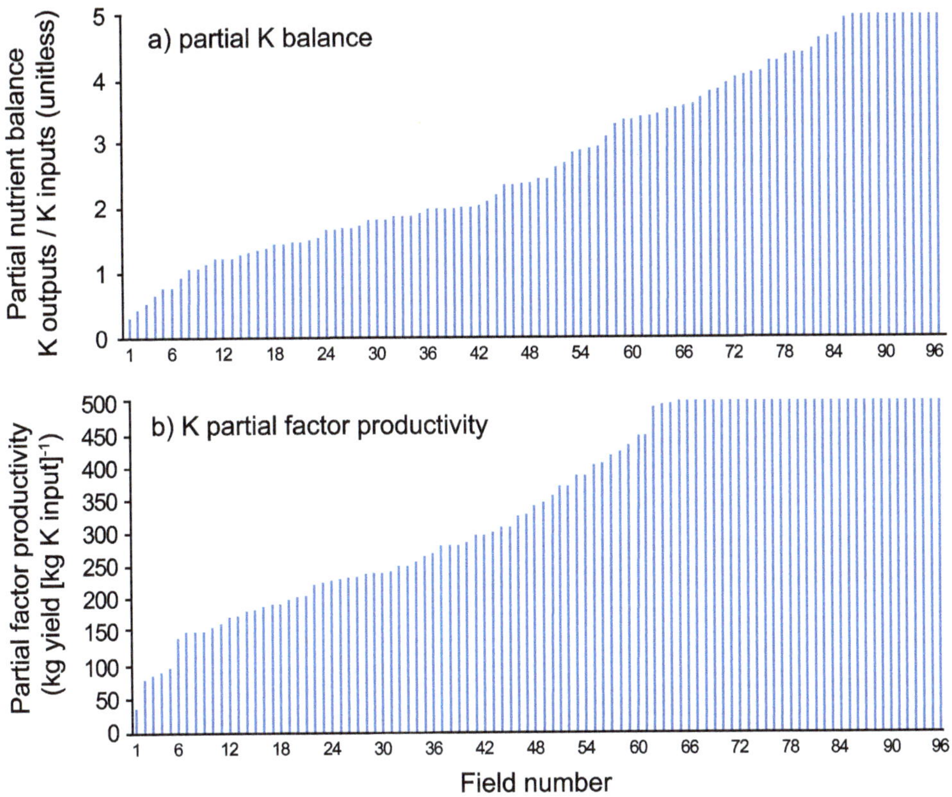

Fig. 11.2 Nutrient performance metrics for K as derived from a survey of Australian farmers' fields: (**a**) partial nutrient balance (PNB_K), and (**b**) partial factor productivity (PFP_K). (adapted from Norton 2017)

critical values (Brennan and Bell 2013). In the south-eastern grain-producing areas, soil K concentrations are generally higher in the lower rainfall areas, but there are regions in the higher rainfall zones near the coast where yield responses to added K are likely. About half the K soil test values on sand and loam soil types in southwestern Victoria are low and these are coincident with the areas where K fertilizer was reported to be applied from the farm survey (Christy et al. 2015). In particular, low soil K concentrations were seen on lighter acid soils under high rainfall conditions and also in formed-in-place Vertosols in both Victoria and Queensland.

11.2.2 Trends in Potassium Removal

Similar to the high PNB_K values reported in the NLWA (2001), Bell and Moody (2005) reported consistently negative K budgets for grain farms in the northern cropping region of Australia. The extent of the deficit was linked to regional crop productivity and the low use of K fertilizer. Potassium removal was highest in chickpea (*Cicer arietinum* L.) crops because of the higher than average grain K concentrations in chickpeas being consistently higher yielding than crops like sorghum (*Sorghum bicolor* (L.) Moench). Typically, K removals were between 8 and 20 kg K ha^{-1} year^{-1} over a five-crop sequence. They also reported low and declining soil K concentrations in many of the summer cropping regions in Queensland. Soil test K values showed stratification with around 1.6–3.2 times more exchangeable K in the topsoil than the subsoil. This depletion has largely gone undetected by soil tests which generally represent only the 0–10 cm topsoil. There are now research projects investigating deep placement of K (and P) as a means of alleviating this deficiency.

There is concern about this depletion trend more widely in regions where K fertilizer has not been traditionally used. Key factors indicating a future K response are cropping on soils that have low to moderate exchangeable and non-exchangeable K reserves, and where K removal in harvested grain and hay is high. Indeed, the consistently low grain K concentrations in some areas coupled with some very low K usage suggested that the soil K status may be approaching low levels—whether due to stratification, presence of high sodium (Na), or the lack of substantial non-exchangeable K reserves.

Potassium fertilizer use in Australia is relatively modest on a world scale, and there is approximately 3 times more K removed in agricultural products than is supplied. While K removal is highest on sugarcane farms, there is a modest deficit for most farms due to low productivity. Regional differences in K use and PNB_K reflect the intensity of production and the inherent K fertility of the regions, although there are inconsistencies in the data available to estimate K balances and nutrient performance indicators. Western Australia uses most K fertilizer for grain production. Victoria uses K mainly on intensive pastures (especially dairy), while in Queensland K is mainly focused on the sugar industry. The data presented here indicate that grain-producing fields in the higher rainfall regions of southern Australia are in significant K deficit, despite the low inherent K fertility. Private, state, and federal agencies are addressing and communicating evidence-based nutrient management strategies to growers to overcome these deficits and improve productivity (Norton 2014b, Norton and vanderMark 2016).

11.3 Southeast Asia

The K balance assessment for Southeast Asia includes four crops most important both economically and for K use: rice (*Oryza sativa* L., maize (*Zea mays* L.), sugarcane, and oil palm (*Elaeis guineensis* Jacq.) in five countries in the region: Indonesia, Malaysia, Philippines, Thailand, and Vietnam. The choice of crops, countries, and crop year included in the K balance estimation were based on the availability of data on fertilizer K application by crop and country. The latest available dataset with the highest number of countries and crops in the region was reported by Heffer (2013) for the crop year 2010–2011.

11.3.1 Data Sources and Limitations

In 2010–2011, the five countries produced a total of 148.5 Mt of rice grain, 23.0 Mt of maize grain, 132.3 Mt of cane sugar, and 218.8 Mt of palm oil (as fresh fruit bunch) (Table 11.3). The total area harvested for these four crops was 57.3 Mha, accounting for 61% of the total crop area (94.9 Mha of arable land and permanent cropland) in the five countries (Table 11.4). For each crop, the percentage of the total crop area was: 38% for rice, 8% for maize, 3% for sugarcane, and 12% for oil palm.

For the estimation of crop K removal, production data for 2010 and 2011 (Table 11.5) was combined with published values of K content in harvested crop biomass. These values were 2.83 g K kg^{-1} for rice grain (Buresh et al. 2010); 2.38 g K kg^{-1} maize grain (Setiyono et al. 2010); 1.1 g K kg^{-1} cane sugar (Dierolf et al. 2001); 3.87 g K kg^{-1} oil palm fresh fruit bunch for Indonesia (Donough et al. 2014);

Table 11.3 Production levels of four crops in five Southeast Asian countries, 2010–2011 season

	Rice production	Maize production	Sugarcane production	Oil palm production
Country	kt grain	kt grain	kt sugar	kt fresh fruit bunch
Indonesia	56,349	6800	15,391	118,000
Malaysia	2526	95	174	91,055
Philippines	16,729	7271	21,913	565
Thailand	30,700	4200	84,026	9160
Vietnam	42,194	4607	10,783	–
Total	148,498	22,973	132,287	218,780

Data source: USDA FAS (2017); oil palm production is reported here as fresh fruit bunch instead of palm oil as reported by USDA. Oil palm fresh fruit bunch is calculated by assuming an oil extraction of 20% (Corley and Tinker 2016)

Table 11.4 Harvested area of four crops in five Southeast Asian countries, 2010–2011 season

	Rice area harvested	Maize area harvested	Sugarcane area harvested	Oil palm area harvested
Country	1000 ha	1000 ha	1000 ha	1000 ha
Indonesia	12,075	2850	435	6801
Malaysia	672	27	4	4202
Philippines	4528	2633	440	45
Thailand	10,667	1000	1259	600
Vietnam	7607	1127	282	–
Total	35,549	7637	2420	11,648

Data sources: USDA FAS (2017) for rice, maize, and oil palm; FAOSTAT (2017) for sugarcane (USDA FAS (2017) does not provide data for sugarcane)

Table 11.5 Inputs of K fertilizer, K removal with crop harvest, and partial nutrient balance intensity ($PNBI_K$) of four crops in five countries in Southeast Asia, 2010–2011

	Rice kt K_2O	Maize kt K_2O	Sugarcane kt K_2O	Oil palm kt K_2O
K applied[a]				
Indonesia	138	100	63	775
Malaysia	55	1	1	989
Philippines	24	6	13	2
Thailand	22	19	81	47
Vietnam	240	26	32	–
Total	479	152	190	1813
K removed with harvested crop[b]				
Indonesia	192	20	20	550
Malaysia	9	0.3	0.2	385
Philippines	57	21	29	3
Thailand	105	12	111	43
Vietnam	144	13	14	–
Total	507	66	175	981
Partial nutrient balance intensity ($PNBI_K$)				
Indonesia	–54	80	43	225
Malaysia	46	0.7	0.8	604
Philippines	–33	–15	–16	–1
Thailand	–83	7	–30	4
Vietnam	96	13	18	–
Total	–28	86	15	832

[a]Values from Heffer (2013)
[b]Estimated using production data for 2010–2011 (USDA FAS 2017) and K concentration in the harvested crop for rice grain (Buresh et al. 2010), maize grain (Setiyono et al. 2010), sugarcane (Dierolf et al. 2001), and oil palm (Tarmizi and Mohd Tayeb 2006, Donough et al. 2014)

and 3.51 g K kg^{-1} oil palm fresh fruit bunch for Malaysia (Tarmizi and Mohd Tayeb 2006). We assumed that the K content of oil palm fresh fruit bunch in the Philippines and Thailand was similar to that of Indonesia. Since oil palm production data were reported by USDA FAS (2017) as palm oil, the equivalent production of fresh fruit bunches was calculated by assuming an oil yield of 20% of fresh fruit bunch biomass (Corley and Tinker 2016).

11.3.2 Trends in Potassium Balance

Rice is the most important cereal crop in Southeast Asia, covering 38% of the total crop area (Table 11.4). In rice, crop removal exceeded K application with a deficit of –33 to –83 kt K_2O in Indonesia, Philippines, and Thailand; whereas K application exceeded crop removal with a K surplus of 46 kt and 96 kt K_2O in Malaysia and

Vietnam, respectively (see $PNBI_K$ values in Table 11.5). Across the five countries, there was a K deficit of –28 kt K_2O in rice-growing areas. Since the calculated $PNBI_K$ values only accounted for grain removal, greater K deficits are expected in areas where rice straw is also removed from the field (for off-farm use). Previous studies reported negative K balances in rice in irrigated long-term experiments in Indonesia, Philippines, and Vietnam (Dobermann et al. 1996) and in rainfed lowland rice in Indonesia (Wihardjaka et al. 1999).

Maize is the second most important cereal crop in Southeast Asia, with an area harvested of 7.64 million ha in the five countries (Table 11.4). Out of the five countries, only the Philippines showed a K deficit (–15 kt K_2O), while the K balance in the other four countries ranged from 0.7 kt K_2O (Malaysia) to 80 kt K_2O (Indonesia) (Table 11.5). Similar to rice, $PNBI_K$ calculations for maize only accounted for grain removal; hence $PNBI_K$ will be further reduced if K removed with maize stover is included. Pasuquin et al. (2014) reported cases of negative K balance in maize in Indonesia, Philippines, and Vietnam when most of the above-ground residues were removed from the field. In Southeast Asia, farmers manage their maize residues in varied ways. In the Philippines, for example, while many farmers retain the residues in their field, some still practice full removal or burning of residues to facilitate land cultivation, especially with the use of non-mechanized tillage implements.

Oil palm occupies the second largest area harvested, with Indonesia and Malaysia being the two largest producers of oil palm in the region (Table 11.4). While oil palm is only grown in four of the five countries, it is the biggest user of applied K fertilizer (69%) among the four crops (Table 11.5). Potassium applications exceeded crop removal (positive $PNBI_K$ values) in Indonesia, Malaysia, and Thailand; while crop removal exceeded applications (negative $PNBI_K$ value) in the Philippines (Table 11.6).

Sugarcane covers the least area among the four crops (Table 11.4), but it is one of the important economic crops in the region and a heavy user of fertilizer K. Table 11.5 shows that annual application of K exceeded crop removal by 0.8–43 kt K_2O in Indonesia, Malaysia, and Vietnam. On the other hand, K deficit is indicated in the Philippines (–16 kt K_2O) and Thailand sugarcane production (–30 kt K_2O). A previous study reported negative K balances in Thailand for yields of 30 t ha^{-1} (of dry cane) and above at a K fertilizer application rate of 94 kg K_2O ha^{-1} (Trelo-ges et al. 2004).

In the five countries in Southeast Asia included in this K budget, the amount of K fertilizer needed to replace crop removal follows the order: oil palm > rice > sugarcane > maize (Table 11.6). To be able to replace crop removal of the four crops in five countries, a total amount of 3.46–5.76 Mt K_2O will be needed at fertilizer use efficiencies of 50% and 30%, respectively. With the current crop area dedicated to these crops in this region, the crops with the largest requirement for K fertilizer are: oil palm in Indonesia and Malaysia, rice in the Philippines and Vietnam, and sugarcane and rice in Thailand. Based on 2010–2011 fertilizer application and the estimated fertilizer requirement to replace crop removal, opportunities for improving K application are evident in most of the countries and crops.

Table 11.6 Estimated fertilizer K requirement to replace crop removal and fertilizer K deficit of the current application practice at 30% (A) and 50% (B) fertilizer use efficiency scenarios in four crops in five Southeast Asian countries (expressed at K_2O)

	Rice		Maize		Sugarcane		Oil palm	
	kt K_2O		kt K_2O		kt K_2O		kt K_2O	
Country	A	B	A	B	A	B	A	B
Fertilizer K requirement to replace crop removal[a]								
Indonesia	641	385	65	39	68	41	1834	1101
Malaysia	29	17	1	0.5	1	0.5	1284	770
Philippines	190	114	70	42	97	58	9	5
Thailand	349	210	40	24	371	223	142	85
Vietnam	480	288	44	26	48	29	–	–
Total	1689	1014	220	132	584	351	3269	1961
Fertilizer K deficit of the current application practice[b]								
Indonesia	–503	–247	35	61	–5	22	–1059	–326
Malaysia	26	38	0.1	0.5	0	1	–295	219
Philippines	–166	–90	–64	–36	–84	–45	–7	–3
Thailand	–327	–188	–21	–5	–290	–142	–95	–38
Vietnam	–240	–48	–18	–0.4	–16	3	–	–
Total	–1210	–535	–68	20	–394	–161	–1456	–148

[a]Using 2010–2011 crop removal data at fertilizer use efficiencies of 30% (A) and 50% (B) during the year of application

[b]Difference between 2010 and 2011 fertilizer K application and estimated fertilizer K requirement to replace crop removal at fertilizer use efficiencies of 30% (A) and 50% (B). A negative value denotes a K application deficit

Fertilizer K deficits of current application practices (Table 11.6) are indicated in: rice for all countries except Malaysia; maize in the Philippines, Thailand, and Vietnam; sugarcane for all countries except Malaysia; and oil palm for all four countries. In the 30% fertilizer use efficiency scenario, there is a predicted annual total K deficit in the five countries of 1.2 Mt K_2O in rice; 0.068 Mt K_2O in maize; 0.39 Mt K_2O in sugarcane; and 1.46 Mt K_2O in oil palm. Current production levels also indicate opportunities for increasing production with intensification through improved crop and nutrient management, which may require a further increase in K fertilizer application.

11.4 China

China is the largest national consumer of fertilizer nutrients, accounting for nearly 30% of global fertilizer use. From 1978 to 2015, total ($N + P_2O_5 + K_2O$) fertilizer consumption increased from 8.8 to 60.2 Mt, with an average annual increase rate of 5.3%. The N, P_2O_5, and K_2O consumption in 2015 was 30, 19, and 11 Mt, respectively.

11.4.1 Potassium Use and Crop Production

Potassium deficiency in crops was initially reported in Southern China in the 1970s (Lin 1989). Potassium did not receive as much attention as N and P in the next decade but now has become a widespread limiting factor in agricultural production, especially in some parts of northeast and north-central China (Liu et al. 2000). Recent research has demonstrated that K deficiencies in intensified agricultural production areas in China continue to be a challenge (He et al. 2009); however, K fertilizer input has increased dramatically over time, rising from 0.38 Mt K_2O in 1980 to 10.6 Mt K_2O in 2015, with an average annual increment of 0.29 Mt K_2O. The total grain production increased from 310 to 657 Mt during the same period, with an average annual increase rate of 2.1%. Fruit and vegetable production increased annually by 6.4% and 9.8%, respectively, in the past 20 years. The unprecedented growth in China's fertilizer consumption and crop production in the last decades prompted researchers to look at nutrient input–output balances. Understanding the surplus/deficit and balance of cropland nutrients provides guidance to the production, distribution, and application of fertilizer.

11.4.2 Potassium Balance Studies

Previous research assessed farmland nutrient balances in China, including the K balance at various scales (Wang et al. 2014; Chuan et al. 2014). These assessments emphasized that the nutrient balance estimations at the national scale would help develop national fertilizer policies, including decisions on investments for fertilizer factories and the exploitation of local resources and minerals to supply nutrients. Potassium balances for the 30 provinces in China from 1961 to 1997 illustrated annual K depletion that increased from 2.9 Mt K (3.5 Mt K_2O) in 1961 to 8.3 Mt K (10.0 Mt K_2O) in 1997 (Sheldrick et al. 2003). Negative $PNBI_K$ (from −17 to −245 kg K ha^{-1} year^{-1}) were also reported from long-term fertilizer experiments in rice-based systems, irrespective of mineral K application and site (Zhang et al., 2010). Negative $PNBI_K$ persisted on wheat–maize rotations even at K application rates as high as 112–300 kg K_2O ha^{-1} (Tan et al. 2012). A winter wheat study in the North China plain between 2005 and 2007 showed negative K balance in all the treatments under different production practices, especially at high-yield levels (Niu et al. 2013). About 79% of the assessed on-farm trials (n = 120) in potato (*Solanum tuberosum* L.) in north-western China had an average negative balance of −102 kg K ha^{-1} (Li et al. 2015). This overwhelming negative balance of K in soils of China started improving (becoming less negative) with the increase in fertilizer K application, along with the in-field retention of crop residues (Wang et al. 2014; Shen et al. 2005). Several studies also show significant soil K surpluses in vegetable production fields in China (Huang et al. 2007; Wang et al. 2008).

11.4.3 Potassium Balances in Grain and Cash Crops

These studies provide critical information on K balances in China but mainly focus on specific experimental sites or short-term observations. Nutrient balances can vary considerably from year to year, and short-term studies providing limited information can be misleading.

Two follow-up studies in the past few years provide a more comprehensive overview on how soil test K levels and crop K responses changed over time and space, and the status of K balances across China. He et al. (2015) used large on-farm experimental datasets to assess spatial and temporal variation in soil test K and crop responses to K between 1990 and 2012. The authors utilized datasets for plant-available K from on-farm experiments, where soil samples from a depth of 0–20 cm were collected and analyzed before sowing. The corresponding crop yield differences were measured between treatments fertilized with N+P+K or with only N+P. The data were spatially desegregated and grouped based on geographical locations, such as northeast (NE), northcentral (NC), northwest (NW), southeast (SE), and southwest (SW) regions of China. In addition, each region was further divided into two sub-groups based on soil utilization pattern, grain crop, and cash crop systems. The grain crop category included wheat, maize, rice, potato, and soybean (*Glycine max* (L.) Merr.), while the cash crop group included vegetables, fruits, rapeseed (*Brassica spp.*), sunflower (*Helianthus annuus* L.), cotton, and sugar crops. Results indicated that soil-extractable K concentrations (exchangeable K) increased with time between 1990 and 2012. For grain crops, exchangeable K concentrations increased only slightly; however, for cash crops, the K concentrations increased dramatically over the period. The trends of increased soil K in cash crops were consistent with increased relative yield for cash crops and the high fertilizer K application rates. The authors suggested that higher soil K concentrations under cash crops increased the average soil K estimations in China, leading to assumptions of surplus K in all soils. Even though extractable K in soils under grain crops increased in the Northcentral, Southeast, and Southwest China in the 2000s as compared with that in the 1990s, the average soil K values were less than the critical concentration of 80 mg L^{-1}, except in the NW region. The results indicated that the exchangeable K continued to show a declining trend with large crop removal associated with higher yields.

The PNB_K was greater than 1.0 for both cash and grain crop categories, suggesting that K removal by crop uptake exceeded K inputs from fertilizer (Fig. 11.3). The PNB_K was higher for cash crops (mean: 2.1; range: 1.1–4.2) than for grain crops (mean: 1.3; range: 1.0–1.5), indicating higher K depletion in cash crops.

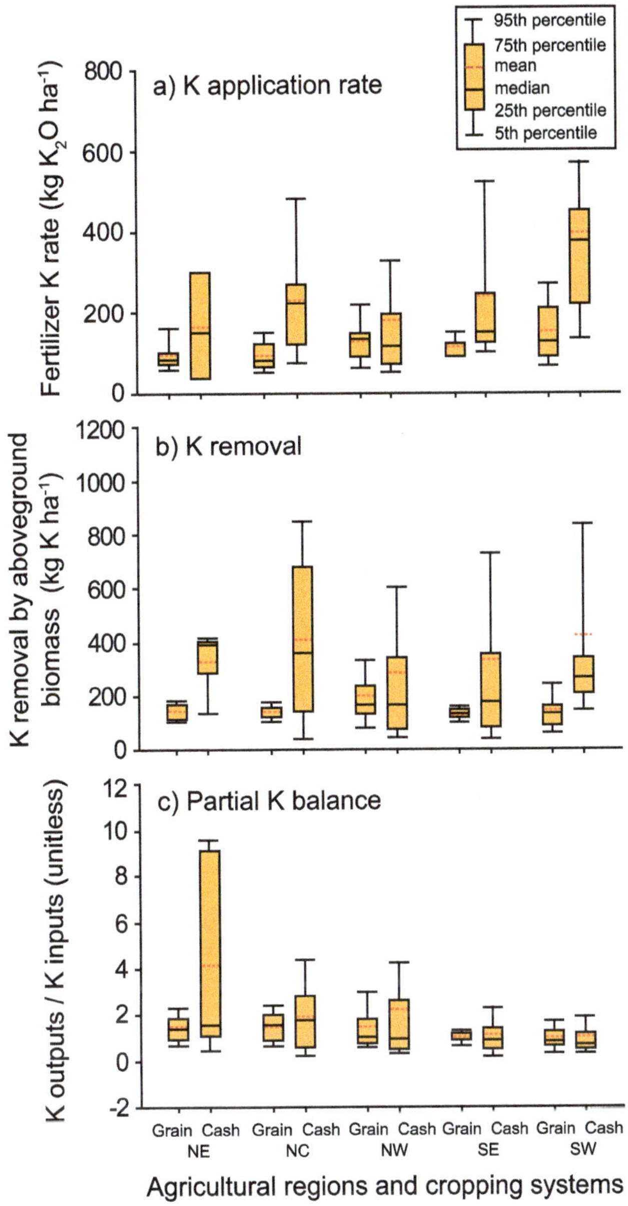

Fig. 11.3 (**a**) Fertilizer application rate (as K_2O), (**b**) K uptake, and (**c**) PNB_K across northeast (NE), north-central (NC), northwest (NW), southeast (SE), and southwest (SW) China for grain and cash crops. (He et al. 2015)

11.4.4 Spatial and Temporal Changes in Potassium Balance

A comprehensive assessment of soil K balances between 1980 and 2010 was undertaken by Liu et al. (2017b) using 31 provincial datasets from the China Agriculture Statistical Report for 1981 and 2011. The K consumption from fertilizers, human and livestock manure, straw return to field, cake manure, deposition, irrigation, seeds, crop removal, and K loss from soils were calculated with provincial data and parameters from refereed literature. The arable crop output was the principal nutrient output pathway in the agricultural systems, which was calculated through economic yield and K requirement per unit of economic yield of various crops. The arable crop outputs were categorized into five groups: cereal crops, fruit and vegetables (fruit, vegetable, and melons), oil crops (oil crops and beans), industrial crops (sugar beet (*Beta vulgaris* L.), sugarcane, fiber crops, cotton and tobacco (*Nicotiana tabacum* L.)) and root crops (sweet potato (*Ipomoea batatas* (L.) Lam.) and potato). The $PNBI_K$ (kg K_2O or kg K_2O ha^{-1}) was estimated by calculating the total K added to farmland from various sources during the year (inputs) and subtracting the total amount of K removed from farmland by crop removal and K loss from leaching and runoff (outputs).

The farmland $PNBI_K$ was significantly different between 1980 and 2010 at the provincial scale. The $PNBI_K$ in China changed from an average deficit of –13.6 kg K_2O ha^{-1} in 1980 to a surplus of 22.9 kg K_2O ha^{-1} in 2010 (Fig. 11.4). In 1980, almost 68% of the farmland area across 19 provinces showed K deficits (very low and low classes); however, 30% of farmland area in 10 provinces had $PNBI_K$ in the moderate class (1–50 kg K_2O ha^{-1}) (Fig. 11.4a). In 2010, 40 and 44% of the farmland was in the low and moderate $PNBI_K$ classes in 9 and 13 provinces, respectively. Meanwhile, 11% of farmland was in the high $PNBI_K$ class in 5 provinces. Only 6% of farmland area showed a $PNBI_K$ in the very high class in four provinces (Fig. 11.4).

Liu et al. (2017a) followed up with a more comprehensive K balance across four decades: 1980–1989, 1990–1999, 2000–2009, and 2010–2015. The data were desegregated into six agricultural regions, based on geographical locations and China's administrative divisions: northeast (NE); north-central (NC); northwest (NW); the middle and lower reaches of the Yangtze River (MLYR); southeast (SE); and southwest (SW).

Mineral fertilizer, organic manure, and other K input resources (atmospheric deposition, irrigation, and seeds) accounted for 33% (4.91 Mt K_2O), 59% (8.85 Mt K_2O), and 8% (1.19 Mt K_2O) of the total K input from 1980 to 2015, respectively. The K input through fertilizers increased from 0.38 Mt K_2O in 1980 to 10.58 Mt K_2O in 2015. At the regional level, K inputs increased in all regions over time. Increases in K inputs were greatest for the SE and least for both the NW and NE (Fig. 11.5a). By 2015, K input in the SE was the highest, followed by the SW, MLYR, NC, NW, and NE.

The average K output from 1980 to 2015 was 14.78 Mt K_2O (146 kg K_2O ha^{-1} $year^{-1}$). Crop removal accounted for 99% (14.63 Mt K_2O) of the total output from 1980 to 2015. Cereals removed 70, 65, 55, and 56% of the total K output in the 1980s, 1990s, 2000s, and 2010s, respectively. The K output of fruit and vegetables and oil crops increased over the years with average annual increments of 0.13 and 0.05 Mt of K_2O, respectively. Potassium outputs increased over time in all regions (Fig. 11.5). The K outputs in MLYR and SE were significantly larger than those in other regions. In the NE, K outputs increased from 1982 to 1989 then stabilized, with some variation, from 1990 on.

Potassium nutrient balance intensity increased from its lowest value in 1984 (–3.23 Mt K_2O) to a relatively stable value of around 2.1 Mt K_2O in 2015. At the

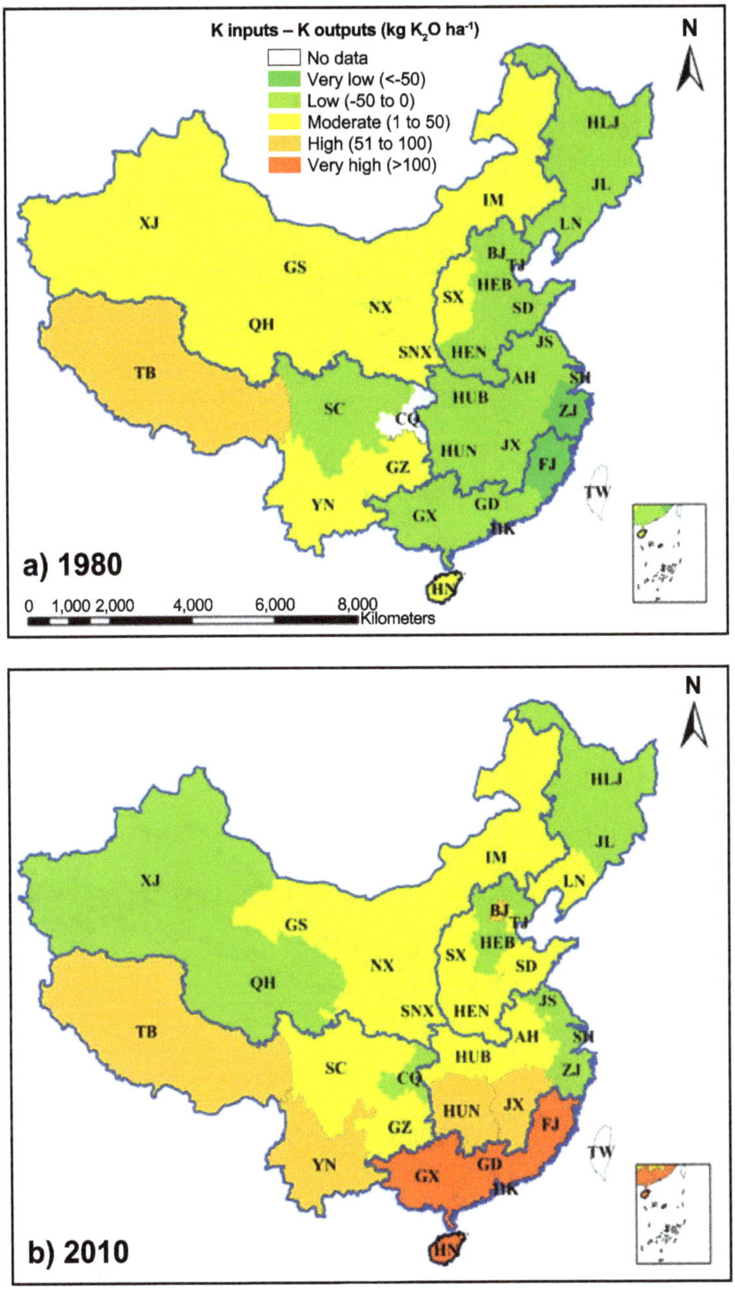

Fig. 11.4 Spatial distributions of $PNBI_K$ in China's agricultural land in (**a**) 1980 and (**b**) 2010. Adapted from Liu et al. (2017b)

national scale, the average $PNBI_K$ from 1980 to 2015 was 0.81 kg K_2O ha^{-1} $year^{-1}$. On average, the $PNBI_K$ changed from its 1980 level by –24.2 kg K_2O ha^{-1} in the 1980s, –5.9 kg K_2O ha^{-1} in the 1990s, 21.3 kg K_2O ha^{-1} in the 2000s, and 19.5 kg K_2O ha^{-1} in the 2010s. The average $PNBI_K$ in the SE region increased steadily

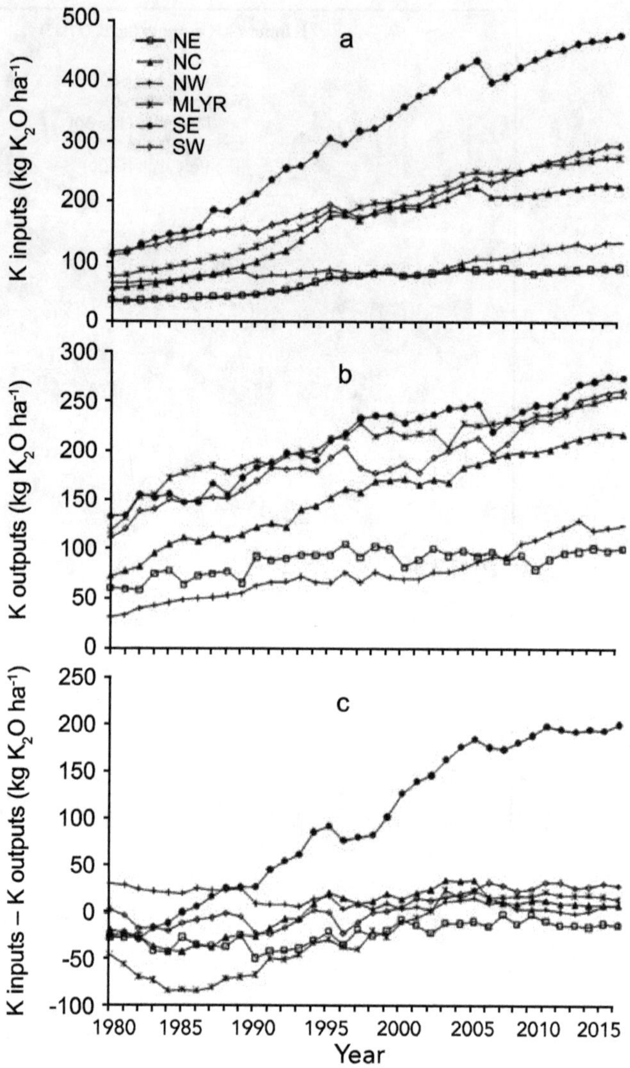

Fig. 11.5 Temporal and regional variation of: (**a**) K inputs, (**b**) K outputs, and (**c**) $PNBI_K$ in six regions of China from 1980 to 2015. (adapted from Liu et al. 2017a)

from –1.5 kg K_2O ha^{-1} in 1980 to 198 kg K_2O ha^{-1} in 2015 and had the greatest K surplus of any region (Fig. 11.5). The average $PNBI_K$ in the 2010s (2010–2015) varied by region: SW (31.3 kg K_2O ha^{-1}) > MLYR (15.9 kg K_2O ha^{-1}) > NC (9.5 kg K_2O ha^{-1}) > NW (8.2 kg K_2O ha^{-1}) > NE (–11.9 kg K_2O ha^{-1}). The NE region stayed in deficit for the entire period of assessment (1980–2015). The positive $PNBI_K$ in the NW region declined over time. In the NC region, $PNBI_K$ increased during the first three decades but declined in the 2010s.

Reliable K fertilizer use and crop output data, along with information from published peer-reviewed literature, allowed estimation of K balance in Chinese agriculture. Recent assessments of K inputs, outputs, and $PNBI_K$ in China showed significant spatial and temporal differences. Widespread negative K balances in soils before the 1980s have now changed toward a more positive balance due to increased use of K fertilizers and retention of crop residues in the field. However, regional and crop-specific disparity in K use provides opportunities to further refine the K balance through site-specific nutrient management for sustainable crop production systems.

11.5 India

India is the second-largest consumer of fertilizer in the world. In nutrient terms, the 2.5 Mt of K_2O consumed in 2014 was 7.7% of the global consumption. Alluvial (Inceptisols and Entisols), Red and Lateritic (Alfisols and Ultisols), and Black (Vertisols) soils are dominant soil types in India. The Alluvial soils in the Indo-Gangetic Plains and elsewhere have ample available and reserve soil K. Despite having high available soil K, the Black soils, mainly in the Central and Western India, often show positive yield responses to added K. This response is most likely due to low K availability because of relatively higher contents of Ca and Mg in these soils. The Red and Lateritic soils in coastal and peninsular India typically have lower K content. Heffer (2013) estimated that over 60% of the total N + P_2O_5 + K_2O is used in cereal crops. Potassium used in cereals is about 50% of the total consumption. An estimate by Tewatia et al. (2017), using crop production statistics of 2015–2016 and crop nutrient uptake information from multiple sources, showed that removal of K_2O by cereals is about half of the total K removed by all crops. The seeming parity between K applied and removed by cereals, however, is misplaced as the removal of K by cereals is about 11 times more than the applied K.

11.5.1 Crop Requirement and Potassium Use

Post Green Revolution, India's agricultural production was largely driven by inorganic fertilizer use. The demand for crop commodities will rise in the future with population increase and changing food habits. Chand (2007) estimated that 262, 19, 54, and 345 Mt of cereals, pulses, oilseeds, and sugarcane, respectively, would be required in 2020, compared to the base year (2004–2005) production of 193, 14, 35, and 262 Mt of the same crops. In 2014–2015, 235, 17, 28, and 363 Mt of cereals, pulses, oilseeds, and sugarcane were produced in India (FAI 2016). A slowdown of crop productivity (Kumar et al. 2004) and decline in PFP (NAAS 2006), however, has been identified as major concerns. Biswas and Sharma (2008) showed that response to applied fertilizer (kg grain [kg of added NPK]$^{-1}$) decreased from 13.4 to 3.7 between 1970 and 2005.

Imbalanced fertilization, more specifically low use of K, has been identified as one of the major reasons for declining PFP in India. Several studies identified inequality between applied K and its removal in harvested crops, which impacts crop productivity, nutrient use efficiencies, and soil K mining (Naidu et al, 2011; Majumdar et al. 2012, 2016; Singh et al. 2014). Long-term cropping with negative $PNBI_K$ has been associated with yield declines in the rice–wheat system in South

Table 11.7 Nutrient uptake from soil reserves under rice–wheat cropping system

	K uptake from the soil reserve[a]		
	Rice	Wheat	System
Location	kg K_2O ha^{-1}	kg K_2O ha^{-1}	kg K_2O ha^{-1}
Sabour	166	95	261
Ranchi	127	78	205
Ludhiana	211	143	354
Palampur	141	85	226
R.S. Pura	189	112	301
Faizabad	177	75	252
Kanpur	146	101	247
Modipuram	173	121	294
Varanasi	130	91	221
Pantnagar	122	98	220
Mean	158	100	258

Tiwari et al. (2006)
[a]rice (*Oryza sativa* L.); wheat (*Triticum aestivum* L.)

Asia (Regmi et al. 2002). Potassium partial nutrient balance intensities were negative even with recommended fertilizer application rates of K and were least negative when farmyard manure was a nutrient source or wheat residues were returned to the field. Although the K-supplying capacity of illite-dominated alluvial soils of the Indo-Gangetic Plains is relatively high (Dobermann et al. 1996), long-term intensive cropping with inadequate application of K have been associated with large negative $PNBI_K$ and depletion of native K reserves (Gami et al. 2001; Yadvinder-Singh et al. 2004). Multi-location studies in the rice–wheat system indicated that soil reserves contributed on an average 258 kg K_2O in K omission plots (Tiwari et al. 2006) (Table 11.7). Several researchers have assessed K balances in systematic ways contributing to our understanding at various scales.

11.5.2 Potassium Balance at the Country Scale

The K partial nutrient balance intensity in most soils of India was largely negative as crop removal far exceeded K additions through manures and fertilizers. Tandon (2004) reported an annual negative PNBI of –9.7 Mt of N+P+K in 1999–2000, of which 19% was N, 12% was P, and 69% was K. Satyanarayana and Tewatia (2009) estimated K balances in major agriculturally important states of India. Besides the major field crops, the authors considered the removal of nutrients by fruits and vegetables in all the states; tea (*Camellia sinensis* (L.) Kuntze), coffee (*Coffea* spp.), jute (*Corchorus capsularis* L.), rubber tree (*Hevea brasiliensis* (Willd. ex A. Juss.) Müll.Arg.), and other plantation crops in states wherever applicable; and the total production values of commodities were multiplied by the nutrient uptake coefficient to arrive at nutrient removal figures. The authors used crop production, state fertilizer

Table 11.8 Regional K additions, removal by harvested crops, and K partial nutrient balance intensity ($PNBI_K$) in India

Region	Addition kt K_2O	Removal kt K_2O	$PNBI_K$ kt K_2O
Eastern Region	517.6	2428.2	–1910.6
Western Region	756.7	4579.0	–3822.3
Northern Region	918.1	3534.0	–2615.9
Southern Region	1118.1	2447.7	–1329.6
India	3310.5	12,988.9	–9678.4

Source: Satyanarayana and Tewatia (2009)

use, and nutrient uptake coefficients from the Fertilizer Association of India (FAI) database (FAI 2008), as well as baseline information and assumptions on nutrient contribution from sources other than fertilizer, such as organic manures, farmyard manure, crop residues, irrigation water, biological nitrogen fixation for individual states while estimating $PNBI_K$. Nutrient additions only through fertilizer was used where information on other nutrient sources was not available. The overall negative K balance was estimated at 9.7 Mt, and the western region (3.8 Mt) had the highest share, followed by northern (2.6 Mt), eastern (1.9 Mt), and southern (1.3 Mt) regions (Table 11.8).

Dutta et al. (2013) estimated $PNBI_K$ in different states of India using the IPNI NuGIS approach (Fixen et al. 2012). The study used the K addition through inorganic and organic sources and the removal by key crops to estimate the K budget. The authors used data for fertilizer and the total amount of recoverable manure used in different states from various sources (DAC 2011; FAI 2007, 2011). The amount of manure consumed in each state was multiplied by a coefficient, based on average K content in recoverable manure, to estimate the K_2O contribution from organic sources. The K_2O removal by crops was calculated by multiplying production with K_2O removal coefficient (GOI 2020; FAI 2007, 2011). The crops considered in this study were rice, wheat, maize, barley, chickpea, pigeon pea (*Cajanus cajan* (L.) Millsp.), green gram (*Vigna radiata* (L.) R.Wilczek), lentil (*Lens culinaris* Medik.), moth bean (*Vigna aconitifolia* (Jacq.) Marechal), groundnut (*Arachis hypogaea* L.), sesame (*Sesamum indicum* L.), mustard (*Brassica juncea* (L.) Czern.), linseed, (*Linum usitatissimum* L.) cotton, and sugarcane. The study reported large negative $PNBI_K$ values in northern, western, and eastern regions (Fig. 11.6). The authors attributed the negative $PNBI_K$ in most states of India to inadequate application of K, high removal of K by intensive cropping, and export of crop residues. They cautioned that negative $PNBI_K$ in soils is not a one-off phenomenon but a recurring one, which will significantly affect the future sustainability of crop production.

Patra et al. (2017), using the national soil health card database, highlighted significant decreases in soil K concentrations in areas of north-western India that were earlier flagged as areas of high negative $PNBI_K$ (Fig. 11.6). Recently, Tewatia et al. (2017) did an extensive analysis of nutrient inputs and outputs for all major crops, including food grains and other cereals, pulses, oilseeds, forages, fiber crops, fruits, vegetables, plantation crops, spices, tubers, and sugarcane. The authors used

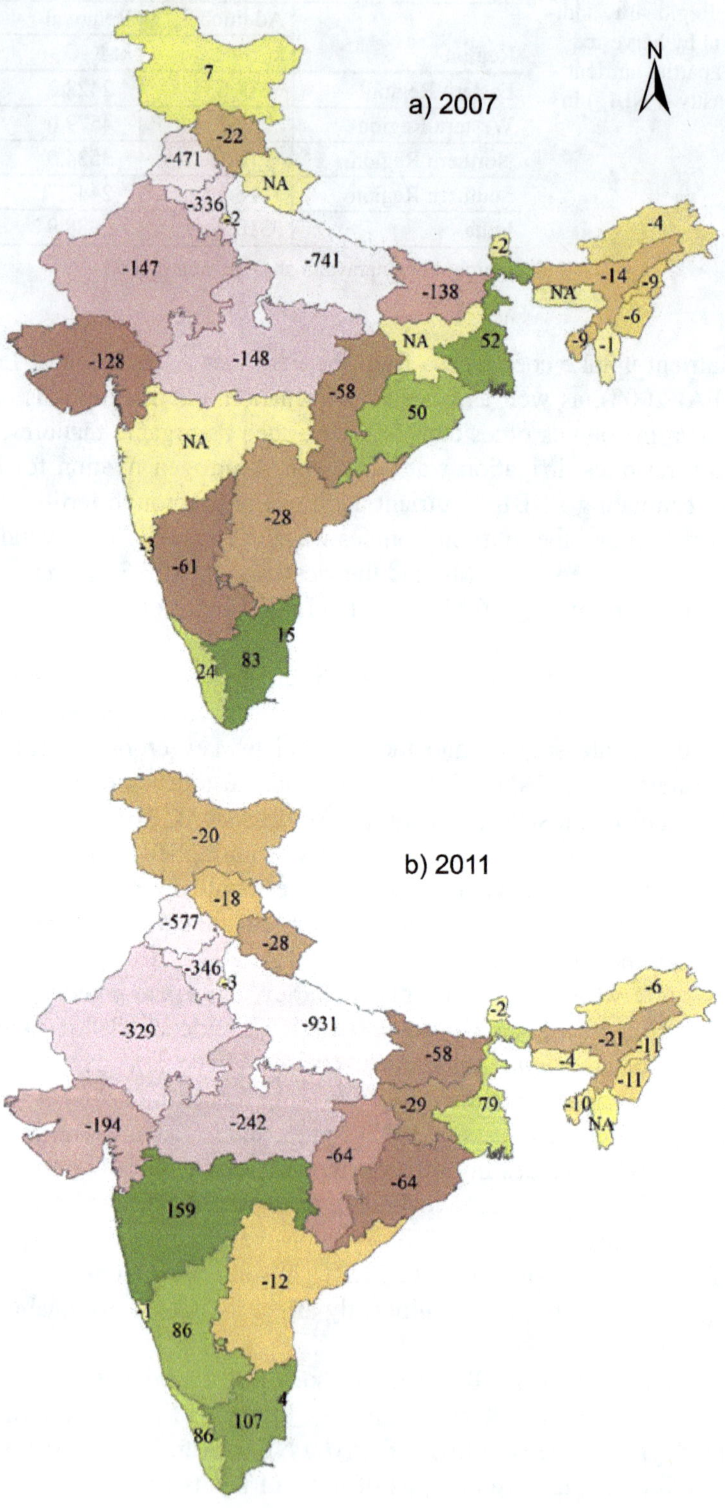

Fig. 11.6 The $PNBI_K$ for (**a**) 2007 and (**b**) 2011 across different states of India. (Dutta et al. 2013)

Table 11.9 Partial nutrient balance intensities (PNBI) in India for N, P_2O_5, K_2O, and total ($N + P_2O_5 + K_2O$). (adapted from Tewatia et al. 2017)

Parameter	N	P_2O_5	K_2O	Total
Country PNBI (Mt)	–2.3	–2.1	–8.6	–13.0
Average per hectare PNBI across crops (kg ha^{-1})	–15.3	–9.4	–42.2	–66.9

Source: Modified from Tewatia et al. (2017)

the FAI database for fertilizer consumption and crop production data for 2015 and 2016, and nutrient uptake coefficients were taken from several sources. Only inorganic fertilizer contributions were considered as inputs. The authors calculated a net negative K balance of –8.6 Mt of K_2O, which was 66% of the net negative $N + P_2O_5 + K_2O$ balance (Table 11.9). The authors further desegregated the K balance information at a crop and land area basis. The average areal negative balance was –42.2 kg K_2O ha^{-1} across all crop groups. It was highest for sugarcane (–121 kg K_2O ha^{-1}), fruits (–71 kg K_2O ha^{-1}), and plantation crops (–58 kg K_2O ha^{-1}), while pulses (–15 kg K_2O ha^{-1}) had the lowest K deficit, most likely due to low productivity levels. The negative balance in food grains was –40 kg K_2O ha^{-1}; however, it has the most profound effect on K balance at the national scale as the major food grain crops (rice, wheat, and maize) cover nearly 80% of the arable land.

11.5.3 Potassium Balance at the Cropping System Scale

Cropping systems strongly influence soil nutrient balance. Two or more crops are generally grown in a sequence within a year in much of India, except where climatic or other factors restrict growing multiple crops. Yadav et al. (1998) illustrated how nutrient removal differs by the number of crops grown in sequence and by the type of crops within the cropping system rotation (Table 11.10).

Careful nutrient and crop management planning is necessary for cropping systems to avoid nutrient imbalances in the soil. Long-term intensive cropping with inadequate application of K results in K mining leading to large negative balances and depletion of native K reserves (Gami et al. 2001; Regmi et al. 2002; Singh et al. 2002; Yadvinder-Singh et al. 2005; Majumdar et al. 2016). Therefore, adequate input of K is essential to prevent further depletion of soil K. Fixation of applied K as non-exchangeable K is also not ruled out in K-depleted soils. Evidences suggested that the non-exchangeable K reserves of the mica-rich soils of the Indo-Gangetic Plains (IGP) were exhausted due to continued lack of K fertilization in intensive cropping, that raised their K fixation capacity (Tiwari et al. 1992; Bijay-Singh et al. 2003). Singh et al. (2018) stated that short-term inconsistencies, if any, in crop response to K or changes in exchangeable K content in the soils of the IGP should not guide the K fertilization decisions. Rather, adequate K replenishment matching with annual K removals is required for the long-term sustainability of the rice–maize system in these soils without impairing soil fertility. Soil K depletion is faster with its

Table 11.10 Nutrient removal in important cropping systems

Crop sequence[a]	Total yield[b] t ha^{-1}	Total crop nutrient removal[c] N kg ha^{-1}	P_2O_5 kg ha^{-1}	K_2O kg ha^{-1}
Maize–wheat–green gram	8.2	306	27	232
Rice–wheat–green gram	11.0	328	30	305
Maize–wheat	7.6	247	37	243
Rice–wheat	8.8	235	40	280
Maize–wheat	7.7	220	38	206
Pigeon pea–wheat	4.8	219	31	168
Pearl millet–wheat–green gram	10.0	278	42	284
Pearl millet–wheat–cowpea	19.9[d]	500	59	483
Soybean–wheat	7.7	260	37	170
Maize–wheat–green gram	9.0	296	47	256
Maize–rapeseed–wheat	8.6	250	41	200

[a]Cowpea (*Vigna unguiculata* (L.) Walp.); green gram (*Vigna radiata* (L.) R.Wilczek); maize (*Zea mays* L.); pearl millet (*Pennisetum glaucum* (L.) R.Br.); pigeon pea (*Cajanus cajan* (L.) Millsp.); rapeseed (*Brassica napus* L.); rice (*Oryza sativa* L.); soybean (*Glycine max* (L.) Merr.); wheat (*Triticum aestivum* L.)
[b]Sum of harvested biomass across all crops in the sequence
[c]Sum of quantities of nutrients removed in harvested crop portions across all crops in the sequence
[d]Includes fodder

Table 11.11 Nutrient removal, partial nutrient balance (PNB), and recovery efficiency of K (RE_K) under different cropping systems

Cropping system[a]	Nutrient removal N kg N ha^{-1}	P kg P ha^{-1}	K kg K ha^{-1}	Partial nutrient balances (PNB) PNB_N	PNB_P	PNB_K	RE_K
Rice–wheat	243	41	268	0.78	0.64	2.25	0.61
Rice–rice	269	47	281	0.74	0.69	2.31	0.68
Rice–chickpea	251	37	185	1.32	0.62	1.95	0.77
Rice–potato	299	47	208	0.88	0.65	1.30	0.72
Maize–wheat	307	64	397	0.93	0.78	2.97	0.54
Pigeon pea–wheat	283	41	201	1.17	0.65	1.63	0.73
Sugarcane–ratoon–wheat	369	156	547	0.92	0.76	3.56	0.69
Sesbania (GM)–rice–wheat	245	39	259	0.63	0.61	1.37	0.71

Singh et al. (2002, 2008, 2015a, b)
[a]Chickpea (*Cicer arietinum* L.); maize (*Zea mays* L.); pigeon pea (*Cajanus cajan* (L.) Millsp.); potato (*Solanum tuberosum* L.); rice (*Oryza sativa* L.); sesbania (*Sesbania bispinosa* (Jacq.) W. Wight); sugarcane (*Saccharum giganteum* (Walter) Pers.); wheat (*Triticum aestivum* L.)

continuous inadequate replenishment. It is further aggravated when crops remove more K than required as luxury consumption. The data presented in Table 11.11 indicate that except for the crop sequences with a pulse or potato as component crops, all other sequences remove more K than N and P. The higher output: input ratio for K compared with N and P indicated insufficient K addition under different cropping systems, leading to depletion of soil K (Tiwari et al. 1992; Singh et al. 2015a, b).

Multi-locational studies (Singh et al. 2014) revealed that omission of K in rice–wheat cropping system resulted in annual mining of 158–349 kg K ha^{-1} from soil reserves, and the authors cautioned that continuous inadequate application of K in this cropping system may not be able to sustain high productivity over time. In a long-term experiment on rice–rice rotations at Gazipur in Central Bangladesh, Mazid Miah et al. (2008) reported that rice grain yield decreased sharply in a clay-loam soil from about 10 t ha^{-1} in 1985 to 6.2 t ha^{-1} in 2000 in the K-omission plots, whereas K application at 50 kg ha^{-1} resulted in positive K balance and maintained rice yields. In another study on rice–wheat rotations in northwest Bangladesh, the application of 54 kg K ha^{-1} increased average grain yield by 25–30% of rice and wheat by 53–86% across a range of demonstration plots on farmers' fields (Mazid Miah et al. 2008). Buresh et al. (2010) showed that near-neutral K balance in rice–wheat cropping systems in India would require 100 and 15% retention of rice and wheat residues, respectively, when irrigation water contributed 125 kg K ha^{-1}. Replacing winter wheat with maize in the rice–wheat cropping system, with typical crop management practices in India, reduces $PNBI_K$ well below –100 kg K ha^{-1}, which becomes even more negative at higher system productivity. At a maize grain yield of 12 t ha^{-1}, there is a net export of about 200 kg K ha^{-1}, which is often difficult to resupply with external K applications until and unless crop residues are fully returned to the field. Residues are often removed for several reasons. Rice straw has several competitive uses. Maize residue is difficult to manage in the field, and N immobilization often occurs because of high C:N ratios in that residue. Nutrient depletion–replenishment studies in the rice-based systems in Bangladesh have also shown negative $PNBI_K$ (Timsina et al. 2013).

On-farm results from the AICRP-IFS (2012) showed negative $PNBI_K$ in farmers' fields across locations and cropping systems (Table 11.12). Potassium partial nutrient balance intensity was influenced by the crops selected in various cropping systems.

The "SR+M" treatment received supplemental application of secondary and micronutrients, along with the same amount of N, P, and K as in "SR" (the state recommendation). The addition of secondary and micronutrients to SR triggered higher removal of K in most cases, suggesting that the application of limiting secondary and micronutrients, even at suboptimal application of major nutrients, may increase crop yield and removal of K.

The deficit in K supply will cause changes in K soil concentrations. Singh et al. (2013) showed that, in the absence of added K fertilizer, average exchangeable and non-exchangeable K decreased by 13–18 mg kg^{-1} and 26–41 mg kg^{-1}, respectively, across 60 on-farm locations during one rice–wheat cropping cycle. Another study assessing spatial variation of different soil K fractions in an intensively cultivated region of West Bengal (Chatterjee et al. 2015) associated the low non-exchangeable K in parts of the study area with high K removal exceeding K application rates in banana (*Musa* spp.) plantations. Non-exchangeable K is not measured in routine soil

Table 11.12 Potassium nutrient balance intensity ($PNBI_K$) for several cropping systems

Cropping system[a]	Treatment	K_2O addition kg K_2O ha^{-1}	K_2O removal kg K_2O ha^{-1}	$PNBI_K$ kg K_2O ha^{-1}
Rice–wheat (24)	FFP	0.0	150.0	−150.0
	SR	74.7	160.0	−85.3
	SR + M	74.7	174.0	−99.3
Rice–rice (24)	FFP	70.6	172.0	−101.5
	SR	66.4	189.0	−135.6
	SR + M	66.4	202.0	−135.6
Pearl millet–mustard (18)	FFP	0.0	104.0	−104.0
	SR	54.0	116.0	−62.1
	SR + M	54.0	122.0	−65.1
Pearl millet–wheat (18)	FFP	0.0	65.0	−65.0
	SR	83.0	95.0	−12.0
	SR + M	83.0	101.0	−18.0
Maize–Bengal gram (24)	FFP	0.0	133.0	−133.0
	SR	20.8	169.0	−148.3
	SR + M	20.8	181.0	−160.3
Rice–green gram(18)	FFP	34.0	129.0	−95.0
	SR	66.4	161.0	−94.6
	SR + M	66.4	176.0	−109.6
Maize–wheat(18)	FFP	21.6	53.0	−31.4
	SR	58.1	89.0	−30.9
	SR + M	58.1	97.0	−38.9
Cotton–pearl millet (18)	FFP	0.0	85.0	−85.0
	SR	83.0	91.0	−8.0
	SR + M	83.0	102.0	−19.0

FFP, farmers' fertilization practice; SR, state recommendation; SR + M, state recommendation with additional micro- and secondary nutrients
Modified from AICRP-IFS (2012)
[a]Bengal gram (*Cicer arietinum* L.); cotton (*Gossypium hirsutum* L.); green gram (*Vigna radiata* (L.) R.Wilczek); maize (*Zea mays* L.); mustard (*Brassica juncea* (L.) Czern.); pearl millet (*Pennisetum glaucum* (L.) R.Br.); rice (*Oryza sativa* L.); wheat (*Triticum aestivum* L.); values in parenthesis are the number of trials

tests in most countries, including India, and any decline in this K fraction generally remains unnoticed.

In conclusion, high-resolution and high-quality data are available to assess K or other nutrient balances in India. Several assessments at national and cropping system scales showed that K balances are largely negative. Low application rates of K in crops combined with high offtake of K at harvest due to straw and other crop residue removal from farms are the major reasons. The on-farm studies showed declines in exchangeable and non-exchangeable K fractions in soils of intensively cultivated areas. Potassium fertilization decisions based on agronomic evidence are required to improve crop yields and farm profitability and to maintain K fertility of Indian soils.

11.6 Sub-Saharan Africa

Soil fertility depletion is considered one of the major factors limiting crop production in sub-Saharan Africa. Crop production in soils of low native fertility with little or no fertilizer application is depleting and unsustainable. Maintaining the productive capacity of these soils through effective nutrient management remains a major challenge. Nutrient mass balance in specific soil–crop systems is a reliable indicator of nutrient depletion or mining.

When estimated at different spatial scales, nutrient balances provide insight into sustainability challenges affecting production systems in the short- and long-term. Access, price, and affordability of fertilizers are major production constraints in Sub-Saharan Africa, especially for K fertilizers. The following sections outline the K balance at various spatial scales in the region and how that affects crop productivity in different production systems.

11.6.1 Potassium Balance at Continental and Country Scales

Analysis of nutrient balances in sub-Saharan Africa shows consistent negative trends that result from the continuous cultivation of crops for many decades with little addition of nutrient inputs in agricultural soils in the region. Annual $PNBI_K$ at the regional level averages less than –15 kg K ha^{-1} and ranges between –5 and –45 kg K ha^{-1} for various countries (Stoorvogel et al. 1993; Van den Bosch et al. 1998) (Fig. 11.7). The losses of K from the soil translate to about 3 Mt K year^{-1}. The average fertilizer use of 18 kg ha^{-1} of N+P+K in sub-Saharan Africa is the lowest in the world, accounting for less than 2% of the world fertilizer consumption. This is far below requirements to prevent nutrient depletion. Potassium inputs from fertilizer are proportionally lower than N and P, as most fertilizers that are recommended and used in crop production in sub-Saharan Africa exclude K. As a result, the amounts of K applied as fertilizer are less than 3 kg ha^{-1}. The application of K through organic resources is also limited and negligible due to low amounts of manures and crop residues that are available. The removal of K in grain and crop residues and K losses through erosion and leaching combine to result in severe nutrient depletion. The average yield of cereal crops which covers 80% of croplands is about 1.5 t ha^{-1}, accounting for removal of 20–30 kg K ha^{-1}. Potassium losses due only to erosion in sub-Saharan Africa range from 3 to 15 kg of N+P+K ha^{-1} year^{-1} (Henao and Baanante 2006).

Despite the overall large negative $PNBI_K$ in sub-Saharan Africa, nutrient balances vary greatly among different cropping systems and agro-ecological zones. The largest losses of 20–50 kg K ha^{-1} year^{-1} occur in the sub-humid savannahs of West Africa and the highlands and sub-humid areas of east Africa and southern Africa, a region with high potential for crop production and high population densities. Moderate K losses of 10–20 kg K ha^{-1} year^{-1} occur in the humid forests and

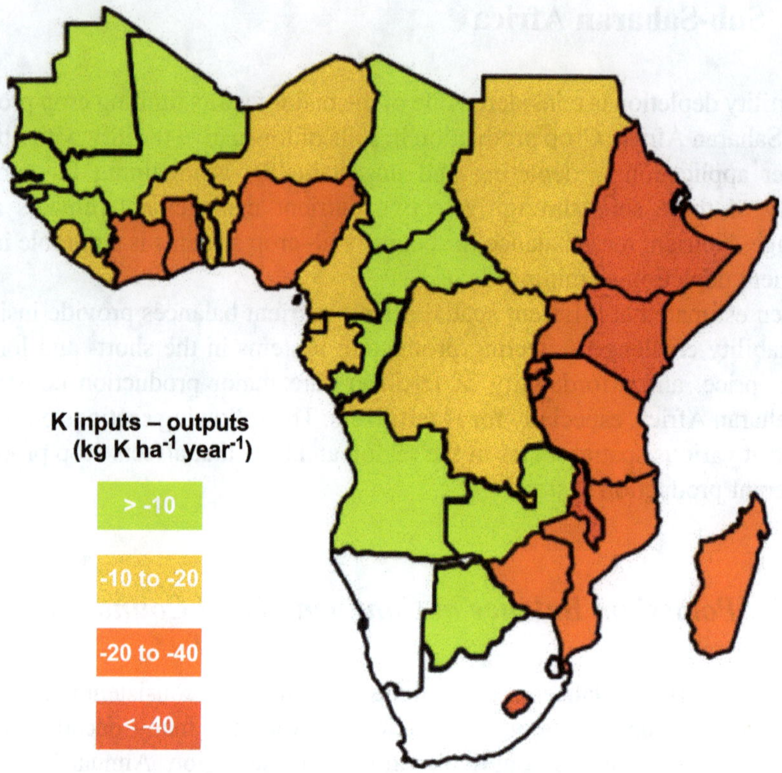

Fig. 11.7 Potassium balances for countries in sub-Saharan Africa. (adapted from Henao and Baanante 2006)

wetlands of south-central Africa and Sudan. The soils in the arid zones in southern and West Africa have the lowest K losses (5–15 kg K ha^{-1} year^{-1}), mainly due to low nutrient removals from low yields.

11.6.2 Potassium Balance at the Regional Scale

Potassium partial nutrient balance intensities for regions within countries are highly variable, similar to the patterns of K balances at the continental level. For example, $PNBI_K$ levels in Kenya are consistently negative, and range from –5 to –34 kg K ha^{-1} year^{-1} (Henao and Baanante 2006) (Table 11.13). The Rift Valley region with the lowest balances is characterized by high potential crop production and intensive cultivation, while the highest nutrient balances in northeastern Kenya is associated with arid conditions that result in both low crop yields and nutrient removals. Relatively higher K fertilizer use in the central Kenya Rift Valley is due to large areas of tea, coffee, and sugarcane production that receive recommended fertilizers containing K.

Table 11.13 Regional K partial nutrient balance intensity ($PNBI_K$) levels in Kenya

	K removal	K added		
	Crop removal	K fertilizer	Manure	$PNBI_K$
Region	kg K ha^{-1} year^{-1}	kg K ha^{-1} year^{-1}	kg K ha^{-1} year^{-1}	kg K ha^{-1} year^{-1}
Central	20	8	3	–9
Coast	21	–	1	–20
Eastern	22	1	1	–20
Nairobi	11	2	1	–8
North Eastern	6	–	1	–5
Nyanza	27	2	1	–24
Rift Valley	40	5	1	–34
Western	25	3	2	–20

Henao and Baanante (2006)

11.6.3 *Potassium Balance at the Farm Scale*

Aggregated nutrient balance estimates at national and regional levels for sub-Saharan Africa have shown overall large negative balances. At such scales, variability is driven by soil-forming factors, such as the underlying geology and position on the landscape, jointly termed the "soilscape" (Deckers 2002). In contrast, variation in soil fertility associated with resource management at the farm level has generally been overlooked, despite evidence that such operations affect K balances and productivity of both crops and livestock (Smaling and Braun 1996; Giller et al. 2006). In addition, there are strong indications that K balances differ widely among farms in different wealth categories and among plots at different distances from homesteads (Shepherd and Soule 1998).

Differences in access to fertilizer by farmers contribute significantly to the variability of fertilizer use. Farmers with more access to fertilizer use greater amounts of mineral fertilizers (Tittonell et al. 2005) and tend to own more cattle and thus manure, thereby using significant quantities of nutrients (Swift et al. 1989; Achard and Banoin 2003). Consequently, K accumulates on farms with more access to resources, often at the expense of thc poorer farms.

Within farms, soil fertility status of different plots on smallholder farms in sub-Saharan Africa may vary considerably due to both inherent factors and different resource management strategies (Tittonell et al. 2005). Smallholder farms consist of multiple plots managed differently in terms of allocation of crops, fertilizers, and labor resources, resulting in within-farm soil fertility gradients caused by management strategies a common feature (Zingore et al. 2007). In most cases, both organic and mineral fertilizer resources are preferentially allocated to the part of the farm used for growing the main food-security crop, often close to the homestead, while plots further away are neglected. Such management decisions culminate in the creation of gradients of decreasing soil fertility with distance from the homestead (Elias et al. 1998; Vanlauwe et al. 2002). Even where small quantities of manure and mineral fertilizers are available, farmers still apply them at high rates by concentrating them in small areas. For instance, farmers in Zimbabwe apply cattle manure at amounts as high as 80 t ha^{-1} $year^{-1}$ by concentrating cattle manure on preferred plots (Mugwira and Murwira 1998). The underlying reasons for targeting of nutrient resources to few fields are not fully understood, but important factors include farm size, distance of different plots from the homestead, restricted availability of fertilizers and manures, availability and efficiency of labor use, risk of theft, and the need to reduce the risk associated with erratic rainfall (Nkonya et al. 2005).

Potassium application in African agriculture is low. Most of the farmland in Africa is dependent on soils that experienced some level of erosion and nutrient mining. The exchangeable K concentrations of the soil are generally low because of the high degree of weathering and a long history of crop production without proper soil management. Farmers' wealth, resource management strategies, and access to K fertilizers all contribute to significant variation in largely negative K balances across African soils. Improved access to K fertilizers and their site-specific management are key to increasing or even just maintaining the productive capacity of these soils.

11.7 North Africa

The North Africa region is dominated by the Mediterranean climate, with rainfall during the colder winter and dry and hot during the summer. The soils of this region are varied but generally low in organic matter. In arable lands, the predominant soils are calcimagnesic, fersialitic, and clay-rich vertisols. Major land areas of North Africa are under arid and semi-arid climate zones. Although lack of sufficient rainfall is the main limitation of crop production, nutrient deficiencies also limit crop productivity. Even though most of the research undertaken in North Africa in the 1980s reported that soils are well supplied with K (Ghanem et al. 1983; El Oumri 1985; Stitou 1985), continuous withdrawal of soil K has led to a decline in soil fertility. Several areas in North Africa now report low exchangeable K and high spatial variability of K in soils (Badraoui et al. 2003). As K availability in soils of most North African countries (Morocco, Algeria, and Tunisia) is perceived to be generally adequate (Azzaoui et al. 1993; Mhiri 2002; Belouchrani et al. 2002), fertilizer K consumption has remained very low, with an average of 12% of the total consumed N, P, and K (FAOSTAT 2018). Potassium partial nutrient balance intensities show negative trends due to crop K removal with none or low K inputs. In some rainfed areas, farmers do not fertilize crops with K, even when it is required, due to the relatively high cost of fertilizer, which has led to visible K deficiency symptoms in crops in the last few decades (Mhiri 2002). For wheat, the main annual crop of the region, fertilizer recommendations are based mainly on N and P, while K application is generally neglected. Research on K also became limited in the last decades with the most recent research programs on K undertaken more than 30 years ago.

Table 11.14 Cumulative K partial nutrient balances ($PNBI_K$) of cropping systems at different locations in Morocco

Crop rotations (location)[a]	Study years	K application	Cumulative K fertilizer addition[b] kg K ha^{-1}	Cumulative K removal[c] kg K ha^{-1}	$PNBI_K$ kg K ha^{-1}
Wheat–maize–forage–cotton–sugar beet	12	–K	0	1645	–1645
(Sidi Kacem, Morocco)	12	+K	1380	1646	–266
Wheat–legumes	15	–K	0	521	–521
(Mohammedia, Morocco)	15	+K	1320	555	765
Wheat–maize–forage–cotton–sugar beet	16	–K	0	9479	–9479
(Tessaout, Morocco)	16	+K	2520	9721	–7201
Wheat–maize–forage–cotton–sugar beet	14	–K	0	5863	–5863
(Souihla, Morocco)	14	+K	2320	6086	–3766
Wheat–maize–forage–cotton–sugar beet	17	–K	0	6457	–6457
(Afourer, Morocco)	17	+K	2640	6450	–3810
Wheat–maize–forage–cotton–sugar beet	10	–K	0	5406	–5406
(Boulaouane, Morocco)	10	+K	2680	5416	–2736
Potato–wheat–onion	10	–K	0	516	–516
(Dar Bouazza, Morocco)	10	+K	1800	455	1345

Azzaoui and Alilou (1990)
[a]Cotton (*Gossypium hirsutum* L.); forage (barley, ryegrass, vetch) legumes (pulses); maize (*Zea mays* L.); onion (*Allium cepa* L.); potato (*Solanum tuberosum* L.); sugar beet (*Beta vulgaris* L.); wheat (*Triticum aestivum* L.)
[b]Sum of K applications across all years
[c]Sum of K removed in harvested crop portions across all years

Limited information showed negative $PNBI_K$ levels in Morocco across locations and cropping systems (Table 11.14). In only two cases (the wheat–legume rotation in Mohammedia and the potato–wheat–onion (*Allium cepa* L.) rotation in Dar-Bouazza) was K added in quantities large enough to produce positive $PNBI_K$. A similar study in the Saharan region of Algeria showed positive $PNBI_K$ levels when K was added in an irrigated wheat production system (Table 11.15). Since the K balance in these studies accounted only for K output by grain removal, greater K deficits would be expected in areas where wheat straw is also removed from the field for off-farm use. Previous research demonstrated that critical levels of K in soils depends on soil types and crop yield performances (Badraoui et al. 2002; Aissa and Mhiri 2002). There are large differences in K use between rainfed and irrigated areas.

Table 11.15 Potassium partial nutrient balance intensity ($PNBI_K$) of irrigated wheat in sandy soils of the Saharan area (South Ouargla) of Algeria

K addition	K removal in grain	$PNBI_K$
kg K ha^{-1}	kg K ha^{-1}	kg K ha^{-1}
0	13.86	−13.86
60	22.13	38.87
120	31.27	88.73
180	38.66	142.34

Halilat (2004)

The expansion of irrigated area in North Africa due to incentive programs such as the Green Plan in Morocco and the National Program for Rural and Agricultural Development (PNDAR) in Algeria intensified production and increased K removal by crops. As a consequence, growers in irrigated areas use more K for industrial crops, fruits, and vegetables. In the citrus sector, which represents one of the main export crops in the region, K plays an important role in fruit quality (Hamza et al. 2012). For irrigated wheat, research undertaken in the Doukkala region of Morocco showed that wheat responded to added K fertilizer only when the yield was higher than 5 t ha^{-1} (Badraoui et al. 2002). The highest consumers of K fertilizer are greenhouse crop production systems. For example, banana production uses on average of 500 kg K_2O ha^{-1}, strawberry production uses 400 kg K_2O ha^{-1}, pepper (*Capsicum* spp.) production uses 140 kg K_2O ha^{-1}, and potato production uses 115 kg K_2O ha^{-1} in northwest Morocco (Zerouali and Mrini 2004).

In Tunisia, the status of K in soils differs among three soil categories (Mhiri 2002): (1) acidic soils in the northwest are generally low in K; (2) alkaline, calcareous, fine-textured soils located in the north are rich in total and exchangeable K, and are considered among the most fertile soils in the country; and (3) alkaline, calcareous coarse-textured soils, located in the center and south of the country and the Sahel, are relatively low to very low in exchangeable K.

In Algeria, arable land is limited to less than 3% of the total geographical area. Approximately 8.7 Mha are available to grow cash crops, forest, pasture, rangelands, and scrub and alfalfa (Laoubi and Yamao 2012). The consumption of K fertilizer in Algeria is low and represents about 11% of the total consumed N, P, and K fertilizer in the country (FAOSTAT 2018).

There are few studies on K mass balance in Morocco, Algeria, and Tunisia. Potassium fertilizers are applied in limited quantities due to perceived high native K in soils. Potassium deficiency in crops, however, is increasing due to higher removal of K in intensive agricultural production systems with improved irrigation facilities. Research evidence of better crop productivity and quality with site-specific K management in the region suggests the need for higher levels of K inputs to crops.

11.8 United States

Potassium budgets for the USA were calculated using the NuGIS model (Fixen et al. 2012). The model calculates K budgets for each county in each state for 1987, 1992, 1997, 2002, 2007, 2010, 2011, and 2012. Briefly, the NuGIS model considers the sum of nutrient removal by up to 21 major crops as the only output: alfalfa (*Medicago sativa* L.), apple (*Malus domestica* Borkh.), barley, common bean (*Phaseolus vulgaris* L.), canola, maize grain, maize silage, cotton, hay (various crops other than alfalfa), orange (*Citrus sinensis* (L.) Osbeck), peanut (*Arachis hypogaea* L.), potato, rice, sorghum, soybean, sugar beet, sugarcane, sunflower, sweet corn (*Zea mays* L.), tobacco, and wheat. The model uses crop-specific K removal rates (g K kg^{-1} yield) and does not consider spatiotemporal variability. The nutrient inputs considered by the model are fertilizer and manure.

A map PNB_K for the contiguous 48 states of the USA is shown in Fig. 11.8. A $PNB_K > 1$ dominates the more arid western half of the USA where soil K levels have historically been higher and where K application rates have generally been lower.

Based on the distribution of PNB_K (Fig. 11.8), the USA was divided into four groups, and trends in PNB_K over time were assessed (Fig. 11.9). Temporal trends were evaluated through linear regression of the log of PNB_K calculated for each regional group as well as for the USA. For the 48 contiguous states considered as a

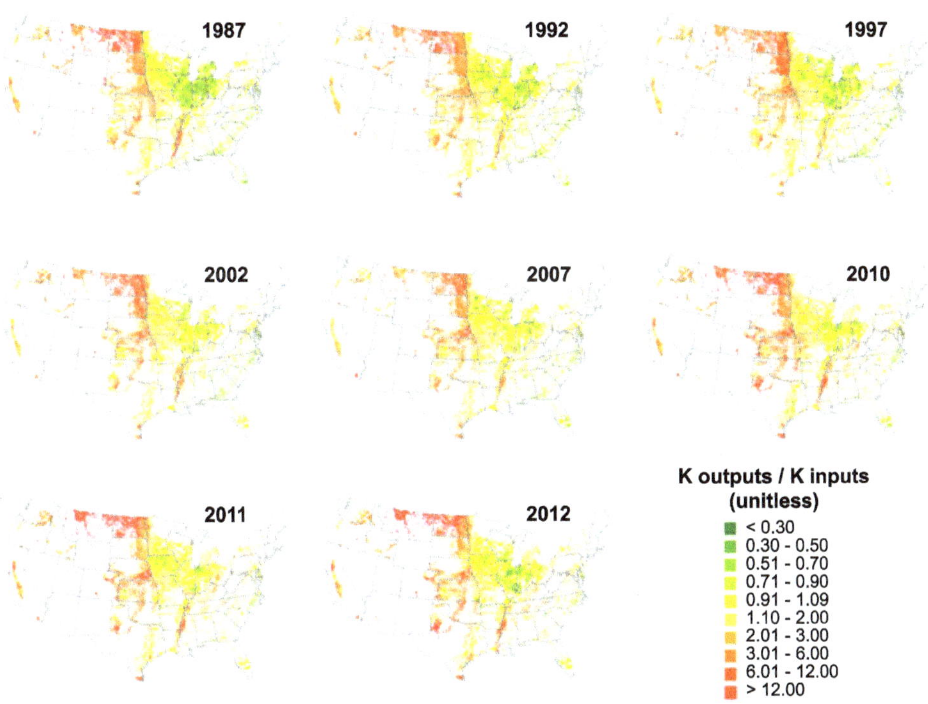

Fig. 11.8 Potassium partial nutrient balance (PNB_K) for the contiguous 48 states of the USA for years 1987, 1992, 1997, 2002, 2007, 2010, 2011, and 2012 as calculated by the NuGIS model

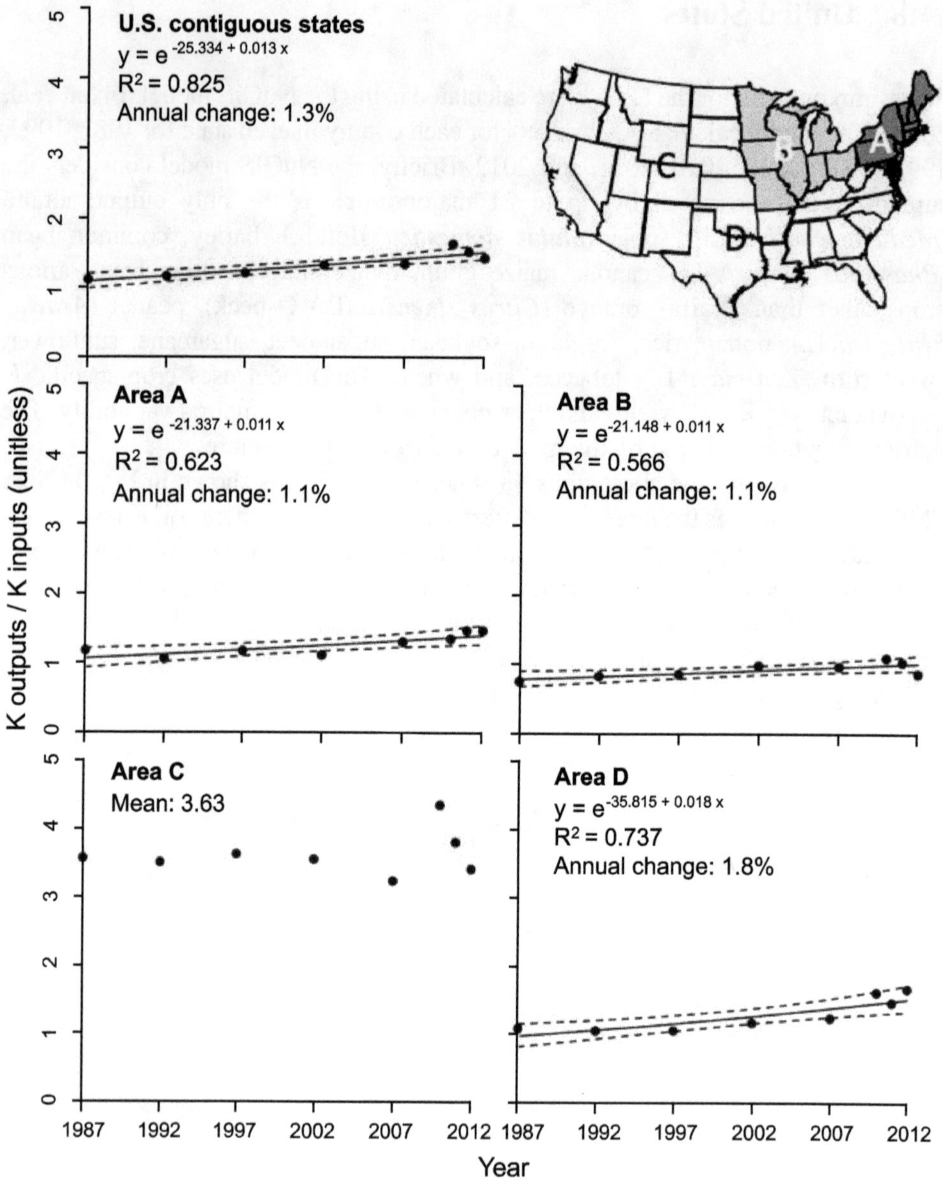

Fig. 11.9 Temporal trends in K removal to use ratios for the contiguous 48 states of the USA and for four groupings of U.S. states, denoted as areas A, B, C, and D. Dotted lines are the 95% confidence intervals for the regression curves

group, PNB_K levels have been increasing at a rate of 1.3% year^{-1}. Similar increases (1.1% year^{-1}) were also observed in the northeastern states (Group A) and the corn and soybean-growing states (Group B). States in Group D had the greatest increase in K removal to use (1.8% year^{-1}). Much of this area is where cotton has traditionally been grown, although cotton production has declined in recent years and is being replaced by other crops, such as soybean. In the more arid western states (Group C), no significant trend emerged over time. Soil K concentrations have historically been high; consequently, farmers have not applied much K fertilizer, resulting in a higher PNB_K, averaging 3.63.

The widely varied K-balance patterns across the USA arise from interactions of climate, soil properties, cropping patterns, market conditions, governmental policies, and many other factors. In general, the combination of rather steady K consumption but increased yields and therefore increased nutrient removal has led to negative K balances that are becoming increasingly negative in much of the USA.

11.9 Brazil

Since the soil nutrient status of Brazilian soils is generally low, supplying additional nutrients is generally necessary to make agriculture effective and profitable. Potassium fertilizer consumption in Brazil is high, with consumption of 5.4 Mt K_2O in 2015, ranking second in the world, with about 90% of it imported. A nutrient budget can represent the amount of a nutrient that is exported from the whole country, a specific location (e.g., state), or a field by a crop relative to the amount of applied nutrient.

The nutrient budgets for Brazil have been calculated for the whole country, states, and main crops based on fertilizer consumed (ANDA 2010–2015), crop production for the eighteen major crops in Brazil, namely banana, common bean (*Phaseolus vulgaris* L.), cassava (*Manihot esculenta* Crantz), castor bean (*Ricinus communis* L.), cocoa (*Theobroma cacao* L.), coffee, cotton, maize, orange, peanut, potato, rice, sorghum, soybean, sugarcane, tobacco, tomato (*Solanum lycopersicum* L.), and wheat (IBGE 2010–2015), and the average nutrient concentration in the harvested portion of the crops (Yamada and Lopes 1998; Cunha et al. 2010, 2011, 2014).

The data from the most recent survey (Francisco et al. 2015) for an average of 4 years (2009–2012) are summarized in Table 11.16. It is important to note that the Midwest region was responsible for 34% of the total K fertilizer consumption. This region provides the bulk of soybean and maize production in the country, with

Table 11.16 Average (2009–2012) annual K partial nutrient balance intensities ($PNBI_K$) and partial nutrient balances (PNB_K) for regions in Brazil

	Crop removal of K	Applied K	$PNBI_K$	PNB_K
Region	Mt K_2O	Mt K_2O	Mt K_2O	
South	0.91	0.96	0.05	0.95
Midwest	1.06	1.29	0.23	0.82
Southeast	0.66	1.02	0.35	0.65
Northeast	0.30	0.43	0.13	0.70
North	0.09	0.09	0.00	1.04
Brazil	3.03	3.79	0.76	0.80

Francisco et al. (2015) using data from ANDA 2010–2013

Table 11.17 Average (2009–2012) annual K partial nutrient balance intensities ($PNBI_K$) and partial nutrient balances (PNB_K) for main crops in Brazil

Crop[a]	Crop removal of K Mt K_2O	Applied K Mt K_2O	$PNBI_K$ Mt K_2O	PNB_K
Soybean	1.64	1.66	0.02	0.99
Maize	0.34	0.52	0.18	0.65
Sugarcane[b]	0.66	0.78	0.12	0.85
Coffee	0.05	0.25	0.20	0.20
Cotton	0.08	0.14	0.06	0.56
Rice	0.06	0.07	0.01	0.86
Common bean	0.06	0.05	–0.01	1.20
Orange	0.03	0.05	0.02	0.60
Wheat	0.02	0.06	0.04	0.33

Francisco et al. (2015)

[a]Common bean (*Phaseolus vulgaris* L.); coffee (*Coffea* spp.); cotton (*Gossypium hirsutum* L.); maize (*Zea mays* L.); orange (*Citrus sinensis* (L.) Osbeck); rice (*Oryza sativa* L.); soybean (*Glycine max* (L.) Merr.); sugarcane (*Saccharum giganteum* (Walter) Pers.); wheat (*Triticum aestivum* L.)

[b]For sugarcane, a 20% deduction was considered for K removal considering the regular disposal of vinasse

plant-available K being inherently low in the soil. In summary, PNB_K for the whole country was 0.8.

Nutrient budgets for nine crops grown between the years of 2009 and 2012 are presented in Table 11.17. Potassium use was higher than crop removal in most crops, with the exception of beans (1.20). Potassium use was most balanced in soybean, which had a PNB_K of 0.99, followed by rice (0.86) and sugarcane (0.85). Coffee had the lowest PNB_K value (0.20).

Soybean, maize, and sugarcane are responsible for about 70% of the K fertilizer consumed in Brazil. The PNB_K levels for these crops are very reasonable, compared to other situations worldwide, which is indicative that Brazil has had fairly appropriate use of K for these crops.

In terms of states (data not shown), there is a wide range of results in terms of K mass balance, with some states presenting very low and others very high PNB_K levels (ranging from 0.54 to 3.17 in the main agricultural states). The results from studies at specific locations with K mass balances allows researchers to focus on crops or regions with the most severe deficits. For example, coffee and citrus were two of the most important crops in Brazil with the lowest levels of PNB_K among all crops considered in the study. Consequently, it is necessary to better study fertilizer practices and management systems that could increase NUE, leading to improved PNB_K levels for both crops.

In order to extend this analysis back to represent PNB for N, P, and K prior to 2009, trends in mass balances of these nutrients between 1988 and 2012 are provided in Fig. 11.10. Nitrogen removal was higher than N input until the late 1990s. Later, N use in the country increased due to the adoption of more intensive cropping systems with higher inputs, especially for sugarcane, orange, coffee, and maize,

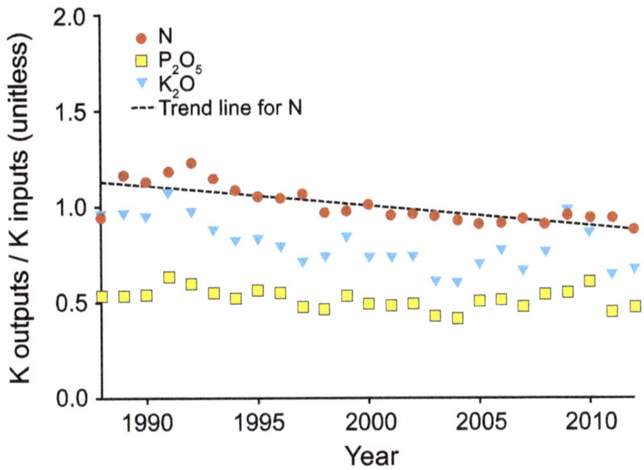

Fig. 11.10 Partial nutrient balances (PNB) for N, P, and K for main crops in Brazil from 1988 to 2012

which resulted in lower N PNB (PNB_N), reaching 0.87 in 2012. Phosphorus PNB (PNB_P) has essentially hovered around 0.60, which may be acceptable when compared to estimates of 0.30 in many situations. Potassium partial nutrient balances had a similar decreasing trend as N, and PNB_K reached 0.67 in 2012. Potassium showed a dramatic increase in PNB_K in 2009 (0.98) that reflected a significant and temporary decrease in K use by farmers due to higher fertilizer prices. The generally steady growth in nutrient use in Brazil in recent years has been effective in improving crop production, increasing average yields. In 1990, the average yield was around 1.70 t ha^{-1} and subsequently increased to 3.44 t ha^{-1} 20 years later.

Nutrient mass balances were estimated periodically to help identify gaps in fertilizer use for specific crops or regions as well as to forecast future nutrient demands. In this context, initiatives aimed at educating farmers and agronomists on how to assess the performance of nutrient inputs are crucial for promoting fertilizer use efficiency, minimizing nutrient loss, and increasing crop production sustainability. With this in mind, a new tool was developed recently by Nutrição de Plantas Ciência e Tecnologia (NPCT) in Brazil which allows the calculation of specific nutrient mass balances at the farm level (NPCT 2020). Many agronomists around the country are using this tool to calculate nutrient mass balances for different crops and farms.

Calculation and publication of nutrient mass balances have helped agricultural stakeholders to better understand the status and efficiency in the use of different plant nutrients. For Brazil, the PNB_K indicates generally appropriate ratios. Between 2009 and 2012, the average PNB_K value was 0.80, which means that overall, 80% of the fertilizer applied to crops was exported as agricultural products. The recently released web tool to calculate nutrient mass balances at the farm level is helping farmers around the country better manage fertilizers. This tool can be adapted for use in other regions.

11.10 Southern Cone of Latin America

The total agricultural area of the five countries of the Southern Cone of Latin America (Argentina, Bolivia, Chile, Paraguay, and Uruguay) is 238 Mha over a total land area of 513.6 Mha. Total arable land represents approximately 22% of the agricultural area. The total area for different crop groups defined by FAOSTAT is 56.5 Mha. Field crops (oil crops, coarse grain, and cereals) lead the area and production in the region.

The main field crop areas of the Southern Cone include the Pampas and Gran Chaco regions of Argentina and Uruguay, the southern highlands of eastern Paraguay, the eastern lowlands of Bolivia, and the central valleys and southern volcanic regions of Chile.

Several soil associations are found in this region. Mollisols are dominant throughout the Pampas–Chaco plains and Uruguay and are among the best suited for agriculture because of their high native fertility. Alfisols are also widespread in the Pampas–Chaco region. Alfisols are generally fertile, with a high concentration of nutrient cations. Ultisols and Oxisols are the main soils in eastern Paraguay. These soils have good physical qualities but require high lime and phosphorus inputs. Vertisols are mainly located in the central-eastern region of Argentina, in the Entre Rios province, as well as in Uruguay. The vertisols have good nutrient concentrations, but challenging soil physical properties that demand careful management. Alluvial soils dominate the eastern lowlands of Bolivia and also have good natural soil fertility. Soils of Chile are quite variable and include Alfisols, Mollisols, Entisols, and Inceptisols in the central region, and Entisols, Inceptisols, Andisols, and Ultisols in the south. Native soil K availability is high in most of the cropping areas of the Southern Cone of Latin America (Barbazán et al. 2012; Sainz Rozas et al. 2013). Thus, fertilizer K consumption in the region has been low, about 335–425 kt K_2O in the last years (summed across the individual country consumptions shown in Fig. 11.11). However, expansion of agriculture into new areas and replacement of pastures by annual crops, mainly soybean, in the last 20 years (Wingeyer et al. 2015) have increased K removal by grains and induced K deficiencies in several regions of Uruguay (Barbazán et al. 2012, 2017), and some areas of Paraguay and Argentina.

In Argentina, the apparent consumption of K as fertilizers reached a peak of 54 kt K_2O (44 kt K) by 2007 (Fig. 11.11). An annual PNB_K has been estimated for the four major grain crops (soybean, corn, wheat, and sunflower) considering K application and grain K removal (García and González Sanjuan 2016). This estimation used average grain K concentration for the four crops and assumed that crop residues were not removed from fields, a typical practice under no-tillage agriculture in Argentina.

The PNB_K has been very high, with K removal exceeding K application by almost 800 times during the period 2012–2014 (Fig. 11.12). Negative K mass balances have reduced soil K availability (0–20-cm depth) to the current soil test K levels of 370–750 mg K kg^{-1} in agricultural fields in the central Pampas (Correndo et al. 2013). This is a 32–62% reduction from the pristine (pre-agriculture) soil test K

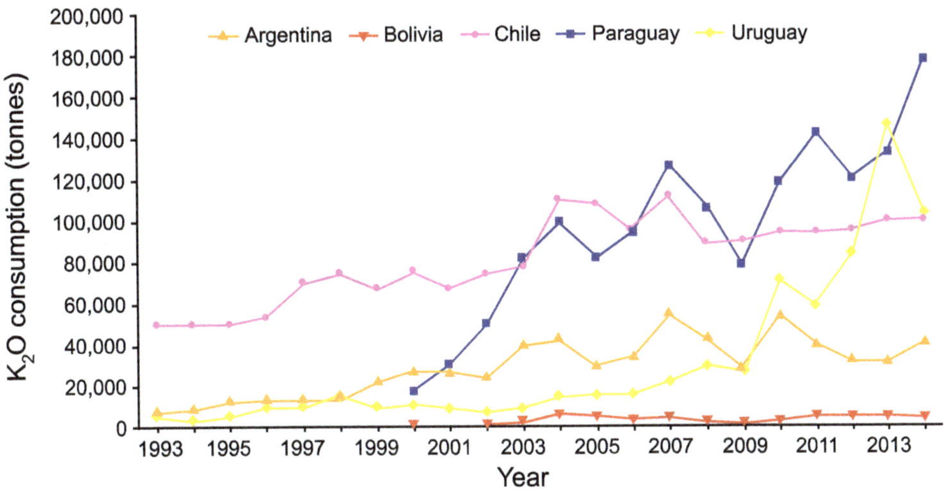

Fig. 11.11 Changes in K_2O fertilizer consumption in the countries of the Latin America Southern Cone region, 1993–2014. [adapted from data of FAOSTAT, IFA, Fertilizar AC (Argentina), APIA (Bolivia), ODEPA (Chile), CAPECO (Paraguay), and MGAP-DIEA (Uruguay)]

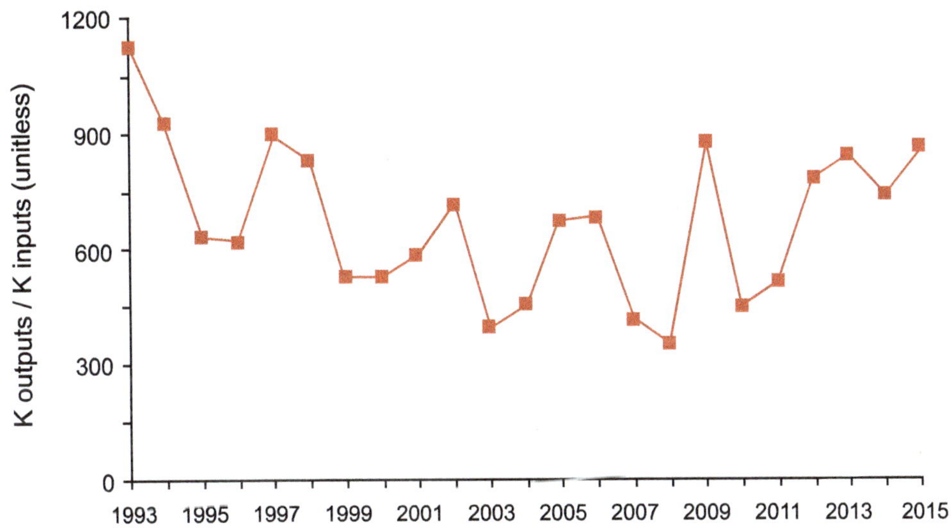

Fig. 11.12 Potassium partial nutrient balance (PNB_K) estimates for the main four field crops in Argentina: soybean, maize, wheat, and sunflower. (adapted from García and González Sanjuan 2016)

concentrations of 990–1140 mg K kg^{-1} (Fig. 11.13). Areas in northwestern Argentina under continuous sugarcane production for more than 50 years without K inputs have decreased soil K concentrations to low enough levels that crops now respond to K fertilization (Pérez Zamora 2015).

Consumption of fertilizer K in Bolivia is very low, with an annual average of approximately 4.4 kt during 2012–2014 (FAOSTAT 2017). Estimation of annual K

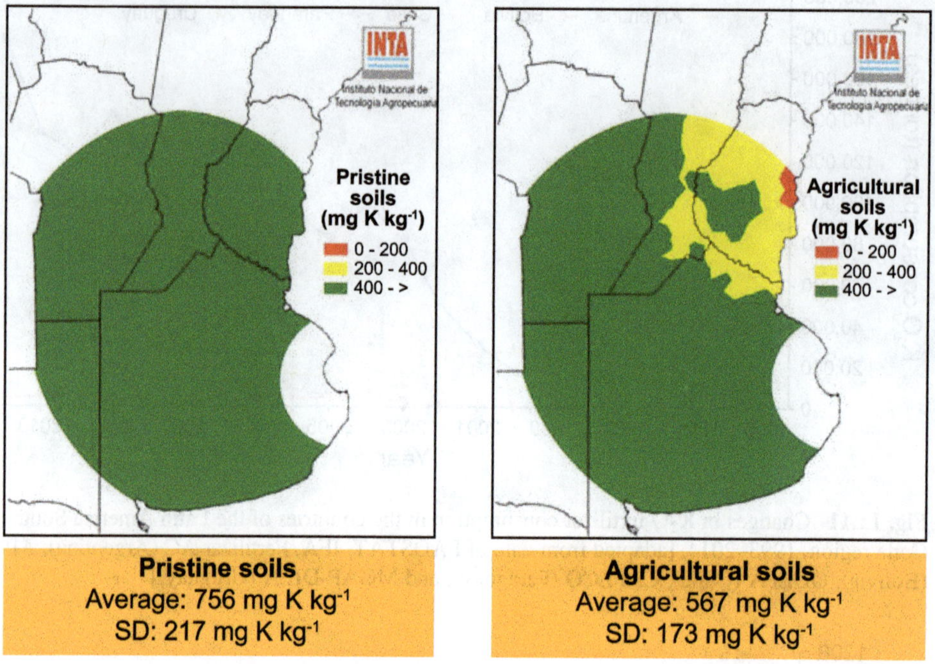

Fig. 11.13 Soil test K concentrations in the Pampas region of Argentina. (Sainz Rozas et al. 2013)

removal in the main harvestable products, using average K concentrations in crops, was 90.7 kt K year^{-1} for the same period, resulting in a $PNB_K > 20$. Continuous removal of soil K without replacement will eventually result in soil fertility declines adversely affecting crop yields and farm profitability.

Mancassola and Casanova (2015) estimated a comprehensive nutrient balance for agricultural production of Uruguay. In their study, nutrient removal was estimated for: field crop production; beef, dairy, and wool production; fruit and vegetable production; and forest production. Nutrient concentrations were gathered from local information and from literature. Nutrient application was estimated from fertilizer imports. In 2010, estimated $PNBI_K$ was approximately –24 kt K_2O (–20 kt K) (Fig. 11.14). These negative $PNBI_K$ levels were observed for most of the production systems because of low K fertilization rates, with the exception of vegetables and fruits. Furthermore, as soybean area increased in the last two decades, $PNBI_K$ in the soil has become more negative due to the high K requirements of soybean. Considering an average grain K content in 3.6 Mt of soybean exports, it is estimated that approximately 63 kt K_2O was removed from the soil in 2014. In addition, agriculture has also expanded to marginal soils in the north-central and eastern regions of the country, where soils with low soil test K levels are common (Fig. 11.15). As a result, K deficiencies in crops have been evident since the early 2000s (Barbazán et al. 2012, 2017).

Most of Paraguay's agricultural production takes place in the eastern half of the country on lateritic soils (mainly Oxisols and Ultisols) and includes soybean as a

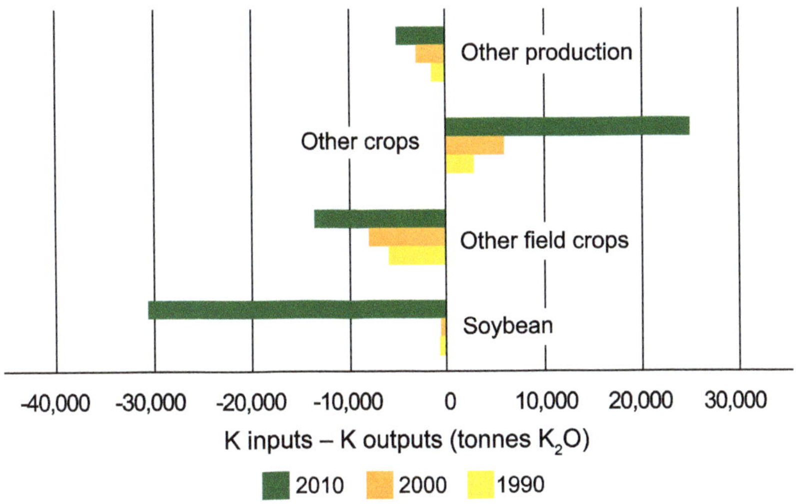

Fig. 11.14 Potassium nutrient balance intensity ($PNBI_K$) by production system in 1990, 2000, and 2010 in Uruguay; other field crops include wheat (*Triticum aestivum* L.), maize (*Zea mays* L.), barley (*Hordeum vulgare* L.), sunflower (*Helianthus annuus* L.), sorghum (*Sorghum bicolor* (L.) Moench), and rice (*Oryza sativa* L.); other crops include fruits, citrus, and vegetables; and other production includes forestry, beef, dairy, and sheep production. (Mancassola and Casanova 2015)

main field crop, along with maize, wheat, sunflower, canola, and others. Causarano Medina (2017) estimated nutrient mass balances for the period between 1996 and 2015, considering K fertilizer imports and K removal with soybean, maize, wheat, rice, sunflower, and canola crop production. Figure 11.16 shows that PNB_K in Paraguay has decreased from 3 to 4 in the period between 1996 and 2003 to a PNB_K between 2 and 3 since 2003. Despite this improvement in PNB_K, it is still much greater than 1, indicating that soil K reserves are being heavily relied upon to provide K for an increasingly productive agriculture. Recent surveys indicate that 58% of soil samples are above soil test K critical levels [> 0.5 cmol(+) kg^{-1} or 195 mg K kg^{-1} (Causarano Medina 2017).

In summary, countries of the Southern Conc of Latin America generally have PNB_K levels > 1, with correspondingly negative $PNBI_K$ values, indicating depletion of soil K. In most of the region, crop production under these negative $PNBI_K$ levels has been sustained from high levels of soil K reserves. However, K deficiencies and positive crop responses to K fertilization have been detected in areas such as Uruguay because of the continuous decline in the exchangeable and non-exchangeable soil K fractions. Site-specific evaluations of soil test K would provide information on the rate of decline of soil K supplies. Lastly, K application decisions based on soil test calibrations would be required to sustain high crop yields and farm profitability and to maintain soil K fertility.

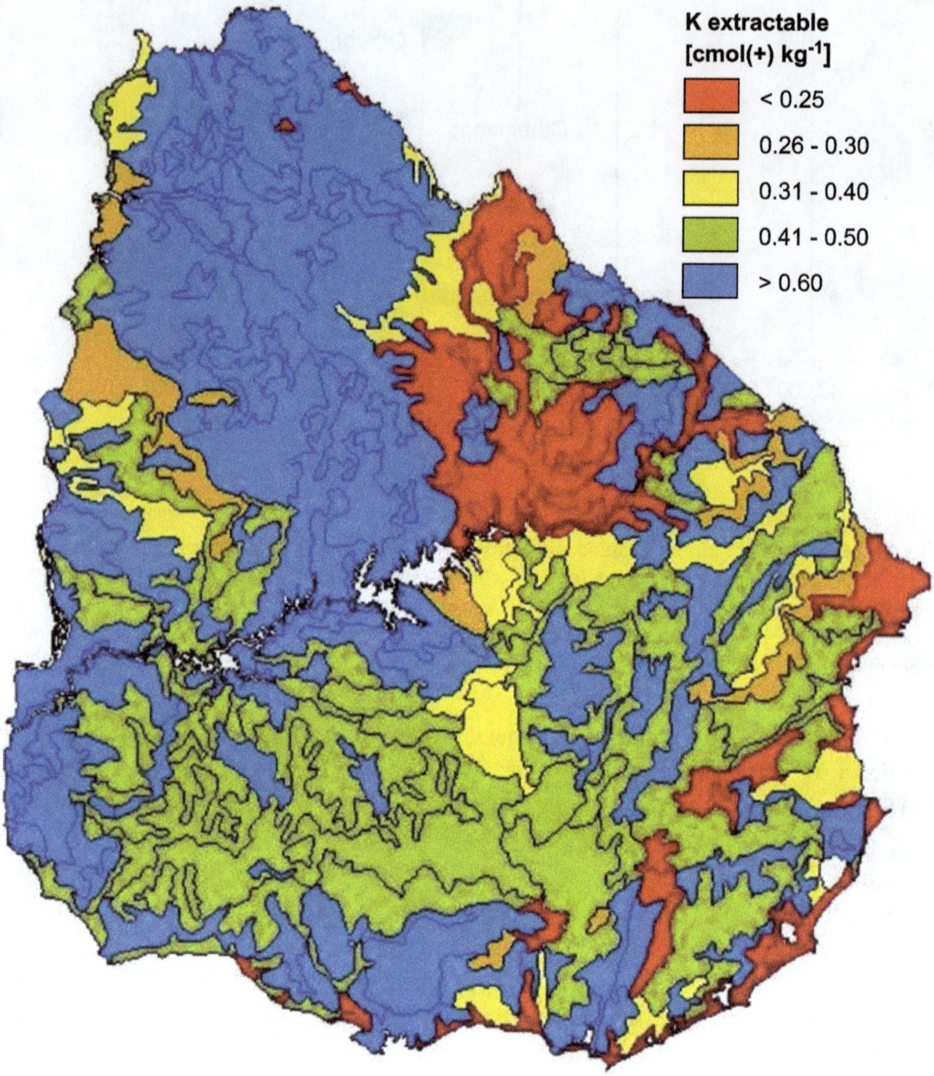

Fig. 11.15 Pristine (pre-agricultural) soil test K concentrations at a depth of 0 to –20 cm, according to the soil recognition guide of Uruguay. (Califra and Barbazan unpublished)

11.11 Conclusion

Estimates of K mass balance provide insight into soil K fertility trends and potential short- and long-term impacts of nutrient management practices in crop production systems. The availability of reliable data to calculate K mass balances differs among geographies. This is reflected in the scale, both spatial and temporal, and resolutions at which K balances are estimated and reported in this chapter. Table 11.18 demonstrates that the quantity of information used to estimate PNB_K and $PNBI_K$ varies among assessments. Common to all, however, are estimates of K removed in

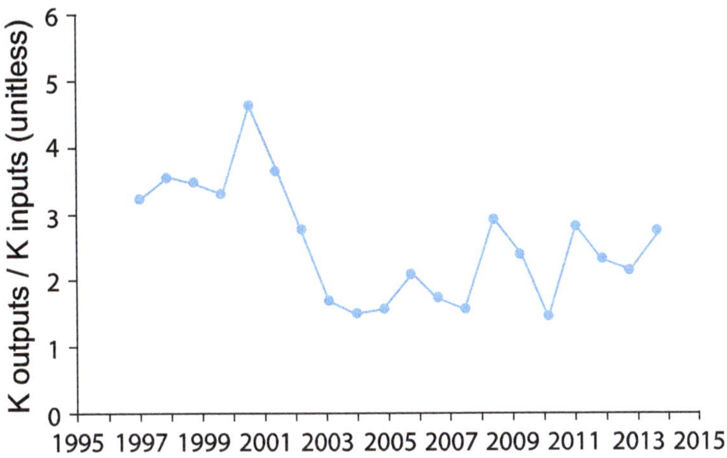

Fig. 11.16 Potassium partial nutrient balance (PNB_K) estimates from 1996 to 2015 for field crop agriculture in Paraguay. (Causarano Medina 2017)

harvested agricultural products and K applied as inorganic fertilizer. Besides the availability and access to quality data, a large diversity of land use and methodological differences in estimating K mass balances are also responsible for divergent metrics reported from various parts of the world. Farmer income levels, access to information on K nutrition of crops, and translation of scientific knowledge to changes in nutrient management practices also play important roles on how K mass balances develop in a specific soil–crop context. This reinforces the concept that nutrient balances are highly contextual and must be interpreted accordingly. Nevertheless, PNB and $PNBI_K$ reported in this chapter, provide a useful global overview of how K is managed in crops, cropping systems, or in mixed enterprises. Nutrient balances are too often reported as snapshots of a year or one cropping cycle, but the trends over time are perhaps the most diagnostic of the sustainability of a given cropping system and its fertilization practices.

Temporal changes in soil K balance over decades at national scales are useful to provide necessary policy guidance for production, distribution and use of K fertilizers. China is such a case. The highly negative $PNBI_K$ levels in soils of China in the 1980s have now changed to generally positive $PNBI_K$ levels, arising from increased awareness at the policy and user level, better access to K fertilizers, and their proper use in the field.

The spectrum of $PNBI_K$ is tilted more toward the negative side globally as more K is removed with each harvest than applied. Potassium applications are typically low in historically high K soils (e.g., countries in the Latin America Southern Cone, North Africa, Western USA), due to an historic lack of crop yield responses. However, crops growing in some soils in these regions are now responding to K fertilizer applications, and crops are more frequently showing deficiency symptoms when no K is applied. A combination of negative $PNBI_K$ and crop response to K fertilizer is an unambiguous indication of K fertility decline.

Table 11.18 Factors considered in K partial nutrient balance (PNB_K) or partial nutrient balance intensity ($PNBI_K$) calculations for the tables and figures in this chapter

		Input						
Fig or table	Mass balance metric (PNB_K or $PNBI_K$)	Inorg. fert.	Org. fert.	Irrigation	Run-off	Erosion	Planted biomass	Atm. deposition
Fig. 11.1	$PNBI_K$	X						
Fig. 11.2	PNB_K	X						
Fig. 11.3	PNB_K	X						
Fig. 11.4	$PNBI_K$	X	X	X			X	X
Fig. 11.5	$PNBI_K$	X	X	X			X	X
Fig. 11.6	$PNBI_K$	X	X					
Fig. 11.7	$PNBI_K$	X	X	X	X	X		X
Fig. 11.8	PNB_K	X	X					
Fig. 11.9	PNB_K	X	X					
Fig. 11.10	PNB_K	X						
Fig. 11.12	PNB_K	X						
Fig. 11.14	$PNBI_K$	X						
Fig. 11.16	PNB_K	X						
Table 11.2	PNB_K, $PNBI_K$	X						
Table 11.5	$PNBI_K$	X						
Table 11.7	$PNBI_K$	X	X	X				
Table 11.8	$PNBI_K$	X						
Table 11.10	$PNBI_K$	X						
Table 11.11	$PNBI_K$	X	X					
Table 11.12	$PNBI_K$	X						
Table 11.13	$PNBI_K$	X						
Table 11.14	PNB_K, $PNBI_K$	X						
Table 11.15	PNB_K, $PNBI_K$	X						

(continued)

Table 11.18 (continued)

Fig or table	Mass balance metric (PNB_K or $PNBI_K$)	Output				
		Harvest	Open burning	Erosion	Run-off	Leaching
Fig. 11.1	$PNBI_K$	X				X
Fig. 11.2	PNB_K	X				
Fig. 11.3	PNB_K	X				
Fig. 11.4	$PNBI_K$	X			X	X
Fig. 11.5	$PNBI_K$	X			X	X
Fig. 11.6	$PNBI_K$	X				
Fig. 11.7	$PNBI_K$	X		X	X	X
Fig. 11.8	PNB_K	X				
Fig. 11.9	PNB_K	X				
Fig. 11.10	PNB_K	X				
Fig. 11.12	PNB_K	X				
Fig. 11.14	$PNBI_K$	X				
Fig. 11.16	PNB_K	X				
Table 11.2	PNB_K, $PNBI_K$	X				
Table 11.5	$PNBI_K$	X				
Table 11.7	$PNBI_K$	X				
Table 11.8	$PNBI_K$	X				
Table 11.10	$PNBI_K$	X				
Table 11.11	$PNBI_K$	X				
Table 11.12	$PNBI_K$	X				
Table 11.13	$PNBI_K$	X				
Table 11.14	PNB_K, $PNBI_K$	X				
Table 11.15	PNB_K, $PNBI_K$	X				

The Indian example is noteworthy in this context. The general perception is Indian soils are inherently rich in K and do not require further K fertilizer applications; however, overwhelming evidence of negative $PNBI_K$ levels at national, regional, and cropping system scales, accompanied by large crop yield losses when no K fertilizer is applied suggest that site-specific K management is necessary to maintain soil K fertility. Negative $PNBI_K$ levels and PNB_K levels greater than unity indicate that a certain portion of crop requirements are derived from native K reserves. Irrespective of the soil K status, this process depletes soil K. How soon the gradual K depletion will impede crop production is a function of the magnitude of the soil K reserve and the level of cropping intensity. Evidence suggests that soils with very high K reserves will eventually become depleted under intensive cropping when adequate K fertilizer is not applied (the PNB_K levels are much greater than 1), giving credence to K mass balance as a reliable indicator of soil K mining.

Acknowledgments The Grains Research and Development Corporation supported the research into Australian farm level nutrient performance indicators reported here and the involvement of Ms Elaina vanderMark and Southern Farming Systems is acknowledged. The estimation of Australian national and regional nutrient performance indicators was done in collaboration with Dr Robert Edis.

References

ABARES (2016) Agricultural commodity statistics 2016, CC by 3.0. https://www.agriculture.gov.au/abares/research-topics/agricultural-commodities/agricultural-commodities-trade-data#2019. Accessed 20 May 2020

ABS (2016) Change in collection scope for ABS rural environment and agricultural commodity collections. In: 4627.0—Land management and farming in Australia, 2014-15. http://www.abs.gov.au/AUSSTATS/abs@.nsf/Lookup/4627.0Main+Features102014-15?OpenDocument. Accessed 20 May 2020

Achard F, Banoin M (2003) Fallows, forage and nutrient transfers by livestock in Niger. Nutr Cycl Agroecosyst 65:183–189. https://doi.org/10.1023/B:FRES.0000019453.19364.70

AICRP-IFS (2012) Annual report 2011-12. Project Directorate for Farming Systems Research (Indian Council of Agricultural Research), Modipuram, Meerut-250 110, pp 224. http://www.iifsr.res.in/AICRP/pdf/AICRP%20on%20IFS_Annual%20Report_2011-12_Final.pdf. Accessed 20 May 2020

Aissa, M, Mhiri M (2002) Détermination du seuil critique du sol en potassium pour le blé dur. Atelier sur la gestion de la fertilisation potassique, acquis et perspectives de la recherche, Tunis 10 Dec. 2002, VI:1–9. https://www.ipipotash.org/uploads/udocs/DETERMINATION%20DU%20SEUIL%20CRITIQUE%20DU%20SOL%20EN%20POTASSIUM%20POUR%20LE%20BLE%20DUR.pdf. Accessed 20 May 2020

ANDA (2010–2015) Anuário estatístico do setor de fertilizantes, São Paulo, Brazil. http://anda.org.br/statistics/. Accessed 20 May 2020

Azzaoui A, Alilou F (1990) Bilan potassique des essais de longues durées de l'INRA. Rapport annuel (1990/91) du Programme Aridoculture, INRA/MIACAID Project 0136

Azzaoui A, El Mourid M, Loudyi B, Ryan J (1993) Fertilité et fertilisation potassique au Maroc: acquis et perspectives d'avenir. Al Awamia 83:241–263. https://www.inra.org.ma/sites/default/files/08314.pdf. Accessed 20 May 2020

Badraoui M, Albani M, Agbani M, Bouabid R, El Gharous M, Karrou M, Zeraouli M (2002) New references for the fertilization of wheat, sugar beet and sunflower in Doukkala and Gharb irrigated perimeters in Morocco. In: Zdruli P, Steduto P, Kapur S (eds) 7th international meeting on soils with Mediterranean type of climate (selected papers), Bari (Italy), CIHEAM 2002, Option Méditerranéennes, Serie A, 50:213–218. http://om.ciheam.org/om/pdf/a50/04002035.pdf. Accessed 20 May 2020

Badraoui M, Agbani M, Bouabid R, Ait Houssa A (2003) Potassium status in soils and crops: fertilizer recommendations and present use in Morocco. In: Johnston AE (ed) Potassium and water management in west Asia and north Africa. The regional workshop of the International Potash Institute, Amman, Jordan, 5–6 November 2001, pp 161–167. https://repository.rothamsted.ac.uk/download/8ba1e172ade49197d9da49b59f921ced78b0f5b3dd93d8fc47163c39c46b3f73/2813429/potassium_and_water_management_in_west_asia_and_north_africa.pdf#page=151. Accessed 20 May 2020

Barbazán M, Bautes C, Beux L, Bordoli M, Califra A, Cano JD, del Pino A, Ernst O, García A, García F, Mazzilli S, Quincke A (2012) Soil potassium in Uruguay: current situation and future prospects. Better Crops 96(4):21–23. http://www.ipni.net/publication/bettercrops.nsf/0/1FFDD7D98D161AA906257ABB005B785A/$FILE/BC%202012-4%20p.%2021.pdf. Accessed 20 May 2020

Barbazán M et al (2017) Assessment of potassium deficiencies in agricultural systems in Uruguay. In: International Plant Nutrition Institute (ed) Abstracts of the frontiers of potassium science conference, Rome, Italy, January 2017. https://www.apni.net/k-frontiers/. Accessed May 29 2020

Bell MJ, Moody P (2005) Chemical fertility and soil health in northern systems. https://grdc.com.au/Media-Centre/Ground-Cover/Ground-Cover-Issue-56/Chemical-fertility-and-soil-health-in-northern-systems. Accessed 20 May 2020

Belouchrani AS, Daoud Y, Derdour H (2002) Study of potassium in Mediterranean red soils of Algeria by biological method. In: Rubio JL, Morgan RPC, Asins S, Andreu A (eds) Proceeding of the 3rd international congress: man and soil at the third millennium. Geoforma ediciones, Logroño, pp 1065–1069

Bijay-Singh, Yadviner-Singh, Imas P, Xie JC (2003) Potassium nutrition of the rice–wheat cropping system. Adv Agron 81:203–259. https://doi.org/10.1016/S0065-2113(03)81005-2

Biswas PP, Sharma PD (2008) A new approach for estimating fertiliser response ratio—the Indian scenario. Indian J Fert 4(7):59–62

Brennan RF, Bell MJ (2013) Soil potassium—crop response calibration relationships and criteria for field crops grown in Australia. Crop Pasture Sci 64(5):514–522. https://doi.org/10.1071/CP13006

Buresh RJ, Pampolino MF, Witt C (2010) Field-specific potassium and phosphorus balances and fertilizer requirements for irrigated rice-based cropping systems. Plant Soil 335:35–64. https://doi.org/10.1007/s11104-010-0441-z

Causarano Medina HJ (2017) Fertilización y balance de nutrientes en la agricultura mecanizada paraguaya. In: Benítez León G et al (eds) IV Congreso Nacional de Ciencias Agrarias. San Lorenzo, Paraguay, 19–21 April 2017, pp 32–35. http://www.agr.una.py/fca/index.php/libros/catalog/book/303. Accessed 20 May 2020

Chand R (2007) Demand for food grains. Econ Polit Wkly 42(52):10–13. https://www.epw.in/journal/2007/52/commentary/demand-foodgrains.html. Accessed 20 May 2020

Chatterjee S, Santra P, Majumdar K, Ghosh D, Das I, Sanyal SK (2015) Geostatistical approach for management of soil nutrients with special emphasis on different forms of potassium considering their spatial variation in intensive cropping system of West Bengal, India. Environ Monit Assess 187:183. https://doi.org/10.1007/s10661-015-4414-9

Christy B, Clough A, Riffkin P, Norton R, Midwood J, O'Leary G, Stott K, Weeks A, Potter T (2015) Managing crop inputs in a high yield potential environment—HRZ of southern

Australia. State of Victoria Department of Economic Development, Jobs, Transport and Resources. Melbourne. https://doi.org/10.13140/RG.2.1.3224.6884

Chuan LM, Zheng HG, Tan CP, Sun SF, Zhang JF (2014) Characteristics of K nutrient input/output and its balance for wheat season in China. Appl Mech Mater 678:720–725. https://doi.org/10.4028/www.scientific.net/AMM.678.720

Corley RHV, Tinker PB (2016) The products of the oil palm and their extraction. In: Corley R, Tinker P (eds) The oil palm. Blackwell, London, pp 445–466. https://doi.org/10.1002/9780470750971.ch13

Correndo A, Ciampitti IA, Rubio G et al (2013) Soil-test potassium dynamics in Mollisols as affected by crop management. In: XVII international plant nutrition colloquium, Istanbul, Turkey

Cunha JF, Casarin V, Prochnow LI (2010) Nutrient budget in Brazilian agriculture in 2008. Informações Agronômicas 130:1–11 (In Portuguese). https://www.npct.com.br/publication/ia-brasil.nsf/CB94A790AA6AB82683257A90000C0822/$File/Page1-11-130.pdf. Accessed 20 May 2020

Cunha JF, Casarin V, Prochnow LI (2011) Nutrient budget in Brazilian agriculture: 1988 to 2010. Informações Agronômicas 135:1–7 (In Portuguese). https://www.npct.com.br/publication/ia-brasil.nsf/9CA193D11CE9775583257A8F005D3F2C/$File/Page1-7-135.pdf. Accessed 20 May 2020

Cunha JF, Francisco EAB, Casarin V, Prochnow LI (2014) Nutrient budget in Brazilian agriculture: 2009 to 2012. Informações Agronômcias 145:1–13 (In Portuguese). https://www.npct.com.br/publication/ia-brasil.nsf/0FAA336F68608D3983257CB30071DE8C/$File/Page1-13-145.pdf. Accessed 20 May 2020

DAC (2011) Agriculture Census Division, Department of Agriculture and Cooperation, Ministry of Agriculture, Government of India. http://inputsurvey.dacnet.nic.in/districttables.aspx. Accessed 20 May 2020

Deckers J (2002) A systems approach to target balanced nutrient management in soilscapes of Sub-Saharan Africa. In: Vanlauwe B, Diels J, Sanginga N, Merckx R (eds) Integrated plant nutrient management in sub-Saharan Africa: from concept to practice. CAB, Wallingford, pp 47–61. https://books.google.ca/books?id=0oP_0c-MjNEC. Accessed 20 May 2020

Dierolf T, Fairhurst T, Mutert E (2001) Soil fertility kit: a toolkit for acid, upland soil fertility management in Southeast Asia. Deutsche Gesellschaft für Technische Zusammenarbeit (GTZ) GmbH; Food and Agriculture Organization; PT Jasa Katom; and Potash & Phosphate Institute (PPI), Potash & Phosphate Institute of Canada (PPIC)

Dobermann, A (2007) Nutrient use efficiency—measurement and management. In: IFA international workshop on fertilizer best management practices. Brussels, Belgium, pp 1–28

Dobermann A, Santa Cruz PC, Cassman KG (1996) Fertilizer inputs, nutrient balance, and soil nutrient supplying power in intensive, irrigated rice systems. I. Potassium uptake and K balance. Nutr Cycl Agroecosyst 46(1):1–10. https://doi.org/10.1007/BF00210219

Donough CR, Cahyo A, Oberthür T, Wandri R, Gerendas J, Rahim GA (2014) Improving nutrient management of oil palms on sandy soils in Kalimantan using the 4R concept of IPNI. Paper presented at the 5th international oil palm conference on the green palm oil for food security and renewable energy, Bali, Indonesia, 17–19 June 2014. http://seap.ipni.net/ipniweb/region/seap.nsf/0/ACD99CDC8F27744A48257D2B0027D3C9/$FILE/05%20OP%20Nutrient%20Mgt%20on%20Sandy%20Soils.pdf. Accessed 20 May 2020

Drew N. (2018) Fertilizer Australia. Personal Communication

Dutta S, Majumdar K, Khurana HS, Sulewski G, Govil V, Satyanarayana T, Johnston A (2013) Mapping potassium budgets across different states of India. Better Crops South Asia 7(1):28–31. http://sap.ipni.net/ipniweb/region/sap.nsf/0/74A459D53F41349E85257C44003364A9/$FILE/Mapping%20Potassium%20Budgets%20Across%20Different%20States%20of%20India_2013.pdf. Accessed 20 May 2020

Edis R, Norton RM, Dassanayake K (2012) Soil nutrient budgets of Australian natural resource management regions. In: Burkitt LL, Sparrow LA (eds) Proceedings of the 5th joint Australian and New Zealand soil science conference: soil solutions for diverse landscapes. Australian Society of Soil Science, Hobart, p 11

El Oumri M (1985) Etude pédologique au l/50.000ème des sols de Abda.Projet intégré de Abda. Dpt. du Milieu Physique. Rabat, INRA, Maro

Elias E, Morse S, Belshaw DRG (1998) Nitrogen and phosphorus balances in Kindo Kiosha farms in southern Ethiopia. Agric Ecosyst Environ 71:93–113. https://doi.org/10.1016/s0167-8809(98)00134-0

EU Nitrogen Expert Panel. (2015) Nitrogen Use Efficiency (NUE) - an indicatro of the utilization of nitrogen in agriculture and food system. Available from: http://www.eunep.com/wp-content/uploads/2017/03/Report-NUE-Indicator-Nitrogen-Expert-Panel-18-12-2015.pdf [04 October 2020]

FAI (2007) Fertiliser statistics 2007. The Fertiliser Association of India, New Delhi, India. https://www.faidelhi.org/statistics/statistical-database. Accessed 20 May 2020

FAI (2008) Fertiliser Statistics 2008. The Fertiliser Association of India, New Delhi, India. https://www.faidelhi.org/statistics/statistical-database. Accessed 20 May 2020

FAI (2011) Fertiliser Statistics 2011. The Fertiliser Association of India, New Delhi, India. https://www.faidelhi.org/statistics/statistical-database. Accessed 20 May 2020

FAI (2016) Fertiliser Statistics 2016. The Fertiliser Association of India, New Delhi, India. https://www.faidelhi.org/statistics/statistical-database. Accessed 20 May 2020

FAOSTAT (2017) Food and Agriculture Organization of the United Nations. http://faostat.fao.org. Accessed 20 May 2020

FAOSTAT (2018) Food and Agriculture Organization of the United Nations. http://faostat.fao.org. Accessed 20 May 2020

Fixen P, Brentrup F, Bruulsema T, Garcia F, Norton R, Zingore S (2015) Nutrient/fertilizer use efficiency: measurement, current situation and trends. In: Drechsel P, Heffer P, Magen H, Mikkelsen R, Wichelns D (eds) Managing water and fertilizer for sustainable agricultural intensification. International Fertilizer Industry Association (IFA), International Water Management Institute (IWMI), International Plant Nutrition Institute (IPNI), and International Potash Institute (IPI). Paris, France. pp 8–37. http://www.iwmi.cgiar.org/Publications/Books/PDF/managing_water_and_fertilizer_for_sustainable_agricultural_intensification.pdf. Accessed 20 May 2020

Fixen P, Williams R, Rund Q. (2012) Nugis: a nutrient use information system (NuGIS) for the U. S. http://www.ipni.net/ipniweb/portal.nsf/0/5D3B7DFAFC8C276885257743. Accessed 04 Oct 2020

Francisco EAB, Cunha JF, Prochnow LI, Cassarin V (2015) A look at the nutrient budget for Brazilian agriculture. Better Crops 99:4–6. http://www.ipni.net/publication/bettercrops.nsf/0/4AA9B3E5A82C5C5F85257E4C0062CB6D/$FILE/BC-2015-2.pdf. Accessed 20 May 2020

Gami SK, Ladha JK, Pathak H, Shah M, Pasuquin E, Pandey S, Hobbs P, Joshy D, Mishra R (2001) Long-term changes in yield and soil fertility in a twenty-year rice-wheat experiment in Nepal. Bio Fert Soils 34:73–78. https://doi.org/10.1007/s003740100377

García FO, González Sanjuan MF (2016) Consumo de fertilizantes en el mundo y en la Argentina. In: Lavado, R (ed) Sustentabilidad de los agrosistemas y uso de fertilizantes. Orientacion Grafica Editora, Buenos Aires, Argentina. http://lacs.ipni.net/ipniweb/region/lacs.nsf/e0f085ed5f091b1b852579000057902e/251e0b2ce526f8b1032580360060025a/$FILE/Garcia%20y%20Gonzalez%20Sanjuan%20-%20Consumo%20de%20fertilizantes%20en%20el%20mundo%20y%20en%20la%20Argentina.pdf. Accessed 20 May 2020

Ghanem H, Amnai L, Azzaoui M, Bouksirat H, El Gharous M, Oubahammou S (1983) Nitrogen, phosphate, potassium and management of arid and semi-arid soils of Morocco (preliminary results of a large-scale research project). In: Nutrient balance and the need of fertilizers in semi-arid and arid regions. In: Proceedings of the 17th colloquium of the International Potash Institute, Rabat and Marrakech, Morocco, pp 259–278. https://www.ipipotash.org/uploads/udocs/nutrient_balances_and_the_need_for_fertilizers_in_semi-arid_and_arid_regions.pdf. Accessed 20 May 2020

Giller KE, Rowe EC, de Ridder N, van Keulen H (2006) Resource use dynamics and interactions in the tropics: scaling up in space and time. Agric Syst 88(1):8–27. https://doi.org/10.1016/j.agsy.2005.06.016

GOI (2020) Web based land use statistics information system, Agriculture Information Division, Ministry of Communication & IT, Govt. of India, New Delhi. https://aps.dac.gov.in/LUS/Index.htm. Accessed 20 May 2020

Gourley CJP, Dougherty WJ, Weaver DM, Aarons SR, Awty IM, Gibson DM, Hannah MC, Smith AP, Peverill KI (2012) Farm-scale nitrogen, phosphorus, potassium and sulfur balances and use efficiencies on Australian dairy farms. Anim Prod Sci 52:929–944. https://doi.org/10.1071/AN11337

Halilat MT (2004) Effect of potash and nitrogen fertilization on wheat under Saharan conditions. Paper presented at the International Potash Institute regional workshop on potassium and fertigation development in west Asia and north Africa, Rabat, Morocco, 24–28 November 2004. https://www.ipipotash.org/uploads/udocs/Effect%20of%20Potash%20and%20Nitrogen%20Fertilization%20on%20Wheat%20under%20Saharan%20Conditions.pdf. Accessed 20 May 2020

Hamza A, Bamouh A, El Guilli M, Bouabid R (2012) Response of clementine citrus var. *Cadoux* to foliar potassium fertilization; Effects on fruit production and quality. Int Potash Insti ifc 31:14–28. https://www.ipipotash.org/publications/eifc-241. Accessed 20 May 2020

He P, Li ST, Jin JY, Wang HT, Li CJ, Wang YL, Cui RZ (2009) Performance of an optimized nutrient management system for double-cropped wheat–maize rotations in north-central China. Agron J 101(6):1489–1496. https://doi.org/10.2134/agronj2009.0099

He P, Yang LP, Xu XP, Zhao SC, Chen F, Li ST, Tu SH, Jin JY, Johnston AM (2015) Temporal and spatial variation of soil available potassium in China (1990–2012). Field Crop Res 173:49–56. https://doi.org/10.1016/j.fcr.2015.01.003

Heffer P (2013) Assessment of fertilizer use by crop at the global level: 2010-2010/11. International Fertilizer Industry Association. Paris, France. https://www.fertilizer.org/images/Library_Downloads/AgCom.13.39%20-%20FUBC%20assessment%202010.pdf. Accessed 20 May 2020

Henao J, Baanante C (2006) Agricultural production and soil nutrient mining in Africa: implications for resource conservation and policy development. IFDC International Center for Soil Fertility and Agricultural Development, Muscle Shoals. https://allafrica.com/download/resource/main/main/idatcs/00010778:649998c684810e9ebb44cc5c59e24454.pdf. Accessed 20 May 2020

Huang SW, Jin JY, Bai YL, Yang LP (2007) Evaluation of nutrient balance in soil–vegetable system using nutrient permissible surplus or deficit rate. Commun Soil Sci Plan 38 (7-8):959–974. https://doi.org/10.1080/00103620701277973

IBGE (Instituto Brasileiro de Geografia e Estatística) (2010–2015). n.23–26. https://www.ibge.gov.br/estatisticas/economicas/agricultura-e-pecuaria.html. Accessed 20 May 2020

IFA (2020) Regional nutrient use efficiency trends and sustainable fertilizer management, synthesis papers. https://www.ifastat.org/nutrientuse- efficiency. Accessed 04 Oct 2020

Kuznets S (1955) Economic growth and income inequality. Am Econ Rev 45:1–28. www.jstor.org/stable/1811581. Accessed 20 May 2020

Ladha JK, Dawe D, Pathak H, Padre AT, Yadav RL, Singh B, Singh Y, Singh Y, Singh P, Kundu AL, Sakal R, Ram N, Regmi AP, Gami SK, Bhandari AL, Amin R, Yadav CR, Bhattarai EM, Das S, Aggarwal HP, Gupta RK, Hobbs PR (2003) How extensive are yield declines in longterm rice wheat experiments in Asia? Field Crop Res 81:159–180. https://doi.org/10.1016/S0378-4290(02)00219-8

Laoubi K, Yamao M (2012) The challenge of agriculture in Algeria: are policies effective? Agric Fish Econ Res 12:65-73. http://home.hiroshima-u.ac.jp/food0709/seika/seika2012-3.pdf. Accessed 20 May 2020

Lassaletta L, Billen G, Grizzetti B, Anglade J, Garnier J (2014) 50 year trends in nitrogen use efficiency of world cropping systems: the relationship between yield and nitrogen input to cropland. Environ Res Lett 9(10):105011. https://doi.org/10.1088/1748-9326/9/10/105011

Li S, Duan Y, Guo T, Zhang P, He P, Johnson A, Scherbakov A (2015) Potassium management in potato production in Northwest region of China. Field Crop Res 174:48–54. https://doi.org/10.1016/j.fcr.2015.01.010

Lin B (1989) Application of chemical fertilizers in China. Beijing Science and Technology Press, Beijing

Liu RL, Jin JY, Wu RG et al (2000) Potassium balance in soil crop system and effectiveness of potash fertilizer in north China. II: yield responses on main crops to potash fertilizer application. Soil Fertil Sci China 1:9–11. (In Chinese)

Liu YX, Ma JC, Ding WC, He WT, Lei QL, Gao Q, He P (2017a) Temporal and spatial variation of potassium balance in agricultural land at national and regional levels in China. PloS One 12(9): e0184156. https://doi.org/10.1371/journal.pone.0184156

Liu YX, Yang JY, He WT, Ma JC, Gao Q, Lei QL, He P, Wu HY, Ullah S, Yang FQ (2017b) Provincial potassium balance of farmland in China between 1980 and 2010. Nutr Cycl Agroecosyst 107(2):247–264. https://doi.org/10.1007/s10705-017-9833-2

Majumdar K, Kumar A, Shahi V, Satyanarayana T, Jat ML, Kumar D, Pampolino M, Gupta N, Singh V, Dwivedi BS, Meena MC, Singh VK, Kamboj BR, Sidhu HS, Johnston A (2012) Economics of potassium fertiliser application in rice, wheat and maize grown in the Indo-Gangetic Plains. Indian J Fert 8(5):44–53. https://repository.cimmyt.org/handle/10883/1573. Accessed 20 May 2020

Majumdar K, Sanyal SK, Dutta SK, Satyanarayana T, Singh VK (2016) Nutrient mining: addressing the challenges to soil resources and food security. In: Singh U, Praharaj C, Singh S, Singh N (eds) Biofortification of food crops, Springer, New Delhi, India, pp 177–198, ISBN: 978-81-322-2714-4. https://doi.org/10.1007/978-81-322-2716-8_14

Mancassola V, Casanova O (2015) Balance de nutrientes de los principales productos agropecuarios de Uruguay para los años 1990, 2000 y 2010. Informaciones Agronomicas de Hispanoamerica 17:2–13

Mazid Miah MA, Saha PK, Islam A, Nazmul Hasan M, Nosov V (2008) Potassium fertilization in rice-rice and rice-wheat cropping system in Bangladesh. Bangladesh J Agric Environ 4:51–67

McLaughlin M, Fillery IR, Till AR (1992) Operation of the phosphorus, sulphur and nitrogen cycles. In: Gifford RM, Barson MM (eds) Australia's renewable resources: sustainability and global change. Bureau of Rural Resources and CSIRO Division of Plant Industry, Canberra, pp 67–110. http://hdl.handle.net/102.100.100/248737?index=1. Accessed 20 May 2020

Mhiri A (2002) Le potassium dans les sols de Tunisie. In: Atelier sur la gestion de la fertilisation potassique: acquis et perspectives de la recherche, II.1-II.13. https://www.ipipotash.org/uploads/udocs/LE%20POTASSIUM%20DANS%20LES%20SOLS%20DE%20TUNISIE.pdf. Accessed 20 May 2020

Mugwira LM, Murwira HK (1998) A review of manure as a soil fertility amendment in Zimbabwe: some perspectives. In: Waddington SR, Murwira HK, Kumwenda JDT, Hikwa D, Tagwira F (eds) Soil fertility research for maize-based farming systems in Malawi and Zimbabwe, Proceedings of the soil fertility network results and planning workshop, Mutare, Zimbabwe 7–11 July 1997. Soil FertNet and CIMMYT-Zimbabwe, Harare, pp 195–201 https://repository.cimmyt.org/bitstream/handle/10883/539/66144.pdf?sequence=1&isAllowed=y. Accessed 20 May 2020

NAAS (2006) Low and declining crop response to fertilizers. Policy paper no. 35, National Academy of Agricultural Sciences (NAAS), New Delhi, p 8. http://naasindia.org/Policy%20Papers/policy%2035.pdf. Accessed 20 May 2020

Naidu LGK, Ramamurthy V, Sidhu GS, Sarkar D (2011) Emerging deficiency of potassium in soils and crops of India. Karnataka J Agric Sci 24:12–19

NAL (2020) Soil nutrient balance. National Agricultural Library (NAL) thesaurus and glossary. https://agclass.nal.usda.gov/glossary.shtml. Accessed 20 May 2020

National Land and Water Resources Audit 2001. Nutrient balance in regional farming systems and soil nutrient status. National Land and Water Resources Audit, Final Report, September 2001. National Heritage Trust, Australia

Niu JF, Zhang WF, Chen XP, Li CJ, Zhang FS, Jiang LH, Liu ZH, Xiao K, Assaraf M, Imas P (2013) Potassium fertilization on maize under different production practices in the North China Plain. Agron J 103(3):822–829. https://doi.org/10.2134/agronj2010.0471

Nkonya E, Kaizzi C, Pender J (2005) Determinants of nutrient balances in a maize farming system in eastern Uganda. Agric Syst 85(2):155–182. https://doi.org/10.1016/j.agsy.2004.04.004

NLWA (2001) Nutrient balance in regional farming systems and soil nutrient status. National Heritage Trust, Canberra. p 89. http://lwa.gov.au/programs/national-land-and-water-resources-audit. Accessed 20 May 2020

Norton RM (2012) Wheat grain nutrient concentrations for south-eastern Australia. In: Yunusa I (ed) Capturing opportunities and overcoming obstacles in Australian agronomy. In: Proceedings of the 16th Australian agronomy conference, Armidale, NSW 14–18 Oct 2012

Norton RM (2014a) Canola seed nutrient concentrations for southern Australia. In: Ware AH, Potter TD (eds) Proceedings of the 18th Australian research assembly on brassicas (ARAB 18). Tanunda, Australia, 29 September – 2 October, 2014. Australian Oilseed Federation, pp 1–6. http://anz.ipni.net/ipniweb/region/anz.nsf/0/0C598E0085BBA369CA257D69000B5FD9/$FILE/Norton%20ARAB%20Canola%20Paper.pdf. Accessed 20 May 2020

Norton RM (2014b) Do we need to revisit potassium? In: Proceedings of the grains research and development corporation advisor updates, Adelaide, Australia, 25–26 February 2014, pp 199–204. https://grdc.com.au/resources-and-publications/grdc-update-papers/tab-content/grdc-update-papers/2014/02/do-we-need-to-revisit-potassium. Accessed 20 May 2020

Norton RM (2017) Potassium removal and use in Australia. Murrell TS, Mikkelsen RL (eds) Frontiers of potassium science, Rome, Italy, 25–27 Jan 2017. International Plant Nutrition Institute, Peachtree Corners, GA, pp 33–44. https://www.apni.net/k-frontiers/. Accessed May 29 2020

Norton RM, Davidson E, Roberts TL (2015) Nitrogen use efficiency and nutrient performance indicators. GPNM task team report and recommendations, position paper from the GPNM's task team workshop, Washington, DC, 8 Dec 2014. http://wedocs.unep.org/bitstream/handle/20.500.11822/10750/Nutrient_use.pdf?sequence=1&isAllowed=y. Accessed 20 May 2020

Norton RM, vanderMark E (2016) Nitrogen performance indicators for southern Australian grain farms. Proceedings of the International Nitrogen Conference, (Ed J Angus), Melbourne Australia, December 2016. http://www.proceedings.com.au/nitrogen-performance-indicators-on-southern-australian-grain-farms/ [04 October 2020]

NPCT (2020) Nutrição de Plantas Ciência e Tecnologia. Av. Independência, 350, Sala 141A, Piracicaba, SP, BRASIL https://www.npct.com.br/npctweb/npct.nsf/article/calculadora. Accessed 20 May 2020

Öborn I, Edward AC, Witter E, Oenema O, Ivarsson K, Withers PJA, Nilsson SI, Stinzing AR (2003) Element balances as a tool for sustainable nutrient management: a critical appraisal of their merits and limitations within an agronomic and environmental context. Eur J Agron 20:211–225. https://doi.org/10.1016/S1161-0301(03)00080-7

OzDSM (2020) Australian Digital Soil Mapping. http://www.ozdsm.com.au/ozdsm_map.php. Accessed 20 May 2020

Pasuquin JM, Pampolino MF, Witt C, Dobermann A, Oberthür T, Fisher MJ, Inubushi K (2014) Closing yield gaps in maize production in Southeast Asia through site-specific nutrient management. Field Crop Res 156:219–230. https://doi.org/10.1016/j.fcr.2013.11.016

Patra AK, Dutta SK, Dey P, Majumdar K (2017) Potassium fertility status of Indian soils: national soil health card database highlights the increasing potassium deficit in soils. Indian J Fert 13 (11):28–33

Pérez Zamora (2015) La fetilización de la caña de azucar. In: Echeverría HE and García FO (eds) Fertilidad de suelos y fertilización de cultivos, ed. INTA. 2a. ed. Buenos Aires, Argentina, pp 609–630. ISBN: 9-789875-215658.

Regmi A, Ladha JK, Pasuquin E, Pathak H, Hobbs P, Shrestha L, Gharti D, Duveiller E (2002) The role of potassium in sustaining yields in a long-term rice–wheat experiment in the Indo-Gangetic Plains of Nepal. Biol Fert Soils 36:240–247. https://doi.org/10.1007/s00374-002-0525-x

Roberts TR, Majumdar K (2017) Global P & K use and balance in world agriculture. Indian J Fert 13(4):32–39

Sainz Rozas H, Eyherabide M, Echeverría HE, Barbieri P, Angelini H, Larrea GE, Ferraris GN, Barraco M (2013) ¿Cuál es el estado de la fertilidad de los suelos argentinos? In: Garcia F, Correndo A (eds) Symposio fertilidad 2013, IPNI Southern Cone-Fertilizar AC, Rosario, Argentina, pp 62–72. https://inta.gob.ar/documentos/¿cual-es-el-estado-de-la-fertilidad-de-los-suelos-argentinos. Accessed 20 May 2020

Satyanarayana T, Tewatia RK (2009) State wise approaches to crop nutrient balances in India. In: Proceedings of the IPI-OUAT-IPNI international symposium on potassium role and benefits in improving nutrient management for food production, quality and reduced environmental damages, Bhubaneswar, Orissa, India, 5–7 Nov 2009, pp 467–485

Setiyono TD, Walters DT, Cassman KG, Witt C, Dobermann A (2010) Estimating maize nutrient uptake requirements. Field Crop Res 118(2):158–168. https://doi.org/10.1016/j.fcr.2010.05.006

Sheldrick WF, Syers JK, Lingard J (2003) Soil nutrient audits for China to estimate nutrient balances and output/input relationships. Agric Ecosyst Environ 94(3):341–354. https://doi.org/10.1016/S0167-8809(02)00038-5

Shen RP, Sun B, Zhao QG (2005) Spatial and temporal variability of N, P and K balances for agroecosystems in China. Pedosphere 15(3):347–355

Shepherd KD, Soule MJ (1998) Soil fertility management in west Kenya: dynamic simulation of productivity, profitability and sustainability at different resource endowment levels. Agr Ecosyst Environ 71(1-3):131–145. https://doi.org/10.1016/S0167-8809(98)00136-4

Singh M, Singh VP, Reddy DD (2002) Potassium balance and release kinetics under continuous rice-wheat cropping system in Vertisol. Field Crops Res 77:81–91. https://doi.org/10.1016/S0378-4290(01)00206-4

Singh VK, Tiwari R, Gill MS, Sharma SK, Tiwari KN, Dwivedi BS, Shukla AK, Mishra PP (2008) Economic viability of site-specific nutrient management in rice-wheat cropping. Better Crops India 2:16–19

Singh VK, Dwivedi BS, Buresh RJ, Jat ML, Majumdar K, Gangwar B, Govil V, Singh SK (2013) Potassium fertilisation in rice–wheat system across northern India: crop performance and soil nutrients. Agron J 105(2):471–481. https://doi.org/10.2134/agronj2012.0226

Singh VK, Dwivedi BS, Tiwari KN, Majumdar K, Rani M, Singh SK, Timsina J (2014) Optimizing nutrient management strategies for rice-wheat system in the Indo-Gangetic Plains of India and adjacent region for higher productivity, nutrient use efficiency and profits. Field Crop Res 164:30–44. https://doi.org/10.1016/j.fcr.2014.05.007

Singh VK, Shukla AK, Dwivedi BS, Singh MP, Majumdar K , Kumar V, Mishra R, Rani M, Singh SK (2015a) Site-specific nutrient management under rice-based cropping systems in Indo-Gangetic Plains: yield, profit and apparent nutrient balance. Agric Res 4:365–377. https://doi.org/10.1007/s40003-015-0179-1

Singh VK, Shukla AK, Singh MP, Mujumdar K, Mishra RP, Rani M, Singh SK (2015b) Effect of site-specific nutrient management on yield, profit and apparent nutrient balance under pre-dominant cropping systems of Upper Gangetic Plains. Indian J Agric Sci 85:335–343

Singh VK, Dwivedi BS, Yadvinder-Singh SSK, Mishra RP, Shukla AK, Rathore SS, Shekhawat K, Majumdar K, Jat ML (2018) Effect of tillage and crop establishment, residue management and K fertilization on yield, K use efficiency and apparent K balance under rice maize system in north-western India. Field Crops Res 224:1–12. https://doi.org/10.1016/j.fcr.2018.04.012

Smaling EMA, Braun AR (1996) Soil fertility research in sub-Saharan Africa: new dimensions, new challenges. Commun Soil Sci Plan 27(3-4):365–386. https://doi.org/10.1080/00103629609369562

Stitou M (1985) Etude pédologique de la région de Settat et Ben Ahmed. Rapport du marché no. 46182/DPN42. DPA, Settat, Rabat

Stoorvogel JJ, Smaling EMA, Janssen BH (1993) Calculating soil nutrient balances in Africa at different scales. Fert Res 35(3):227–235. https://doi.org/10.1007/BF00750641

Swift MJ, Frost PGH, Campbell BM, Hatton JC, Wilson KB (1989) Nitrogen cycling in farming systems derived from savanna: perspectives and challenges. In: Clarholm M, Bergström L (eds) Ecology of arable land—perspectives and challenges, Developments in plant and soil sciences, vol 39. Springer, Dordrecht, pp 63–76. https://doi.org/10.1007/978-94-009-1021-8_7

Tan DS, Jin JY, Jiang LH, Huang SW, Liu ZH (2012) Potassium assessment of grain producing soils in north China. Agr Ecosyst Environ 148:65–71. https://doi.org/10.1016/j.agee.2011.11.016

Tandon HLS (2004) Fertilizers in Indian agriculture—from 20th to 21st century. Fertilizer Development and Consultation Organisation, New Delhi

Tarmizi AM, Mohd Tayeb D (2006) Nutrient demands of Tenera oil palm planted on inland soils of Malaysia. J Oil Palm Res 18:204–209. http://palmoilis.mpob.gov.my/publications/joprv18june-tarmizi.pdf. Accessed 20 May 2020

Tewatia RK, Rattan RK, Bhende S, Kumar L (2017) Nutrient use and balances in India with special reference to phosphorus and potassium. Indian J Fert 13(4):20–29

Timsina J, Singh VK, Majumdar K (2013) Potassium management in rice–maize systems in South Asia. J Plant Nutr Soil Sci 176(3):317–330. https://doi.org/10.1002/jpln.201200253

Tittonell P, Vanlauwe B, Leffelaar PA, Rowe EC, Giller KE (2005) Exploring diversity in soil fertility management of smallholder farms in western Kenya. II. Within-farm variability in resource allocation, nutrient flows and soil fertility status. Agr Ecosyst Environ 110 (3-4):166–184. https://doi.org/10.1016/j.agee.2005.04.003

Tiwari KN, Dwivedi BS, Subba Rao A (1992) Potassium management in rice-wheat system. In: Pandey RK et al (eds) Rice–wheat cropping system: Proceedings of the rice–wheat workshop, Modipuram, Meerut. Project Directorate for Cropping Systems Research, Modipuram, Meerut, India, pp 93–114

Tiwari KN, Sharma SK, Singh VK., Dwivedi BS, Shukla AK (2006) Site-specific nutrient management for increasing crop productivity in India: results with rice-wheat and rice-rice system. PDCSR Modipuram and PPIC India Programme, Gurgaon, pp. 92

Trelo-ges V, Limpinuntana V, Patanothai A (2004) Nutrient balances and sustainability of sugarcane fields in a mini-watershed agroecosystem of Northeast Thailand. Southeast Asian Stud 41 (4):473–490. https://pdfs.semanticscholar.org/c154/3228177ddfb17d998d1458761c3a7dc6faa2.pdf?_ga=2.40050337.2057332575.1576787027-249372282.1576787027. Accessed 20 May 2020

USDA FAS (2017) Production, supply and distribution database. United States Department of Agriculture Foreign Agriculture Service. https://apps.fas.usda.gov/psdonline/app/index.html#/app/home. Accessed 20 May 2020

Van den Bosch H, de Jager A, Vlaming J (1998) Monitoring nutrient flows and economic performance in African farming systems (NUTMON): II. Tool development. Agr Ecosyst Environ 71(1-3):49–62. https://doi.org/10.1016/S0167-8809(98)00131-5

Vanlauwe B, Diels J, Lyasse O, Aihou K, Iwuafor ENO, Sanginga N, Merckx R, Deckers J (2002) Fertility status of soils of the derived savanna and northern Guinea savanna and response to major plant nutrients, as influenced by soil type and land use management. Nutr Cycl Agroecosyst 62:139–150. https://doi.org/10.1023/A:1015531123854

Wang HJ, Huang B, Shi XZ, Darilek JL, Yu DS, Sun WX, Zhao YC, Chang Q, Öborn I (2008) Major nutrient balances in small-scale vegetable farming systems in peri-urban areas in China. Nutr Cycl Agroecosyst 81(3):203–218. https://doi.org/10.1007/s10705-007-9157-8

Wang XL, Feng AP, Wang Q, Wu CQ, Liu Z, Ma ZS, Wei XF (2014) Spatial variability of the nutrient balance and related NPSP risk analysis for agro-ecosystems in China in 2010. Agric Ecosyst Environ 193:42–52. https://doi.org/10.1016/j.agee.2014.04.027

Wihardjaka A, Kirk GJD, Abdulrachman S, Mamaril CP (1999) Potassium balances in rainfed lowland rice on a light-textured soil. Field Crop Res 64(3):237–247. https://doi.org/10.1016/S0378-4290(99)00045-3

Wingeyer AB, Amado TJC, Pérez-Bidegain M, Studdert GA, Varela CHP, García FO, Karlen DL (2015) Soil quality impacts of current South American agricultural practices. Sustainability 7 (2):2213–2242. https://doi.org/10.3390/su7022213

Yadav RL, Prasad K, Gangwar KS (1998) Prospects of Indian agriculture with special reference to nutrient management under irrigated systems. In: Swarup A et al (eds) Long term fertilizer management through integrated plant nutrient supply. Indian Institute of Soil Science, Bhopal, India, pp 1–325

Yadvinder-Singh, Bijay-Singh, Ladha JK, Khind CS, Khera TS, Bueno CS (2004) Effects of residue decomposition on productivity and soil fertility in rice–wheat rotation. Soil Sci Soc Am J 68:854–864. https://doi.org/10.2136/sssaj2004.8540

Yadvinder-Singh, Bijay-Singh, Timsina J (2005) Crop residue management for nutrient cycling and improving soil productivity in rice-based cropping systems in the tropics. Adv Agron 85:269–407. https://doi.org/10.1016/S0065-2113(04)85006-5

Yamada T, Lopes AS (1998) Nutrient budget in Brazilian agriculture. Informações Agronômicas 84:1–8. (In Portuguese). https://www.npct.com.br/publication/ia-brasil.nsf/issue/BRS-1998-84. Accessed 20 May 2020

Zerouali M, Mrini M (2004) Fertilité des sols et fertilisation potassique des principales cultures dans la region du Gharb (Maroc): développement de la fertigation. In: Badraoui M, Bouabid R, Ait-Houssa A (eds) International Potash Institute (IPI) regional workshop on potassium and fertigation development in west Asia and north Africa; Rabat, Morocco, 24–28 Nov 2004, pp 1–9. https://www.ipipotash.org/uploads/udocs/Fertilite%20des%20Sols%20et%20Fertilisation%20Potassique%20des.pdf. Accessed 20 May 2020

Zhang HM, Xu MG, Shi XJ, Li ZH, Huang QH, Wang XJ (2010) Rice yield, potassium uptake and apparent balance under long-term fertilization in rice-based cropping systems in southern China. Nutr Cycl Agroecosyst 88(3):341–349. https://doi.org/10.1007/s10705-010-9359-3

Zhang X, Davidson EA, Mauzerall DL, Searchinger TD (2015) Managing nitrogen for sustainable development. Nature 528:51–57. https://doi.org/10.1038/nature15743

Zingore S, Murwira HK, Delve RJ, Giller KE (2007) Influence of nutrient management strategies on variability of soil fertility, crop yields and nutrient balances on smallholder farms in Zimbabwe. Agric Ecosyst Environ 119:112–126. https://doi.org/10.1016/j.agee.2006.06.019

12

Soil Potassium (K) Fertilization Strategies

Michael J. Bell, Antonio P. Mallarino, Jeff Volenec, Sylvie Brouder, and David W. Franzen

Abstract Placement strategies can be a key determinant of efficient use of applied fertilizer potassium (K), given the relative immobility of K in all except the lightest textured soils or high rainfall environments. Limitations to K accessibility by plants caused by immobility in the soil are further compounded by the general lack of K-stimulated root proliferation in localized soil zones enriched with K alone, compared with root proliferation due to concentrated N and P. Further, effects of K fixation reactions in soils with certain clay mineralogies and the declining concentration and activity of soil solution K with increasing clay content can also limit plant K acquisition. Variation in root system characteristics among crops in a rotation sequence and fluctuating soil moisture conditions in fertilized soil horizons in rain-fed systems increase the complexity of fertilizer placement decisions to ensure efficient K recovery and use. This complexity has resulted in extensive exploration of fertilizer K application strategies, with this chapter focusing on K applications to the soil. Issues discussed include comparisons of broadcast versus banded applications, depth of fertilizer placement, and the impacts of co-location of K with other nutrients. While research findings are often specific to the crop, soil, and seasonal conditions under which they are conducted, we attempt to identify strategies that most consistently deliver improved crop recovery and utilization of fertilizer K.

M. J. Bell (✉)
School of Agriculture and Food Sciences, The University of Queensland, Brisbane, QLD, Australia
e-mail: m.bell4@uq.edu.au

A. P. Mallarino
Department of Agronomy, Iowa State University, Ames, IA, USA
e-mail: apmallar@iastate.edu

J. Volenec · S. Brouder
Department of Agronomy, Purdue University, West Lafayette, IN, USA
e-mail: jvolenec@purdue.edu; sbrouder@purdue.edu

D. W. Franzen
Soil Science Department, North Dakota State University, Fargo, ND, USA
e-mail: david.franzen@ndsu.edu

12.1 Introduction

Plants typically accumulate potassium (K) in similar quantities to nitrogen (N), with the potential for luxury accumulation of plant K resulting in greater K accumulation than N in some situations. The scale of crop K requirements and the time-critical nature of K uptake, with maximum uptake rates often well in advance of biomass accumulation, means that soil K availability and appropriate fertilizer application methods are critical to ensure adequate crop K nutrition. While foliar applications of K are practiced in the culture of some crops including cotton (Coker et al. 2009) and some horticultural crops (Jifon and Lester 2011) fertilizer K applications are typically limited to supplementing K uptake from the soil, with the quantity of foliar K supplied relatively small compared to total crop K accumulation. Foliar application of K is discussed explicitly in Chap. 13 of this book, while this chapter focuses on soil K fertilization strategies.

12.2 Factors Affecting Root Access to Zones of K Enrichment

The main factors affecting the efficiency of applied K recovery involve: (1) the interactions between crop root systems and the soil physical and chemical properties that affect the movement of K to plant roots and (2) the replenishment of depleted soil solution K concentrations in response to plant K uptake. Plant root factors are discussed in detail in Chap. 4, but relate primarily to the temporal coincidence of active crop roots and K-enriched soil profile layers, the proportion of the crop root system that is in the enriched zone, the extent to which those roots can deplete soil solution K concentration and/or exploit non-exchangeable soil K. The mobility of K through the soil profile, and hence the possible expansion of the zone of K enrichment beyond the original fertilized soil volume, is an important factor to consider in order to understand the interactions between soil-applied K and crop root systems that collectively determine plant K uptake.

Important physical properties such as pore size and pore continuity influence the diffusion path length (tortuosity) and hence the rate of diffusive resupply of soil solution K depleted by plant uptake. Chemical factors include those that influence the impact of applied K on the activity of K in the soil solution (i.e., the K buffer capacity—BC_K), which is a function of the number and specificity of potential K sorption sites (estimated by measurement of cation exchange capacity) and the presence of clay minerals that can fix (usually temporarily) some of the applied K. These chemical factors also influence the relative importance of diffusion and mass flow in meeting crop K requirements.

12.2.1 Crop Root Distribution

Underlying genetics determine the potentially different patterns of root distribution among species and genotypes. Some of those differences are fundamental, such as the contrast in root system morphology and distribution between fibrous-rooted monocots like wheat (*Triticum aestivum* L.) and barley (*Hordeum vulgare* L.) and tap-rooted dicots like cotton (*Gossypium spp.*) and grain legumes, as illustrated by Gulick et al. (1989). In that study, barley exhibited much higher root length densities than cotton, especially in the uppermost 12 cm of the soil columns. Cotton was characterized by low root length densities throughout the soil profile. Finer-level differences between genotypes can also have important implications for functional traits like accessing water stored deep in the soil profile (e.g., Liakat et al. 2015) and potentially for responding to nutrient-rich patches like fertilizer bands or layers in cropped fields. Further, substantive K acquisition differences among cultivars have been linked to the root surface area (Brouder and Cassman 1990), although such studies remain sparse. However, the contribution of these differences to root system function and the crop it supports will be largely determined by the interaction of soil characteristics and seasonal conditions in the field (Rich and Watt 2013). As a result, the combination of soil type and seasonal conditions, modified by management inputs such as tillage system and irrigation, can have major impacts on the root distribution of the same genotype of a given species. Statistical analyses of many studies suggest that 70% of the root mass of many crop species is usually found in the upper 30 cm of the soil profile (Jackson et al. 1996), with irrigation (Gan et al. 2009) and the adoption of reduced or zero tillage systems (Williams et al. 2013) tending to increase the density of roots in the upper horizons.

This zone of high root density tends to coincide with the zones of greatest nutrient enrichment, including fertilizer application, microbial activity, and nutrient cycling; in natural systems surface concentration of nutrients is expected as aboveground residues accumulate and decompose largely without mechanical soil incorporation. Therefore, the acquisition of nutrients from these layers is clearly important. However, there is limited evidence that applications of K fertilizer alone impact the root distribution within the soil profile (Brouder and Cassman 1994), except perhaps for the situation where K application contributed to a reduced severity of crop water stress and an extended period of biomass accumulation and root growth (Grzebisz et al. 2013). An example is shown from a soybean (*Glycine max* (L.) Merr.) crop in Indiana, USA in Fig. 12.1. The uppermost 5 cm of soil has at least 1.5-fold the root length density (RLD) of any other soil profile segment, and the RLD of soybean in this study was very low (~ 1 cm cm^{-3}) below 20 cm (Fig. 12.1b). Contrasting rates of K addition resulted in large differences in soil test K in the 0–5, 5–10 cm soil profile increments, but these differences did not impact RLD. Unlike N and P that stimulate localized proliferation of fine roots, localized high concentrations of K do not appear to enhance root growth (Rengel and Damon 2008). The only recent reports of root proliferation in a fertilizer K-enriched soil volume was a report for maize by Perna and Menzies (2010) that showed some evidence of root proliferation when 6–12% of

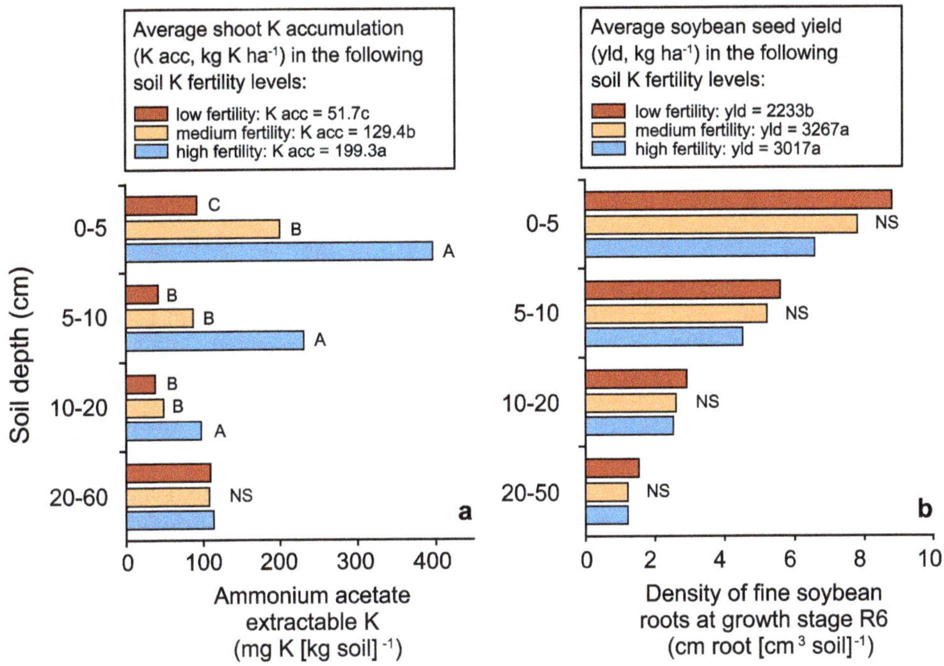

Fig 12.1 Distribution of (**a**) soil exchangeable K (ammonium acetate extraction) and (**b**) soybean roots by soil depth in plots previously fertilized for 6 years with varied rates of K (0–900 kg ha^{-1} cumulative additions). Low, medium, and high correspond to soil test K levels less than, approximately equal to, and greater than 104 mg kg^{-1} measured at 0–20 cm one year after the last K addition. Four replicates per category were assessed over 2 years at planting (soil) and R6 (roots). Within each graph and depth increment, bars followed by a different letter are significantly different ($P \leq 0.05$). Legends show a 2-year mean total aboveground K accumulation (A) and seed yields (B). Adapted from Fernandez et al. (2008) and Navarrete-Ganchozo (2014) (graph **a**) and Fernández et al. (2009) (graph **b**)

the available root volume was enriched with K. Ma et al. (2007) also suggested that there was some evidence of roots proliferating in K-enriched compartments of split-pot systems relative to unfertilized compartments in the same pot, but effects were quite inconsistent.

Collectively, the literature suggests that large changes in rooting patterns are unlikely to occur in response to the application of fertilizer K alone. This then suggests that either the volume of K-enriched soil needs to be large enough to encompass a significant proportion of the crop root system or that other strategies to enhance root activity in the fertilized zone, such as irrigation strategies that ensure synchronous availability of soil moisture and K, are enacted. The minimal effect of K on root proliferation may partly explain, along with salt effects, the usually small effects of starter K applications (fertilizer applied in or near the seeding row at planting) on early crop growth and yield, compared to effects of N and P (Kaiser et al. 2005; Mallarino et al. 2010).

12.2.2 Mobility of K in Soil in Soil Profiles

Fertilizer K applied to soils initially enriches the soil solution K pool, which is then depleted by plant uptake, by rapid adsorption onto exchange sites on clay or organic matter surfaces, or by more gradual fixation in wedge and interlayer positions of weathered micas, vermiculite, and high-charge smectite (e.g., Goulding 1987, Chap. 7). The most important factors affecting the mobility of K in the soil are: (1) the cation exchange capacity (CEC) of the soil and the specificity of exchange sites for K (both determined by the clay and organic matter content as well as the type of clay present); (2) the presence of K-fixing minerals; (3) the formation of sparingly soluble reaction products in bands containing K and other nutrients; and (4) the seasonal moisture dynamics—specifically the frequency and duration of wetting and drying cycles (affecting K fixation and release) and the extent of through drainage or leaching (Luo and Jackson 1985; Sparks and Huang 1985). In light-textured soils with low CEC and organic matter content, there is a limited capacity to adsorb significant amounts of K on the exchange complex. In these soils, a large proportion of applied K will remain in the soil solution and may be subject to leaching into deeper soil horizons. In terms of increasing the volume of K-enriched soil in the crop root zone, this can be beneficial to crop growth; but in high rainfall environments, it may result in leaching of K too deeply for access by plant roots. In contrast, soils with even moderate CEC will typically adsorb K leached from crop residues or applied as fertilizer, such that K is considered as sparingly mobile or effectively immobile in the soil profile. This is illustrated in Fig. 12.2 (adapted from Bell et al. 2009), with fertilizer incorporated into the top

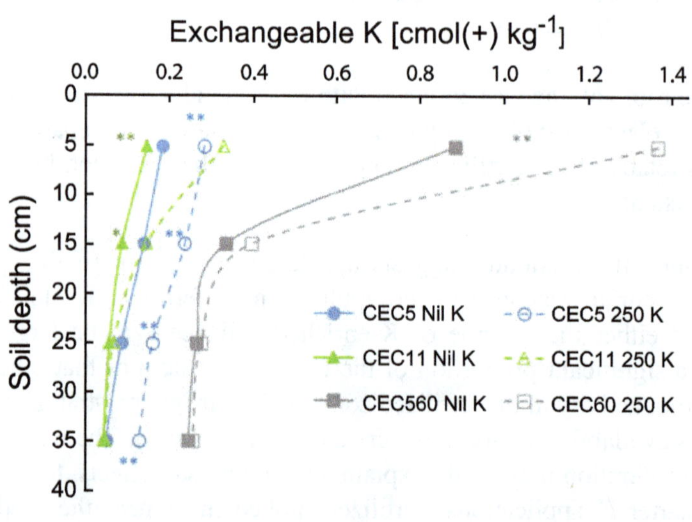

Fig. 12.2 Examples of the mobility of K through the soil profile after an application of 250 kg ha^{-1} of K (as KCl) applied into soils with contrasting cation exchange capacities (5–60 cmol(+) kg^{-1}) in northeast Australia (Bell et al. 2009; Halpin et al. 2019). Fertilizer was incorporated into the top 10 cm of each field profile using mechanical tillage, and soil samples were collected in 10 cm increments to 40 cm at 12 months and 750–950 mm of rainfall after fertilizer application. Exchangeable K (cmol(+) kg^{-1}) data are plotted at the midpoint of each depth increment

10 cm of the soil showing no redistribution into deeper soil layers in a Vertisol with CEC of 60 cmol(+) kg^{-1}, movement into the next 10 cm profile increment in the Oxisol with CEC of 11 cmol(+) kg^{-1} and leaching beyond the depth of measurement in a Chromosol with CEC of 4 cmol(+) kg^{-1}.

Because of their permeability with regards to K, low CEC soils require careful management of K applications to ensure efficient use by crops rather than leaching losses in high rainfall environments. However, the situation in higher CEC soils represents real challenges in ensuring that applied or recycled K enriches the zone where crop roots are most active. Particularly in reduced or zero-tillage cropping systems, where physical mixing of profile layers has been minimized or eliminated entirely, plant-available K reserves become increasingly concentrated near the soil surface in a situation commonly described as K stratification (Grant and Bailey 1994; Mallarino and Borges 2006). The impact of K stratification will depend on the frequency with which those K-enriched topsoil layers are rewet and can support active root growth during the period of crop K accumulation. In situations where in-season rainfall events are infrequent and the crop is reliant on subsoil reserves of moisture and nutrients for extended periods, such as the northern Australian grain growing regions on Vertisol soils (Bell et al. 2010), root activity and K acquisition from these topsoil layers is minimal.

12.2.3 Movement of K to Plant Roots

In most situations, the concentration of K in the soil solution is low, primarily due to the propensity for rapid adsorption of K onto exchange surfaces. As a result, the extent to which mass flow contributes to K supply to the plant root is limited to typically <5% of overall plant uptake (Jungk 2001), although this proportion can be higher when soil solution K concentration is high. Examples include when high rates of K are applied to light-textured soils with low CEC (Rosolem et al. 2003), or when soils are irrigated with wastewaters containing elevated K concentrations (Arienzo et al. 2009). In most conditions, K supply is dominated by diffusion through the soil solution along a concentration gradient established between the plant root surface and the undepleted soil solution (Barber 1985). The efficiency of diffusion is determined by a variety of soil and seasonal factors, including: (1) the moisture content of the soil—effective diffusion rates increase with increasing volumetric water content; (2) the impedance or tortuosity of the diffusion path—effective diffusion slows as clay content increases or as soil structure is degraded; (3) the concentration gradient established between the rhizosphere and surrounding undepleted soil—soil solution K concentrations typically decrease as BC_K increases, lessening the potential concentration gradients; and (4) the soil temperature—diffusion rates increase with increasing temperature due to lower viscosity of the soil water (Barber 1995).

These factors obviously interact with the crop root system, with root density and inter-root competition in a given soil volume affecting the root depletion profile and hence the uptake of K per unit of root length (Jungk 2001; Mengel et al. 2001). Species differences in root hair length and mycorrhizal colonization will also affect the volume of soil K depletion.

12.3 Fertilizer K Application Strategies in Soil

Potassium fertilizer strategies typically involve applications that are broadcast onto the soil surface or placed in discrete bands (alone or in combination with other nutrients) in the topsoil layers, with or without subsequent incorporation with tillage implements. The latter application method is particularly prevalent in reduced or no-till systems where soil structure and retention of surface residue cover are important management considerations. There is also some use of K in "starter" fertilizer programs, where small amounts of nutrients (typically compound fertilizers containing N–P or N–P–K, maybe with micronutrient additives) are placed in the seeding row or in bands immediately beside or below the seeding trench to ensure early contact between developing roots and the nutrient source. However, the amount of K applied in this approach is often limited by the risk of salt-induced damage to the developing seedlings and their root systems. Although benefits of this application method have been observed occasionally, they have been linked to very specific circumstances (e.g., delayed planting of full-season maize hybrids in the upper US Midwest (Bundy and Andraski 1999). Generally, the impact on the yield of this application method has been shown to be inconsistent and sometimes detrimental (Mallarino et al. 2010).

While broadcast applications to the soil surface are typically more cost-effective in terms of rate of land area treated, their efficiency in supplying K to the crop depends on the extent to which either tillage or rainfall/irrigation can redistribute K deeper into the soil profile where roots can access the applied fertilizer. In light-textured soils, the low water-holding capacity, high internal drainage rates, and low capacity to adsorb K on the exchange complex can facilitate the redistribution of broadcast K into deeper profile layers (e.g., the low CEC soil in Fig. 12.2). However, in some situations, K can be leached completely from the crop root zone (Alfaro et al. 2003; Askegaard et al. 2004). More typically, broadcast K redistributes down the profile slowly, and the rate of leaching is greatly exceeded by crop uptake and deposition on the soil surface in crop residues. The result is "stratification" of soil K, where the concentration of labile K in the top 5–10 cm typically exceeds that in the soil layers immediately below (e.g., 10–20 cm and beyond), with the extent dependent on root distribution and K removal rates (Saarela and Vuorinen 2010). This situation is exacerbated by minimum and no-till systems, and K fertilization of perennial species where the physical redistribution of K throughout the cultivated layer by plows or discs has been minimized or discontinued entirely (Robins and Voss 1991; Holanda et al. 1998; Vyn et al. 2002).

While the existence of soil profiles with stratified labile K is not necessarily a constraint to crop K acquisition, provided those topsoil layers are moist for extended periods and characterized by an extensive network of active roots (Fernandez et al. 2008; Fernández et al. 2009), intensive cropping can deplete the upper layers of the subsoil. This results in an increased reliance on optimal conditions in the topsoil layers throughout the period of maximum K uptake (e.g., Cassman et al., 1989). Even in growing areas where growing season rainfall for grain production ensures moist top soils for K uptake, significant periods of temporal drought may limit K availability. This problem is accentuated in growing areas relying on stored soil moisture, and in many situations, K application strategies other than broadcasting have been adopted. These include banding (Bordoli and Mallarino 1998; Borges and Mallarino 2001) as well as occasional tillage operations designed to "redistribute" stratified K reserves (Yin and Vyn 2004).

In reduced- and no-tillage systems, the favored alternative to broadcast application is to apply K in bands. The strategies that determine effective K fertilizer banding have been developed by considering a number of key principles. These include: (1) not placing high fertilizer concentrations close to the seed row to avoid high salt concentrations that have a negative impact on germination and seedling establishment (Gelderman 2007); (2) band placement that maximizes root interception and crop K acquisition (e.g., below and/or beside the plant line, or in the planting hill); and (3) co-locating other nutrients with K to encourage root proliferation in and around the fertilizer band (Officer et al. 2009). Bands can be particularly effective in situations where the rate of root development and access to a larger soil volume is constrained by cold soils or high soil strength/compaction (Oborn et al. 2005), with the higher soil solution concentration in the vicinity of the band allowing rapid K uptake. Banding K deeper than the common 5–10 cm depth is sometimes beneficial in conditions where the topsoil is frequently dry but deeper soil layers have moisture (Bordoli and Mallarino 1998). However, the effectiveness of applying K into deeper soil profile layers will represent a compromise between placing fertilizer into soil layers that are moist enough to allow K acquisition for a greater part of the growing season and also having sufficient root length density to enable a significant amount of K uptake. In the example in Table 12.1, broadcast application of K to alfalfa resulted in the highest yield and greatest K recovery when compared to K injected at discrete soil depths; adequate rainfall likely enabled uptake of surface-applied K and precluded K recovery from greater soil depths even in a deep-rooted species like alfalfa.

Another consideration with banded K applications is that as the crop grows and the K demand increases, the proportion of plant K that can be supplied from a localized fertilizer band enriching a small soil volume diminishes. This suggests that where banded K applications are necessary (e.g., no-till systems on heavier textured soils), strategies to enhance K diffusion into larger soil volumes or to encourage a greater proportion of the crop root system to develop in the proximity of the bands (e.g., co-location with other nutrients like N and P; Ma et al. 2011) need to be considered. An example of the latter is shown in Fig. 12.3 where the addition of P to a band of KCl fertilizer, either alone or in combination with N, enhances the uptake of rubidium (RB) tracer added in the band. While the effects of N are transitory and

Table 12.1 Influence of K placement depth on tissue K concentrations, K recovery, and yield of alfalfa. (Petersen and Smith 1973)

Depth of K placement	Tissue K	K recovery	Yield
cm	g kg^{-1}	%	Mg ha^{-1}
Check, No K	10.7	–	4.46
Broadcast	21.7	28	5.06
7.5	16.5	15	4.92
722.5	14.5	9	4.66
37.5	12.2	4	4.59
52.5	11.6	4	4.72
67.5	12.0	5	4.82
82.5	11.4	–	4.12
LSD[a]	2.6		0.51

Potassium as K_2SO_4 was surface broadcast or injected as a solution into the silt loam soil at specific depths using a Leur-Lok syringe. The fertilization rate of 224 kg K/ha was applied on April 15 and the yield of this 2-year-old alfalfa stand determined on June 3

[a]LSD: least significant difference at $p < 0.05$

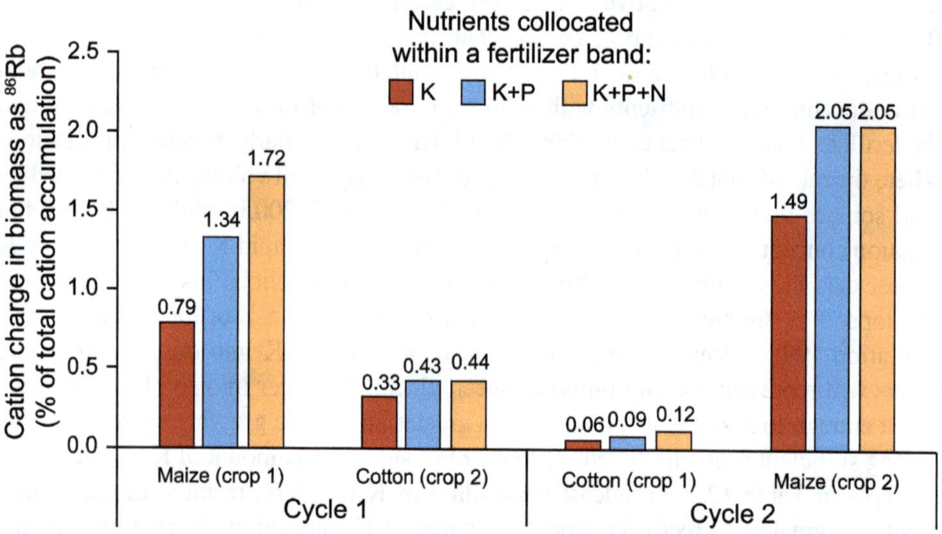

Fig. 12.3 Impact of co-locating P or (N and P) on the exploitation of KCl bands by maize and cotton plants grown in a Vertisol soil over two experimental cycles with contrasting crop sequences. Effects were assessed by quantifying the uptake of Rb (applied as RbCl mixed into a KCl band), expressed as a percentage of the total cation accumulation in plant biomass, cmol(+) kg^{-1} dry matter. (MJ Bell and PW Moody, unpublished data)

limited to the first crop in each sequence, the effects of P co-location with K are more persistent.

Greater crop nutrient recovery has sometimes been recorded when fluid forms of nutrients have been deployed at similar nutrient rates compared to granular products. The improved availability of fluid P over granular P to crops grown on highly calcareous soils in South Australia is a good example (Lombi et al. 2004). The mechanism for this response was shown to be increased P diffusion away from the point of fertilizer injection, thus enhancing the volume of soil enriched with P and so accessible to plant roots. However, there are no reports of similar advantages for fluid forms of K fertilizer over granules, perhaps because of the generally greater solubility of K fertilizers. Choice of a fluid K formulation would be based on factors related to ease of application and the ability to blend different products rather than an expected increase in K use efficiency.

12.4 Quantifying Fertilizer K Recovery

There has been less research focused on the efficient recovery of applied K fertilizer by crops and the utilization of that K in the production of crop or forage biomass and harvestable yield than there has for nutrients that are more mobile and/or cause off-site impacts in the atmosphere or adjacent water bodies (i.e., N and P). While concerns about excessive K applications after land application of wastewaters do arise (Arienzo et al. 2009), most scientists consider excessive K application as reducing the profitability of crop production and an inefficient use of a natural resource, but not having off-site impacts on the environment. The K fertilizer placement method can have an impact on the K recovered by plants and what is removed from the field at harvest or recycled to the soil, but studies focusing on this issue are scarce. Research with corn (*Zea mays* L.) and soybean has demonstrated that banded K fertilizer almost always greatly increases the K uptake during vegetative growth periods relative to broadcast K application for several tillage systems (Mallarino et al. 1999; Borges and Mallarino 2000; Borges and Mallarino 2003), although the persistence of these effects through to maturity was not measured. However, the impact of increased K uptake with banding on net K removal will depend greatly on the crop species (Oltmans and Mallarino 2015) and the crop part harvested, such as biomass removed in forage or silage production compared to harvested grain.

The metrics used to quantify fertilizer K recovery by crops and the efficiency of use to increase yield are discussed in detail in Chap. 5. Interestingly, the use-efficiency data (mainly for cereal crops) suggest applied K is used less efficiently to produce additional grain yield than either N or P, which may reflect the lower critical grain K concentrations for these species and the fact that most crop K is returned to the field in roots and residues in these species. This is supported by published fertilizer K recoveries in crop biomass that are more in line with reported values for N and P (Fixen et al. 2015).

Reported fertilizer K recovery figures may underestimate crop uptake of applied K, as they are generally based on an assumption that only the additional crop K uptake in the fertilized treatments, when compared to a 0 K control, are due to fertilizer recovery. Given the impact of K fertilizer on soil solution K concentrations, especially in the vicinity of bands, and hence the likely improved efficiency of diffusive supply across a stronger concentration gradient to a plant root, there may well be some unaccounted preferential fertilizer K exploitation in the fertilized layers and some sparing of soil K reserves elsewhere in the soil profile.

The phenomenon of preferential fertilizer exploitation by crop roots has been commonly observed for P through the use of radioactive P isotopes, but there seems to have been little published work on the topic for K. There are real opportunities to re-examine the use of tracers like Rb to provide more accurate determinations of fertilizer K uptake and better assess the efficiency of different K application strategies. Strategies could include either enriching a K fertilizer band (Hafez and Rains 1972) or simply by using the relative abundance of K and Rb in unfertilized and fertilized treatments (Hafez and Stout 1973). The example of using Rb-enriched KCl bands provided in Fig. 12.3 illustrates the insights that can be obtained from using such techniques. In that study, biomass K concentration and plant uptake were similar in the banded treatments with K alone as in those with added P, or N and P. However, the Rb tracer data clearly illustrates more extensive exploitation of the fertilizer band, presumably sparing K reserves in the bulk soil, when these other nutrients were co-located with K in the fertilizer band.

12.5 Crop Characteristics Influencing K Application Strategy

To optimize recovery of applied K there must be a spatial coincidence of active roots and enriched K layers or patches. Several studies by Barber and collaborators, summarized in Barber (1995), have suggested that optimal K recovery required fertilizer K to be mixed through a greater proportion of the root zone than for P. However, the implications for fertilizer application strategy will vary with the physiological characteristics of the root cells, with the inherent root distribution of the different plant species or genotype and with the continuity of moisture availability in the fertilized soil layer.

A recent review by Fan et al. (2016) concluded that at least half of the total root mass of agricultural crops grown in temperate regions could be found in the top 20 cm of the soil profile, while Gan et al. (2009) suggested that these proportions may be conservative for a range of winter cereal, oilseed, and pulse crops (i.e., >75% of roots in the top 20 cm). These reports showed slightly shallower root distributions in temperate systems than the broader global analysis of Jackson et al. (1996), suggesting that effective K fertilizer strategies in temperate environments should be able to focus on the upper part of the soil profile—a zone that is relatively easily accessible to most fertilizer application/tillage equipment. However, the applicability of these results to rain-fed cropping systems in the more variable rainfall environments of the tropics and subtropics (Bell et al. 2009), or to flood-irrigated cotton on heavy clay soils (Lester and Bell 2015) is questionable. In such environments, either extended dry periods or excessive moisture and low oxygen availability limit root activity and nutrient acquisition from the uppermost zones of the soil.

An additional complication is apparent in no-till systems. While the proportional allocation of root biomass in the topsoil can be pronounced, the spatial heterogeneity of the root distribution may limit the effectiveness of exploiting this zone for K. For example, an analysis by White and Kirkegaard (2010) suggested that 20–30% of wheat roots at 20 cm were confined to pores and cracks wider than normal root diameters, with this proportion rapidly increasing to 60% by 60 cm and effectively 100% at depths of 80–90 cm. This "clumping" of roots around existing pores and root channels, rather than being distributed through the bulk soil, may have significant implications for the acquisition of a relatively immobile nutrient like K. Theoretically, new roots will exploit the same (previously depleted) soil volume around these channels, while homogenous fertilizer K distribution would be much less effective at replenishing depleted K soil around such biopores. These effects would likely be more pronounced in subsoils (i.e., >20 cm depth), where the interaction between the distribution of K bands, crop row spacings, biopore density, and soil water availability may explain the lack of consistent response to deep bands in the literature.

Given the limited evidence of increased root density in response to soil K enrichment in zones/patches, it could be assumed that enrichment of as much of the active root zone as possible would be a desirable strategy. Such an approach requires either redistribution of surface broadcast K into deeper layers with soil water movement (in light-textured soils), or through soil inversion/tillage—including occasional strategic tillage operations in otherwise no-till systems (Dang et al. 2015). The effectiveness of this general approach to K replenishment in the entire rooting zone will be determined by soil properties that regulate the extent to which K application increases soil solution K activity (e.g., K buffering) and the extent to which fertilizer K is fixed into slowly available forms by clay minerals. An example of where mixing K through a greater proportion of the root zone has been very effective is shown in Fig. 12.4, with crop biomass (15%) and K content (55%) increasing with the degree of profile mixing at the same rate of K application.

An alternative approach is employed when fertilizer K is banded, with very small soil volumes enriched. The effectiveness of this K application strategy could be considered risky in some soils, given the greatly reduced volume of fertilized soil and hence the smaller chance that enough roots will encounter K-enriched soil to optimize crop K uptake. However, co-location of K with other nutrients that do cause root proliferation such as P (Barber 1995; Ma et al. 2011) can be used to increase root density around the fertilizer band and enhance recovery of banded K (e.g., Fig. 12.3). There are limited reports of the benefits of this approach in the literature, although Brouder and Cassman (1994) were able to demonstrate enhanced K uptake in cotton in response to root proliferation in zones where K had been co-located with NH_4^+-N. A possible limitation with a strategy of nutrient co-location in bands is the potential to precipitate insoluble K minerals (pool 13—neoformed K

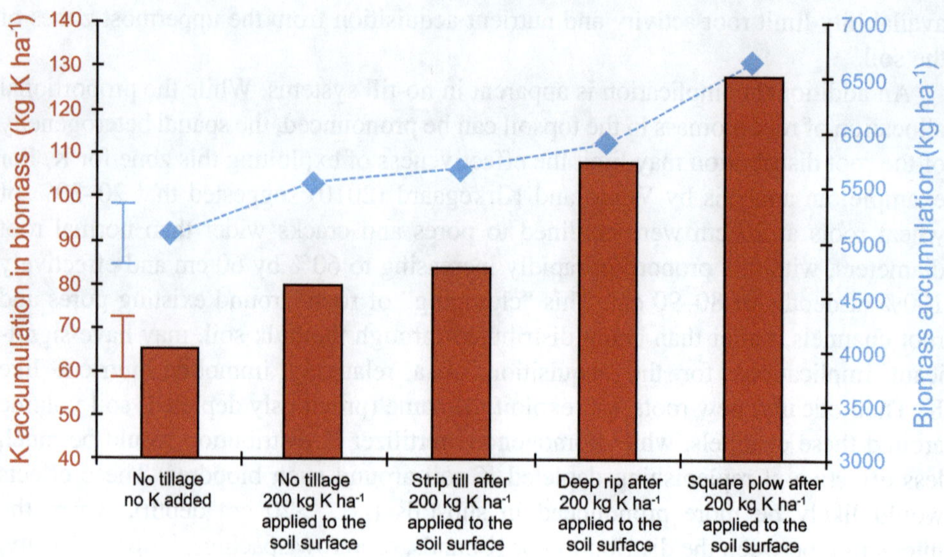

Fig. 12.4 Impact of various degrees of profile mixing on biomass production (diamond symbols) and K accumulation (solid bars) by a peanut crop grown on an Oxisol in NE Australia (Bell et al. 2009). Treatments range from a surface broadcast application with no incorporation, through to aggressive tillage with profile inversion (square plow) to a depth of 25 cm

minerals, discussed in Chap. 7), as a result of radical changes in the pH and ionic strength of the soil solution over short periods. Circumstantial evidence of this phenomenon has been recorded in Australian studies (MJ Bell and DW Lester, unpublished data), but further definitive research is needed.

Finally, the K accumulation dynamics of different crop species may also influence the fertilizer application strategies. Different K application strategies may be suitable for crops in which the intensity of K demand varies within the growing season (e.g., due to the duration of the period of rapid K uptake or internal redistribution). As an example, crop K accumulation in a uniculm species like maize occurs mainly in a sharply defined period early in the growing season, well in advance of maximum dry matter accumulation (Welch and Flannery 1985). However, in species with a greater reliance on staggered tiller addition (e.g., grain sorghum), or in less determinate species such as cotton (Mullins and Burmester 1990) or soybean (Hanway 1985), K uptake occurs at lower rates over a longer period, mirroring dry matter accumulation. These differences in crop K demand may influence the choice of application method (banding vs. broadcast) and the timing of K application relative to crop establishment (Chap. 13), including the use of supplementary foliar K applications during periods of rapid K uptake (maize) or redistribution (boll loading in cotton).

12.6 Soil Characteristics Influencing K Application Strategy

The most obvious soil characteristic influencing K application strategy is the plant-available K status of the soil itself. There are consistent reports of negative K balances in many agricultural systems (Oborn et al. 2005; Rengel and Damon 2008; Bell et al. 2010) suggesting that depletion of native soil K reserves is widespread in agricultural lands. The first pre-condition for growers to commence a K fertilizer program will be a determination of the plant-available K status and an assessment of the likelihood of an economic response. This will involve the collection of soil samples representative of the available K status within the crop zone, followed by appropriate sample processing, analysis, and result interpretation. As mentioned previously, sampling strategies will need to consider the heterogeneity of K both vertically and horizontally due to differential K enrichment/depletion of profile layers and application of K bands—especially in reduced or zero-till systems. Research on K fertility assessments for different placement strategies and under different tillage systems is clearly required.

As outlined in Chap. 8, most commercial laboratory tests typically estimate plant-available K as the soil solution and adsorbed K pools measured as exchangeable K, although the latter may also contain some K from secondary phyllosilicate minerals, depending on mineralogy and extraction method. However, the dissolution of structural K may reduce the fertilizer K requirement (Moody and Bell 2006), while fixation or release of K from secondary phyllosilicate minerals during the growing season may increase or reduce the fertilizer K requirement, respectively. Identifying different K pools in the soil for which a fertilizer recommendation is being developed is the first step to determine a successful K fertilizer strategy. Unfortunately, the current lack of quantitative diagnostic tests to link the presence of these K pools to their likely release rates under varying rhizosphere conditions can make the decision to apply fertilizer K challenging. The development of simple laboratory indices for K fixation (Chap. 8) or the development of soil classification and landscape indices linked to the likely presence of mineralogy conferring particular K behavior (Chap. 7) may provide some benefits in this regard.

In soils with low BC_K [e.g., those with low CEC and low clay content—Barber (1985)], even small application rates of K dispersed through the soil volume can significantly increase the concentration of K in the soil solution (e.g., the Oxisol soil in Fig. 12.5) and ensure the development of stronger concentration gradients and more rapid diffusion rates of K to plant roots. In these situations, fertilizer K dispersed through the cultivated layer would be expected to result in high efficiency of recovery of applied K, because such applications would ensure exposure of a large proportion of the root zone to elevated soil solution K. In such soils, comparable RE_K values might be expected from banded applications only where soil structure and porosity were such that diffusive supply could efficiently occur over larger distances (i.e., strongly structured and with high permeability). In high rainfall

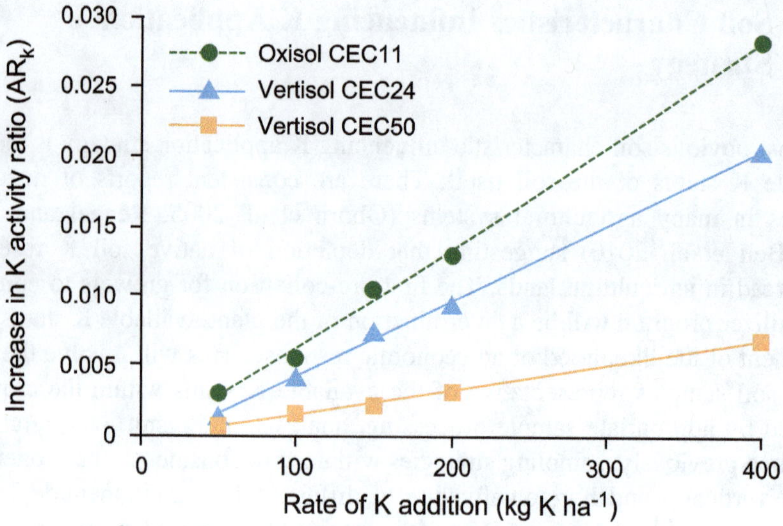

Fig. 12.5 The increase in activity of K in the soil solution (AR_K) in response to varying rates of K fertilizer application for an Oxisol (CEC 11 cmol(+) kg^{-1}) and two Vertisols (CEC 24 and CEC 50 cmol(+) kg^{-1}). Data calculated from Bell et al. (2009)

environments a subset of these soils with very low CEC (<5 cmol(+) kg^{-1}) may experience leaching losses of K, and in these cases split applications of broadcast K may be an appropriate way to ensure fertilizer K is available to meet crop demands and minimize luxury consumption of K by plants.

Conversely, in soils where BC_K is high (e.g., high CEC and high clay contents, such as in the CEC 50 Vertisol in Fig. 12.5), a much higher rate of applied K would need to be dispersed through the soil volume to generate significant increases in either soil solution K concentration or AR_K. As an illustration, the rate of applied K needed to achieve a specified change in AR_K in the CEC 50 Vertisol would be *ca.* 4 times that required to achieve the same impact in the Oxisol if the K were dispersed through the same soil volume. However, if the applied K is concentrated in fertilizer bands there is a much higher effective K application rate in a small soil volume and a much more substantial impact on soil solution K and AR_K. In these soils, banded applications should provide the opportunity for higher fertilizer K recovery efficiencies provided that sufficient root proliferation can be generated in the vicinity of the fertilizer band, or that the spatial density of banding is sufficient to ensure that a greater proportion of the crop root system has access to zones of elevated solution K. Clearly, more research is needed to explore the trade-offs between BC_K in different soil types and the effectiveness of banded or dispersed fertilizer K application strategies.

Similarly, soil physical properties are also likely to affect the efficacy of K application strategies within BC_K classes, with issues such as poor soil structure (e.g., in sodic soils) or compaction likely to reduce the effective diffusion path length, and hence the efficiency of K supply to roots. In such situations, an appropriate response may be to increase the density of K fertilizer bands to ensure a greater number of enriched K patches to compensate for the restricted diffusion path lengths around each band. In a similar vein, soils and cropping systems where moisture availability is seasonally limited will also experience reduced K diffusion rates (Mengel et al. 2001), potentially increasing the frequency of K responses provided crop demand is not substantially decreased simultaneously. Such conditions may prompt use of either higher K application rates (to ensure stronger concentration gradients) or placement strategies that ensure fertilizer K is placed where soil moisture status is more favorable for longer in the growing season (e.g., by placing K bands deeper in the soil profile).

There is little published information about how fertilizer K application strategies could be modified for soils where K fixation is significant. As noted by Blake et al. (1999), the recovery of applied K is typically lower on soils with significant K-fixing capacity (i.e., only 70% of that recorded on comparable non-K fixing soils in long-term fertilizer trials). A common application strategy is simply to increase fertilizer K rates to compensate for the lower recoveries. Theoretically, large rates of K addition would be needed to saturate the K-specific fixation sites before application rates that matched crop removal could be safely adopted (Mengel 2007). For nutrients like P, where strong precipitation or fixation reactions can reduce the fraction of the applied nutrient that is available for crop uptake in some soils, a strategy of minimizing the interaction between the fertilizer and the bulk soil by banding has been successfully used to slow the decline in plant-available nutrient and to improve crop recovery. The effectiveness of such a strategy for K in soils with significant fixation capacity or specific tillage systems may offer some benefits, and it may already be occurring in situations where the advantages of banding over broadcast K have been recorded (Bordoli and Mallarino 1998; Borges and Mallarino 2001). This is an area requiring further research.

A tentative framework for considering the impact of the key soil properties of BC_K and K fixation on the choice of fertilizer K application strategy is presented in Fig. 12.6. While hypothetical, it is based on the concepts discussed in this section and could provide the sort of framework upon which to base broader investigations of K application strategies.

12.7 Conclusions

The relative immobility of K in many soil profiles dictates that agricultural K inputs must be managed to ensure a coincidence of K-enriched soil with a significant proportion of the active root system during periods of high K demand. This creates challenges for fertilizer application strategies and equipment, particularly in systems where soil inversion and other forms of aggressive tillage are no longer practiced. A better understanding of the capacity of crop and pasture root systems to utilize K-rich patches (typically fertilizer bands) will be a key prerequisite for developing successful K management strategies, as will an understanding of the potential benefits that can be achieved through the co-location of different nutrients with K in bands to

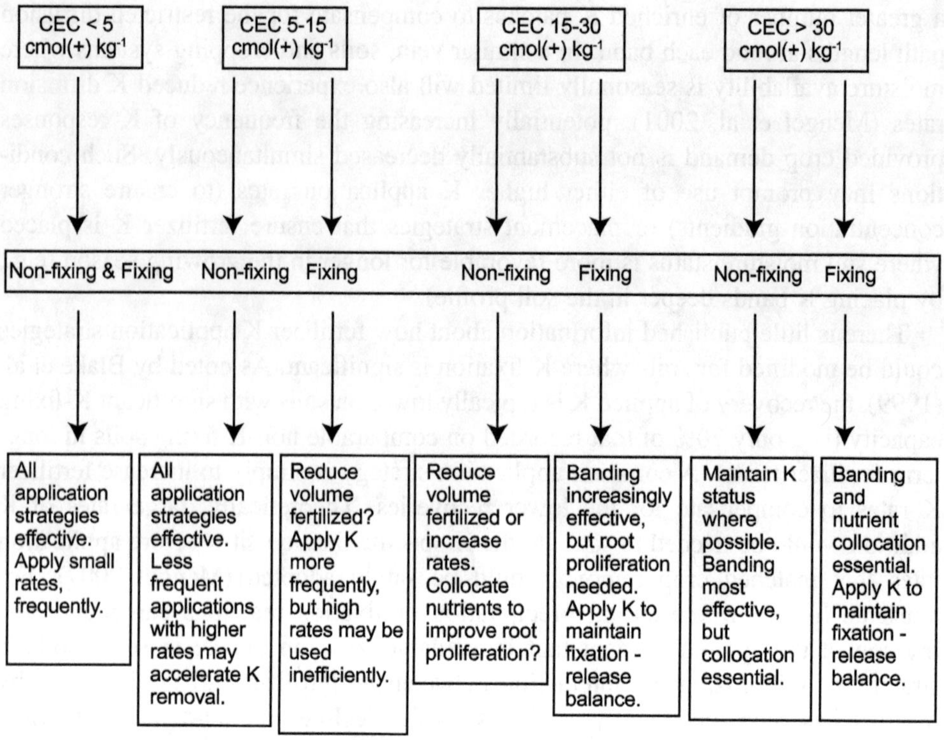

Fig. 12.6 A tentative framework for allowing for the likely impact of key soil properties on choice of banding or broadcast K fertilizer application strategy in the soil. The cation exchange capacity (CEC) classes are tentative at this stage, but a framework along these lines could produce a useful guide for agricultural land managers

encourage root proliferation. The appropriate placement strategy (e.g., depth, band spacing) will also be critical, and it will vary with soil type (e.g., water holding capacity in rainfed situations), climate (temperature and the frequency of effective rainfall events), and irrigation availability and method. However, in soils where CEC is moderate (10–15 cmol(+) kg^{-1}) to high (>15 cmol(+) kg^{-1}), there may be real advantages in applying large rates of K less frequently across a crop rotation, rather than applying lower rates on a crop-by-crop basis.

Conversely, in light-textured soils with low CEC and limited capacity to adsorb K, redistribution of applied K into deeper profile layers is possible if there is sufficient drainage, so a broader range of placement options appear to be available. The challenges in these systems relate more to ensuring that K remains in the crop root zone through the periods of peak K demand, and that leaching losses do not reduce the efficiency of K recovery and use. In these situations, K fertilizer management is likely to be on a crop-by-crop basis, possibly even requiring split applications within a crop season where the potential for leaching losses is high.

Successful soil K placement strategies will therefore need to reflect the interaction of plant, soil, and environmental factors. Development of effective strategies will require an improved understanding of the availability of fertilizer K added to soil, plant root system characteristics for different species in a rotation sequence, the response of roots to dispersed or concentrated patches of K (and other nutrients), and the dynamics of K accumulated in crop biomass and returned to the field in residues. This improved understanding will facilitate optimization of soil K placement strategies that may achieve more efficient use of the fertilizer K resource.

References

Alfaro MA, Jarvis SC, Gregory PJ (2003) Potassium budgets in grassland systems as affected by nitrogen and drainage. Soil Use Manag 19:89–95. https://doi.org/10.1111/j.1475-2743.2003.tb00286.x

Arienzo M, Christen EW, Quale W et al (2009) A review of the fate of potassium in the soil-plant systems after land application of wastewaters. J Hazard Mater 164:415–422. https://doi.org/10.1016/j.jhazmat.2008.08.095

Askegaard M, Eriksen J, Johnston AE (2004) Sustainable management of potassium. In: Schjonning P, Elmholt S, Christensen BT (eds) Managing soil quality: challenges in modern agriculture. CABI, Wallingford, UK, pp 85–10

Barber, SA (1985) Potassium availability at the soil-root interface and factors influencing potassium uptake. In: Munson RD (ed) Potassium in agriculture. American Society of Agronomy, Madison, WI, pp 309–324

Barber SA (1995) Soil nutrient bioavailability: a mechanistic approach, 2nd edn. Wiley, New York

Bell MJ, Moody PW, Harch GR et al (2009) Fate of potassium fertilisers applied to clay soils under rainfed grain cropping in south-east Queensland, Australia. Aust J Soil Res 47(1):60–73. https://doi.org/10.1071/sr08088

Bell, MJ, Moody PW, Klepper K et al (2010) The challenge to sustainability of broadacre grain cropping systems on clay soils in northern Australia. Paper presented at the proceedings of the 19th world congress of soil science; soil solutions for a changing world, Brisbane, Australia, 1–6 August

Blake L, Mercik S, Koerschens M et al (1999) Potassium content in soil, uptake in plants and the potassium balance in three European long-term field experiments. Plant Soil 216(1–2):1–14. https://doi.org/10.1023/A:1004730023746

Bordoli JM, Mallarino AP (1998) Deep and shallow banding of phosphorus and potassium as alternatives to broadcast fertilization for no-till corn. Agron J 90(1):27–33. https://doi.org/10.2134/agronj1998.00021962009000010006x

Borges R, Mallarino AP (2000) Grain yield, early growth, and nutrient uptake of no-till soybean as affected by phosphorus and potassium placement. Agron J 92(2):380–388. https://doi.org/10.2134/agronj2000.922380x

Borges R, Mallarino AP (2001) Deep banding phosphorus and potassium fertilizers for corn managed with ridge tillage. Soil Sci Soc Am J 65(2):376–384. https://doi.org/10.2136/sssaj2001.652376x

Borges R, Mallarino AP (2003) Broadcast and deep-band placement of phosphorus and potassium for soybean managed with ridge tillage. Soil Sci Soc Am J 67(6):1920–1927. https://doi.org/10.2136/sssaj2003.1920

Brouder SM, Cassman KG (1990) Root development of two cotton cultivars in relation to potassium uptake and plant growth in a vermiculitic soil. Field Crops Res 23(3-4):187–203. https://doi.org/10.1016/0378-4290(90)90054-f

Brouder SM, Cassman KG (1994) Cotton root and shoot response to localized supply of nitrate, phosphate and potassium: split-pot studies with nutrient solution and vermiculitic soil. Plant Soil 161(2):179–193. https://doi.org/10.1007/bf00046389

Bundy LG, Andraski TW (1999) Site-specific factors affecting corn response to starter fertilizer. J Prod Agric 12(4):664–670. https://doi.org/10.2134/jpa1999.0664

Cassman KG, Roberts BA, Kerby TA et al (1989) Soil potassium balance and cumulative cotton response to annual potassium additions on a vermiculitic soil. Soil Sci Soc Am J 53(3):805–812. https://doi.org/10.2136/sssaj1989.03615995005300030030x

Coker, DL, Oosterhuis DM, Brown RS (2009) Cotton yield response to soil- and foliar-applied potassium as influenced by irrigation. J Cotton Sci 13:1–10. https://www.cotton.org/journal/2009-13/1/upload/JCS13-1.pdf. Accessed 13 May 2020

Dang YP, Moody PW, Bell MJ et al (2015) Strategic tillage in no-till farming systems in Australia's northern grains-growing regions: II. Implications for agronomy, soil and environment. Soil Till Res 152:115–123. https://doi.org/10.1016/j.still.2014.12.013

Fan J, McConkey B, Wang H et al (2016) Root distribution by depth for temperate agricultural crops. Field Crops Res 189:68–74. https://doi.org/10.1016/j.fcr.2016.02.013

Fernandez FG, Brouder SM, Beyrouty CA et al (2008) Assessment of plant-available potassium for no-till, rainfed soybean. Soil Sci Soc Am J 72(4):1085–1095. https://doi.org/10.2136/sssaj2007.0345

Fernández F, Brouder S, Volenec J et al (2009) Root and shoot growth, seed composition, and yield components of no-till rainfed soybean under variable potassium. Plant Soil 322(1):125–138. https://doi.org/10.1007/s11104-009-9900-9

Fixen, P, Brentrup F, Bruulsema TW et al (2015) Nutrient/fertilizer use efficiency: measurement, current situation and trends. In: Drechsel P, Heffer P, Magen H et al (eds) Managing water and fertilizer for sustainable agricultural intensification, 1st edn. International Fertilizer Industry Association (IFA), International Water Management Institute (IWMI), International Plant Nutrition Institute (IPNI), and International Potash Institute (IPI), Paris, France, pp 8–38

Gan YT, Campbell CA, Janzen HH et al (2009) Root mass for oilseed and pulse crops: growth and distribution in the soil profile. Can J Plant Sci 89(5):883–893. https://doi.org/10.4141/CJPS08154

Gelderman, R (2007) Fertilizer placement with seed—a decision aid. In: North Central extension – industry soil fertility conference, Des Moines, IA, pp 46–51

Goulding, KWT (1987) Potassium fixation and release. Paper presented at the Methodology in soil-K research. Proceedings of 20th colloquium of the International Potash Institute, Baden bei Wien, Austria

Grant CA, Bailey LD (1994) The effect of tillage and KCl addition on pH, conductance, NO_3-N, P, K and Cl distribution in the soil profile. Can J Soil Sci 74(3):307–314. https://doi.org/10.4141/cjss94-043

Grzebisz W, Gransee A, Szczepaniak W et al (2013) The effects of potassium fertilization on water-use efficiency in crop plants. J Plant Nutr Soil Sci 176(3):355–374. https://doi.org/10.1002/jpln.201200287

Gulick SH, Cassman KG, Grattan SR (1989) Exploitation of soil potassium in layered profiles by root systems of cotton and barley. Soil Sci Soc Am J 53(1):146–153. https://doi.org/10.2136/sssaj1989.03615995005300010028x

Hafez A, Rains DW (1972) Use of rubidium as a chemical tracer for potassium in long-term experiments in cotton and barley. Agron J 64(4):413–417. https://doi.org/10.2134/agronj1972.00021962006400040002x

Hafez AAR, Stout PR (1973) Use of indigenous soil-rubidium absorbed by cotton plants in determining labile soil-potassium pool sizes. Soil Sci Soc Am Proc 37(4):572–579. https://doi.org/10.2136/sssaj1973.03615995003700040030x

Halpin NV, Bell MJ, Rehbein WE, Moody PW (2019) Potassium fertilisation strategies for rotational grain legume crops—Implications for the subsequent sugarcane crop. In: Proceedings of 41st conference of the Australian Society of Sugar Cane Technologists. https://www.assct.com.au/conference/past-conferences/172-2019-assct-conference. Accessed 13 May 2020

Hanway JJ (1985) Potassium nutrition of soybeans. In: Munson RD (ed) Potassium in agriculture. American Society of Agronomy, Madison, WI, pp 753–764

Holanda FSR, Mengel DB, Paula MB et al (1998) Influence of crop rotations and tillage systems on phosphorus and potassium stratification and root distribution in soil profile. Commun Soil Sci Plant Anal 29:2383–2394. https://doi.org/10.1080/00103629809370118

Jackson RB, Canadell J, Ehleringer JR et al (1996) A global analysis of root distributions for terrestrial biomes. Oecologia 108(3):389–411. https://doi.org/10.1007/bf00333714

Jifon JL, Lester GE (2011) Effect of foliar potassium fertilization and source on cantaloupe yield and quality. Better Crops 95(1):13–15. http://www.ipni.net/publication/bettercrops.nsf/0/4A87D50760EB9CA18525797D0061579A/$FILE/Better%20Crops%202011-1%20p13-15.pdf. Accessed 13 May 2020

Jungk A (2001) Nutrient movement at the soil-root interface: it's role in nutrient supply to plants. Revista de la Ciencia del Suelo y Nutrición Vegetal (J Soil Sci Plant Nutr) 1(1):1–18. https://bibliotecadigital.infor.cl/handle/20.500.12220/12427. Accessed 13 May 2020

Kaiser DE, Mallarino AP, Bermudez M (2005) Corn grain yield, early growth, and early nutrient uptake as affected by broadcast and in-furrow starter fertilization. Agron J 97(2):620–626. https://doi.org/10.2134/agronj2005.0620

Lester, DW, Bell MJ (2015) Cotton root systems and recovery of applied P and K fertilisers. Paper presented at the science securing cotton's future, 2nd Australian cotton research conference, Toowoomba, 8–10 September

Liakat AM, Luetchens J, Nascimento J et al (2015) Genetic variation in seminal and nodal root angle and their association with grain yield of maize under water-stressed field conditions. Plant Soil 397:213–225. https://www.jstor.org/stable/43872870. Accessed 13 May 2020

Lombi E, McLaughlin MJ, Johnston C et al (2004) Mobility and lability of phosphorus from granular and fluid monoammonium phosphate differs in a calcareous soil. Soil Sci Soc Am J 68 (2):682–689. https://doi.org/10.2136/sssaj2004.6820

Luo JX, Jackson ML (1985) Potassium release on drying of soil samples from a variety of weathering regimes and clay mineralogy in China. Geoderma 35:197–208. https://doi.org/10.1016/0016-7061(85)90037-0

Ma Q, Rengel Z, Bowden B (2007) Heterogeneous distribution of phosphorus and potassium in soil influences wheat growth and nutrient uptake. Plant Soil 291(1-2):301–309. https://doi.org/10.1007/s11104-007-9197-5

Ma Q, Rengel Z, Siddique KHM (2011) Wheat and white lupin differ in root proliferation and phosphorus use efficiency under heterogeneous soil P supply. Crop Pasture Sci 62(6):467–473. https://doi.org/10.1071/CP10386

Mallarino AP, Borges R (2006) Phosphorus and potassium distribution in soil following long-term deep-band fertilization in different tillage systems. Soil Sci Soc Am J 70(2):702–707. https://doi.org/10.2136/sssaj2005.0129

Mallarino AP, Bordoli JM, Borges R (1999) Phosphorus and potassium placement effects on early growth and nutrient uptake of no-till corn and relationships with grain yield. Agron J 91 (1):37–45. https://doi.org/10.2134/agronj1999.00021962009100010007x

Mallarino AP, Bergmann EN, Kaiser DE (2010) A look at effects of in-furrow applications of potassium on corn. Fluid J 18(4):12–15. http://www.fluidfertilizer.com/pastart/pdf/F10-A3.pdf. Accessed 13 May 2020

Mengel K (2007) Potassium. In: Barker AV, Pilbeam DJ (eds) Handbook of plant nutrition. CRC, Taylor & Francis, Boca Raton, FL, pp 91–120

Mengel K, Kirkby EA, Kosegarten H et al (2001) Principles of plant nutrition, 5th edn. Kluwer, Dordrecht

Moody PW, Bell MJ (2006) Availability of soil potassium and diagnostic soil tests. Aust J Soil Res 44(3):265–275. https://doi.org/10.1071/SR05154

Mullins GL, Burmester CH (1990) Dry matter, nitrogen, phosphorus, and potassium accumulation by four cotton varieties. Agron J 82(4):729–736. https://doi.org/10.2134/agronj1990.00021962008200040017x

Navarrete-Ganchozo RJ (2014) Quantification of plant-available K in a corn-soybean rotation: a long-term evaluation of K rates and crop K removal. PhD dissertation, Purdue Univ., West Lafayette, IN

Oborn I, Andrist-Rangel Y, Askekaard M et al (2005) Critical aspects of potassium management in agricultural systems. Soil Use Manag 21:102–112. https://doi.org/10.1111/j.1475-2743.2005.tb00114.x

Officer SJ, Dunbabin VM, Armstrong RD et al (2009) Wheat roots proliferate in response to nitrogen and phosphorus fertilisers in Sodosol and Vertosol soils of south-eastern Australia. Aust J Soil Res 47:91–102. https://doi.org/10.1071/SR08089

Oltmans RR, Mallarino AP (2015) Potassium uptake by corn and soybean, recycling to the soil and impact on soil test potassium. Soil Sci Soc Am J 79:314–327. https://doi.org/10.2136/sssaj2014.07.0272

Perna, J, Menzies NW (2010) Shoot and root growth and potassium accumulation of maize as affected by potassium fertilizer placement. In: Gilkes RP, Nattaporn (eds) Proceedings of 19th world congress of soil sciences: soil solutions for a changing world, Brisbane, Australia. International Union of Soil Sciences, Crawley, pp 5096–5099.

Petersen LA, Smith D (1973) Recovery of K_2SO_4 by alfalfa after placement at different depths in a low fertility soil. Agron J 65(5):769–772. https://doi.org/10.2134/agronj1973.00021962006500050028x

Rengel Z, Damon PM (2008) Crops and genotypes differ in efficiency of potassium uptake and use. Physiol Plant 133(4):624–636. https://doi.org/10.1111/j.1399-3054.2008.01079.x

Rich SM, Watt M (2013) Soil conditions and cereal root system architecture: review and considerations for linking Darwin and Weaver. J Exp Bot 64(5):1193–1208. https://doi.org/10.1093/jxb/ert043

Robins SG, Voss RD (1991) Phosphorus and potassium stratification in conservation tillage systems. J Soil Water Conserv 46(4):298–300

Rosolem CA, da Silva RH, de Fatima Esteves JA (2003) Potassium supply to cotton roots as effected by potassium fertilization and liming. Pesquisa Agropecuária Brasileira 38(5):635–641. https://doi.org/10.1590/S0100-204X2003000500012

Saarela I, Vuorinen M (2010) Stratification of soil phosphorus, pH and macro-cations under intensively cropped grass ley. Nutr Cycl Agroecosyst 86:367–381. https://doi.org/10.1007/s10705-009-9298-z

Sparks DL, Huang PM (1985) Physical chemistry of soil potassium. In: Munson RD (ed) Potassium in agriculture. American Society of Agronomy, Madison, WI, pp 201–276

Vyn TJ, Galic DM, Janovicek KJ (2002) Corn response to potassium placement in conservation tillage. Soil Till Res 67(2):159–169. https://doi.org/10.1016/s0167-1987(02)00061-2

Welch LF, Flannery RL (1985) Potassium nutrition of corn. In: Munson RD (ed) Potassium in agriculture. American Society of Agronomy, Madison, WI, pp 647–664

White RG, Kirkegaard JA (2010) The distribution and abundance of wheat roots in a dense, structured subsoil—implications for water uptake. Plant Cell Environ 33(2):133–148. https://doi.org/10.1111/j.1365-3040.2009.02059.x

Williams JD, McCool OK, Reardon CL et al (2013) Root:shoot ratios and belowground biomass distribution for Pacific Northwest dryland crops. J Soil Water Conserv 68(5):349–360. https://doi.org/10.2489/jswc.68.5.349

Yin XH, Vyn TJ (2004) Residual effects of potassium placement for conservation-till corn on subsequent no-till soybean. Soil Till Res 75(2):151–159. https://doi.org/10.1016/s0167-1987(03)00155-7

13

Foliar Application of Potassium (K)

V. K. Singh, B. S. Dwivedi, S. S. Rathore, R. P. Mishra, T. Satyanarayana, and K. Majumdar

Abstract Potassium (K) demand by crops is almost as high as that of nitrogen (N) and plays a crucial role in many plant metabolic processes. Insufficient K application results in soil K mining, deficiency symptoms in crops, and decreased crop yields and quality. Crop K demands vary with crop types, growth patterns, nutrient needs at different physiological stages, and productivity. Science-based K application in crops needs to follow 4R Nutrient Stewardship to ensure high yield, improved farm income, and optimum nutrient use efficiency. Studies around the world report widespread K deficiency, ranging from tropical to temperate environments. Long-term experiments indicate significant yield responses to K application and negative K balances where K application is either omitted or applied suboptimally. Limited understanding of K supplementation dynamics from soil non-exchangeable K pools to the exchangeable and solution phases and over-reliance on native K supply to meet crop demand are major reasons for deficit of K supply to crops. Research on optimum timing of K fertilizer application in diverse climate–soil–crop systems is scarce. The common one-time basal K management

V. K. Singh (✉) · S. S. Rathore
Division of Agronomy, ICAR-Indian Agricultural Research Institute, New Delhi, India

B. S. Dwivedi
Division of Soil Science and Agricultural Chemistry, ICAR-Indian Agricultural Research Institute, New Delhi, India
e-mail: head_ssac@iari.res.in

R. P. Mishra
ICAR-Indian Institute of Farming Systems Reserach, Meerut, Uttar Pradesh, India

T. Satyanarayana
Division of Agronomy and Advisory Services, K Plus S Middle East FZE, Dubai, UAE

K+S Minerals and Agriculture GmbH, Kassel, Germany
e-mail: Satya.Talatam@k-plus-s.com

K. Majumdar
African Plant Nutrition Institute and Mohammed VI Polytechnic University, Ben Guerir, Morocco
e-mail: k.majumdar@apni.net

practice is often not suitable to supply adequate K to the crops during peak demand phases. Besides, changes in crop establishment practices, residue retention, or fertigation require new research in terms of rate, time, or source of K application. The current review assesses the synchrony of K supply from indigenous soil system and from external sources vis-à-vis plant demand under different crops and cropping systems for achieving high yield and nutrient use efficiency.

13.1 Introduction

Potassium (K) is required by plants in large quantities, equal to or more than nitrogen (N), and plays a key role in many metabolic processes. The arable lands are deficient in K globally, which include three-fourth of the paddy soils of China and two-third of the wheat belt of Southern Australia. Additionally, the export of agricultural products and leaching of K, particularly in sandy soils, contribute to the lowering of soil K content (Rengel and Damon 2008). Soils on which K deficiency occurs vary widely, and include acid sandy soils, waterlogged soils, and saline soils (Mengel and Kirkby 2001). Globally, the annual above-ground parts of crops (phytomass) contain 75, 14, and 60 million tons of N, P, and K, respectively, which are being utilized for other purposes (e.g., heating, animal feed, biofuels). In India, animal dung (as fuel cakes) and crop residues are used as a source of bioenergy for cooking and heating without recycling the K-rich ash or sludge back to farming land that receives only low, if any, input of K fertilizers (Hasan 2002). Globally, K is applied at a much lower level to replenish only 35% of the K removed by crops (Smil 1999). Hence widespread K deficiency is observed in major production regions of the world. Potassium-deficient plants, besides producing low yields, become susceptible to drought, excess water, high and low temperatures, and to pests, diseases, and nematodes. Soil K availability is largely governed by soil mineralogical composition. Extent and pathways of weathering of primary K-bearing minerals and the dynamic equilibrium between soil K fractions give rise to soils of varying K-supplying capacity. Most soils of great alluvial flood plains in Asia were considered to have high K fertility due to the abundant presence of K-rich clay minerals (Dobermann et al. 1998; De Datta and Mikkelsen 1985), and K was rarely found a limiting factor in crop production (Bajwa 1994). Later studies, however, indicated continuous soil K depletion due to higher K withdrawal than its supplement (Dobermann et al. 1998; Bijay-Singh et al. 2003; Yadvinder-Singh et al. 2005; Singh et al. 2013, 2014).

Recent studies conducted in intensively cultivated areas of India showed imbalanced N use, optimal to suboptimal P use, and complete neglect of K application by the farmers (Dwivedi et al. 2001; Singh et al. 2014, 2015a; Syers 2003). Timsina et al. (2013) associated soil K fertility depletion with high nutrient demand and excessive extraction of K in intensive production systems of Asia. Such depletion was further aggravated by the general practice of removing crop residues from the field for other competitive uses. This has led to widespread K deficiency in many soils, including the fine-textured soils that originally had high soil K contents. The examples include alluvial illitic soils of India (Singh et al. 2015b), lowland rice soils of Java (Sri Adiningsih et al. 1991), and vermiculitic clay soils of Central Luzon, Philippines (Dobermann and Oberthür 1997). Evidence from long-term experiments in different cropping systems in India and elsewhere showed significant yield responses to K application, and negative K balances where K application is either omitted or applied suboptimally (Dwivedi et al. 2017). Depletion of soil K has been considered as a possible cause of yield decline of rice and wheat in the long-term rice–wheat systems (RWS) of Indo-Gangetic Plains (IGP) of South Asia (Ladha et al. 2003; Regmi et al. 2002). The K content in crops depends on soil type, crop, and fertilizer input; however, concentrations in the range of 0.4–4.3% have been reported (Askegaard et al. 2004). Öborn et al. (2005) concluded in a literature survey that crop K concentrations are often well below (<2.5–3.5%) what is needed to avoid deficiency. For many crops, the critical K concentration is in the range 0.5–2% in dry matter (Leigh and Wyn Jones 1984).

Potassium demand of crops varies with crop type, growth pattern, nutrient needs at different physiological stages, and crop productivity. High K demand of crops is associated with its high extraction from soils, which may lead to declining K fertility unless the extracted K is replenished through external sources. The use efficiency of applied K varies with cropping systems (Singh et al. 2014, 2015b), soil indigenous supplying capacity, source, rate, time, and method of K application. These factors, along with the variable K availability in soils, needs to be considered while formulating K management strategies in cropping systems. The current review assesses the synchrony of K supply through soil and external sources vis-à-vis plant demand. Aspects considered include soil characteristics, indigenous K supply, residue management, crop growth behavior, uptake pattern, and the importance of synchronizing soil K supply and plant need for sustainable high crop productivity and farm income.

13.2 Why the Emphasis on Potassium?

Improving nutrient use efficiency is a global concern. However, among the primary nutrients, K often gets less attention compared with N and P. Fixen et al. (2015) reported that world partial factor productivity (PFP) for K increased between 1983 and 2007, approaching 145 kg production kg^{-1} K (Fig. 13.1). In general, PFP for K tended to increase in Africa, North America, Europe, and EU15, whereas a downward trend was observed in Latin America, India, and China. These researchers also synthesized global ranges for partial factor productivity (PFP), agronomic efficiency (AE), and recovery efficiency (RE) of K for rice–wheat systems, and reported PFP of K varying from 75 to 200 kg kg^{-1} K. Similarly, AE and RE varied from 8 to 20 kg kg^{-1} K and 30–50%, respectively.

The average RE of K in on-farm trials conducted before 1998 as summarized by Dobermann (2007) that fell in the range of 38–51%. In China, the ranges of RE and AE (Jin 2012) observed in field trials between 2002 and 2006 were 25–32% and

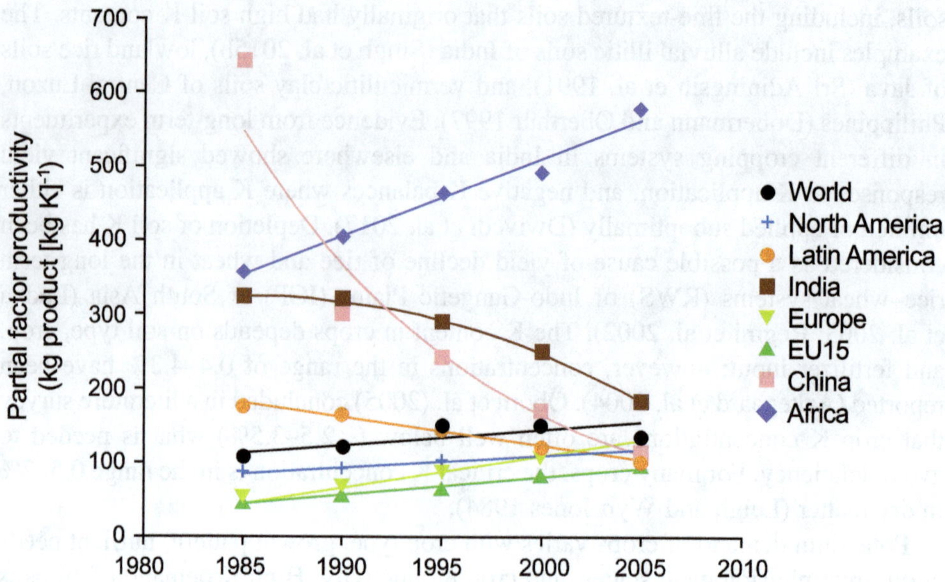

Fig. 13.1 Partial factor productivity of K in different global regions during 1983–2007. (Fixen et al. 2015)

8–12 kg kg^{-1} K for cereal crops. The RE was relatively higher (34–44%) for field trials on winter wheat in Northcentral China, while the AE values were in the range of 8–10 kg kg^{-1} K (He et al. 2012). The researchers observed that the K application rates probably exceeded the optimum for the soil K supply of individual site-year that led to lower AE values. Dobermann (2007) suggested that AE levels of 10–20 kg kg^{-1} K were realistic targets for cereals on soils that do not have high available K reserves.

In the Indian context, the need for enhancing nutrient use efficiency is felt more than ever before for at least two reasons: (i) increasing production of food grains and other crops will require more nutrients, whereas there is less likelihood of significant increase in fertilizer consumptions in foreseeable future due to economic and environmental constraints; and (ii) research established inefficient fertilizer use as one of the most important causes of soil health deterioration and lowering farm income (Tiwari 2002; Tiwari et al. 2006; Dwivedi and Meena 2015). Enhancing K use efficiency (KUE) is of particular significance, as India meets its entire K requirement through imports. Besides, the substantial increase in the retail price of K fertilizers, consequent to the implementation of nutrient-based subsidy since 2010, necessitates judicious use of K fertilizers and calls for enhancement in its use efficiency which seldom exceeds 60% KUE.

Plants acquire K from the soil solution as K^+ ions. Depending on soil type, 90–98% of soil K is relatively unavailable. The K-bearing minerals (especially feldspar and mica) are the source of soil K, which release K very slowly to the more available forms. Readily available K is composed of soil solution K and exchangeable K. In general, solution K ranges from 1 to 2% and exchangeable K from 1 to 10%. An equilibrium exists between non-exchangeable, exchangeable, and soil solution K. Because of this equilibrium, some of the soluble K applied as fertilizer gets temporarily converted to the non-exchangeable K, thus decreasing KUE.

13.3 The Need to Synchronize Potassium Supply with Plant Demand

In most annual crops, basal application of the full dose of K fertilizer is a common practice, though it may lead to low KUE as K demand of crops varies with crop growth pattern, nutrient needs at different physiological stages, and productivity level. For instance, both rice and wheat require large quantities of K, and a sustained supply is necessary until the heading or reproductive stage is over. In fact, most of the applied K before/at the time of sowing got exhausted from the soil through its further transformations to other K pools, resulting in its lower availability during the crop reproductive phase. Thus, an unsynchronized K supply often affects economic yield adversely.

Potassium is highly mobile in plants and differences between genotypes in the efficiency of K utilization have been attributed to differences in their capacity to translocate K at a cellular and whole plant level (Fig. 13.2). Under K deficiency, cytosolic K activity is maintained at the expense of vacuolar K activity (Leigh 2001, Memon et al. 1985), even though vacuolar (but not cytosolic) K activity is regulated differently in the root and leaf cells (Cuin et al. 2003). The capacity to translocate K between organs may also be an important mechanism for efficient K utilization within the plant (Dunlop and Tomkins 1976).

Potassium is not a constituent of any of the cell organelle, but it has a substantial regulatory role in the growth and development of plants. Since K is involved in >60

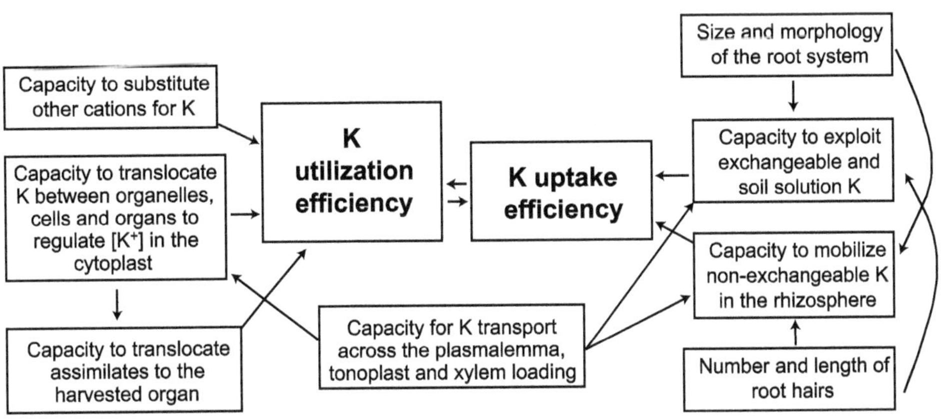

Fig. 13.2 Potential mechanisms of K uptake and utilization efficiency and their interactions that influence the expression of the K-efficient phenotype. (Pettigrew 2008)

enzymatic systems in plants, often referred to as the "the regulator." Potassium helps crop plants in the synthesis of carbohydrates, regulates the opening and closing of stomata and affects root growth which is required for efficient water use. The K stimulates active sites of enzymes for reaction. The K amount in the cell determines activation and rates of chemical reaction of enzymes. Therefore, the rate of a given enzymatic reaction is regulated by K entering the cell. When K moves into the guard cells around the stomata, the cells accumulate water and swell, causing the pores to open and allowing smooth gas exchange. When the water supply is short, K is pumped out of the guard cells. Under water stress, stomata close tightly to prevent excessive loss of water. Accretion of K in plant roots produces a gradient of osmotic pressure that draws water into the roots. Plants deficient in K are thus less able to absorb water and are more prone to stress when water is scarce.

Potassium involvement in ATP production and activation of enzymes is undoubtedly more significant in regulating the rate of photosynthesis than is the role of K in stomatal activity. The loading of photosynthates in sink is regulated by K. The plant transport system uses energy in the form of ATP. If K is inadequate, less ATP is available, and the transport system breaks down. This causes photosynthates to build up in the leaves, and the rate of photosynthesis is reduced. As a result, the normal development of energy storage organs, such as grain, is impeded. An adequate K supply helps to keep all of these processes and transportation systems functioning normally. Translocation of nitrate, phosphate, calcium (Ca), magnesium (Mg), and amino acids is decreased under short supply of K. During the reproductive stages of crop plants, K and Mg in source leaves play a critical role by ensuring an adequate supply of sucrose and K, Mg, N, P, and S to the grain filling, fruits, and tubers (Kirkby and Römheld 2004). The enzyme nitrate reductase catalyzes the formation of proteins, and K is likely responsible for its activation and synthesis. When plants are deficient in K, proteins are not synthesized despite an abundance of available N. The importance of high harvest index as a mechanism of K utilization efficiency has been widely documented for a number of species, including wheat (Damon et al. 2007, Woodend and Glass 1993, Zhang et al. 1999). Even short periods of K deficiency, especially during critical developmental stages, can cause serious losses. Potassium improves the physical quality, disease resistance, and shelf life of fruits and vegetables and the feeding value of grain and forage crops, and fiber quality of cotton. Quality can also be affected in the field before harvesting such as when K reduces lodging. Thus, K is required by the plants throughout the life cycle, and the synchronized supply of K with demand ensures not only high productivity but also the quality of the produce.

The critical K demand stages of some of the important crops are given in Table 13.1.

Multi-locational studies conducted under cereal-based systems in IGP indicated that high K demand of crops is associated with its high extraction from soils, which may lead to declining K fertility unless replenished through external sources (Dwivedi et al. 2017; Singh et al. 2014, 2018). Other factors that influence the efficiency of applied K in the crops are the cropping system followed, indigenous K supplying capacity of soil, and source, rate, time, and method of K application.

Table 13.1 Critical potassium demand stages in different crops

Crop[a]	Critical potassium demand crop stages
Rice	Initial growth (15–45 days after transplanting) and panicle emergence
Wheat	Initial crop growth (30 days after sowing), tillering, and reproductive stage
Maize	Initial crop growth (VE to V3), knee-high (V5 to V7) stage, and reproductive stage (tasseling, silking, and cob formation)
Cotton	Boll formation and fiber formation
Sugarcane	Grand growth phase (45–90 days after sowing) and sugar formation stage (120–170 days after sowing)

[a]Cotton (*Gossypium hirsutum* L.); maize (*Zea mays* L.); rice (*Oryza sativa* L.); sugarcane (*Saccharum giganteum* (Walter) Pers.); wheat (*Triticum aestivum* L.)

13.4 Managing Potassium to Synchronize Supply with Plant Demand

Soil K remains in four pools, which differ in availability to crop plants. These are soil solution K, exchangeable K, non-exchangeable K (positioned in interlayers of clay minerals, especially those of the 2:1 type), and structural K (Moody and Bell 2006; Majumdar and Sanyal 2015). Potassium that is dissolved in soil water (water-soluble) plus that held on the exchange sites on clay particles (exchangeable K) is considered readily available for plant use. The soluble and exchangeable K are the forms of K measured by the routine soil testing procedure. Plants absorb K from the soil solution pool exclusively, which is in a dynamic equilibrium with the exchangeable and, to a minor extent, the non-exchangeable pools. Soil solution plays a pivotal role in providing the pathway for K uptake from the soil to the plant roots. This pool is very low in K concentration, representing only a few percent of total crop demand at any given time (McLean and Watson 1985). In such soils, the exchangeable K pool can make a considerable contribution (80–100%) to available K for plants (Hinsinger and Jaillard 2002) in some cases. Exchangeable K can be rapidly released from exchange sites on the surfaces of clay minerals and organic matter to replenish the K-depleted soil solution (Steingrobe and Claassen 2000). Non-exchangeable K can also be released into soil solution, when the soil solution K concentration dropped below 3.5 mM (Springob and Richter 1998). But non-exchangeable K release from interlayer sites of clay minerals is a sluggish process and is mostly vital in contributing to the renewal of the soil solution and exchangeable pools in the long run (Pal et al. 2001a, 2002). The release of structural K into soil solution can be affected only by weathering of clay minerals; hence, it is a slow process with negligible effect during a single crop cycle (Pal et al. 2001b). Crop uptake synchrony is largely determined by the fate of K in the soil–plant system, and the K cycle is a powerful depiction to understand this relationship.

The rhizosphere environment has a profound effect on the native K supply. For example, solution K concentrations remain high in flooded rice soils because large amounts of soluble Fe^{2+}, Mn^{2+}, and NH_4^+ ions brought into solution displace cations from the clay complex, and exchangeable K is released. The displacement and release of K from the exchange complex, however, ceases on return to aerobic conditions during succeeding crops like wheat, maize, etc. (Timsina et al. 2010). In fields with adequate drainage, K and other basic cations can be lost via leaching. The leaching losses of K can be substantial in highly permeable soils with low cation-exchange capacities. Yadvinder-Singh et al. (2005) found that leaching losses of K were 22% and 16% of the applied K, respectively, in sandy loam and loamy soil maintained at submerged moisture regimes. In Bangladesh, such losses were as high as 0.1–0.2 kg K ha^{-1} day^{-1} (Timsina and Connor 2001). In RWS of South Asia, a common practice is to apply the full basal dose of K fertilizer at the puddling of rice and at the sowing of wheat. In well-drained soils having low cation-exchange capacity, basal application of K to rice should be avoided. As both rice and wheat require large quantities of K, a sustained supply is necessary through the heading stage or the reproductive stage is over. On coarse-textured soils, split application of K fertilizer in both rice and wheat may give higher nutrient use efficiency than its single application due to a reduction in leaching losses and luxury consumption of K (Tandon and Sekhon 1988). Tiwari et al. (1992) cited several references showing distinct benefits of split K applications. In Indian Punjab, Kolar and Grewal (1989) reported an average grain yield advantage of 250 kg ha^{-1} by split application of K (half at transplanting + half at the tillering stage of rice), compared with a single application at transplanting.

On-farm studies conducted in RWS in the IGP indicated that the initial non-exchangeable soil K concentration before rice planting ranged from 1228 to 3145 mg kg^{-1} across 60 farmers' fields, and the yield gain from applied K was relatively constant across the range of non-exchangeable K (Fig. 13.3) (Singh et al. 2013). The relatively small difference in yield gain from applied K across the exchangeable soil K range of 60–162 mg kg^{-1} raises concerns about the effectiveness of soil testing based only on the assessment of exchangeable soil K to detect the probable crop response to applied K for RWS in northern India. Non-exchangeable soil K might be particularly important in the illite-dominated soils of the IGP, and release and plant uptake of K from this soil fraction might mask the K supplied from the exchangeable K fraction (Bijay-Singh et al. 2003).

Other studies in rice/maize systems in the IGP showed that application of 75 kg K_2O ha^{-1} to each crop in two-splits, basal and panicle emergence in rice and basal and pre-silking in maize, significantly improved K uptake and grain yield as compared to a single full basal application (Table 13.2).

Surface residue retention in zero-till winter maize in rice/maize systems increased K uptake, indicating better synchrony of K supply with crop demand. The increased K uptake by crops growing in plots with residue could be due to increased K input from crop residues (Table 13.3). Other authors also showed increased K availability in upper soil layers when residues are retained on the surface under no-till systems (Franzluebbers and Hons 1996). The surface-retained residues decompose slowly (Kushwaha et al. 2000; Balota et al. 2004) and may reduce rapid leaching of K through the soil profile, which is more likely with the incorporation of crop residue into the soil. Increased number of macropores and better aggregation in zero-till conditions increased root growth below the 15-cm soil depth, helping K acquisition

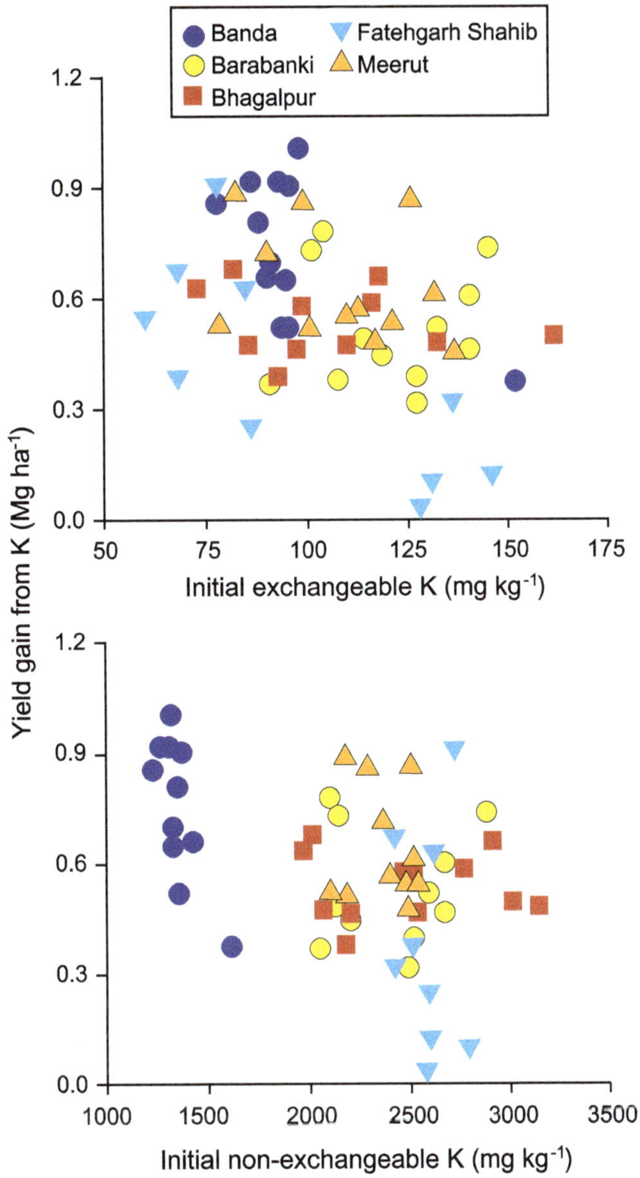

Fig. 13.3 Relationships of exchangeable K and non-exchangeable K with yield gain in rice from applied K across five locations in northern India

from deeper soil layers by rice and maize (Singh et al. 2018; Chakraborty et al. 2008). Surface-retained residues also reduce heat flux in and out of the soil by reflecting a large part of the solar radiation, as compared to no residues on the soil surface (Singh et al. 2018). Residues buffered soil temperature by 3.5–10° C during winter (early growth of maize) and by 0.8–4.8° C during March–April (reproductive stage of maize). An increase in soil temperature in the winter helps stimulate root and shoot growth, while a reduction in soil temperature during the grain filling stage alleviates the adverse impact of heat stress on maize productivity (Gupta et al. 2010; Wilhelm et al. 2004; Acharya et al. 1998). Researchers have observed higher moisture content in surface soil layers where crop residues are retained, increased

Table 13.2 Effect of split K fertilizer application (basal and/or pre-silking) and residue retention (removed or retained) on yield, K agronomic efficiency (AE_K), and K recovery efficiency (RE_K) under rice–maize system

			Yield		AE_K		RE_K	
K rate			Rice	Maize	Rice	Maize	Rice	Maize
kg K_2O ha^{-1}	Timing	Residue	Mg ha^{-1}	Mg ha^{-1}	kg grain (kg K)$^{-1}$		kg K (kg K)$^{-1}$	
0	–	–	7.91	7.8	–	–	–	–
75	B	rem	8.38	8.4	7.58	9.68	0.524	0.502
75	B+P	rem	8.69	8.90	12.58	17.74	0.608	0.584
75	B	ret	8.86	9.37	15.32	25.32	0.450	0.411
75	B+P	ret	9.10	9.54	19.19	28.06	0.533	0.528

B, basal (at sowing); P, pre-silking (V8-V10); rem, residue removed; ret, residue retained

Table 13.3 Interactive effects of crop establishment methods and residue management on total K uptake in maize (5-year cumulative averages)

	Cumulative K uptake in maize[a] (kg ha^{-1})	
	Residue applied (t ha^{-1})	
	0	4
Transplanted flooded rice/conventional till maize	859 bB	930 bA
Conventional till, direct-seeded rice/conventional till maize	921 bB	1018 aA
Zero-till, direct seeded rice/zero-till maize	999 aB	1124 aA

[a]Within a given level of residue application, cumulative K uptake quantities of different cropping systems that are followed by different lowercase letters are significantly different ($p < 0.05$); within a given cropping system, cumulative K uptake quantities associated with the two levels of residue application that are followed by different capital letters are significantly different ($p < 0.05$)

the thermal capacity of the soil, and reduced the soil temperature regime (Govaerts et al. 2009; Verhulst et al. 2010).

Potassium is required in large quantities by cotton, with peak uptake rates ranging from 3 to 5 kg K ha^{-1} day^{-1} (Halevy 1976). The developing cotton bolls are the largest K sink in the cotton plant (Howard et al. 1998). The K requirement dramatically increases during boll formation of cotton and, therefore, K application becomes crucial during the reproductive stage for higher yield (Abaye 2009). The rate of K uptake is slow during the seedling stage, about 10% of the total, but increases rapidly at flowering and reaches a maximum of 4.6 kg ha^{-1} day^{-1} between 72 and 84 days after planting (Halevy 1976). Mullins and Burmester (1990) reported maximum accumulation of K in cotton at the start of flowering, and the K uptake was highest during mid-bloom and then declined rapidly as the boll matured. They reported a maximum K uptake rate of 2.5–3.9 kg K ha^{-1} day^{-1} at flowering from 63 to 98 days after planting. During the critical period of simultaneous boll set, growth, and development, inadequate K uptake by the roots causes boll abortion and shedding (Pettigrew 2008). Obviously, lack of synchrony between the plant demand and soil K supply results in decreased fiber quality and lowered yields. In Pakistan, application of 100 kg K_2O ha^{-1} as two equal splits, and 200 kg K_2O ha^{-1} as four

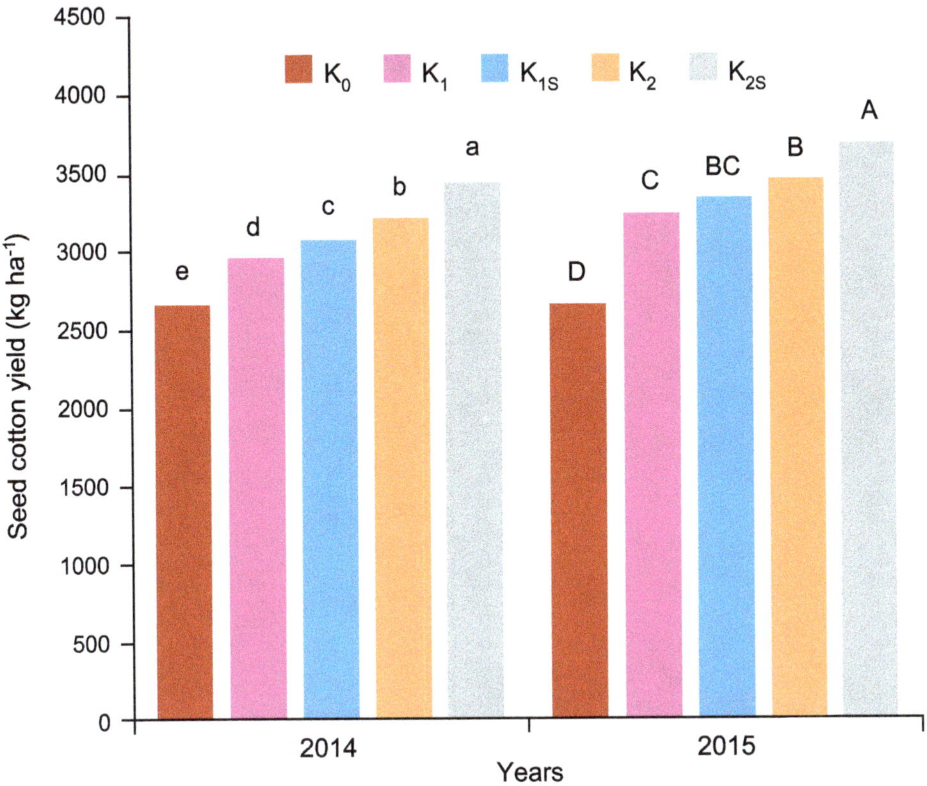

Fig. 13.4 Effect of K fertilization rates and application time on seed cotton yield in Pakistan. K_0 = No-K; K_1 = 100 kg K_2O ha^{-1} single basal application; K_{1s} = 100 kg K_2O ha^{-1} (50% at sowing and 50% at 45 days after sowing; K_2 = 200 kg K_2O ha^{-1} basal application; K_{2s} = 200 kg K_2O ha^{-1} (25% at sowing; 25% at 30 days after sowing; 25% at 45 days after sowing; and 25% at 60 days after sowing). Columns with the same letters do not significantly differ (LSD=0.05). Adapted from Muhammad et al. (2016)

equal splits produced 3.6% and 7% higher seed-cotton yield, respectively, compared to a single K fertilizer dose applied at sowing (Muhammad et al. 2016) (Fig. 13.4). The K requirement of cotton can be met by pre-plant soil application and/or mid-season side-dressing of K.

Foliar K applications offer the opportunity to correct the deficiency, especially at later growth stages when soil application may not be effective. The K concentration in leaf petioles of cotton improved significantly at 70, 90, and 110 days after sowing when 2% KNO_3 sprays were made at critical crop demand stages (Jyothi et al. 2016, Table 13.4). Studies conducted in Tennessee, USA, reported increased cotton yield following foliar K fertilization when applied in a no-till system (Abaye 2009). Three to four foliar applications of K (2.0% w/v K_2SO_4) are recommended during peak boll development at 7–10-day intervals, beginning about 2 weeks after flower initiation.

In sugarcane-based cropping systems of northwest India, application of 120 kg K_2O ha^{-1} to both the initially planted crop as well the succeeding ratoon crop had the

Table 13.4 Potassium concentration in cotton leaf petioles as influenced by foliar KNO_3 application

Fertilizer rate	Days after sowing						Harvest
	70		90		110		
	Before	After	Before	After	Before	After	
	g kg^{-1}	g kg^{-1}	g kg^{-1}	g kg^{-1}	g kg^{-1}	g kg^{-1}	g kg^{-1}
Recommended soil dose	32.3	31.6	31.1	30.4	23.9	23.1	8.1
Recommended dose + 2% KNO_3 spray	45.1	47.6	41.6	44.6	37.4	33.9	11.6
SEM[a]	0.6	0.3	0.4	0.4	0.5	0.2	0.2

Jyothi et al. (2016)

[a]Standard error of the mean

Table 13.5 Effect of split application of 120 kg K ha^{-1} on yield and potassium recovery efficiency (RE_K) of sugarcane and wheat under sugarcane–ratoon–wheat system

Potassium rate and time of application	Yield			RE_K		
	Sugarcane plant crop	Sugarcane ratoon	Wheat	Sugarcane plant crop	Sugarcane ratoon crop	Wheat
kg K_2O ha^{-1}	Mg ha^{-1}			kg K (kg K)$^{-1}$		
0	49.4	61.7	3.7	–	–	–
120 to sugarcane plant crop as basal	56.4	68.2	3.5	0.513	–	0.514
60 to plant crop as basal and 60 to plant crop at grand growth phase	67.4	80.6	3.9	0.678	–	0.521
60 to plant crop as basal and 60 to ratoon as basal	55.7	81.3	4.1	0.546	0.597	0.589
60 to ratoon as basal and 60 to ratoon at grand growth phase	68.6	92.3	4.8	0.668	0.736	0.602

Singh et al. (2008)

highest biomass productivity, compared with K application to planted crop only (Table 13.5). Additionally, K application in two splits (basal and at grand growth stage) resulted in higher yield, K use efficiency, and juice quality (Singh et al. 2008). In Florida, the recommended application rate has been as high as 450 kg K_2O ha^{-1} for the first growth cycle (ratoon), then reduced to 270 kg K_2O ha^{-1} for the second and third ratoons (Rice et al. 2006). Hunsigi (2011) suggested application of up to 350 kg K_2O ha^{-1} and 117 kg K_2O ha^{-1} for the first and second ratoons of sugarcane, respectively. Singh et al. (2008) settled on a standard rate of 150 kg K_2O ha^{-1} in their experimental studies in India, while De Oliveira et al. (2016) concluded that 98 kg K_2O ha^{-1} would be sufficient to obtain a stable sugarcane yield of 80 t ha^{-1}.

13.5 Fertigation for Synchronized Potassium Supply

Fertigation provides an excellent opportunity to optimize nutrient application to crops, with co-benefits of greater nutrient and water use efficiencies and economic return compared with soil application. The nutrient application rate and time can be more precisely regulated in fertigation to match crop demand at different growth stages. Another advantage of applying K through fertigation is that it minimizes the risk applied K being removed from the solution through clay fixation. Synchronized application of K with plant demand and applying the fertilizer K precisely in the root zone reduces the residence time of the K fertilizer in the soil, thus minimizing the contact of the K ions with the soil volume. However, when plant K demands are

Table 13.6 Potassium recommendation (kg K_2O ha^{-1}) at different stages of crops under micro-irrigation-based fertigation

	Transplanting to 6 leaf stage	6 leaf to fruit setting stage	Fruit setting stage to fruit development stage	Fruit setting stage to ripening stage	Total
Crops[a]	kg K_2O ha^{-1}	kg K_2O ha^{-1}	kg K_2O ha^{-1}	kg K_2O ha^{-1}	kg K_2O ha^{-1}
Potato	16	58	196	–	270
Tomato	58	35	45	172	310
Bell pepper	58	35	45	172	310
Onion	60	32	171	32	295
Red cabbage	32	35	173	–	240
Carrot	120	171	171	–	400
Lettuce	122	35	–	–	90
Cucumber	121	35	35	–	125
Watermelon	10	25	75	–	120

Unpublished data from authors

[a]Bell pepper (*Capsicum annuum* L.); carrot (*Daucus carota* subsp. *sativus* (Hoffm.) Arcang.); cucumber (*Cucumis sativus* L.); lettuce (*Lactuca sativa* L.); onion (*Allium cepa* L.); potato (*Solanum tuberosum* L.); red cabbage (*Brassica oleracea* L.); tomato (*Solanum lycopersicum* L.); watermelon (*Citrullus lanatus* (Thunb.) Matsum. & Nakai)

high, supplying the entire plant K requirement via fertigation will require a continuous nutrient supply throughout the growing season to ensure optimum yield and quality (Kafkafi and Tarchitzky 2011). Gupta et al. (2015) suggested drip irrigation at 80% evapotranspiration (ET) and fertigation with 60% of the recommended NPK for improving yield and water and nutrient use efficiencies in tomato. Fertilizers with high solubility are used for fertigation. Several of the common K fertilizers such as potassium chloride (KCl; MOP), potassium sulfate (SOP), monopotassium phosphate (MKP), and potassium nitrate (KNO_3) are excellent sources for fertigation. Potassium recommendations for different crops grown under micro-irrigation-based fertigation are given in Table 13.6.

13.6 Use of Decision Support Tools

The fertilizer decision support tool, Nutrient Expert®, developed by the International Plant Nutrition Institute (IPNI) along with several national and international partners, has been successfully used to estimate site-specific K recommendation for major cereal crops and cropping systems. On-farm studies in different agro-ecologies in India showed that the tool-based K recommendations for rice–wheat, rice–rice, rice–maize, and maize–wheat systems, based on the nutrient demand of a

Table 13.7 Effect of Nutrient Expert® (NE)-based K inputs on grain yield and K uptake under cereal-based cropping systems

Cropping system	Crop[a]	K use			Grain yield			K uptake	
		FFP	NE	Diff. (NE-FFP)	FFP	NE	Yield gain over FFP	FFP	NE
		kg K_2O ha^{-1}			Mg ha^{-1}		%	kg K ha^{-1}	
Rice–wheat	Rice	40	67	+27	5.1	5.83	14	104	139
	Wheat	0	64	+64	3.66	4.18	14	89	101
Rice–rice	Rice (m)	60	61	+1	4.28	4.98	16	81	106
	Rice (w)	52	46	+6	4.67	5.38	15	91	124
Maize–wheat	Maize	19	44	+25	2.1	3.5	67	57	78
	Wheat	0	63	+63	2.21	3.51	59	52	81

FFP, farmer fertilizer practice; NE, Nutrient Expert® recommendations
Unpublished data from authors
[a]Maize (*Zea mays* L.); rice (*Oryza sativa* L.); rice (m), rice grown during the monsoon season; rice (w), winter rice; wheat (*Triticum aestivum* L.)

Table 13.8 A comparison of the effects of a single basal K application (State recommendation) with split K applications (recommended by Nutrient Expert®) for two different maize yield targets on: grain yield, K partial factor productivity (PFP_K), and grain produced per rupee invested

K recommendation	Fertilizer K rate			Grain yield	PFP_K	Grain produced per rupee invested
	Basal	40 days after sowing	Total			
	kg K ha^{-1}			Mg ha^{-1}	kg grain (kg K)$^{-1}$	kg grain (rupee invested)$^{-1}$
State	62	–	62	4.10	66.1	1.68
NE using a 6 Mg ha^{-1} yield target	38	25	63	5.59	88.7	2.26
NE using a 7 Mg ha^{-1} yield target	40	27	67	6.19	92.4	2.35

NE, Nutrient Expert; State, state recommendation
Unpublished data from authors

high-yielding cereal crop and the soil nutrient supplying capacity, increased crop yield and K uptake over the existing farmers' fertilizer practices (Table 13.7).

The Nutrient Expert considers the yield targets along with soil indigenous nutrient supplying capacity to determine the K application rate. It also recommends the time of application to meet K demand at the right physiological growth stages of the crop. Such synchronized K supply to the crops led to higher K uptake and K use efficiencies. A recent study conducted in IGP indicated that the K recommendations from the Nutrient Expert varied with changing yield targets. The yield gain and efficiency parameters (partial factor productivity and economic KUE) were more with highest yield targets (7 t ha^{-1}) (Table 13.8).

13.7 Future Thrusts

The timing of K fertilizer application to crops is site- and crop-specific. Typically, K fertilizers are applied at the time of land preparation as K is relatively immobile in the soil. It presupposes that the basally applied K will be held by the soil matrix and will continue to meet the K demand throughout the crop growth period. However, the variability in soil types in terms of cation-holding capacities and texture may change the mobility of K ions within the soil. For example, a highly weathered red and lateritic soil would most likely not have enough K-holding capacity, and the fast percolation of water in such soils can facilitate the loss of K from the root zone. Under such situations, a split application of K is appropriate to ensure an adequate supply at the right physiological stages. Research is necessary to outline the best time of application of K fertilizers in the crop–soil context that integrates the soil characteristics, duration of the crop-growing period, physiological stages of high K demand, and the influence of seasonal differences on physiological growth stages.

It is well recognized that K plays a significant role in the transport of photosynthates to economic plant parts during the reproductive phase. So splitting K applications to match the demand at the initial growth stage and at the reproductive phase can ensure better productivity and KUE. Evidence from recent research in rice nutrition showed significant gains when the basal K application is skipped, and K is applied at the time of maximum tillering and panicle initiation. This particularly makes sense for puddled transplanted rice, where submergence releases soil matrix-held K into a solution that can meet the K demand of early growth, and the external K application could be done at the reproductive stage. More research is needed, not only for rice but also for other crops, to outline suitable fertilizer-splitting strategies for K, and the proportion of K fertilizer to be applied in each split.

Developing a better fertilizer-splitting strategy for crops has another practical advantage, particularly in areas where access to K fertilizers is limited. Many farmers often skip K application entirely when they cannot apply fertilizer K at the time of land preparation due to the unavailability of K fertilizer in the market. A better understanding of the benefits of late-applied K on yield, quality, and nutrient use efficiencies will provide farmers incentive to apply K at later growth stages for better return on investment. Another important area of research is optimizing K application in cropping systems where multiple crops are grown in sequence. Significant gains, both in crop yield and KUE could be achieved by developing application strategies that are based on the K requirement of the component crops grown in sequence and the crop growing ecology. For example, in a rice (anaerobic)–maize (aerobic) cropping system, better utilization efficiency could likely be achieved by applying the lower proportion of the total system K requirement to rice and the rest higher proportion to maize. Similar research on intercropping is necessary wherein very little guidance is available regarding K application to match crop demand.

Crop establishment under zero-tillage and the return of crop residues also have a profound impact on K dynamics in soils that influence the optimal timing and rate of K fertilizer application. However, in most cases, protocols of K fertilizer application in conventional tillage systems are replicated in no-till systems, with limited understanding of K release from crop residues and their utilization by crops or loss from the system. Significant knowledge gaps are also evident in farming systems where crops receive nutrients through fertigation or through foliar fertilization. What should be the rates or timing of K application in such systems where expected nutrient use efficiencies are significantly higher than soil-applied nutrients need further research to accrue full benefits of investments made on deploying foliar or fertigation systems.

In general, the right time of K fertilizer application rates for major crops like rice, wheat, maize, plantation, and cash crops are well defined. But much remains to be done to understand how to synchronize K application with plant demand for cropping systems, intercrops, and minor crops that fit into specific cropping systems where significant variabilities challenge timing decisions.

13.8 Conclusion

Potassium is a major nutrient required for many physiological functions in the plants. Adequate K application based on specific crop requirements (quantity) and synchronizing it with the high K demand stages of the crops (timing) are essential for optimum crop growth and productivity. Matching K application with the crop demand also ensures the effective utilization of applied K nutrients by the plants leading to higher KUE and return on investment. Despite high K requirements, its applications in crops are limited and a negative partial balance of K is common in most of the intensively cultivated regions of the world. This chapter, besides summarizing the benefits of adequate and synchronized K input on yield, produce quality, and KUE, also highlights significant knowledge gaps in defining the most appropriate time of K fertilizer application in specific climate–soil–crop combinations. Intensive agriculture systems around the world will benefit by bridging these knowledge gaps to produce more from less lands with the efficient use of K fertilizers. The foregoing discussions established the significance of judicious K input for enhancing productivity and use efficiency of K and other nutrients in cropping systems. The myths of adequacy of soil K in the alluvial soils, such as in the Indo-Gangetic Plain, have been negated by profuse crop responses to K fertilization in these regions. Studies on synchronizing K supplies with crop demand are, however, scarce despite their significance in improving K fertilizer recommendations. Future investigations should involve crop- and soil-specific comparison of split application vis-à-vis conventional one-time application with respect to crop yields, use efficiency, native K mining, and losses from root zone.

References

Abaye AO (2009) Potassium fertilization of cotton. Virginia Cooperative Extension, 418-025. Available from: https://www.pubs.ext.vt.edu/418/418-025/418-025.html. Accessed May 21 2020

Acharya CL, Kapur OC, Dixit SP (1998) Moisture conservation for rainfed wheat production with alternative mulches and conservation tillage in the hills of northwest India. Soil Till Res 46:153–163. https://doi.org/10.1016/S0167-1987(98)00030-0

Askegaard M, Eriksen J, Johnston AE (2004) Sustainable management of potassium. In: Schjønning P, Elmholt S, Christensen BT (eds) Managing soil quality—challenges in modern agriculture. CAB, Wallingford, UK, pp 85–102

Bajwa MI (1994) Soil potassium status, potash fertilizer usage and recommendations in Pakistan. Potash Review, No. 3, International Potash Institute, Berne, Switzerland

Balota EL, Colozzi A, Andrade DS, Dick RP (2004) Long-term tillage and crop rotation effects on microbial biomass and C and N mineralization in a Brazilian Oxisol. Soil Till Res 77:137–145. https://doi.org/10.1016/j.still.2003.12.003

Bijay-Singh, Yadviner-Singh, Imas P, Xie JC (2003) Potassium nutrition of the rice–wheat cropping system. Adv Agron 81:203–259. https://doi.org/10.1016/S0065-2113(03)81005-2

Chakraborty D, Shantha N, Aggarwal P, Gupta VK, Tomar RK, Garg RN, Sahoo RN, Sarkar A, Chopra UK, Sundara Sarma KS, Kalra N (2008) Effect of mulching on soil and plant water status and the growth and yield of wheat (*Triticum aestivum* L.) in a semi-arid environment. Agric Water Manag 95:1323–1334. https://doi.org/10.1016/j.agwat.2008.06.001

Cuin TA, Miller AJ, Laurie SA, Leigh RA (2003) Potassium activities in cell compartments of salt-grown barley leaves. J Exp Bot 54:657–661. https://doi.org/10.1093/jxb/erg072

Damon PM, Osborne LD, Rengel Z (2007) Canola genotypes differ in potassium efficiency during vegetative growth. Euphytica 156:387–397. https://doi.org/10.1007/s10681-007-9388-4

De Datta SK, Mikkelsen DS (1985) Potassium nutrition of rice. In: Munson RD (ed) Potassium in agriculture. ASA/CSSA/SSSA, Madison, WI, pp 665–699. https://doi.org/10.2134/1985.potassium.c30

De Oliveira RI, De Medeiros MRFA, Freire CS, Freire FJ, Neto DES, De Oliveira ECA (2016) Nutrient partitioning and nutritional requirement in sugarcane. Aust J Crop Sci 10:69–75

Dobermann A (2007) Nutrient use efficiency—measurement and management. In: Fertilizer best management practices: general principles, strategy for their adoption and voluntary initiatives vs regulations. Proceedings of IFA international workshop on fertilizer best management practices, 7–9 Mar 2007, Brussels, Belgium. International Fertilizer Industry Association, Paris, France, pp 1–28

Dobermann A, Oberthür T (1997) Fuzzy mapping of soil fertility—a case study on irrigated rice land in the Philippines. Geoderma 77:317–339. https://doi.org/10.1016/S0016-7061(97)00028-1

Dobermann A, Cassman KG, Mamaril CP, Sheehy JE (1998) Management of phosphorus, potassium, and sulfur in intensive, irrigated lowland rice. Field Crops Res 56:113–138. https://doi.org/10.1016/S0378-4290(97)00124-X

Dunlop J, Tomkins B (1976) Genotypes differences in potassium translocation in rye-grass. In: Wardlaw IF, Passioura JB (eds) Transport and transfer process in plants. Academic Press, New York, pp 145–152. https://doi.org/10.1016/B978-0-12-734850-6.50018-6

Dwivedi BS, Meena MC (2015) Soil testing service—retrospect and prospects. Ind J Fertil 11 (10):110–122

Dwivedi BS, Shukla AK, Singh VK, Yadav RL (2001) Results of participatory diagnosis of constraints and opportunities based trials from the state of Uttar Pradesh. In: Rao S, Srivastava S (eds) Development of farmers' resource-based integrated plant nutrient supply systems: experience of a FAO-ICARIFFCO collaborative project and AICRP on soil test crop response correlation. Indian Institute of Soil Science, Bhopal, India, pp 50–75

Dwivedi BS, Singh VK, Shekhawat K, Meena MC, Dey A (2017) Enhancing use efficiency of phosphorus and potassium under different cropping systems of India. Ind J Fertil 13(8):20–41

Fixen P, Brentrup F, Bruulsema T, Garcia F, Norton R, Zingore S (2015) Nutrient/fertilizer use efficiency: measurement, current situation and trends. In: Drechsel P, Heffer P, Magen H, Mikkelsen R, Wichelns D (eds) Managing water and fertilizer for sustainable agricultural intensification. IFA, IWMI, IPNI, IPI, Paris, France, pp 8–38. https://pdfs.semanticscholar.org/4efc/c3c517d07573dfcbecb7b6ca98874879dba6.pdf. Accessed May 21 2020

Franzluebbers AJ, Hons FM (1996) Soil-profile distribution of primary and secondary plant available nutrients under conventional and no tillage. Soil Till Res 39:229–239. https://doi.org/10.1016/S0167-1987(96)01056-2

Govaerts B, Sayre KD, Goudeseune B, De Corte P, Lichter K, Dendooven L, Deckers J (2009) Conservation agriculture as a sustainable option for the central Mexican highlands. Soil Till Res 103:222–230. https://doi.org/10.1016/j.still.2008.05.018

Gupta R, Gopal R, Jat ML, Jat RK, Sidhu HS, Minhas PS, Malik RK (2010) Wheat productivity in Indo-Gangetic plains of India: terminal heat effects and mitigation strategies. PACA Newsl 15:1–3. https://www.researchgate.net/publication/279059151_Wheat_productivity_in_Indo-Gangetic_plains_of_India_during_2010_terminal_heat_effects_and_mitigation_strategies_PACA_Newslett. Accessed May 21 2020

Gupta AJ, Chattoo MA, Singh L (2015) Drip irrigation and fertigation technology for improved yield, quality, water and fertilizer use efficiency in hybrid tomato. J AgriSearch 2(2):94–99

Halevy J (1976) Growth rate and nutrient uptake of two cotton cultivars grown under irrigation. Agron J 68:701–705. https://doi.org/10.2134/agronj1976.00021962006800050002x

Hasan R (2002) Potassium status of soils in India. Better Crops Int 16(2):3–5

He P, Jin J, Wang H, Cui R, Li C (2012) Yield responses and potassium use efficiency for winter wheat in North-Central China. Better Crops 96(3):28–30

Hinsinger PH, Jaillard B (2002) Root-induced release of interlayer potassium and vermiculitization of phlogopite as related to potassium depletion in the rhizosphere of ryegrass. J Soil Sci 44:525–534. https://doi.org/10.1111/j.1365-2389.1993.tb00474.x

Howard DD, Gwathmey CO, Roberts RK, Lessman GM (1998) Potassium fertilization of cotton produced on a low K soil with contrasting tillage systems. J Prod Agric 11:74–79. https://doi.org/10.2134/jpa1998.0074

Hunsigi G (2011) Potassium management strategies to realize high yield and quality of sugarcane. Karnataka J Agric Sci 24:45–47

Jin J (2012) Changes in the efficiency of fertilizer use in China. J. Sci Food Agric 92:1006–1009

Jyothi TV, Hebsur NS, Sokolowski E, Bansal SK (2016) Effects of soil and foliar potassium application on cotton yield, nutrient uptake, and soil fertility status. Int Potash Inst e-ifc 46:13–21

Kafkafi U, Tarchitzky J (2011) Fertigation: a tool for efficient fertilizer and water management. International Fertilizer Industry Association, Paris

Kirkby EA, Römheld V (2004) Micronutrients in plant physiology: functions, uptake and mobility. The International Fertiliser Society, York, UK. https://fertiliser-society.org/store/micronutrients-in-plant-physiology-functions-uptake-and-mobility/. Accessed May 21 2020

Kolar JS, Grewal HS (1989) Response of rice to potassium. Int Rice Res Newsl 14:33–40

Kushwaha CP, Tripathi SK, Singh KP (2000) Variations in soil microbial biomass and N availability due to residue and tillage management in a dryland rice agroecosystem. Soil Till Res 56:153–166. https://doi.org/10.1016/S0167-1987(00)00135-5

Ladha JK, Dawe D, Pathak H, Padre AT, Yadav RL, Singh B, Singh Y, Singh Y, Singh P, Kundu AL, Sakal R, Ram N, Regmi AP, Gami SK, Bhandari AL, Amin R, Yadav CR, Bhattarai EM, Das S, Aggarwal HP, Gupta RK, Hobbs PR (2003) How extensive are yield declines in long-term rice wheat experiments in Asia? Field Crops Res 81:159–180

Leigh RA (2001) Potassium homeostasis and membrane transport. J Plant Nutr Soil Sci:164193–164198. https://doi.org/10.1002/1522-2624(200104)164:2<193::AID-JPLN193>3.0.CO;2-7

Leigh RA, Wyn Jones RG (1984) A hypothesis relating critical potassium concentrations for growth to the distribution and functions of this ion in the plant cell. New Phytol 97:1–13. https://doi.org/10.1111/j.1469-8137.1984.tb04103.x

Majumdar K, Sanyal SK (2015) Soil and fertilizer phosphorus and potassium. In: Rattan RK, Katyal JC, Dwivedi BS, Sarkar AK, Bhattacharyya T, Tarafdar JC, Kukal SS (eds) Soil science: an introduction. Indian Society of Soil Science, New Delhi, pp 571–600. ISBN: 81-903797-7-1

McLean EO, Watson ME (1985) Soil measurements of plant available potassium. In: Munson RD (ed) Potassium in agriculture. ASA, CSSA, SSSA, Madison, WI, pp 277–308. https://doi.org/10.2134/1985.potassium.c10

Memon AR, Saccomani M, Glass ADM (1985) Efficiency of potassium utilization by barley varieties: The role of subcellular compartmentation. J Exp Bot 36:1860–1876. www.jstor.org/stable/23691362

Mengel K, Kirkby EA (2001) Principles of plant nutrition, 5th edn. Springer. 849 pp. https://doi.org/10.1007/978-94-010-1009-2

Moody PW, Bell MJ (2006) Availability of soil potassium and diagnostic soil tests. Aust J Soil Res 44:265–275. https://doi.org/10.1071/SR05154

Muhammad DB, Afzal MN, Tariq M, Wakeel A (2016) Impact of potassium fertilization dose, regime, and application methods on cotton development and seed-cotton yield under an arid environment. Int Potash Inst e-ifc 45:3–10

Mullins GL, Burmester CH (1990) Dry matter, nitrogen, phosphorus, and potassium accumulation by four cotton varieties. Agron J 82:729–736. https://doi.org/10.2134/agronj1990.00021962008200040017x

Öborn I, Andrist-Range Y, Askekaard M, Grant CA, Watson CA, Edwards AC (2005) Critical aspects of potassium management in agricultural systems. Soil Use Manag 21:102–112. https://doi.org/10.1111/j.1475-2743.2005.tb00414.x

Pal Y, Gilkes RJ, Wong MTF (2001a) Mineralogy and potassium release from some Western Australian soils and their size fractions. Aust J Soil Res 39:813–822. https://doi.org/10.1071/SR00031

Pal Y, Gilkes RJ, Wong MTF (2001b) Soil factors affecting the availability of potassium to plants for Western Australian soils: A glasshouse study. Aust J Soil Res 39:611–625. https://doi.org/10.1071/SR00030

Pal Y, Gilkes RJ, Wong MTF (2002) Mineral sources of potassium to plants for seven soils from south-western Australia. Aust J Soil Res 40:1357–1369. https://doi.org/10.1071/SR02014

Pettigrew WT (2008) Potassium influences on yield and quality production for maize, wheat, soybean and cotton. Physiol Plant 133:670–681. https://doi.org/10.1111/j.1399-3054.2008.01073.x

Regmi AP, Ladha JK, Pasquin E, Pathak H, Hobbs PR, Shrestha LL, Gharti DB, Duveiller E (2002) The role of potassium in sustaining yields in a long-term rice-wheat experiment in the Indo-Gangetic Plains of Nepal. Biol Fert Soils 36:240–247. https://doi.org/10.1007/s00374-002-0525-x

Rengel Z, Damon PM (2008) Crops and genotypes differ in efficiency of potassium uptake and use. Physiol Plant 133:624–636. https://doi.org/10.1111/j.1399-3054.2008.01079.x

Rice RW, Gilbert RA, McCray JM (2006) Nutritional requirements for Florida sugarcane. In: Gilbert, RA, Rice RW, McCray JM (eds) Sugarcane handbook. Univ Florida Extension Pub #SS-AGR-228. https://edis.ifas.ufl.edu/sc028. Accessed May 21 2020

Singh VK, Tiwari R, Gill MS, Sharma SK, Tiwari KN, Dwivedi BS, Shukla AK, Mishra PP (2008) Economic viability of site-specific nutrient management in rice-wheat cropping. Better Crops-India 2:16–19

Singh VK, Dwivedi BS, Buresh RJ, Jat ML, Majumdar K, Gangwar B, Govil V, Singh SK (2013) Potassium fertilization in rice–wheat system across Northern India: Crop performance and soil nutrients. Agron J 105:471–481. https://doi.org/10.2134/agronj2012.0226

Singh VK, Dwivedi BS, Tiwari KN, Majumdar K, Rani M, Singh SK, Timsina J (2014) Optimizing nutrient management strategies for rice–wheat system in the Indo-Gangetic Plains of India and adjacent region for higher productivity, nutrient use efficiency and profits. Field Crops Res 164:30–44. https://doi.org/10.1016/j.fcr.2014.05.007

Singh VK, Shukla AK, Dwivedi BS, Singh MP, Majumdar K, Kumar V, Mishra R, Rani M, Singh SK (2015a) Site-specific nutrient management under rice-based cropping systems in Indo-Gangetic Plains: Yield, profit and apparent nutrient balance. Agric Res 4:365–377. https://doi.org/10.1007/s40003-015-0179-1

Singh VK, Shukla AK, Singh MP, Mujumdar K, Mishra RP, Rani M, Singh SK (2015b) Effect of site-specific nutrient management on yield, profit and apparent nutrient balance under pre-dominant cropping systems of Upper Gangetic Plains. Indian J Agric Sci 85:335–343

Singh VK, Dwivedi BS, Yadvinder-Singh SSK, Mishra RP, Shukla AK, Rathore SS, Shekhawat K, Majumdar K, Jat ML (2018) Effect of tillage and crop establishment, residue management and K fertilization on yield, K use efficiency and apparent K balance under rice maize system in north-western India. Field Crops Res 224:1–12. https://doi.org/10.1016/j.fcr.2018.04.012

Smil V (1999) Crop residues: Agriculture's largest harvest: Crop residues incorporate more than half of the world's agricultural phytomass. BioScience 49:299–308. https://doi.org/10.2307/1313613

Springob G, Richter J (1998) Measuring interlayer potassium release rates from soil materials. II. A percolation procedure to study the influence of the variable 'solute K' in the <1...10 μM range. J Plant Nutr Soil Sci 161:323–329. https://doi.org/10.1002/jpln.1998.3581610321

Sri Adiningsih J, Santoso D, Sudjadi M (1991) The status of N, P, K and S of lowland rice soils in Java. In: Blair G, Lefroy RDB (eds) Sulfur fertilizer policy for lowland and upland rice cropping systems in Indonesia. Australian Centre for International Agricultural Research, Melbourne

Steingrobe B, Claassen N (2000) Potassium dynamics in the rhizosphere and K efficiency of crops. J Plant Nutr Soil Sci 163:101–106. https://doi.org/10.1002/(SICI)1522-2624(200002)163:1<101::AID-JPLN101>3.0.CO;2-J

Syers JK (2003) Potassium in soils: current concepts. In: Johnston AE (ed) Feed the soil to feed the people :The role of potash in sustainable agriculture. Proceedings of IPI Golden Jubilee Congress 1952-2002, Basel, Switzerland, 8–10 Oct 2002. International Potash Institute, Basel, Switzerland, pp 301–310

Tandon HLS, Sekhon GS (1988) Potassium research and agricultural production in India. Fertilizer Development and Consultation Organisation, New Delhi. ISBN: 10:8185116059

Timsina J, Connor DJ (2001) Productivity and management of rice-wheat cropping systems: Issues and challenges. Field Crops Res 69:93–132. https://doi.org/10.1016/S0378-4290(00)00143-X

Timsina J, Jat ML, Majumdar K (2010) Rice-maize systems of South Asia: Current status, future prospects and research priorities for nutrient management. Plant Soil 335:65–82. https://doi.org/10.1007/s11104-010-0418-y

Timsina J, Singh VK, Majumdar K (2013) Potassium management in rice–maize systems in South Asia. J Plant Nutr Soil Sci 176:317–330. https://doi.org/10.1002/jpln.201200253

Tiwari KN (2002) Nutrient management: Issues and strategies. Fert News 47:23–122

Tiwari KN, Dwivedi BS, Subba Rao A (1992) Potassium management in rice-wheat system. In: Pandey RK et al (eds) Rice–wheat cropping system: Proceedings of the rice–wheat workshop, Modipuram, Meerut. Project Directorate for Cropping Systems Research, Modipuram, Meerut, India, pp 93–114

Tiwari KN, Sharma SK, Singh VK., Dwivedi BS, Shukla AK (2006) Site-specific nutrient management for increasing crop productivity in India: Results with rice-wheat and rice-rice system. PDCSR Modipuram and PPIC India Programme, Gurgaon, pp 92

Verhulst N, Govaerts B, Verachtert E, Castellanos-Navarrete A, Mezzalama M, Wall P, Deckers J, Sayre KD (2010) Conservation agriculture, improving soil quality for sustainable production systems? In: Lal R, Stewart BA (eds) Advances in soil science: food security and soil quality. CRC, Boca Raton, FL, pp 137–208, ISBN: 10:143980057X

Wilhelm WW, Johnson JMF, Hatfield JL, Voorhees WB, Linden DR (2004) Crop and soil productivity response to corn residue removal: A review of the literature. Agron J 96:1–17. https://doi.org/10.2134/agronj2004.1000

Woodend JJ, Glass ADM (1993) Genotype-environment interaction and correlation between vegetative and grain production measures of potassium use-efficiency in wheat (*T. aestivum* L.) grown under potassium stress. Plant Soil 151:39–44. https://doi.org/10.1007/BF00010784

Yadvinder-Singh, Bijay-Singh, Timsina J (2005) Crop residue management for nutrient cycling and improving soil productivity in rice-based cropping systems in the tropics. Adv Agron 85:269–407. https://doi.org/10.1016/S0065-2113(04)85006-5

Zhang G, Chen J, Tirore EA (1999) Genotypic variation for potassium uptake and utilization efficiency in wheat. Nutr Cycl Agroecosyst 54:41–48. https://doi.org/10.1023/A:1009708012381

14

Potassium (K) Fertilizer Recommendations: A Futuristic Approach

Jeffrey J. Volenec, Sylvie M. Brouder, and T. Scott Murrell

Abstract Potassium (K) fertilizer recommendations for annual crops in the USA are generally founded in soil test results. The goal of this chapter is to highlight additional plant-related traits that may impact crop responses to K fertilization. This includes the role of tissue testing, the influence of luxury consumption, genetic improvement of K use efficiency, genotype × environment × management interactions on K uptake and yield, response to foliar K fertilization, intraplant K cycling, fungal associations and K uptake, the influence of K on crop quality, and the role of K in abiotic stress tolerance. Recognizing the potential role of these plant factors may help reconcile response inconsistencies based solely on soil test information, and improve future K recommendations. Finally, we hope to highlight knowledge gaps and opportunities for additional integrated soil–plant K research.

14.1 Introduction

The impact of potassium (K) on growth and yield has been previously compiled and summarized for a wide array of agronomic and horticultural crop species (Munson 1985). In addition, the chemistry of K in soils, the K cycle, and K soil testing methods and corresponding recommendations are discussed at length elsewhere in this book, and as such, these details will not be discussed here. Collectively, these results and other research findings have been translated into an array of Extension publications used to guide soil fertility/plant nutrition practices of farmers (e.g., Vitosh et al. 1995; Buchholz et al. 2004). For K, this translation process continues in

J. J. Volenec (✉) · S. M. Brouder
Department of Agronomy, Purdue University, Indiana, USA
e-mail: jvolenec@purdue.edu; sbrouder@purdue.edu

T. S. Murrell
Department of Agronomy, Purdue University, Indiana, USA

African Plant Nutrition Institute and Mohammed VI Polytechnic University, Ben Guerir, Morocco
e-mail: tmurrell@purdue.edu

an effort to reconcile state-specific K recommendations and to resolve inconsistencies between these recommendations and incorporate findings from emerging research. The goal of this chapter is to broaden our understanding of K in crop performance by building on traditional concepts, expanding knowledge into nontraditional issues, and highlighting knowledge gaps and opportunities for additional research.

14.2 Soil and Tissue Testing

14.2.1 Soil Testing for Potassium

Traditionally, soil testing is used to predict yield responses to K fertilizer application, and this approach can work well (Slaton et al. 2010). However, the relationship between soil test K and seed yield can be inconsistent and not accurately predict crop response to fertilizer application. For example, Clover and Mallarino (2013) conducted K fertilizer response trials for maize (*Zea mays* L.) and soybean (*Glycine max* L. Merr.) yield on over 50 site-years in Iowa. The 16 site-years responsive to K fertilizer application had soil test K levels <173 mg K kg^{-1}. However, another 18 site-years in this study also had soil test K levels below this "critical level" (<173 mg K kg^{-1} soil) and these were unresponsive to K fertilizer applications. Other work in Iowa with maize showed grain yield unresponsive to fertilizer K applications at 22 of 28 sites, even though soil test K concentrations ranged from 85 to 172 mg K kg^{-1} soil (Mallarino and Higashi 2009). Intensive sampling campaigns have revealed unforeseen temporal and spatial variation in soil test K estimates that can preclude an accurate assessment of soil test K, and negate an association with yield from being realized (Randall et al. 1997; Borges and Mallarino 1998; Lissbrant et al. 2010).

14.2.2 Tissue Testing for Potassium

Tissue testing can supplement soil testing and augment efforts to predict crop responses to K fertilizer application, but results from tissue K testing can vary with species, stage of plant growth, and the tissue used for analysis (McNaught 1958; Page and Talibudeen 1982). Macy (1936) introduced the concept of "critical percentage" of each nutrient, above which there is "luxury consumption" and below which there is "poverty adjustment." Results from a multistate K nutrition trial with maize (Hanway 1962; Fig. 14.1) identified a critical percentage of 10.3 g K kg^{-1} DM. Tissue K concentrations at most sites in this trial were above this critical concentration and would be considered in the luxury consumption category. Tyner (1947) estimated the critical leaf K concentration for maize to be 13.0 g K kg^{-1} DM. This generally agrees with more recent results where the critical concentration

Fig. 14.1 Relationship between aboveground dry weight and tissue K concentrations of maize at silking. Plant mass data were adjusted relative to the highest yield at each location-year. Linear-plateau regression was used to identify the critical K concentration for maize tissues in this study (10.3 g kg^{-1} DM) above which yield did not increase (e.g., luxury consumption). (Adapted from Hanway 1962)

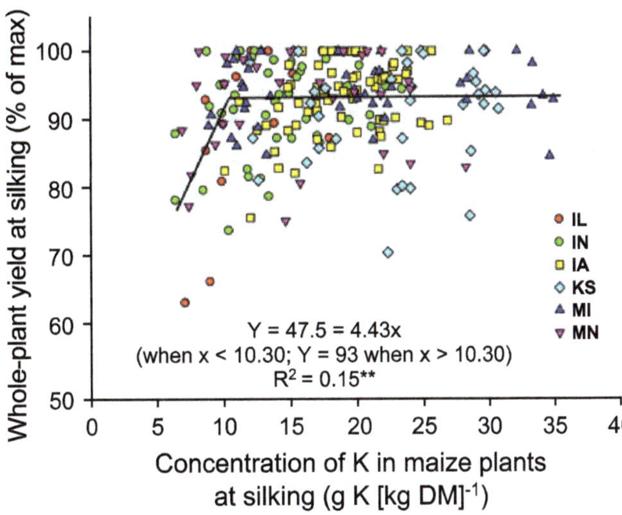

of K in maize vegetative tissues was reported to be in the range of 10.2–14.2 g K kg^{-1} DM (Page and Talibudeen 1982; Clover and Mallarino 2013; Stammer and Mallarino 2018) and the critical concentration identified in Fig. 14.1. While significant relationships were evident between soil test K levels and K concentrations in whole plants (at V5–V6), stems, and ear leaves, it was only the latter tissue whose K concentrations were associated with grain yield critical concentration (12.3 g K kg^{-1} DM, Mallarino and Higashi 2009). Like maize, leaf tissues, rather than whole-plant samples are more closely associated with soybean grain yield (Slaton et al. 2010; Clover and Mallarino 2013; Stammer and Mallarino 2018). The yield was reduced as leaf K concentrations of soybean sampled at ~R2 (late flowering) declined below a critical concentration range of 15.6–20.0 g K kg^{-1} DM. Extension recommendations (Vitosh et al. 1995; Kaiser et al. 2016; Brown 2017) generally identify higher critical K concentrations for maize ear (18.0–30.0 g K kg^{-1} DM) and soybean leaves (17.0–25.0 g K kg^{-1} DM) than the studies reported here. Opportunity clearly exists to broaden the implementation of tissue testing in K fertility recommendations by enhancing our understanding of critical K concentrations in select tissues of major crop species.

14.2.3 Luxury Consumption of Potassium

Using tissue testing for K fertility management is complicated by a phenomenon known as luxury consumption (Hanway and Weber 1971). Luxury consumption occurs when fertilization increases tissue nutrient concentrations without a corresponding increase in biomass (Macy 1936). For example, modest rates of K fertilization (up to 180 kg K ha^{-1} $year^{-1}$) increased yields, whereas higher K application rates (up to 270 kg K $year^{-1}$) failed to further increase but resulted in high tissue K concentrations in alfalfa (*Medicago sativa* L.) (de Campos Bernardi et al. 2013). Luxury K consumption also has been reported in soybean and maize where K concentrations in vegetative tissues of nearly twice the critical value necessary for high grain yield have been reported (Clover and Mallarino 2013).

In a long-term maize K nutrition study, Qiu et al. (2014) reported that fertilizing with 225 kg K ha^{-1} failed to increase grain and stover yields and grain K concentrations over the intermediate K fertilizer rate (113 kg K ha^{-1}), but did result in higher stover K concentrations. This luxury consumption reduced all measures of potassium use efficiency (KUE) including K harvest index, agronomic efficiency (kg kg^{-1}), partial factor productivity (kg kg^{-1}), and K recovery efficiency (%). In addition to reduced efficiencies, extremely high rates of K fertilization as KCl can reduce growth or even kill plants (Page and Talibudeen 1982). Smith and colleagues (Rominger et al. 1976; Smith et al. 1981) applied up to 3000 kg K ha^{-1} to alfalfa (*Medicago sativa* L.) and reported very high tissue K (up to 60 g K kg^{-1} DM) and Cl (up to 80 g Cl kg^{-1} DM) concentrations that were associated with reduced yield/plant death due to the high Cl concentrations. These extremely high rates of K fertilization are uneconomical, but also illustrate the negative agronomic consequences that can occur with extreme luxury consumption.

14.3 Factors Influencing Potassium Uptake

14.3.1 Genetics and Potassium Uptake

Interest in yield-K relationships has evolved from simple yield responses to issues focused on genetic (G), environmental (E), and management (M) strategies and their interactions (G × E × M) that alter K uptake and use in crops. This information underpins the current interest in 4R Nutrient Stewardship Programs (http://www.nutrientstewardship.com/). Unfortunately, less is known regarding the G × E × M effects on K nutrition than is known for nitrogen (N) and phosphorus (P), in part, because the potential environmental issues associated with N and P have stimulated more research on these nutrients. Early work by Kleese et al. (1968) reported significant G effects for seed K of wheat (*Triticum aestivum* L.), barley (*Hordeum vulgare* L.) and soybean, and little G × E effects despite large apparent (but unreported) differences in soil fertility between the two locations of the trial. They concluded that a single environment might suffice for screening genotypes for differences in mineral accumulation. Differences in soil K levels does not always alter seed K concentrations. Mallarino and Higashi (2009) observed no effect of a fourfold range in soil test K on maize grain K concentrations, although soil test K levels did alter K concentrations of vegetative tissue. Forage K concentrations of sorghum (*Sorghum bicolor* L.) hybrids exhibited a nearly twofold range (Gorz et al. 1987). General combining ability and additive gene action was more important than specific combining ability in determining herbage K concentrations of sorghum. Kaiser et al. (2014) analyzed leaf nutrient concentrations of flag leaves from spring wheat at heading in 18 site-years. They reported significant G, E, and G × E interaction effects for K (and most other nutrients). Flag leaf K concentrations ranged from 13.5 to 19.2 g K kg^{-1} DM in the 14 genotypes analyzed. Six genotypes responded consistently across environments, while the other eight, and especially the varieties Albany and Vantage, contributed extensively to the G × E effects. Because of this interaction, these authors suggested that critical concentrations/sufficiency ranges should be developed for individual genotypes or groups of varieties. This G × E for tissue K concentrations may contribute to the extensive variation often observed in tissue K-yield relationships, including those illustrated in Fig. 14.1.

14.3.2 Potassium Uptake and Yield

While direct genetic selection for increased KUE has not been reported, indirect improvement of KUE with selection for higher yield is viewed as a desirable outcome (Rengel and Damon 2008). Ciampitti and Vyn (2014) assembled a literature-based dataset to examine the relationship between maize grain yield and uptake of N, P, and K. These data were parsed into groups representing different eras of maize improvement (1880–1960, 1961–1975, 1976–1985, 1986–1995, 1996–2005, and 2006–2012) with each group containing from 59 to 455 individual observations. Data were further separated into that originating from the United States (US) and data from 31 other countries ("world") exclusive of the US. Regression of plant K uptake and grain yield revealed a linear relationship inclusive of both old and new hybrids that were independent of geography (Fig. 14.2). The slope of this line

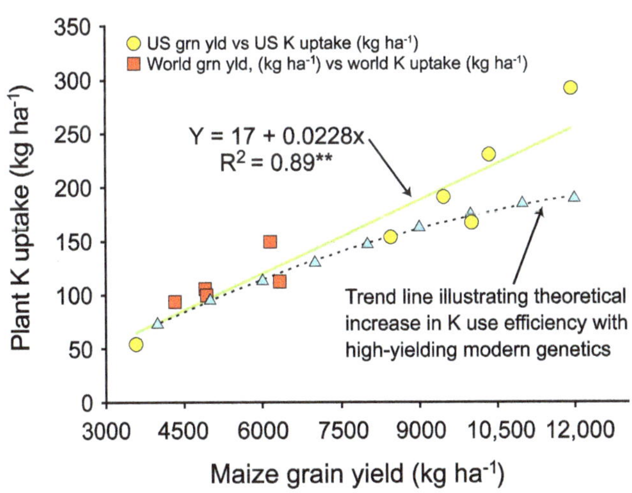

Fig. 14.2 Relationship between maize grain yield and plant K uptake. Data points represent results from historical periods (1880–1960 [US only]; 1961–1975; 1976–1985; 1986–1996; 1996–2005; 2006–2012) and include from 59 to 455 observations per point. The linear regression includes all data. The dotted line illustrates a theoretical trend if K use efficiency had improved during this period of selection for high maize yields. (Adapted from Ciampitti and Vyn 2014)

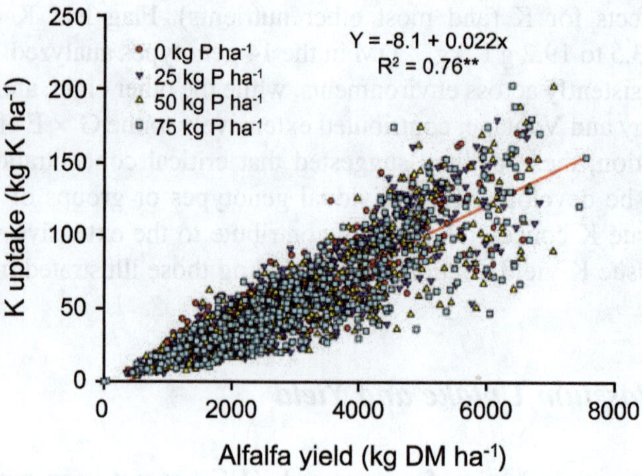

Fig. 14.3 Herbage K accumulation of alfalfa scales with alfalfa yield. Plots were fertilized with five K (0, 100, 200, 300, and 400 kg K ha^{-1} year^{-1}) and four P (0, 25, 50, and 75 kg P ha^{-1} year^{-1}) application rates and harvested four times in each of 7 years. Fertilizer applications were split with one-half the total annual amount applied after the first harvest in May and the other half applied after the final herbage harvest in September. (Adapted from Berg et al. 2005, 2007)

indicates an uptake of 22.8 g K kg^{-1} of grain produced; a value that has remained effectively unchanged for over a century.

A departure from linearity where modern hybrids produced high yields with less K uptake was not observed (Fig. 14.2, dotted line). This suggests that new hybrids had similar KUE as old hybrids; all hybrids required proportional amounts of K to produce grain. This conclusion is supported by estimates of the physiological efficiency of K (PE, grain yield per K content ratio at maturity) that remained largely unchanged irrespective of hybrid yield/era (Ciampitti and Vyn 2014). In a similar study, Morgounov et al. (2013) reported that spring wheat yield doubled between 1930s era varieties and those released after 1986. Seed N concentrations declined slightly, whereas grain K concentrations remained constant in response to selection for high yield. As with maize, K uptake scaled with grain yield, and KUE remained unchanged by selection for high grain yield. By comparison, total K uptake (kg ha^{-1}) by cotton (*Gossypium hirsutum* L.) was unaffected as the lint yield of varieties released between the 1970s and 2006 increased (Rochester and Constable 2015). As a result, K uptake efficiency was improved with selection for yield from approximately 10–13 kg lint kg^{-1} K uptake.

The relatively conserved relationship described for K uptake-grain yield also is observed with K removal-forage yield (Berg et al. 2005, 2007). Analysis of alfalfa samples obtained from four annual harvests of 80 plots receiving 4 P (0–75 kg P ha^{-1}) and 5 K (0–400 kg K ha^{-1}) fertilizer rates over 7 years (28 total harvests, 2240 observations) revealed that, like maize (Fig. 14.2), uptake of K by alfalfa scales with herbage yield (Fig. 14.3). The R^2 value of 0.76 indicates that irrespective of harvest date, stand age, and P-fertilization rate, forage yield remained the principle determinant of alfalfa K uptake. The slope of the regression line (0.022 kg K removed kg^{-1}

DM agrees well with the "book value" of 0.021 kg K removed kg^{-1} DM previously reported (Vitosh et al. 1995).

14.3.3 Potassium Uptake and Management

Genotype × Environment × Management (G × E × M) interactions can modify K uptake and KUE relationships. Parvej et al. (2016) compiled seed K and grain yields for soybean grown at 100 site-years in North America. This included K-fertilized and unfertilized plots and soil test K levels that ranged from 30 to 408 mg K kg^{-1} soil that together, resulted in a wide range in seed K concentrations (12.7–24.3 g K kg^{-1}). Despite these differences, regression of seed K uptake and grain yield in soybean resulted in similar slopes for both fertilized and unfertilized treatments; approximately 18 g of K is removed per kg of seed (Fig. 14.4). The regression line for the "fertilized" plots was consistently above the "unfertilized" line (Note: intercepts were statistically similar) suggesting slightly higher seed K uptake by fertilized plants. This removal value agrees with results from Navarrete-Ganchozo (2014) who analyzed K removal of soybean receiving several K fertilizer rates/timings at five locations in Indiana USA over 7 years ($n = 1049$) and reported 19 g K removed kg^{-1} seed. In both studies, the nature of the yield-seed K removal relationship was virtually identical across vast differences in yield and environments suggesting few G × E × M interactions for this attribute. As with forage K accumulation in alfalfa, yield *per se* is the main determinant of seed K uptake in this species.

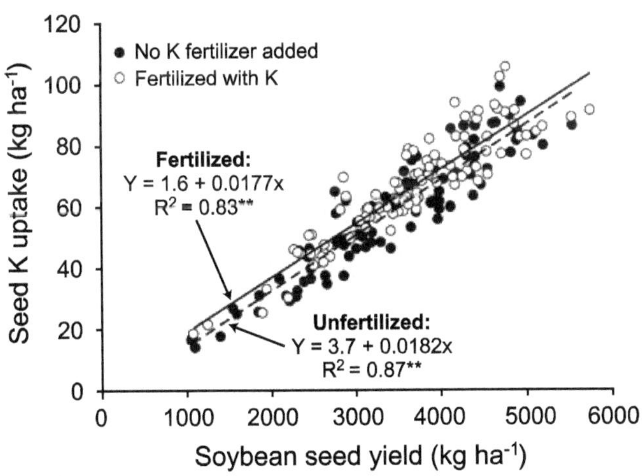

Fig. 14.4 The relationship between seed yield and K uptake of soybean. Data represent 100 site-years located in North America collected primarily from 2002 to 2015 that included paired plots that were fertilized with K or left unfertilized. Thirty-three site-years were irrigated, while the rest were rainfed sites. Slopes of the linear regressions were statistically similar for the K-fertilized versus unfertilized plots. (Adapted from Parvej et al. 2016)

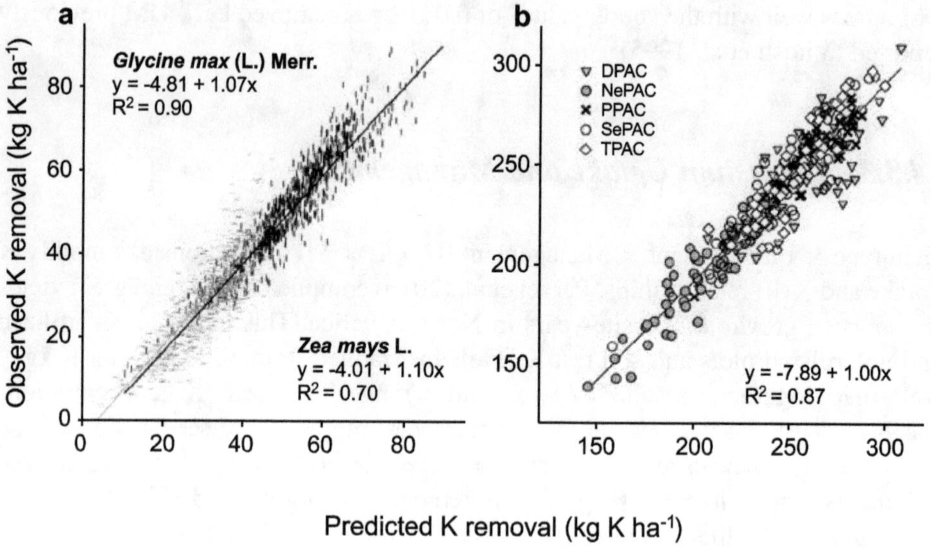

Fig. 14.5 Comparison of observed K removal (grain yield × grain [K]) and K removal based on grain yield and "book values" for grain K (3.9 and 18.3 g K kg^{-1} grain DM for maize and soybean, respectively) across K fertilizer rates and species at 5 locations over 6 years. Locations include the Davis Purdue Ag Center (DPAC), the Northeast Purdue Ag Center (NePAC), Pinney Purdue Ag Center (PPAC), the Southeast Purdue Ag Center (SePAC), and Throckmorton Purdue Ag Center (TPAC). (Adapted from Navarrete-Ganchozo 2014)

Interestingly, grain K concentrations are so conserved that G × E × M interactions often are inconsequential. For example, across species, K rates and application timings, and environments, the so-called "book values" for grain K concentrations were nearly as accurate at predicting K removal in grain as were lab-measured K concentrations and grain yields (Fig. 14.5). In all cases, the slopes of the relationship between measured and predicted K removal ranged from 1.0 to 1.1 and R^2-values were 0.70–0.90. The option to use book values for grain K concentration estimates has enabled K use and K nutrient budgets to be determined at various spatial resolutions ranging from local to national scales (http://nugis.ipni.net/Publication/).

Parvej et al. (2016) also explored the relationship between soil test K and seed K concentrations. Analysis including all site-years and fertilization groups revealed that seed K concentrations declined at soil test K levels below approximately 170 g K kg^{-1} soil; however, R^2-values indicate that soil test K only explained 24 and 40% of the variability in seed K concentration for plots with and without K fertilizer, respectively. This illustrates the challenge of using soil testing in a site-specific manner to predict seed K concentrations.

14.4 Alternative Potassium Management Strategies

14.4.1 Foliar Fertilization with Potassium

An alternative management strategy for K fertilization is foliar fertilization. Unfortunately, most foliar fertilization research has used solutions containing N, P, K, and often micronutrients and not K alone (Garcia and Hanway 1976; Orlowski et al. 2016). Thus, positive responses of these fertilizer blends cannot be attributed solely to K (or any other macronutrient). Results of most studies with an array of species reveal inconsistent responses of yield to foliar fertilization, and when positive, modest yield increases (Thom et al. 1981; Giskin and Efron 1986; Umar et al. 1999; Ling and Silberbush 2002; Hu et al. 2008) with yield increases often attributed to N or P, and not K in the nutrient blend.

Among agronomic species, soybean has been the most studied with respect to foliar fertilization. Early multi-site/year studies showed inconsistent yield response of soybean to foliar fertilization (Haq and Mallarino 1998, 2000). When it occurred, the magnitude of yield enhancement generally did not offset the application costs of the foliar fertilizer. These authors also reported little consistent impact of foliar K on oil and protein concentrations in soybean seed (Haq and Mallarino 2005). Mallarino et al. (2001) reported a significant ($P < 0.10$) yield response of soybean to foliar fertilization in only 2 of 18 small plot trials and 1 of 8 larger strip trials. The responsive plots had relatively high soil test P and K (>30 mg P kg^{-1} and 133–213 mg K kg^{-1} soil), while plots with low P and K (9–13 kg P kg^{-1} soil; 97 mg K kg^{-1} soil) were surprisingly unresponsive to foliar applications. Moreira et al. (2017) also observed a yield response of soybean to foliar fertilization with KNO_3 under certain environmental conditions and attributed this primarily to the N in the foliar spray. Although the application of KNO_3 did not increase seed yield in this study, seed K concentrations were elevated slightly, suggesting that foliar K uptake was possible. However, other studies reveal no impact of foliar K application on seed K concentrations and grain yield, but high rates of foliar K resulted in increased damage to leaf tissues (Syverud et al. 1980).

Applying foliar K fertilizers with fungicides to soybean also has been evaluated. Foliar application of K as 0-0-30 or 0-0-62 (K_2O) did not increase grain yield or ear leaf K concentrations over control plots (Shetley et al. 2015). Soil fertility levels were high at all locations and this may have precluded a positive response to foliar-applied K. Nelson et al. (2005, 2010) compared foliar K application to soybean to pre-plant K applications in soils with low-to-medium K availability. Consistent grain yield increases occurred in response to soil-applied K, and these were accompanied by higher tissue K concentrations. Foliar K application resulted in slightly higher yields in some site-years, especially if applied at vegetative growth stages, but soil-applied K resulted in the highest yields and profitability. These authors concluded that foliar K application to soybean is not a substitute for pre-plant K application. In a 57 site-year study in the Midwest USA, foliar application of a blend of nutrients including K did not significantly increase soybean yield and had a low probability of enhancing profitability (Orlowski et al. 2016). However, only five site-years in this study had soil test K levels below 100 mg K kg^{-1} soil, lessening the likelihood of a response to K application, including foliar applications. Nevertheless, the low probability of a positive yield response along with modest yield increases when they do occur indicates that foliar fertilization is an unreliable and likely unprofitable strategy for fertilizing crops with K.

14.4.2 Potassium Application Methods, Including Fertigation

A recent meta-analysis revealed that the K placement method has no effect on yield when K fertilizer was placed between 0 and 10 cm, and only a modest effect on yield when placed at depths greater than 10 cm (Nkebiwe et al. 2016). This conclusion is supported by a large number of multi-site studies where the yield of soybean and maize were unaffected by K placement and timing of application (e.g., Vyn and Janovicek 2001; Yin and Vyn 2002). The sub-soiling effect that occurs simultaneously with the deep placement of fertilizer can by itself increase yield. Mullins et al. (1994) showed that cotton lint yield was similar between the deep placement of K and subsoil tilled plots without additional K fertilization. This indicates that confounding between the effects of tillage associated with placement versus the effect of the fertilizer *per se* can obscure the underlying cause of yield increases when they occur.

An alternative to soil and foliar fertilization is fertigation; inclusion of K (and often N) in the irrigation water supplied to crops. These systems generally use drip or similar irrigation systems to provide water and nutrients near the root zone with the goal of increasing both water and nutrient use efficiency. Because of the infrastructure involved, this approach has been largely limited to high-value crops. Neilsen et al. (1999) reviewed fertigation practices in fruit trees. They concluded that fertigation of K resulted in similar depth of K movement into the soil (60–75 cm) but increased lateral K movement in soil when compared to broadcast K applications. Fertigation often increased tissue K concentrations and occasionally increased yield. The grain yield of rice was similar among application methods (fertigation, broadcasting, banding) (Ali et al. 2005). However, the timing of K application did impact yield. Applying K before transplanting or 50 days after transplanting (DAT) reduced rice grain yield for all application methods when compared to K applications made 25 DAT.

14.4.3 Recycling Potassium in Plants

The mobility of K in plants is a commonly observed phenomenon; under K-limited conditions, K is mobilized to younger, meristematic tissues where growth and development are occurring (Hoagland 1932). As a result, deficiency symptoms

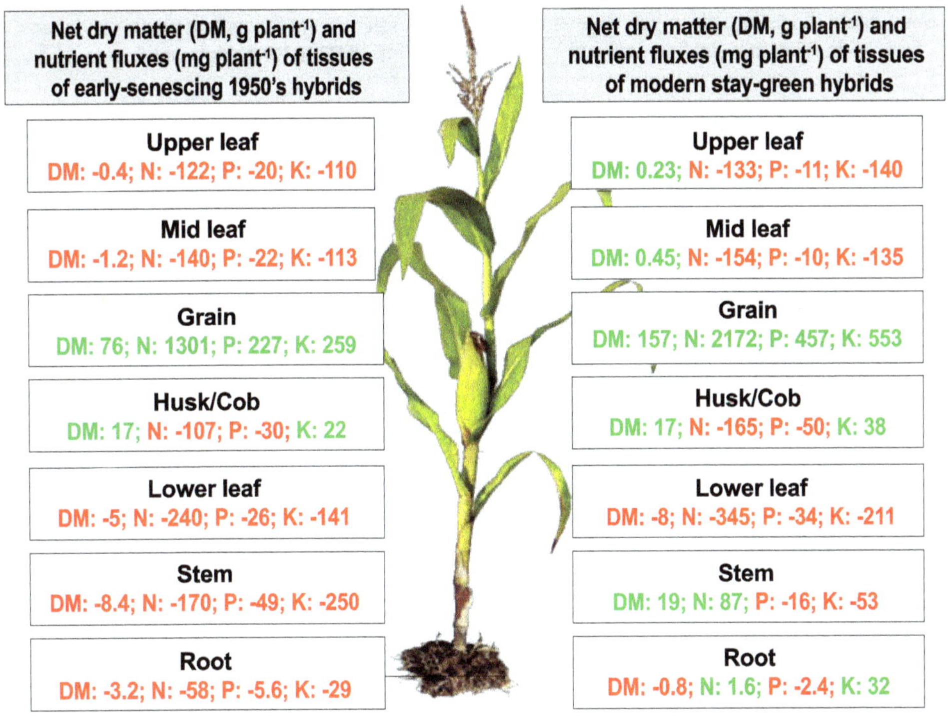

Fig. 14.6 Net changes in dry matter (DM) and macronutrients on a per-plant basis of maize tissues between silking and maturity. Early-senescing hybrids released in the 1950s (left) are compared to modern stay-green hybrids released in the 2000s (right). Net gains in DM and nutrients are identified in green text, while net losses are highlighted in red text. (Adapted from Ning et al. 2013)

usually appear predominately in older vegetative tissues. In addition to this well-characterized K partitioning within the plant, there is a growing awareness that nutrient recycling from vegetative tissues to grain can partially meet the nutrient needs of developing seeds. This might occur because post-anthesis K uptake is low relative to other macronutrients. Van Duivenbooden et al. (1996) surveyed the literature (50–100 experiments) and found that wheat had the lowest post-anthesis K uptake (4% of total aboveground K), while maize, sorghum, millet, and rice had generally similar values (14–18%). Post-anthesis uptake of N and P tended to be higher than K, ranging from 18–35% for N and 10–47% for P. Remobilization of K from vegetative tissues to the seed may meet K needs during grain fill, negating the need for additional soil K uptake.

We estimated net changes in dry matter (DM), and mass of N, P, and K between silking and maturity for maize using data published by Ning et al. (2013) where mobilization patterns of senescing hybrids from the 1950s was compared to that of modern, stay-green hybrids (Fig. 14.6). As expected, grain was a net accumulator of DM, N, P, and K irrespective of hybrid group. Modern hybrids double the net accumulation of DM, P, and K, whereas net N accumulation increased about 67% over that reported for 1950s era hybrids. Net partitioning of DM and macronutrients to husks/cobs and were generally similar for both hybrid groups with net losses of N

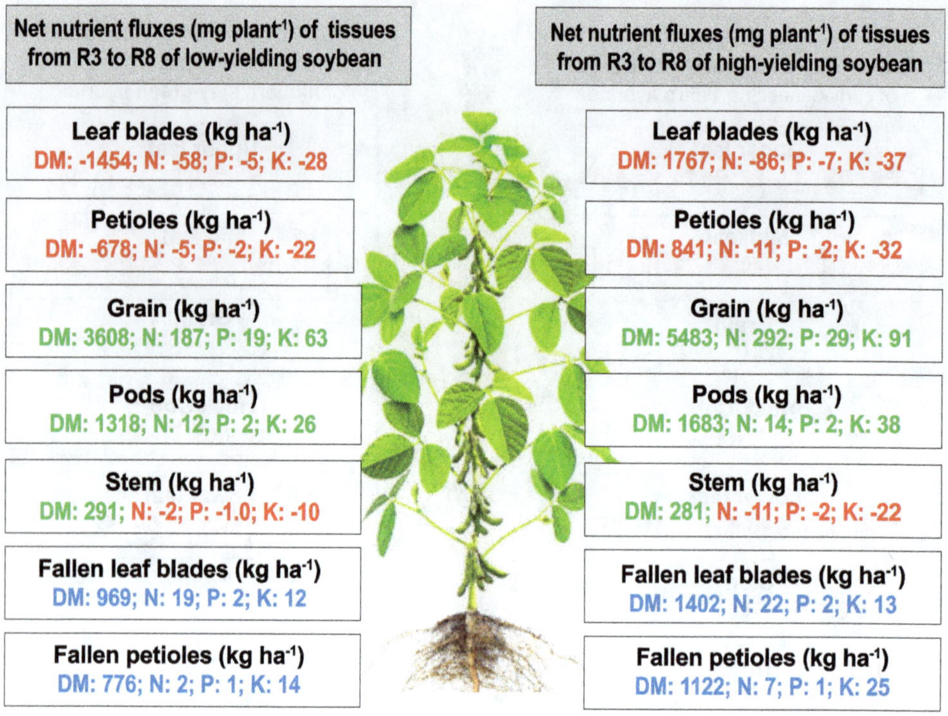

Fig. 14.7 Net changes in dry matter (DM) and macronutrients on a per-hectare basis of soybean tissues between growth stages R3 and R8. Patterns for relatively low-yielding plots are shown on the left, while trends for high-yielding soybean plots are on the right. Net gains in DM and nutrients are identified in green text, net losses are highlighted in red text, while loss to the soil are highlighted in blue. (Adapted from Gaspar et al. 2017a, b)

and P, and net gains in DM and K. However, the early-senescing hybrids of the 1950s exhibited net losses in DM and N, P, and K from all other tissues with especially large losses in K from stems. By comparison, these tissues of the stay-green hybrids had net gains in DM in most tissues between silking and maturity. In addition, stems and roots of these hybrids lost far less, and in some cases had net gains in DM and N, P, and K than the 1950s era plants. For K, these reductions in net K loss from roots and stems were offset, in part, by greater net K mobilization from leaf tissues. Summed over all tissues, net K losses of stay-green hybrids (539 mg plant^{-1}) was 97% of the K accumulated in grain (553 mg plant^{-1}). By comparison, only 40% of the K lost by vegetative tissues of the 1950's era plants was accounted for by net accumulation in grain (259 mg/643 mg) indicating less efficient K recycling and loss of K from the plant system.

A similar analysis with low-yielding versus high-yielding plots of modern soybean lines revealed net gains in DM and masses of N, P, and K that scaled with grain yield (Fig. 14.7). Grain yield and N, P, and K in grain of high-yielding plots increased approximately 50% over that of the low-yielding plots between R3 and R8 growth stages. Irrespective of yield, leaf blades and petioles exhibited net losses of DM, N, P, and K during this period. Net gains of K in grain (63 and 91 kg ha^{-1})

could be accounted for by net K losses in blades, petioles, and stems indicating that K recycling in the plant could meet grain K needs. By comparison, and irrespective of yield level, net losses of N and P could only account for approximately 40% of these nutrients in the grain, suggesting that internal recycling alone could not meet the N and P needs of soybean. In this study, leaf blades and petioles that fell to the ground between R3 and R8 returned up to 38 kg K ha^{-1} to the soil, with lesser amounts of N and a negligible amount of P returned to the soil.

14.4.4 Crop Residues and Potassium Nutrition

Leaching of K from post-harvest residues can contribute significantly to soil test K and the cycling of K in agroecosystems. Oltmans and Mallarino (2015) reported 34 and 55 kg K ha^{-1} in soybean and maize residue, respectively, immediately after grain harvest. Two months post-harvest, approximately 50% of this K remained in maize residue, whereas only 19% of this K remained in soybean residue. In both species, reductions in stover K increased with post-harvest precipitation. Soil test K increased from fall to spring on average 27 and 24 mg K kg^{-1} for soybean and maize, respectively, and these values were positively associated with residue K losses (Fig. 14.8). This residue K input can impact crop yield. Singh et al. (2018) assembled apparent K budgets for rice-maize cropping systems in India for a 5-year

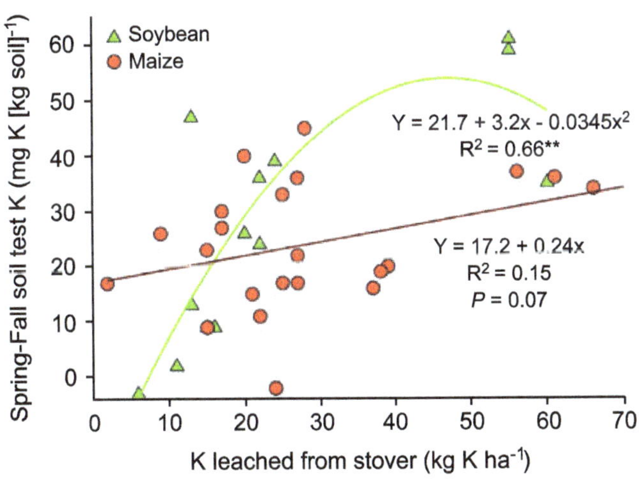

Fig. 14.8 Impact of K leaching from stover of maize and soybean on soil test K (STK) levels. Stover was collected periodically after harvest in fall, K concentrations determined, and mass of K leached from stover calculated based on residue mass. Changes in STK between grain harvest in fall and the following April averaged 27 and 24 mg K kg^{-1} for soybean and maize, respectively. (Adapted from Oltmans and Mallarino 2015)

period. In addition to fertilizing plots with K (620 kg K ha^{-1} for the 5-year period), they measured K inputs from alternative sources including crop residue (618–678 kg K ha^{-1}) and roots/stubble (95–152 kg K ha^{-1}). Only plots receiving K from all three of these sources maintained a positive K balance and high yields.

14.4.5 Fungal Associations and Potassium Nutrition

Symbiotic associations between plants and mycorrhizal fungi that enhance nutrient uptake have been well-documented for some nutrients, especially P. Less is known regarding the role of these fungi in augmenting plant K uptake. Rosendahl (1943) reported greater K uptake and enhanced plant growth of pine seedlings grown in orthoclase-amended sand when inoculated with *Boletus felleus*, a fungus known to form mycorrhizal associations with conifers. Recent reviews (Garcia and Zimmermann 2014; Dominguez-Nuñez et al. 2016) summarized the role of mycorrhizae in improving K nutrition of several plant species, especially under K-limited conditions as is often found in forest ecosystems. These authors also indicated that this symbiosis enhanced general abiotic stress tolerance. While the fundamental mechanisms controlling this symbiosis are poorly understood, overexpression of specific K channel genes in the genome of the ectomycorrhizal fungi *Hebeloma cylindrosporum* increased K accumulation in shoots of pine seedlings (Guerrero-Galán et al. 2018). These genes may serve as molecular targets for enhancing K uptake under low soil K conditions via this symbiotic association. Additional work is needed to more fully characterize the role and quantify the metabolic costs of mycorrhizae in providing K to plants.

Fungal endophytes form mutualistic associations with some plant species resulting in enhanced growth and abiotic stress tolerance. A classic example of this mutualism is endophyte-infected tall fescue (*Schedonorus arundinaceus* (Schreb.) Dumort.), where P uptake is enhanced in infected plants under low soil P conditions (Malinowski and Belesky 2000). Less is known regarding how endophyte infection impacts K uptake. Rahman and Saiga (2005) reported greater shoot growth and K uptake in tall fescue plants infected with the endophyte *Neotyphodium coenophialum* when compared to uninfected plants. Responses varied with soil type and tall fescue ecotype. Malinowski and Belesky (2000) also reported increased K absorption rates in one of two tall fescue ecotypes. Additional research, especially under low soil K conditions, is needed to fully understand the role of these fungal endophytes on K uptake. Further, it is unclear if the endophyte enhances shoot growth that then leads to greater K uptake (sink driven) or if greater K uptake occurs that leads to faster shoot growth (source driven).

14.5 Impact of Potassium on Crop Quality

14.5.1 Potassium Nutrition and Crop Quality

Given the large impact that K nutrition can have on physiological function and yield of crops, it is not surprising that crop composition also can be altered under K-deficient conditions. Usherwood (1985) and Mengel (1997) presented an overview of this topic previously. Rather than review the numerous recent papers published on this topic, the impact K deficiency can have on crop quality will be illustrated with key examples for crop categories.

14.5.2 Cereals

Unlike N and P, few reports exist describing the impact of K on grain quality of cereals. A recent review of N, P, and K on wheat performance (Duncan et al. 2018) noted the little consistent effect of K on grain protein concentration. These authors noted that balanced nutrition with N, P, and K maintained protein concentrations in yield-responsive environments. However, few of the 32 field studies reviewed assessed the impact of K on wheat grain quality *per se*. Holland et al. (2019) reported significant declines in crude protein concentration (%N) of wheat grain with even modest rates (12.5–25 kg K ha^{-1}) of K fertilizer that was associated with increased grain yield. By comparison, the yield of triticale did not increase until a much higher K rate (100 kg K ha^{-1}) was applied and only then did grain protein decline in one of two studies. This suggests that the decline in grain protein concentrations is a result of dilution of grain N as K-stimulated yield.

Rice (*Oryza sativa* L.) grain quality measured as gel consistency, amylose concentration, gelatinization temperature, and protein concentration can be altered by K fertilization (Bahmaniar and Ranjbar 2007). Some of these attributes interacted with K fertilization rate, plant genotype, and N fertilizer application making it challenging to develop a unified N and K fertilizer program targeting grain quality for rice cultivars used in this study.

In a multi-year study conducted at five locations, Brouder examined the impact of K on maize grain yield and quality (Fig. 14.9). Plots were fertilized with up to 180 kg K ha^{-1} annually or 360 kg K ha^{-1} biennially, and large differences in grain yield were observed at some locations (Navarrete-Ganchozo 2014). Despite a three-fold difference in grain K concentration, there was no significant impact of concentrations of starch, protein, or oil in grain. Maize grain amino acid concentrations were generally higher in plants provided adequate K (Usherwood 1985).

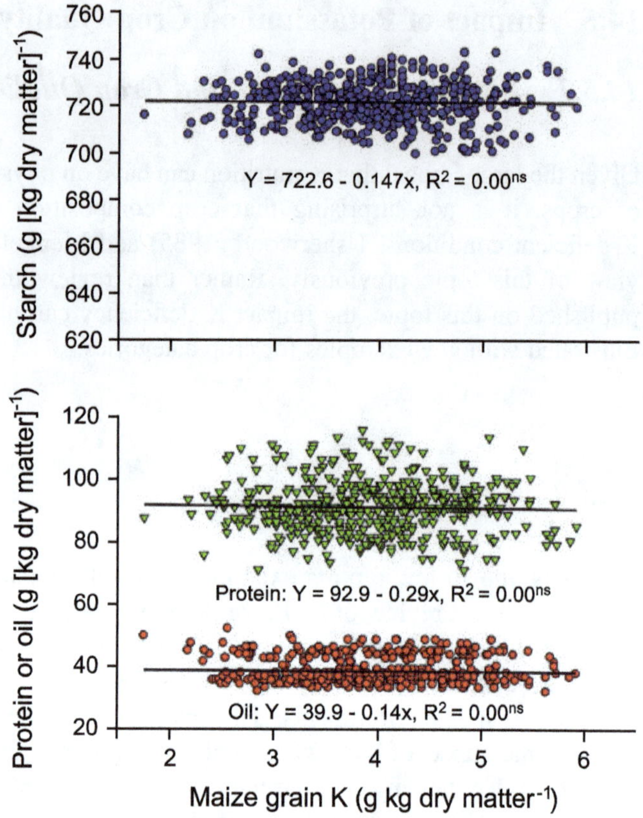

Fig. 14.9 Relationship between grain K concentration and starch, protein, and oil concentration of maize grain. Data were obtained from 17 site-years in Indiana USA. Plants were fertilized with 0–180 kg K ha^{-1} annually or 0–360 kg K ha^{-1} biennially. Grain K concentrations were determined with an inductively coupled plasma spectrophotometer, while grain starch, protein, and oil were analyzed with a near-infrared reflectance spectrophotometer. (Adapted from unpublished data from S. Brouder, Purdue University)

14.5.3 Oilseeds

Bailey and Soper (1985) reported little consistent impact of K on the oil and protein composition of the seed of rape, flax, sunflower, and safflower. This agrees with more recent results with canola where seed oil and protein concentrations were unaffected by K fertilization rates that, in some environments, increased yield nearly twofold (Brennan and Bolland 2007). Oil and protein concentrations in soybean seeds also are often not responsive to K fertilizer application, especially if initial soil test K levels are adequate (Krueger et al. 2013). In some studies, seed oil concentrations increase with K, while seed protein concentrations declined (Yin and Vyn 2003). However, recent studies reveal that both protein and oil concentrations of soybean can increase in response to increasing K fertility (Abbasi et al. 2012; Bellaloui et al. 2013), including increases in specific fatty acids (Krueger et al. 2013). Potassium also can interact with P in determining oil and protein concentrations of soybean. Abbasi et al. (2012) applied P as single superphosphate and K as K_2SO_4 to a soil containing 3.4 mg kg^{-1} available P and 67 mg kg^{-1} exchangeable K. In addition to increasing soybean grain yield and nodulation, both grain oil and protein concentrations increased with K fertilizer application, with the highest concentrations achieved only when both P and K were adequate (Fig. 14.10). It is

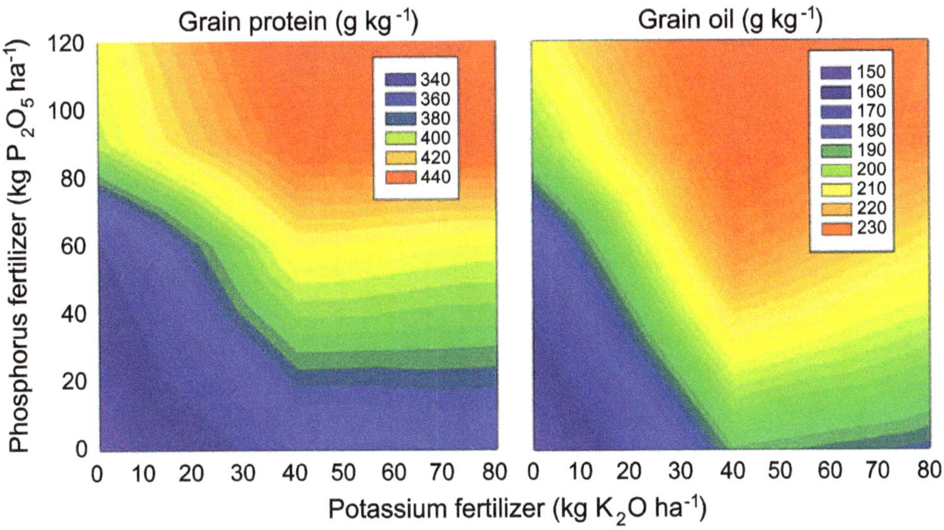

Fig. 14.10 Impact of potassium (K) and phosphorus (P) fertilizer application on the concentrations of protein and oil in soybean seed. Prior to fertilization, the clay loam soil (Humic Lithic Eutrudept) contained 3.7 and 67 mg kg^{-1} P and K, respectively. (Adapted from Abbasi et al. 2012)

not clear if the addition of sulfur with the K to this soil containing 10 g kg^{-1} organic C also contributed to the changes in seed oil and protein.

14.5.4 Forage

Numerous plant species can be used as forage, including silage produced from plants normally used for grain production like maize, sorghum, and small grains. Besides yield, K can alter plant morphology (e.g., leaf:stem ratio) and composition of cell walls that together, can impact forage intake, digestibility, and ultimately animal performance. Lissbrant et al. (2009) reported higher in vitro dry matter disappearance (IVDMD) and protein concentrations for low-yielding, K-deficient alfalfa when compared to forage from plants fertilized with K (Table 14.1). The greater digestibility was associated with lower concentrations of neutral detergent fiber (NDF), acid detergent fiber (ADF), and lignin of the K-deficient plants. However, the yield of both digestible nutrients (DNY) and protein ha^{-1} were greater in K-fertilized plants because of their greater forage yield. Similar trends for reduced forage protein concentrations with increased K fertilization have been reported (Macolino et al. 2013), but concentrations of NDF, ADF, and lignin were not increased in their study conducted in a high-K soil.

Most studies on forage grass quality have focused on N nutrition, with few studies exploring K impacts on forage grass quality. Balasko (1977) reported small, but significant increases in IVDMD in 4 of 8 winter harvests of tall fescue when N and P fertilizers were supplemented with K. Forage yield increased in 3 of 4 harvests where

Table 14.1 Impact of K fertilization on average yield per harvest and forage quality of alfalfa

K	Yield	IVDMD	DNY	Protein	Protein	NDF	ADF	Lignin
kg ha^{-1} year^{-1}	kg ha^{-1} harvest^{-1}	g kg^{-1}	kg ha^{-1}	g kg^{-1}	kg ha^{-1}	g kg^{-1}		
0	2392c	806a	1875c	192a	450c	368e	268d	72.2e
100	2895b	794b	2255b	187b	531b	389d	288c	75.4d
200	3137a	788c	2436a	185bc	572a	399c	298b	76.8c
300	3176a	785cd	2464a	183c	573a	404b	302b	78.0b
400	3254a	782d	2517a	180d	580a	411a	309a	78.9a

In vitro dry matter disappearance (IVDMD), digestible nutrient yield (DNY), protein concentration and yield, and concentrations of neutral detergent fiber (NDF), acid detergent fiber (ADF), and lignin, expressed on a dry matter basis. Data were averaged over four *P* application rates on plots harvested four times annually for 7 years (112 observations per mean). Means within a column followed by the same letter are statistically similar ($P > 0.05$). (Adapted from Lissbrant et al. 2009)

IVDMD increased. Similarly, when stargrass (*Cynodon* spp.) was adequately fertilized with P, yield responses to K fertilization were evident in 2 of 3 years and these higher yields were accompanied by increases in forage IVDMD, but reductions in forage protein (Pant et al. 2004). By comparison, Malhi et al. (2005) reported similar concentrations of protein, NDF, and ADF even in environments where timothy (*Phleum pretense* L.) forage yields were enhanced by K fertilization. Wheat and triticale (*Triticosecale* spp. Wittm. ex A. Camus) forage composition can be altered by K fertilization (Holland et al. 2019). Significant increases in dry matter disappearance, metabolizable energy, water-soluble carbohydrates, and reductions in NDF, ADF, and protein were observed in some environments and growth stages as a result of K fertilization.

In addition, because of luxury consumption discussed previously, forage can accumulate high-K concentrations leading to mineral imbalances that result in potentially fatal livestock diseases like hypomagnesemia and milk fever (Kayser and Isselstein 2005; Lunnan et al. 2018); thus, there is a potential anti-quality issue with forage K nutrition. The ratio of magnesium and calcium to K in forage, and limiting excess K also is a consideration for plants used for ruminant livestock feed.

14.5.5 Fiber

In cotton, K deficiency reduced most measures of fiber quality including fiber elongation, 50% span length, uniformity ratio, micronaire, fiber maturity, and perimeter in all genotypes studied (Pettigrew et al. 1996). This confirms earlier findings of Cassman et al. (1990) where most fiber attributes of cotton were positively associated with soil, leaf, and fiber K concentrations. Analysis of fiber composition of industrial hemp (*Cannabis sativa* L.) grown for six site-years in Canada revealed no effect of K fertilization (0–200 kg K ha^{-1}) on cellulose, hemicellulose, or lignin concentrations (Aubin et al. 2015). However, fiber yield also was not influenced by K application, leading the authors to suggest that high initial soil test K levels (~200 kg K ha^{-1}) may have prevented responses to K fertilization.

14.5.6 Tubers and Tuberous Roots

Starch synthase, a key enzyme in starch synthesis, requires K for proper activation (Murata and Akazawa 1968). Thus, it is not surprising that K influences the growth and composition of starch-rich tubers. Potato (*Solanum tuberosum* L.) quality is often influenced by both rate of K fertilization and K source. Tuber specific gravity, a trait positively associated with processing quality, was lower with K fertilization; however, specific gravity losses were less when K_2SO_4 rather than KCl is used as a K source (Panique et al. 1997). Stanley and Jewell (1989) reported a reduction in tuber dry matter with increasing K. Tubers of K-fertilized plants had lower reducing sugars that cause discoloration of fried food like chips and fries. Sensory evaluation of French fries made from K-fertilized tubers scored higher with consumers in several categories including color and texture. Other work on chipping potatoes revealed that chip yield was greatest, and residual chip oil content lowest for K_2SO_4-fertilized plants when compared to plants provided KCl as a fertilizer source (Kumar et al. 2007).

By comparison, increases in yield and quality of sweet potato (*Ipomoea batatas* L.) were not influenced by K source. In general, K fertilization increased the frequency of large tuberous roots in this species and slowed weight losses in storage; an index of quality (Nicholaides et al. 1985). In a 13 site-year study where the addition of K to N- and P-fertilized soil more than doubled tuber yields, John et al. (2013) reported increased tuber yield and a slight increase in tuber starch concentration for cassava (*Manihot esculenta* Crantz) with K fertilizer application. They also observed altered starch rheological characteristics including increased amylose content, granule size, pasting temperature, viscosity, and swelling volume with K fertilization. Concentrations of cyanogenic glycosides, a serious anti-quality attribute of this species, were reduced with K fertilizer application. By comparison, Obigbesan (1977) reported that cultivar and environment were more important that K fertilization in determining levels of cyanogenic glycosides in this species.

14.5.7 Fruits and Vegetables

Because this category contains numerous species whose K fertilizer practices are managed in diverse ways (source, timing, application methods, . . .), it is not possible to comprehensively represent general trends and effects here. An early review (Greenwood et al. 1980) indicated that K fertilization decreased the quality of some species (e.g., carrot, cauliflower, turnips), while improving the quality of others (spinach, parsnips). In a more recent review, Lester et al. (2010) summarized the impact of K nutrition on the quality of over 20 fruit/vegetable species. Fruit quality, measured as either compositional (e.g., sugar, acidity, vitamins, carotene, . . .) or physical attributes (e.g., color, firmness, texture, shelf-life, . . .) generally improved with K fertilization. In studies where results were not in agreement, this was attributed to variation in fertilizer application method and/or K source (Mikkelsen 2017). The underlying mechanisms involved in K-enriched fruit quality are emerging. For example, in pear (*Pyrus communis* L.) K regulates the expression of key genes involved in sugar and sorbitol metabolism in both sources and sink tissues that ultimately enhances sugar accumulation in and quality of pear fruits (Shen et al. 2017, 2018).

14.5.8 Human Nutrition and Health

A vast body of evidence regarding the role of K in human health and nutrition has been published. A few key examples are highlighted here. The positive impact of K on reducing blood pressure and related diseases has been a recurring research topic. Dietary K supplementation has been shown to reduce blood pressure in hypertensive individuals, particularly in high-sodium consumers not currently receiving hypertensive drug treatment, and those in the lowest category of K intake (Filippini et al. 2017). Adequate dietary K intake (~90 mmol day^{-1}) is recommend to achieve blood pressure control. In a meta-analysis, D'Elia et al. (2011) reported that high dietary K intake is associated with low rates of stroke and might also reduce the risk of coronary heart disease and total cardiovascular disease. These results support recommendations for higher consumption of K-rich foods to prevent vascular diseases. Increased fruit and vegetable intake in the range commonly consumed is associated with a reduced risk of stroke (He et al. 2006). The protective effects of fruit and vegetables on stroke prevention have a strong biological basis, including the fact they are rich sources of dietary K. The positive effects of K on human health extend beyond blood pressure and heart disease to include reduction in osteoporosis (Lambert et al. 2015), insulin-resistant diabetes (Ekmekcioglu et al. 2016), kidney disease (Zhang et al. 2019), ulcerative colitis and Crohn's disease (Khalili et al. 2016), and obesity and metabolic syndrome (Cai et al. 2016).

14.6 Plant Stress Tolerance and Potassium Nutrition

14.6.1 Potassium and Abiotic Stress Tolerance

Potassium has long been identified as a nutrient that is critical in alleviating the detrimental effects of abiotic stresses in plants. Several reviews have discussed the role of K in tolerance to drought, salinity, chilling, and freezing temperatures, flooding, and stresses associated with climate change (Ahmad et al. 2018; Amtmann et al. 2018; Anschütz et al. 2014; Cakmak 2005; Kant and Kafkafi 2002; Oosterhuis et al. 2014; Sardans and Peñuelas 2015; Wang et al. 2013). Rather than reiterate the details found in these reviews, a few key examples of how K functions in plants and imparts stress tolerance are outlined here.

The role of K in abiotic stress tolerance is exemplified by the regulation of stomatal aperture that influences both water loss from leaves and CO_2 uptake in photosynthesis. Humble and Raschke (1971) used X-ray microprobe analyses to demonstrate the K accumulation in guard cells and subsequent water influx that ultimately opened stomata. This response was specific to K and highlights the key role it has in both transpiration and photosynthesis, processes central to plant water relations, and dry weight accumulation. For example, Pervez et al. (2004) reported increases in both photosynthesis and transpiration with increasing K fertilizer

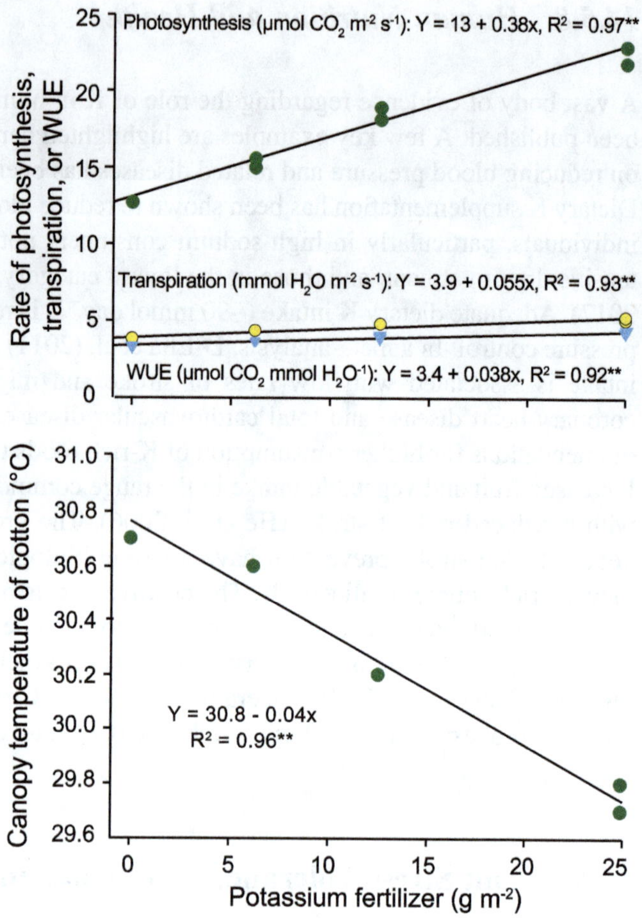

Fig. 14.11 Influence of K fertilizer application on rates of photosynthesis, transpiration, water use efficiency (WUE), and canopy temperature of cotton. Data are averaged over four cotton cultivars and two K sources. (Adapted from Pervez et al. 2004)

applications to cotton (Fig. 14.11). They also observed improved instantaneous water use efficiency (WUE, the ratio of photosynthesis to transpiration) and lower canopy temperatures with K fertilization. Lower leaf temperatures result from evaporative cooling associated with the latent heat of vaporization as liquid water evaporates from the leaf surface; a process critical to temperature regulation in plants.

Potassium also influences the flooding tolerance of plants. Dwivedi et al. (2017) reported that K fertilization mitigated submergence-induced stress in rice. This included reducing membrane damage during flooding and improved post-flooding recovery of photosynthesis. This result confirms earlier findings with barley where the loss of membrane integrity and tissue K during waterlogging were associated with flooding intolerance (Zeng et al. 2014).

Long-term survival of perennial plants also is enhanced by K fertilization. Lissbrant et al. (2010) used cluster analysis to categorize plots from a 7-year-old P and K fertility study with alfalfa into groups that varied as a function of forage yield in May of the final yield of the study. The highest yielding plots in the High cluster had tissue K concentrations between 18.2 and 26.2 g K kg^{-1} dry matter (Table 14.2),

Table 14.2 Herbage K concentrations, plant survival, and herbage yield of alfalfa analyzed using cluster analysis

Year	Herbage K		Plant populations		Yield	
	High	Very Low	High	Very Low	High	Very Low
	g K kg^{-1} DM		Plants m^{-2}		kg DM ha^{-1}	
1	26.2	15.7^{**}	259	226	5676	4946
2	21.3	8.7^{**}	178	228^{**}	6453	5872^{*}
3	23.2	8.1^{**}	134	160	5795	4639^{**}
4	18.2	7.5^{**}	112	$82^{\dagger}$	3984	2789^{**}
5	21.0	11.7^{**}	68	61	4051	1327^{**}
6	19.4	$-^{\ddagger}$	56	11^{**}	4754	0^{**}
7	21.4	$-^{\ddagger}$	45	6^{**}	4390	0^{**}

Plots were grouped into six clusters based on yield in May of Year 7. Only data for the highest (High) and the lowest (Very Low) yielding clusters are shown. (Adapted from Lissbrant et al. 2010)
†, *, and ** indicate significant difference between the High and Very Low clusters in this year
‡ Tissue not available for analysis

whereas tissue K concentrations ranged from 7.5 to 15.7 g K kg^{-1} for the Very Low cluster. Plant populations declined with time in all clusters and averaged 9 and 51 plants m^{-2} for the Very Low and High clusters, respectively, in Years 6 and 7. Forage yields declined to 0 kg ha^{-1} for the Very Low cluster at the end of the study. Plants died during summer rather than winter (Berg et al. 2018); a finding that is contrary to the general understanding that K enhances winter survival of alfalfa. Additional work is necessary to identify the underlying cause(s) for the death of these alfalfa plants in summer.

14.6.2 Potassium and Biotic Stress Tolerance

Fewer reviews have summarized the role of K in biotic stress tolerance. In many cases, K application reduces disease incidence, but in some studies, disease prevalence is unaffected or even increases with K fertilization (Huber and Arny 1985). Others found insufficient data to conduct a thorough quantitative review of the role of K on disease incidence (Veresoglou et al. 2013). Amtmann et al. (2008) reported inconsistencies associated with plant- and experiment-specific interactions of K with plant physiological traits (growth, cell wall structure, metabolites, hormones, ...) that are components of the disease-resistance response. In this study, response to K varied with pathogen group; reductions in fungal diseases more common with K application than were diseases caused by bacteria and viruses.

A meta-analysis of the role of K in insect resistance also revealed mixed results that were influenced by the insect group being studied (Butler et al. 2012). While insect populations generally showed a negative response to K fertilization, data were too sparse to identify significant effects. As with diseases, individual studies describe a positive impact of K fertilizer application on the reduction of insects, while other studies report no impact of K on insect stress (Kitchen et al. 1990; Myers and Gratton 2006; Myers et al. 2005). Closer collaboration among soil fertility/plant nutrition and entomologists/pathologists would advance our understanding of the role of K in biotic stress tolerance.

References

Abbasi MK, Tahir MM, Azam W, Abbas Z, Rahim N (2012) Soybean yield and chemical composition in response to phosphorus–potassium nutrition in Kashmir. Agron J 104:1476–1484. https://doi.org/10.2134/agronj2011.0379

Ahmad Z, Anjum S, Waraich EA, Ayub MA, Ahmad T, Tariq RMS, Ahmad R, Iqbal MA (2018) Growth, physiology, and biochemical activities of plant responses with foliar potassium application under drought stress–a review. J Plant Nutr 41:1734–1743. https://doi.org/10.1080/01904167.2018.1459688

Ali A, Zia MS, Hussain F, Salim M, Mahmood IA, Shahzad A (2005) Efficacy of different methods of potassium fertilizer application on paddy yield, K uptake and agronomic efficiency. Pak J Agric Sci 42:27–32

Amtmann A, Troufflard S, Armengaud P (2008) The effect of potassium nutrition on pest and disease resistance in plants. Physiol Plant 133:682–691. https://doi.org/10.1111/j.1399-3054.2008.01075.x

Amtmann A, Armengaud P, Volkov V (2018) Potassium nutrition and salt stress. In: Annual Plant Reviews. Wiley, New Jersey, pp 328–379. https://doi.org/10.1002/9781119312994.apr0151

Anschütz U, Becker D, Shabala S (2014) Going beyond nutrition: regulation of potassium homoeostasis as a common denominator of plant adaptive responses to environment. J Plant Physiol 171:670–687. https://doi.org/10.1016/j.jplph.2014.01.009

Aubin MP, Seguin P, Vanasse A, Tremblay GF, Mustafa AF, Charron JB (2015) Industrial hemp response to nitrogen, phosphorus, and potassium fertilization. Crop Forage Turfgrass Manage 1:1–10. https://doi.org/10.2134/cftm2015.0159

Bahmaniar MA, Ranjbar GA (2007) Response of rice (*Oryza sativa* L.) cooking quality properties to nitrogen and potassium application. Pak J Biol Sci 10:1880–1884

Bailey LD, Soper RJ (1985) Potassium nutrition of rape, flax, sunflower, and safflower. In: Munson RD (ed) Potassium in agriculture. American Society of Agronomy, Madison WI, pp 765–798

Balasko JA (1977) Effects of N, P, and K fertilization on yield and quality of tall fescue forage in winter. Agron J 69:425–428. https://doi.org/10.2134/agronj1977.00021962006900030023x

Bellaloui N, Yin X, Mengistu A, McClure AM, Tyler DD, Reddy KN (2013) Soybean seed protein, oil, fatty acids, and isoflavones altered by potassium fertilizer rates in the Midsouth. Amer J Plant Sci 4:976–988. https://doi.org/10.4236/ajps.2013.45121

Berg WK, Cunningham SM, Brouder SM, Joern BC, Johnson KD, Santini J, Volenec JJ (2005) Influence of phosphorus and potassium on alfalfa yield and yield components. Crop Sci 45:297–304. https://doi.org/10.2135/cropsci2005.0297

Berg WK, Cunningham SM, Brouder SM, Johnson KD, Joern BC, Volenec JJ (2007) The long-term impact of phosphorus and potassium fertilization on alfalfa yield and yield components. Crop Sci 47:2198–2209. https://doi.org/10.2135/cropsci2006.09.0576

Berg WK, Lissbrant S, Cunningham SM, Brouder SM, Volenec JJ (2018) Phosphorus and potassium effects on taproot C and N reserve pools and long-term persistence of alfalfa (*Medicago sativa* L.). Plant Sci 272:301–308. https://doi.org/10.1016/j.plantsci.2018.02.026

Borges R, Mallarino AP (1998) Variation of early growth and nutrient content of no-till corn and soybean in relation to soil phosphorus and potassium supplies. Comm Soil Sci Plant Anal 29:2589–2605. https://doi.org/10.1080/00103629809370136

Brennan RF, Bolland MDA (2007) Influence of potassium and nitrogen fertiliser on yield, oil and protein concentration of canola (*Brassica napus* L.) grain harvested in South-Western Australia. Austr J Exp Agric 47:976–983. https://doi.org/10.1071/EA06114

Brown C (2017) Agronomy guide for field crops publication 811. Ministry of agriculture, food and rural affairs, Queen's printer for Ontario, Toronto, p 433. http://www.omafra.gov.on.ca/english/crops/pub811/pub811.pdf. Accessed 13 May 2020

Buchholz DD, Brown JR, Garret JD, Hanson RG, Wheaton HN (2004) Soil test interpretations and recommendations handbook. Univ of Missouri-College of Agric, Div Plant Sci, Columbia, MO

Butler J, Garratt MPD, Leather SR (2012) Fertilisers and insect herbivores: a meta-analysis. Ann Appl Biol 161:223–233. https://doi.org/10.1111/j.1744-7348.2012.00567.x

Cai X, Li X, Fan W, Yu W, Wang S, Li Z, Scott EM, Li X (2016) Potassium and obesity/metabolic syndrome: a systematic review and meta-analysis of the epidemiological evidence. Nutrients 8:183. https://doi.org/10.3390/nu8040183

Cakmak I (2005) The role of potassium in alleviating detrimental effects of abiotic stresses in plants. J Plant Nutr Soil Sci 168:521–530. https://doi.org/10.1002/jpln.200420485

Cassman KG, Kerby TA, Roberts BA, Bryant DC, Higashi SL (1990) Potassium nutrition effects on lint yield and fiber quality of Acala cotton. Crop Sci 30:672–677. https://doi.org/10.2135/cropsci1990.0011183X003000030039x

Ciampitti IA, Vyn TJ (2014) Understanding global and historical nutrient use efficiencies for closing maize yield gaps. Agron J 106:2107–2117. https://doi.org/10.2134/agronj14.0025

Clover MW, Mallarino AP (2013) Corn and soybean tissue potassium content responses to potassium fertilization and relationships with grain yield. Soil Sci Soc Amer J 77:630–642. https://doi.org/10.2136/sssaj2012.0223

de Campos Bernardi, AC, Rassini JB, MendonÓa FC, de Paula Ferreira R (2013) Alfalfa dry matter yield, nutritional status and economic analysis of potassium fertilizer doses and frequency. Inter J Agron Plant Prod 4:389–98.

D'Elia L, Barba G, Cappuccio FP, Strazzullo P (2011) Potassium intake, stroke, and cardiovascular disease: a meta-analysis of prospective studies. J Amer College Cardio 57:1210–1219. https://doi.org/10.1016/j.jacc.2010.09.070

Dominguez-Nuñez JA, Benito B, Berrocal-Lobo M, Albanesi A (2016) Mycorrhizal fungi: role in the solubilization of potassium. In: Potassium solubilizing microorganisms for sustainable agriculture. Springer, New Delhi, pp 77–98

Duncan EG, O'Sullivan CA, Roper MM, Biggs JS, Peoples MB (2018) Influence of co-application of nitrogen with phosphorus, potassium and Sulphur on the apparent efficiency of nitrogen fertiliser use, grain yield and protein content of wheat. Field Crops Res 226:56–65. https://doi.org/10.1016/j.fcr.2018.07.010

Dwivedi SK, Kumar S, Bhakta N, Singh SK, Rao KK, Mishra JS, Singh AK (2017) Improvement of submergence tolerance in rice through efficient application of potassium under submergence-prone rainfed ecology of Indo-Gangetic plain. Func Plant Biol 44:907–916. https://doi.org/10.1071/FP17054

Ekmekcioglu C, Elmadfa I, Meyer AL, Moeslinger T (2016) The role of dietary potassium in hypertension and diabetes. J Physiol Biochem 72:93–106. https://doi.org/10.1007/s13105-015-0449-1

Filippini T, Violi F, D'Amico R, Vinceti M (2017) The effect of potassium supplementation on blood pressure in hypertensive subjects: a systematic review and meta-analysis. Inter J Cardio 230:127–135. https://doi.org/10.1016/j.ijcard.2016.12.048

Garcia RL, Hanway JJ (1976) Foliar fertilization of soybeans during the seed-filling period. Agron J 68:653–657. https://doi.org/10.2134/agronj1976.00021962006800040030x

Garcia K, Zimmermann SD (2014) The role of mycorrhizal associations in plant potassium nutrition. Front Plant Sci 5:337. https://doi.org/10.3389/fpls.2014.00337

Gaspar AP, Laboski CA, Naeve SL, Conley SP (2017a) Dry matter and nitrogen uptake, partitioning, and removal across a wide range of soybean seed yield levels. Crop Sci 57:2170–2182. https://doi.org/10.2135/cropsci2016.05.0322

Gaspar AP, Laboski CA, Naeve SL, Conley SP (2017b) Phosphorus and potassium uptake, partitioning, and removal across a wide range of soybean seed yield levels. Crop Sci 57:2193–2204. https://doi.org/10.2135/cropsci2016.05.0378

Giskin M, Efron Y (1986) Planting date and foliar fertilization of corn grown for silage and grain under limited moisture. Agron J 78:426–429. https://doi.org/10.2134/agronj1986.00021962007800030005x

Gorz HJ, Haskins FA, Pedersen JF, Ross WM (1987) Combining ability effects for mineral elements in forage sorghum hybrids. Crop Sci 27:216–219. https://doi.org/10.2135/cropsci1987.0011183X002700020017x

Greenwood DJ, Cleaver TJ, Turner MK, Hunt J, Niendorf KB, Loquens SMH (1980) Comparison of the effects of phosphate fertilizer on the yield, phosphate content and quality of 22 different vegetable and agricultural crops. J Agric Sci 95:457–469. https://doi.org/10.1017/S0021859600039502

Guerrero-Galán C, Delteil A, Garcia K, Houdinet G, Conéjéro G, Gaillard I, Sentenac H, Zimmermann SD (2018) Plant potassium nutrition in ectomycorrhizal symbiosis: properties and roles of the three fungal TOK potassium channels in *Hebeloma cylindrosporum*. Envir Microbiol 20:1873–1887. https://doi.org/10.1111/1462-2920.14122

Hanway JJ (1962) Northcentral regional potassium studies. III. Field studies with corn. Res Bull (Iowa Agriculture and Home Economics Experiment Station) 34(503):Article 1

Hanway JJ, Weber CR (1971) Accumulation of N, P, and K by soybean (*Glycine max* (L.) Merrill) plants. Agron J 63:406–408. https://doi.org/10.2134/agronj1971.00021962006300030017x

Haq MU, Mallarino AP (1998) Foliar fertilization of soybean at early vegetative stages. Agron J 90:763–769. https://doi.org/10.2134/agronj1998.00021962009000060008x

Haq MU, Mallarino AP (2000) Soybean yield and nutrient composition as affected by early season foliar fertilization. Agron J 92:16–24. https://doi.org/10.2134/agronj2000.92116x

Haq MU, Mallarino A (2005) Response of soybean grain oil and protein concentrations to foliar and soil fertilization. Agron J 97:910–918. https://doi.org/10.2134/agronj2004.0215

He FJ, Nowson CA, MacGregor GA (2006) Fruit and vegetable consumption and stroke: meta-analysis of cohort studies. Lancet 367:320–326. https://doi.org/10.1016/S0140-6736(06)68069-0

Hoagland DR (1932) Mineral nutrition of plants. Annu Rev Biochem 1:618–636. https://doi.org/10.1146/annurev.bi.01.070132.003154

Holland JE, Hayes RC, Refshauge G, Poile GJ, Newell MT, Conyers MK (2019) Biomass, feed quality, mineral concentration and grain yield responses to potassium fertiliser of dual-purpose crops. New Zeal J Agric Res 62:476–494. https://doi.org/10.1080/00288233.2018.1512505

Hu Y, Burucs Z, Schmidhalter U (2008) Effect of foliar fertilization application on the growth and mineral nutrient content of maize seedlings under drought and salinity. Soil Sci Plant Nutr 54:133–141. https://doi.org/10.1111/j.1747-0765.2007.00224.x

Huber DM, Arny DC (1985) Interactions of potassium with plant disease. In: Munson RD (ed) Potassium in agriculture. American Society of Agronomy, Madison WI, pp 467–488

Humble GD, Raschke K (1971) Stomatal opening quantitatively related to potassium transport: evidence from electron probe analysis. Plant Physiol 48:447–453. https://doi.org/10.1104/pp.48.4.447

John KS, Ravindran CS, George J, Nair MM, Suja G (2013) Potassium: a key nutrient for high tuber yield and better tuber quality in cassava. Better Crops–South Asia 2013:26–27

Kaiser DE, Wiersma JJ, Anderson JJ (2014) Genotype and environment variation in elemental composition of spring wheat flag leaves. Agron J 106:324–336. https://doi.org/10.2134/agronj2013.0329

Kaiser DE, Rosen CJ, Lamb JA (2016) Potassium for crop production. Univ. of Minn. St. Paul MN https://extension.umn.edu/phosphorus-and-potassium/potassium-crop-production. Accessed 13 May 2020

Kant S, Kafkafi U (2002) Potassium and abiotic stresses in plants. In: Pasricha NS, Bansal SK (eds) Potassium for sustainable crop production. Potash Institute of India, Gurgaon, pp 233–251

Kayser M, Isselstein J (2005) Potassium cycling and losses in grassland systems: a review. Grass Forage Sci 60:213–224. https://doi.org/10.1111/j.1365-2494.2005.00478.x

Khalili H, Malik S, Ananthakrishnan AN, Garber JJ, Higuchi LM, Joshi A, Peloquin J, Richter JM, Stewart KO, Curhan GC, Awasthi A (2016) Identification and characterization of a novel association between dietary potassium and risk of Crohn's disease and ulcerative colitis. Front Immun 7:554. https://doi.org/10.3389/fimmu.2016.00554

Kitchen NR, Buchholz DD, Nelson CJ (1990) Potassium fertilizer and potato leafhopper effects on alfalfa growth. Agron J 82:1069–1074. https://doi.org/10.2134/agronj1990.00021962008200060008x

Kleese RA, Rasmusson DC, Smith LH (1968) Genetic and environmental variation in mineral element accumulation in barley, wheat, and soybeans. Crop Sci 8:591–593. https://doi.org/10.2135/cropsci1968.0011183X000800050025x

Krueger K, Goggi AS, Mallarino AP, Mullen RE (2013) Phosphorus and potassium fertilization effects on soybean seed quality and composition. Crop Sci 53:602–610. https://doi.org/10.2135/cropsci2012.06.0372

Kumar P, Pandey SK, Singh BP, Singh SV, Kumar D (2007) Influence of source and time of potassium application on potato growth, yield, economics and crisp quality. Potato Res 50:1–13. https://doi.org/10.1007/s11540-007-9023-8

Lambert H, Frassetto L, Moore JB, Torgerson D, Gannon R, Burckhardt P, Lanham-New S (2015) The effect of supplementation with alkaline potassium salts on bone metabolism: a meta-analysis. Osteoporosis Inter 26:1311–1318. https://doi.org/10.1007/s00198-014-3006-9

Lester GE, Jifon JL, Makus DJ (2010) Impact of potassium nutrition on postharvest fruit quality: melon (*Cucumis melo* L) case study. Plant Soil 335:117–131. https://doi.org/10.1007/s11104-009-0227-3

Ling F, Silberbush M (2002) Response of maize to foliar vs. soil application of nitrogen–phosphorus–potassium fertilizers. J Plant Nutr 25:2333–2342. https://doi.org/10.1081/PLN-120014698

Lissbrant S, Stratton S, Cunningham SM, Brouder SM, Volenec JJ (2009) Impact of long-term phosphorus and potassium fertilization on alfalfa nutritive value–yield relationships. Crop Sci 49:1116–1124. https://doi.org/10.2135/cropsci2008.06.0333

Lissbrant S, Brouder SM, Cunningham SM, Volenec JJ (2010) Identification of fertility regimes that enhance long-term productivity of alfalfa using cluster analysis. Agron J 102:580–591. https://doi.org/10.2134/agronj2009.0300

Lunnan T, Øgaard AF, Krogstad T (2018) Potassium fertilization of timothy-based cut grassland—effects on herbage yield, mineral composition and critical K concentration on soils with different K status. Grass Forage Sci 73:500–509. https://doi.org/10.1111/gfs.12341

Macolino S, Lauriault LM, Rimi F, Ziliotto U (2013) Phosphorus and potassium fertilizer effects on alfalfa and soil in a non-limited soil. Agron J 105:1613–1618. https://doi.org/10.2134/agronj2013.0054

Macy P (1936) The quantitative mineral nutrient requirement of plants. Plant Physiol 11:749–764. https://doi.org/10.1104/pp.11.4.749

Malhi SS, Loeppky H, Coulman B, Gill KS, Curry P, Plews T (2005) Fertilizer nitrogen, phosphorus, potassium, and sulphur effects on forage yield and quality of timothy hay in the Parkland region of Saskatchewan. Can J Plant Nutr 27:1341–1360. https://doi.org/10.1081/PLN-200025834

Malinowski DP, Belesky DP (2000) Adaptations of endophyte-infected cool-season grasses to environmental stresses: mechanisms of drought and mineral stress tolerance. Crop Sci 40:923–940. https://doi.org/10.2135/cropsci2000.404923x

Mallarino AP, Higashi SL (2009) Assessment of potassium supply for corn by analysis of plant parts. Soil Sci Soc Amer J 73:2177–2183. https://doi.org/10.2136/sssaj2008.0370

Mallarino AP, Haq MU, Wittry D, Bermudez M (2001) Variation in soybean response to early season foliar fertilization among and within fields. Agron J 93:1220–1226. https://doi.org/10.2134/agronj2001.1220

McNaught KJ (1958) Potassium deficiency in pastures: I. potassium content of legumes and grasses. New Zeal J Agric Res 1:148–181. https://doi.org/10.1080/00288233.1958.10431069

Mengel K (1997) Impact of potassium on crop yield and quality with regard to economical and ecological aspects. In: Johnston AE (ed) Food security in the WANA region, the essential need for balanced fertilization. International Potash Institute, Basle, pp 157–174

Mikkelsen R (2017) The importance of potassium management for horticultural crops. Indian J Fert 13(11):82–86

Moreira A, Moraes LAC, Schroth G, Becker FJ, Mandarino JMG (2017) Soybean yield and nutritional status response to nitrogen sources and rates of foliar fertilization. Agron J 109:629–635. https://doi.org/10.2134/agronj2016.04.0199

Morgounov AI, Belan I, Zelenskiy Y, Roseeva L, Tomoskozi S, Bekes F, Abugalieve A, Cakmak I, Vargas M, Crossa J (2013) Historical changes in grain yield and quality of spring wheat varieties cultivated in Siberia from 1900 to 2010. Can J Plant Sci 93:425–433. https://doi.org/10.4141/cjps2012-091

Mullins GL, Reeves DW, Burmester CH, Bryant HH (1994) In-row subsoiling and potassium placement effects on root growth and potassium content of cotton. Agron J 86:136–139. https://doi.org/10.2134/agronj1994.00021962008600010025x

Munson RD (1985) Potassium in agriculture. American Society of Agronomy, Madison, WI

Murata T, Akazawa T (1968) Enzymic mechanism of starch synthesis in sweet potato roots. I. Requirements of potassium ions for starch synthetase. Arch Biochem Biophys 126:873–879. https://doi.org/10.1016/0003-9861(68)90481-5

Myers SW, Gratton C (2006) Influence of potassium fertility on soybean aphid, *Aphis glycines* Matsumura (Hemiptera: Aphididae), population dynamics at a field and regional scale. Environ Entom 35:219–227. https://doi.org/10.1603/0046-225X-35.2.219

Myers SW, Gratton C, Wolkowski RP, Hogg DB, Wedberg JL (2005) Effect of soil potassium availability on soybean aphid (Hemiptera: Aphididae) population dynamics and soybean yield. J Econ Entom 98:113–120. https://doi.org/10.1093/jee/98.1.113

Navarrete-Ganchozo RJ (2014) Quantification of plant-available potassium (K) in a corn-soybean rotation: A long-term evaluation of K rates and crop K removal effects (Doctoral dissertation, Purdue University). (Order No. 3636478). Available from Agricultural & Environmental Science Database; Dissertations & Theses @ CIC Institutions; ProQuest Dissertations & Theses Global. (1615794935). Retrieved from https://search.proquest.com/docview/1615794935?accountid=13360. Accessed 13 May 2020

Neilsen GH, Neilsen D, Peryea F (1999) Response of soil and irrigated fruit trees to fertigation or broadcast application of nitrogen, phosphorus, and potassium. HortTechnol 9:393–401. https://doi.org/10.21273/HORTTECH.9.3.393

Nelson KA, Motavalli PP, Nathan M (2005) Response of no-till soybean [*Glycine max* (L.) Merr.] to timing of preplant and foliar potassium applications in a claypan soil. Agron J 97:832–838. https://doi.org/10.2134/agronj2004.0241

Nelson KA, Motavalli PP, Stevens WE, Dunn D, Meinhardt CG (2010) Soybean response to preplant and foliar-applied potassium chloride with strobilurin fungicides. Agron J 102:1657–1663. https://doi.org/10.2134/agronj2010.0065

Nicholaides JJ, Chancy HF, Mascagni HJ, Wilson LG, Eaddy DW (1985) Sweet potato response to K and P fertilization. Agron J 77:466–470. https://doi.org/10.2134/agronj1985.00021962007700030024x

Ning P, Li S, Yu P, Zhang Y, Li C (2013) Post-silking accumulation and partitioning of dry matter, nitrogen, phosphorus and potassium in maize varieties differing in leaf longevity. Field Crops Res 144:19–27. https://doi.org/10.1016/j.fcr.2013.01.020

Nkebiwe PM, Weinmann M, Bar-Tal A, Muller T (2016) Fertilizer placement to improved crop nutrient acquisition and yield. A review and meta-analysis. Field Crop Res 196:389–401. https://doi.org/10.1016/j.fcr.2016.07.018

Obigbesan GO (1977) Investigations on Nigerian root and tuber crops: effect of potassium on starch yields, HCN content and nutrient uptake of cassava cultivars (*Manihot esculenta*). J Agric Sci 89:29–34. https://doi.org/10.1017/S0021859600027167

Oltmans RR, Mallarino AP (2015) Potassium uptake by corn and soybean, recycling to soil, and impact on soil test potassium. Soil Sci Soc Am J 79:314–327. https://doi.org/10.2136/sssaj2014.07.0272

Oosterhuis DM, Loka DA, Kawakami EM, Pettigrew WT (2014) The physiology of potassium in crop production. Adv Agron 126:203–233. https://doi.org/10.1016/B978-0-12-800132-5.00003-1

Orlowski JM, Haverkamp BJ, Laurenz RG, Marburger D, Wilson EW, Casteel SN, Conley SP, Naeve SL, Nafziger ED, Roozeboom KL, Ross WJ (2016) High-input management systems effect on soybean seed yield, yield components, and economic break-even probabilities. Crop Sci 56:1988–2004. https://doi.org/10.2135/cropsci2015.10.0620

Page MB, Talibudeen O (1982) Critical potassium potentials for crops: 2. Potentials for wheat, maize, peas, beans and sugar beet in their early growth on a sandy loam. Eur J Soil Sci 33:771–778. https://doi.org/10.1111/j.1365-2389.1982.tb01806.x

Panique E, Kelling KA, Schulte EE, Hero DE, Stevenson WR, James RV (1997) Potassium rate and source effects on potato yield, quality, and disease interaction. Amer Potato J 74:379–398. https://doi.org/10.1007/BF02852777

Pant HK, Mislevy P, Rechcigl JE (2004) Effects of phosphorus and potassium on forage nutritive value and quantity: environmental implications. Agron J 96:1299–1305. https://doi.org/10.2134/agronj2004.1299

Parvej M, Slaton NA, Fryer MS, Roberts TL, Purcell LC (2016) Postseason diagnosis of potassium deficiency in soybean using seed potassium concentration. Soil Sci Soc Amer J 80:1231–1243. https://doi.org/10.2136/sssaj2016.02.0030

Pervez H, Ashraf M, Makhdum MI (2004) Influence of potassium nutrition on gas exchange characteristics and water relations in cotton (*Gossypium hirsutum* L.). Photosynthetica 42:251–255. https://doi.org/10.1023/B:PHOT.0000040597.62743.5b

Pettigrew WT, Heitholt JJ, Meredith WR (1996) Genotypic interactions with potassium and nitrogen in cotton of varied maturity. Agron J 88:89–93. https://doi.org/10.2134/agronj1996.00021962008800010019x

Qiu S, Xie J, Zhao S, Xu X, Hou Y, Wang X, Zhou W, He P, Johnston AM, Christie P, Jin J (2014) Long-term effects of potassium fertilization on yield, efficiency, and soil fertility status in a rain-fed maize system in Northeast China. Field Crops Res 163:1–9. https://doi.org/10.1016/j.fcr.2014.04.016

Rahman MH, Saiga S (2005) Endophytic fungi (*Neotyphodium coenophialum*) affect the growth and mineral uptake, transport and efficiency ratios in tall fescue (*Festuca arundinacea*). Plant Soil 272:163–171. https://doi.org/10.1007/s11104-004-4682-6

Randall GW, Iragavarapu TK, Evans SD (1997) Long-term P and K applications: I. Effect on soil test incline and decline rates and critical soil test levels. J Prod Agric 10:565–571. https://doi.org/10.2134/jpa1997.0565

Rengel Z, Damon PM (2008) Crops and genotypes differ in efficiency of potassium uptake and use. Physiol Plant 133:624–636. https://doi.org/10.1111/j.1399-3054.2008.01079.x

Rochester IJ, Constable GA (2015) Improvements in nutrient uptake and nutrient use-efficiency in cotton cultivars released between 1973 and 2006. Field Crops Res 173:14–21. https://doi.org/10.1016/j.fcr.2015.01.001

Rominger RS, Smith D, Peterson LA (1976) Yield and chemical composition of alfalfa as influenced by high rates of K topdressed as KC1 and K_2SO_4 1. Agron J 68:573–577. https://doi.org/10.2134/agronj1976.00021962006800040010x

Rosendahl RO (1943) The effect of mycorrhizal and nonmycorrhizal fungi on the availability of difficultly soluble potassium and phosphorus. Soil Sci Soc Amer J 7:477–479. https://doi.org/10.2136/sssaj1943.036159950007000C0080x

Sardans J, Peñuelas J (2015) Potassium: a neglected nutrient in global change. Glob Ecol Biogeogr 24:261–275. https://doi.org/10.1111/geb.12259

Shen C, Wang J, Shi X, Kang Y, Xie C, Peng L, Dong C, Shen Q, Xu Y (2017) Transcriptome analysis of differentially expressed genes induced by low and high potassium levels provides insight into fruit sugar metabolism of pear. Front Plant Sci 8:938. https://doi.org/10.3389/fpls.2017.00938

Shen C, Li Y, Wang J, Al Shoffe Y, Dong C, Shen Q, Xu Y (2018) Potassium influences expression of key genes involved in sorbitol metabolism and its assimilation in pear leaf and fruit. J Plant Growth Reg 37:883–895. https://doi.org/10.1007/s00344-018-9783-1

Shetley J, Nelson KA, Stevens WG, Dunn D, Burdick B, Motavalli PP, English JT, Dudenhoeffer CJ (2015) Corn yield response to pyraclostrobin with foliar fertilizers. J Agric Sci 7:18. https://doi.org/10.5539/jas.v7n7p18

Singh VK, Dwivedi BS, Singh SK, Mishra RP, Shukla AK, Rathore SS, Shekhawat K, Majumdar K, Jat ML (2018) Effect of tillage and crop establishment, residue management and K fertilization on yield, K use efficiency and apparent K balance under rice-maize system in North-Western India. Field Crops Res 224:1–12. https://doi.org/10.1016/j.fcr.2018.04.012

Slaton NA, Golden BR, DeLong RE, Mozaffari M (2010) Correlation and calibration of soil potassium availability with soybean yield and trifoliolate potassium. Soil Sci Soc Amer J 74:1642–1651. https://doi.org/10.2136/sssaj2009.0197

Smith D, Dobrenz AK, Schonhorst MH (1981) Response of alfalfa seedling plants to high levels of chloride-salts. J Plant Nutr 4:143–174. https://doi.org/10.1080/01904168109362909

Stammer AJ, Mallarino AP (2018) Plant tissue analysis to assess phosphorus and potassium nutritional status of corn and soybean. Soil Sci Soc Amer J 82:260–270. https://doi.org/10.2136/sssaj2017.06.0179

Stanley R, Jewell S (1989) The influence of source and rate of potassium fertilizer on the quality of potatoes for French fry production. Potato Res 32:439–446. https://doi.org/10.1007/BF02358499

Syverud TD, Walsh LM, Oplinger ES, Kelling KA (1980) Foliar fertilization of soybeans (*Glycine max* L.). Comm Soil Sci Plant Anal 11:637–651. https://doi.org/10.1080/00103628009367069

Thom WO, Miller TC, Bowman DH (1981) Foliar fertilization of rice after midseason. Agron J 73:411–414. https://doi.org/10.2134/agronj1981.00021962007300030007x

Tyner EH (1947) The relation of corn yields to leaf nitrogen, phosphorus, and potassium content. Soil Sci Soc Amer J 11:317–323. https://doi.org/10.2136/sssaj1947.036159950011000C0059x

Umar S, Bansal SK, Imas P, Magen H (1999) Effect of foliar fertilization of potassium on yield, quality, and nutrient uptake of groundnut. J Plant Nutr 22:1785–1795. https://doi.org/10.1080/01904169909365754

Usherwood NR (1985) The role of potassium in crop quality. In: Munson RD (ed) Potassium in agriculture. American Society of Agronomy, Madison WI, pp 489–513

Van Duivenbooden N, deWit CT, Van Keulen H (1996) Nitrogen, phosphorus and potassium relations in five major cereals reviewed in respect to fertilizer recommendations using simulation modelling. Fert Res 44:37–49. https://doi.org/10.1007/BF00750691

Veresoglou SD, Barto EK, Menexes G, Rillig MC (2013) Fertilization affects severity of disease caused by fungal plant pathogens. Plant Path 62:961–969. https://doi.org/10.1111/ppa.12014

Vitosh ML, Johnson JW, Mengel DB (1995) Tri-state fertilizer recommendations for corn, soybeans, wheat and alfalfa. Ext Bull E-2567. Michigan State Univ East Lansing

Vyn TJ, Janovicek KJ (2001) Potassium placement and tillage system effects on corn response following long-term no till. Agron J 93:487–495. https://doi.org/10.2134/agronj2001.933487x

Wang M, Zheng Q, Shen Q, Guo S (2013) The critical role of potassium in plant stress response. Inter J Mol Sci 14:7370–7390. https://doi.org/10.3390/ijms14047370

Yin X, Vyn TJ (2002) Soybean responses to potassium placement and tillage alternatives following no-till. Agron J 94:1367–1374. https://doi.org/10.2134/agronj2002.1367

Yin X, Vyn TJ (2003) Potassium placement effects on yield and seed composition of no-till soybean seeded in alternate row widths. Agron J 95:126–132. https://doi.org/10.2134/agronj2003.1260

Zeng F, Konnerup D, Shabala L, Zhou M, Colmer TD, Zhang G, Shabala S (2014) Linking oxygen availability with membrane potential maintenance and K+ retention of barley roots: implications for waterlogging stress tolerance. Plant Cell Envir 37:2325–2338. https://doi.org/10.1111/pce.12422

Zhang Y, Chen P, Chen J, Wang L, Wei Y, Xu D (2019) Association of low serum potassium levels and risk for all-cause mortality in patients with chronic kidney disease: a systematic review and meta-analysis. Ther Apher Dial 23:22–31. https://doi.org/10.1111/1744-9987.12753

15

Benefits of Potassium (K) Intake and its Role in Human Health

Michael Stone and Connie Weaver

Abstract Potassium (K) is the most abundant cation in intracellular fluid where it plays a key role in maintaining cell function. The majority of K consumed (60–100 mmol day^{-1}) is lost in the urine, with the remaining excreted in the stool, and a very small amount lost in sweat. Little is known about the bioavailability of K, especially from dietary sources. Less is understood on how bioavailability may affect health outcomes. Potassium is an essential nutrient that has been labeled a shortfall nutrient by recent Dietary Guidelines for Americans Advisory Committees. Increases in K intake have been linked to improvements in cardiovascular and other metabolic health outcomes. There is growing evidence for the association between K intake and blood pressure (BP) reduction in adults; hypertension (HTN) is the leading cause of the cardiovascular disease (CVD) and a major financial burden (US$53.2 billion) to the US public health system and has a significant impact on all-cause morbidity and mortality worldwide. Evidence is also accumulating for the protective effect of adequate dietary K on age-related bone loss and glucose control. Understanding the benefit of K intake from various sources may help to reveal how specific compounds and tissues influence K movement within the body, and further the understanding of its role in health.

15.1 Potassium Intake Needs

Potassium (K) is an essential nutrient, that has been labeled a shortfall nutrient by recent Dietary Guidelines for Americans Advisory Committees (National Academies of Sciences and Medicine 2019; DeSalvo et al. 2016). Physiologically, K is the most abundant cation in intracellular fluid where it plays a key role in cell function, maintaining intracellular fluid (ICF) volume and transmembrane electrochemical gradients (Stone et al. 2016). Because K is a major intracellular ion, it is widely

M. Stone (✉) · C. Weaver (✉)
Department of Nutrition Science, College of Health and Human Sciences, Purdue University, Indiana, USA
e-mail: stone59@purdue.edu; weavercm@purdue.edu

distributed in foods once derived from living tissues. Potassium concentrations are generally highest in fruits and vegetables, but can also be quite high in cereals, grains, dairy, and meat (DeSalvo et al. 2016; Stone et al. 2016). The evolution of dietary practices in the USA over the last several decades, and more recently worldwide, has seen a higher intake of low nutrient density convenience foods, coupled with decreased consumption of fruits and vegetables, leading to a diet lower in K and higher in sodium (Na) (Weaver 2013). The average intake of K of the US adults participating in the National Health and Nutrition Examination Survey (NHANES) 2013–2014 was 2668 mg K day^{-1}, below the adequate intake (AI) of 3000 mg day^{-1} set forth by the 2019 DRI committee, and well below the previous AI of 4700 mg K day^{-1} (Institute of Medicine 2005; National Academies of Sciences and Medicine 2019). This chapter gives a comprehensive overview of K as a nutrient, the physiology of how it moves through the body including K bioavailability and excretion, and how this may affect vascular pressure, glucose metabolism, the movement and storage of calcium (bone) throughout the body, and the health consequences of these relationships.

15.1.1 Dietary Reference Intakes

The reference values for the intake of any nutrient are referred to as the Dietary Reference Intakes (DRIs) and include: the Estimated Average Requirement (EAR), or intake level at which 50% of the population have adequate intakes; the Recommended Dietary Allowance (RDA), based on the EAR, is sufficient to meet the requirements of nearly the entire population (98%); Adequate Intake (AI), used in lieu of an RDA when there is insufficient evidence to set an EAR and thus an RDA; and the Tolerable Upper Intake Level (UL), the estimated maximum intake that poses no health risk, developed from a "NOAEL" with a safety factor applied (Fulgoni 2007; Lupton et al. 2016; Millen et al. 2016; Institute of Medicine 2005). Dietary reference intakes are quantitative values established by review committees commissioned by the National Academy of Sciences, Engineering, and Medicine (NASEM), Health and Medicine Division (formerly the Institute of Medicine), after a review of the appropriate research surrounding any nutrient's role in eliminating nutritional deficiencies, and reducing the risk of chronic disease. Basic concepts of establishing the proper level of intake for each nutrient are that the needs of healthy (non-diseased) individuals are met, nutrients are grouped by physiological functionality, and age groupings are revised to reflect changes of biological patterns (e.g., gender, growth, pregnancy, etc.) (Lupton et al. 2016; Millen et al. 2016; Institute of Medicne 2005). Chronic disease endpoints are only considered when a sufficient body of knowledge has been established. To this point, the recent Dietary Reference Intake report for sodium (Na) and K was the first to establish a chronic disease risk reduction (CDRR) level for Na, a new DRI intended to help differentiate between nutrient intakes necessary for adequacy vs. those which may improve health (National Academies of Sciences and Medicine 2019).

15.1.2 *Potassium Intakes Worldwide*

Recommended K intakes in various countries worldwide often utilize the guidelines set by the North American DRIs or World Health Organization (WHO) (Strohm et al. 2017; World Health Organization 2012). Despite this, few countries meet these recommendations and large global variation in K consumption exists (Weaver et al. 2018).

The most recent WHO recommendations for K intake come from guidelines published in 2012, examining key chronic disease endpoints related to blood pressure (BP), stroke, CVD, coronary heart disease (CHD), blood lipids, and catecholamines (World Health Organization 2012). Based primarily off one large systematic review with meta-analysis (Aburto et al. 2013), the WHO set recommendations to consume at least 90 mmol (~3500 mg) of K day^{-1} to reduce BP, cardiovascular disease (CVD), stroke, and coronary heart disease (World Health Organization 2012; Weaver et al. 2018).

Current recommendations for the USA and Canada were recently revised by the National Academy of Sciences, Health, and Medicine Division. According to the 2019 DRI guidelines for K, lack of a sensitive biomarker and limitations across K bioavailability and retention studies offer insufficient evidence to establish EAR and RDA levels for adequacy or deficiency (National Academies of Sciences and Medicine 2019). Because of this, the committee set AIs using intake data from two nationally representative surveys, NHANES and Canadian Community Health Survey (CCHS). The highest median K intake across the two surveys was selected for each DRI group and set as the AI. For adults, the data that informed the K AIs were from healthy, normotensive individuals without a self-reported history of CVD. In contrast to the 2005 DRI report, adult AIs were separated by sex, with a K intake of 3400 mg day^{-1} for men and 2600 mg day^{-1} for women (National Academies of Sciences and Medicine 2019). This is remarkably lower than the AIs established in 2005, set at 4700 mg day^{-1} for adults 18 and older (Institute of Medicine 2005). Because observational data looking at increased K intakes and CVD (and associated disease) risk are mixed (Newberry et al. 2018; National Academies of Sciences and Medicine 2019), CDRR intake level for K could not be established. Blood pressure was considered for a surrogate marker for CVD risk reduction, based on findings that show a reduction in BP with increased supplemental K intake (Newberry et al. 2018), but given the lack of clear evidence supporting K intake alone in the reduction of CVD and related mortality, the committee decided against this.

Actual K requirements would vary with an individual's genetics, Na intake, and status of various health-related biomarkers. Potential benefits of increasing K consumption may include decreases in vascular pressure, optimal kidney function, improvement in glucose control, and possible bone benefit (He and MacGregor 2008; Weaver 2013).

15.2 Internal Balance of Potassium

15.2.1 Potassium Tissue Movement

About 90% of dietary K is passively absorbed in the small intestine. In the proximal small intestine (duodenum, jejunum) K^+ absorption primarily follows water absorption, while distally (ileum) movement is more influenced by changes in transepithelial electrical potential difference. In the colon, K is both excreted, in exchange for Na, as well as reabsorbed via H^+/K^+ ATPases (Meneton et al. 2004). Total body K is estimated to be approximately 43 mmol K kg^{-1} in adults, with only 2% of this found in the extracellular fluid. Most of the body K content is found in the intracellular space of skeletal muscle. Potassium is the primary intercellular cation and plays a key role in maintaining cell function, having a marked influence on transmembrane electrochemical gradients (Palmer 2015; Stone et al. 2016). The gradient of K^+ across the cell membrane determines cellular membrane potential, which, based on the normal ratio of intracellular to extracellular K^+, is −90 mV. This potential difference is maintained in large part by the ubiquitous ion channel, the sodium-potassium (Na^+/K^+) ATPase pump. Transmembrane electrochemical gradients cause the diffusion of sodium (Na^+) out of the cell and K^+ into the cell. This process is reversed, and cellular potential difference is held constant, via the aforementioned Na^+/K^+ ATPase pumps. When activated, the Na^+/K^+ ATPase pump exchanges two extracellular K^+ ions for three intracellular Na^+ ions, influencing membrane potential based on physiological excitation or inhibition. These channels are partially responsible, along with the Na^+/K^+ chloride symporter, and sodium-calcium exchanger, for maintaining the potential difference across the resting cell membrane as well. Both resting membrane potential and the electrochemical difference across the cell membrane are crucial for normal cell biology, especially in muscle, cardiac, and nervous tissue (Palmer 2015; Unwin et al. 2011; Stone et al. 2016; Stone and Weaver 2018).

Distribution of K under normal physiological conditions is referred to as internal balance. In healthy individuals, the blood K concentration ranges between 3.5 and 5.5 mM, with numerous homeostatic mechanisms in place for maintenance within this narrow margin. Changes in plasma concentrations of K^+ alter the electrochemical gradient and can lead to physiological dysfunction. In hyperkalemia, when K concentrations exceed 5.5 mM, membrane depolarization can lead to muscle weakness, paralysis, and cardiac dysrhythmias (e.g., sinus bradycardia, ventricular tachycardia, ventricular fibrillation). Conversely, hypokalemia, when K plasma concentration is below 3.5 mM, can cause membrane hyperpolarization, interfering with normal nerve and muscle function leading to muscle weakness and decreases in smooth muscle contraction (Stipanuk 2006). Hypokalemia can also cause both atrial and ventricular cardiac dysrhythmias, as well as lead to paralysis and if left untreated, death. Total body K is found intercellularly (98%) primarily in the muscle (70%) and to some extent all other tissues. Distribution and metabolism of K are influence by hormones (insulin, aldosterone, catecholamines), acidemia, and fluid balance.

In response to the dietary consumption of a high K meal, insulin enhances the cellular uptake of K^+. Insulin, released from pancreatic beta-cells, increases K uptake via the stimulation of Na^+/K^+ ATPase activity in skeletal and cardiac muscle, fat tissue, liver, bone, and red blood cells, attenuating the rise in plasma K^+ following consumption (Greenlee et al. 2009). Potassium uptake is also influenced by the stimulation of both α and β2 adrenergic receptors by the circulating stress hormones catecholamines (epinephrine, norepinephrine) (Palmer 2015; Unwin et al. 2011). Mechanistically, the insulin-mediated regulatory pathway leads to Na^+/K^+ ATPase activation via stimulation of cell surface tyrosine kinase receptors (insulin substrate receptor-1; IRS1), which also stimulates the translocation of intracellular glucose transport proteins (GLUT4 in muscle) facilitating the influx of glucose into the cell. Downstream activation of signaling cascades involving IRS1-phosphatidylinositide-3-kinase (PI3-K) and protein kinase A (PKA) facilitate both K and glucose uptake (Unwin et al. 2011; Stone et al. 2016). Catecholamine binding to β2 adrenergic receptors activates pathways mediated by cyclic adenosine-mono-phosphate (cAMP) and PKA to increase Na^+/K^+ ATPase activity and cellular K^+ uptake. In contrast, stimulation of α1 and α2 adrenergic receptors, primarily through increased circulating levels of the stress hormone norepinephrine, lead to activation of hepatic calcium-dependent K^+ channels and increased plasma K concentration via K release from the liver. Aldosterone, which has a marked effect on renal handling of K, may also influence the transmembrane distribution of K^+ via stimulation of cellular Na^+ uptake through activation of Na^+/H^+ or $Na^+/K^+/Cl^-$ transporters and subsequently Na^+/K^+ ATPases (Unwin et al. 2011; Stipanuk 2006). While hormones play an important role in the movement of K^+ within the body, the concentration of other ions (inorganic and organic) is also influential in maintaining proper internal balance (Stone et al. 2016).

Metabolic acidosis caused by inorganic anions (mineral acidosis) can also stimulate the K^+ movement. The effect of acidemia on enhancing cellular K loss is not related to direct K^+-H^+ ion exchange, but rather via action on transporters which normally regulate skeletal muscle pH (Aronson and Giebisch 2011; Stone et al. 2016). The decrease in extracellular pH reduces the rate of Na^+/H^+ exchange and inhibits Na^+/bicarbonate (HCO_3^-) cotransport. The fall in intracellular Na^+ reduces Na^+/K^+ ATPase activity, leading to decreased K^+ influx, cellular K^+ losses, and possible hyperkalemia (Palmer 2015; Stone et al. 2016). Additionally, a fall in extracellular HCO_3^- increases inward flux of Cl^- via upregulation of Cl^-/HCO_3^- exchange, increasing K^+/Cl^- cotransport and subsequent K^+ efflux. In metabolic acidosis via organic anion (e.g., lactic acid) accumulation, loss of K from the cell is much smaller. Accumulation here, through the movement of both anions and H^+ through monocarboxylate transporters (MCT; MCT1, MCT4), leads to a lower intracellular pH, stimulating the movement of Na^+ via Na^+/H^+ and Na^+/HCO_3^- transporters. An increase of intracellular Na^+ maintains Na^+/K^+ ATPase activity, limiting the efflux of K^+. Generally, metabolic acidosis (inorganic or organic) causes greater K^+ efflux than respiratory acidosis, HCO_3^- being the primary anion accumulating in the cell to balance the influx of hydrogen ions (Perez et al. 1981; Stone et al. 2016). Movement of cellular K varies similarly in response to different types of physiological alkalosis as well. In respiratory alkalosis, K^+ influx is reduced compared to metabolic alkalosis, due to the efflux of cellular HCO_3^- (Stone et al. 2016).

15.2.2 Renal Potassium Handling

The majority of consumed K is excreted in the urine, with the remaining excreted in the stool, and, under homeostatic conditions, a variable amount in sweat (Shils and Shike 2006). Potassium has a higher ratio of dietary intake to extracellular pool size; recall only 2% of the total body K^+ is distributed in the extracellular fluid (ECF) with the remaining distributed in the intracellular fluid (ICF) of various tissues. To meet the challenge of a high K meal, the K homeostatic system is very efficient at clearing plasma K via an increase in renal K excretion. When dietary K intake increases or decreases, the kidneys modulate excretion accordingly, ensuring the maintenance of plasma K^+ concentration (Stone et al. 2016). In addition, with the administration of acute K loads, only approximately half of the dose appears in the urine after 4–6 h, suggesting that extrarenal tissues (e.g., muscle, liver, adipose) play an important role in K homeostasis as well via insulin and catecholamine uptake (Youn 2013; Bia and DeFronzo 1981; Stone et al. 2016). Excessive extrarenal K losses are usually small but can occur in individuals with diarrhea, severe burns, or excessive and prolonged sweating (Stone et al. 2016; Stone and Weaver 2018).

Potassium is freely filtered by the glomerulus of the kidney, with most of it being reabsorbed (70–80%) in the proximal convoluted tubule (PCT) and loop of Henle. Under physiological homeostasis, the delivery of K to the nephron remains constant. Conversely, the secretion of K by the distal nephron is variable and depends on intracellular K concentration, luminal K concentration, and cellular permeability (Palmer 2015; Stone et al. 2016). Two major factors of K secretion/loss involve the renal handling of Na and mineralocorticoid activity. Reabsorption in the proximal tubule is primarily passive and proportional to reabsorption of solute and water, accounting for ~60% of filtered K (Penton et al. 2015; Ludlow 1993; Stone et al. 2016). Within the descending limb of Henle's loop, a small amount of K^+ is secreted into the luminal fluid, while in the thick ascending limb (TAL), reabsorption occurs together with Na^+ and Cl^-, both trans- and paracellularly. This leads to the K concentration of the fluid entering the distal convoluted tubule to be lower than plasma levels (~2 mM), facilitating eventual secretion (Ludlow 1993; Stone et al. 2016). Similar to reabsorption in the proximal tubule, paracellular diffusion in Henle's loop is mediated via solvent drag, while transcellular movement occurs primarily through the apical sodium-potassium-chloride ($Na^+/K^+/2Cl^-$) cotransporter (Ludlow 1993; Stone et al. 2016). The renal outer medullary K channel (ROMK), also located on the apical membrane, mediates the recycling of K from the cell to the lumen, sustaining the activation of the $Na^+/K^+/2Cl^-$ cotransporter and K reabsorption in the ascending limb. The movement of K through ROMK induces a positive lumen voltage potential, increasing the driving force of paracellular cation (e.g., Ca^{2+}, Mg^{2+}, K^{+}) reabsorption as well. Na^{+}/K^{+} ATPase pumps located basolaterally throughout the loop, maintain low levels of intracellular Na^{+} and further provide a favorable gradient for K^{+} reabsorption (Palmer 2015; Unwin et al. 2011; Stone et al. 2016; Stone and Weaver 2018).

Major regulation of K excretion begins in the late distal convoluted tubule (DCT) and progressively increases through the connecting tubule and cortical collecting duct. In the early DCT luminal, Na^{+} influx is mediated by the apical sodium chloride cotransporter (NCC) and continues into the late DCT via the epithelial Na^{+} channel (ENaC) (Meneton et al. 2004; Stone et al. 2016). Both are expressed apically and are the primary means of Na reabsorption from the luminal fluid. Sodium reabsorption leads to an electrochemical potential that is more negative than peritubular capillary fluid. This charge imbalance is matched by an increase in the aforementioned paracellular reabsorption of Cl^{-} from the lumen, as well as increases in Na^{+}/K^{+} ATPase and ROMK activity. Increased distal delivery of Na increases Na reabsorption, leading to a more negative luminal/plasma potential gradient and an increase in K secretion (Stone et al. 2016).

Most K excretion is mediated by principal cells in the collecting duct. Principal cells possess basolateral Na^{+}/K^{+} ATPases, which facilitate the movement of K from the blood and into the cell. The high cellular concentration of K provides a favorable gradient not only for the movement of K into the tubular lumen but for the reabsorption of Na as well. Movements of K and Na occur through the ROMK and ENaC channels, respectively. In conditions of K depletion, reabsorption of K occurs through H^{+}/K^{+} ATPases, located on the apical membrane of α-intercalated cells in the collecting duct, thus, providing a mechanism in which K depletion increases K reabsorption (Meneton et al. 2004; Stone and Weaver 2018).

Two primary types of K channels have been identified in the cortical collecting duct, the aforementioned ROMK, as well as the maxi-K channel (also known as the BK large conductance K^{+} channel). The ROMK is known to be the major K secretory pathway, characterized by activity during the low conductance of normal physiologic renal fluid excretion. Conversely, the maxi-K channel is quiescent in basal conditions and becomes activated during periods of increased tubular flow, increasing K secretion in a flow-dependent manner (e.g., hypervolemia, high arterial pressure) (Palmer 2015).

15.2.3 Interactions with Sodium Balance

Sodium and K^{+} are the primary electrolytes found in body fluids and work in concert to maintain normal fluid balance. There are no known receptors capable of detecting fluctuations of Na^{+} within the body, however physiological mechanisms that control extracellular fluid volume effectively control Na^{+} balance, influencing K^{+} movement as well. Perturbations in extracellular fluid volume lead to the recruitment of mechanisms that influence both the volume and pressure of circulation (cardiac and arterial pressure). Vascular pressure receptors (baroreceptors) sense changes in stretch or tension in vascular beds. Receptors that respond to low-pressure found in the central venous portion of the vascular tree respond to changes in blood volume, while high-pressure receptors located in the arterial circulation respond to changes in blood pressure (Stipanuk 2006). With hypovolemia (low fluid volume) baroreceptors are activated in the vasculature of the pulmonary vein and/or walls of the cardiac atria and send efferent signals to the central nervous system (CNS) to induce both a sympathetic and hormonal response. Hormonally this causes increased release of arginine vasopressin (AVP; antidiuretic hormone) from the posterior pituitary gland, which increases the permeability of the collecting ducts of the kidneys to water, facilitating water reabsorption and increased fluid volume. AVP also increases the reabsorption of Na^+ and Cl^- in the TAL and collecting duct, overall decreasing Na and water loss. As part of a reflex response to a fall in systemic pressure, sympathetic neurons that innervate the afferent/efferent arterioles of the glomerulus release the neurotransmitter norepinephrine, causing an increase in renal vasculature resistance and a decrease in fluid filtration. The decrease in renal blood flow leads to an overall reduction in filtration and Na loss. Stimulated $\alpha1$ and $\alpha2$ adrenergic receptors in the proximal tubular cells of the kidney also increase the activity of basolaterally located Na^+/K^+ ATPase and apical Na^+/H^+ exchanger, respectively, increasing reabsorption of Na^+ from the PCT luminal fluid.

In addition to affecting renal hemodynamics, stimulation of $\alpha1$ adrenergic receptors induce the release of renin from the juxtaglomerular cells of the kidney afferent and efferent arterioles. Renin, as part of the renin–angiotensin–aldosterone hormonal axis, is a proteolytic enzyme that when released into circulation is responsible for cleaving the hepatically produced protein angiotensinogen into angiotensin 1. Angiotensin 1 undergoes further cleavage into angiotensin 2 (ANG-2), catalyzed by angiotensin-converting enzyme (ACE) which is produced primarily by the epithelial cells of the lungs. Angiotensin 2 is a vasoactive hormone, increasing total peripheral vascular resistance in response to low blood volume thus normalizing total pressure. In the CNS ANG-2 stimulates the release of AVP from the posterior pituitary, and increases thirst and salt appetite. Angiotensin 2 also has direct and indirect effects on renal Na loss. Directly ANG-2 increases vascular resistance of the efferent arterioles, decreasing renal plasma flow. Angiotensin 2 also has direct effects on the tubular transport system, increasing expression of the Na^+/K^+ ATPase and Na^+/HCO_3^- exchanger in the basolateral and apical membrane of the proximal kidney, respectively, and the Na^+/H^+ exchanger and ENaC in the distal tubules (Gumz et al. 2015). Overall decreasing loss and increasing Na reabsorption. Indirectly ANG-2 in circulation stimulates the release of the mineralocorticoid aldosterone from the adrenal cortex. Aldosterone is secreted in response to low plasma Na (hypovolemia), high plasma K, and increases in ANG-2. Aldosterone increases K secretion by stimulating an increase in luminal Na reabsorption. Aldosterone directly increases renal cellular uptake of Na via apical stimulation of ENaC and ROMK expression and increased activity of basolateral Na^+-K^+ ATPases (Shils and Shike 2006). Increased reabsorption of Na^+ also increases the potential difference across the tubular cell, enhancing the secretion of K^+ from the cell into the more electronegative lumen.

15.3 Potassium Bioavailability

Potassium is found in most plant and animal tissues, with fruits and vegetables having a higher nutrient density than cereals and animal foods. Potassium is intrinsically soluble and quickly dispersed in the luminal water of the upper digestive tract. The small intestine is the primary site of K absorption, with approximately 90% of dietary K being absorbed by passive diffusion (Demigne et al. 2004; Stone et al. 2016). Little is known about the bioavailability of K, with the majority of work being centered on the assessment of urinary K losses after K salt supplementation (Melikian et al. 1988; Bechgaard and Shephard 1981; Betlach et al. 1987; Stone et al. 2016).

15.3.1 Kinetic Modeling and Potassium Bioavailability

Many different models of K movement within the body have been proposed, each developed to fit various areas of biological interest. The complexity of each model varies, from early recommendations by the International Commission on Radiological Protection for evaluation of radio potassium exposure limiting the body to one large mixed pool of K to more complex anatomically related compartmentalization (ICRP 1975, 2007; Valentin 2002; Stone et al. 2016). In one of the earliest schemes, Ginsburg and Wilde (1954) constructed a five-compartment model, mathematically derived from murine data looking at tissue groupings (muscle/testes, brain/RBC, bone, lung/kidney/intestine, liver/skin/spleen) and their K exchange between a common compartment of ECF (Ginsburg 1962; Ginsburg and Wilde 1954; Stone et al. 2016). Utilizing $^{42}K^+$ intravenous (IV) injections, researchers noted a wide spectrum of tracer exchange rates between tissues, with kidneys being the fastest (equilibrium with plasma at 2 min) and muscle and brain being the slowest ($\geq$600 min) (Ginsburg 1962). Based on this model, the total K mass of the four primary tissue compartments should be equivalent to total body K. However, findings revealed that this was not the case, the total sum only accounting for 73% of K mass. Investigators concluded that exchange rates/pools may be heterogeneous across both organs and organ groups, making the idea of grouping tissue compartments even more complex, and the internal movement of K more nuanced. Later, Leggett and Williams (1986) proposed a more anatomically specific model based on the quantitative movement of K through mathematically derived compartments within a physiologically relevant framework (Stone et al. 2016). Their model, similar to previous depictions, identifies plasma/ECF as the primary feeding compartment, with equilibrium distribution of K, regional blood flow rates, and K tissue extraction fractions, all influencing K exchange. The model also describes K exchange from plasma/ECF to tissues as a relatively rapid and uniform process; skeletal muscle being the only exception, with slower exchange due to its role as the main site of K storage. This concept is confirmed by earlier studies looking at exchange rates of total body K using measures of whole-body counting of radioactivity, IV administration of ^{42}K, and $^{40}K/^{42}K$ ratios (Edmonds and Jasani 1972; Jasani and Edmonds 1971; Stone et al. 2016). These early works revealed that after absorption, most body K exchanges rapidly with a half-life of less than 7 h, while a small portion thought to be contained primarily in skeletal muscle exchanges more slowly (~70 h) (Jasani and Edmonds 1971; Surveyor and Hughes 1968). A better understanding of kinetic modeling and K movement throughout the body may help to reveal how specific tissues influence K bioavailability, and further the understanding of its role in health (Stone et al. 2016).

In a recent study conducted by Macdonald and colleagues (2016), researchers aimed to assess and compare the bioavailability of K from potato (*Solanum tuberosum* L.) sources (non-fried white potatoes, French fries) and a K supplement (potassium gluconate). Thirty-five healthy men and women (age of 29.7 ± 11.2 year, body mass index of 24.3 ± 4.4 kg m^{-2}) were randomized to nine, five-day interventions of additional K equaling: 0 mmol (control at phase 1 and repeated at phase 5), 20 mmol (1500 mg), 40 mmol (3000 mg), 60 mmol (4500 mg) K day^{-1} consumed as K^+ gluconate or potato, and 40 mmol K^+ day^{-1} from French fries. Bioavailability of K was determined from the area under the curve (AUC) of serial blood draws and 24-h urinary excretion assessed after a test meal of varying K dose given on the fourth day. Investigators found increases in serum K AUC with increasing dose regardless of source, while 24-h urine K concentration also increased with dose but was greater with potato compared to supplement. Blood pressure (BP) was also assessed throughout the study but resulted in no significant findings. These outcomes reveal the need for a full K balance study, looking at intakes from a variety of dietary sources and complete losses (urine and feces), to fully understand K bioavailability differences between dietary K and supplements and their subsequent health effects (Stone et al. 2016).

15.4 Potassium and Hypertension

Hypertension, or high blood pressure, is the leading cause of cardiovascular disease and a major contributing risk factor for the development of stroke, coronary heart disease, myocardial infarction, heart failure, and end-stage renal disease, amounting to a US public health financial burden of $53.2 billion (Roger et al. 2012; Benjamin et al. 2018). Approximately one in three American adults ≥20 years (~86 million) are estimated to have HTN, while nearly 60 million are at risk for developing HTN (BP greater than 120/80 mmHg) (Benjamin et al. 2018). Approximately 90% of US adults older than 50 year are at risk for the development HTN, with systolic rises being the most prevalent (Svetkey et al. 2004). Hypertension is a leading cause of morbidity and mortality worldwide and second only to smoking as a preventable cause of death in the US (Lopez and Mathers 2006; Stone et al. 2016; Stone and Weaver 2018).

15.4.1 Mechanisms of Arterial Pressure Control

Regulation of systemic arterial pressure is the most important role of the cardiovascular system. Arterial pressure is a result of cardiac output (heart rate × stroke volume), or blood being pumped from the heart into the systemic circulation, and total peripheral vascular resistance, or the degree to which the systemic vasculature is in a state of constriction or dilation (Mohrman and Heller 2010). Blood pressure is regulated by both short-term and long-term mechanisms. In the short-term, arterial baroreceptors, located predominately in the walls of the aorta and the carotid arteries, respond to sensory inputs of increased stretch in the vasculature sending afferent signals to the medullary cardiovascular center in the CNS. Subsequently, the CNS integration process is such that increased input from the arterial baroreceptor reflex, caused by increases in arterial pressure, will cause a decrease in the tonic activity of cardiovascular sympathetic nerves and an increase in cardiac parasympathetic nerve activity. The result of this negative feedback system being an overall decrease in BP. Conversely, a decrease in mean arterial pressure would increase sympathetic and decrease parasympathetic neural activity (Ekmekcioglu et al. 2016). If arterial pressure remains elevated for several days the baroreceptor reflex will gradually adjust to this new pressure set point and cease firing. Because of this, it is not considered a good mechanism for long-term control. Long-term pressure regulation is closely tied to the prevalence and potential causes of hypertension. Long-term regulation is theorized to be primarily dependent on the way the kidneys handle Na (e.g., extracellular osmolarity) and regulate blood volume. Arterial pressure has a marked effect on urinary output rate and total body fluid volume. A disturbance that leads to an increase in arterial pressure will in turn cause an increase in urinary output, decreasing total fluid volume and bring arterial pressure back to a homeostatic level. Again conversely, a decrease in arterial pressure would lead to fluid volume expansion. Similar to short-term regulation, long-term regulation works as a negative feedback loop, utilizing modulation of fluid volume as a means for pressure regulation.

As discussed previously, the kidneys play a major role in regulating electrolyte balance and the osmolarity of blood plasma. Plasma is filtered within the glomerular capillaries before entering the renal tubules of the nephron. The rate at which this process occurs is referred to as the glomerular filtration rate (GFR) and is influenced by both the hydrostatic and oncotic aspects of arterial pressure. Increased blood volume and pressure will increase GFR, and when the body is at physiological steady-state, arterial pressure must remain at a level that ensures urinary output equal fluid intake (Mohrman and Heller 2010). Filtered fluid enters the renal tubules where it is either reabsorbed and reenters the cardiovascular system, or is excreted as urine.

As stated earlier, the kidneys regulate blood osmolarity primarily via modulation of total body water rather than total solutes, although some fluid reabsorption occurs because Na^+ is actively pumped out of the renal tubules. The previously discussed hormonal influences of arginine vasopressin (antidiuretic hormone; AVP) and the renin–angiotensin–aldosterone axis stimulate both water and Na^+ reabsorption in response to low fluid volume/low blood pressure. The resulting increase in BP, and overall dysregulation of this long-term control mechanism, may explain the incidence of hypertension to some degree, although the majority of primary HTN remains idiopathic.

Systemic HTN is defined as an elevation of systolic BP (vascular pressure during cardiac muscle contraction) above 140 mm Hg and diastolic blood pressure (vascular pressure during cardiac muscle relaxation) above 90 mm Hg. Secondary HTN can be traced to a preexisting comorbidity such as kidney disease, obesity and/or diabetes, and various forms of cancer. Primary or essential HTN ("essential" to drive blood through the vasculature) often has no diagnosable cause, leaving only the symptom of high BP to be treated either pharmacologically, or through lifestyle modification (e.g., exercise and diet).

15.4.2 Potassium and Arterial Pressure

Theorized mechanisms of how K^+ influences vascular health include effects on the renin–angiotensin–aldosterone system, reduction in adrenergic tone, increased Na^+ excretion (natriuresis), and increases in vasodilation. Short-term increased consumption of K^+ may improve the function of endothelial cells, a monolayer of cells within the vasculature that control the tone of the underlying vascular smooth muscle. Elevated serum K^+, within the physiological range, may induce endothelial hyperpolarization via a stimulation of Na^+/K^+ ATPase pumps and the activation of plasma membrane K^+ channels, leading to subsequent vasodilation via efflux of Ca^{2+} from vascular smooth muscle cells (Haddy et al. 2006; Ekmekcioglu et al. 2016). Increased K^+ intake may also enhance vasodilation and improve BP regulation via inhibition of sympathetic neural transmission and reduced sensitivity to catecholamine-induced vasoconstriction, increased endothelial nitric oxide release, alteration of baroreceptor sensitivity, and increased Na^+ excretion (Haddy et al. 2006; He et al. 2010; Stone and Weaver 2018).

In relation to Na^+, increases in K^+ intake can lead to increased Na^+ excretion which may improve the overall fluid volume and BP control. As described previously, active Na^+ and K^+ reabsorption and excretion are primarily regulated by the epithelial Na^+ channels (ENaC; Na reabsorption) and the renal outer medullary K^+ channel (ROMK; K excretion) transporters of the kidney. Na^+ is also actively reabsorbed in DCT by the Na^+/Cl^- cotransporter (NCC), which determines the delivery of Na^+ to the downstream ENaC and ROMK, and directly influences the reabsorption of Na^+ and excretion of K^+. Assessed in animal models, increased K^+ feeding increases extracellular K^+ concentration leading to a decrease in NCC activity (via a phosphorylation-dephosphorylation mechanism) reducing Na^+ reabsorption, and increasing urinary loss (Veiras et al. 2016). Prospective human population studies show that higher fruit and vegetable intake (and assumed increased dietary K^+) increases Na^+ excretion, which may lead to improvements in fluid balance and BP control (Cogswell et al. 2016). In contrast, low K^+ intake may lead to excessive Na^+ retention independent of fluid dynamics. In animal models, inadequate K^+ upregulates the Na^+/H^+ exchanger in the PCT, leading to increased Na^+ reabsorption and fluid expansion (Soleimani et al. 1990). Potassium depletion may also lead to increased activity of the NCC, increasing Na^+ and fluid reabsorption in the distal kidney, and promoting arterial pressure dysregulation. While the influence of both K^+ and Na^+, and the complex physiological relationship between the two are intimately tied to fluid balance and arterial pressure, the mechanisms behind this are still unknown (Stone and Weaver 2018).

15.4.3 Epidemiological Data

Numerous epidemiological studies suggest diet as a key component in BP control, with some studies showing lower BP in populations consuming higher amounts of fruits and vegetables (INTERSALT 1988; Young et al. 1995; Elford et al. 1990). Dietary patterns shown to lower BP include increased K and reduced Na intake, increases in fruit and vegetable consumption, as well as other foods rich in antioxidants (Appel et al. 1997; Svetkey et al. 1999). A population study conducted by Khaw et al. in St. Lucia, West Indies suggested an increase in K by ~700–1200 mg/day (20–30 mmol/day) resulted in a 2–3 mmHg reduction in systolic blood pressure (SBP) (Khaw and Rose 1982). In adults, a 2-mmHg reduction in BP can reduce CHD and stroke mortality rates by 4 and 6%, respectively (Stamler 1991). The INTERSALT study, a worldwide epidemiologic study (n = 10,079 men and women aged 20–59 year from 32 countries) that looked at the relationship between 24 h. Na excretion and BP provided evidence of K intake as an important factor affecting population BP, independent of Na, among diverse population groups (Stamler 1991). The American Heart Association has estimated that increasing K intake may decrease HTN incidence in Americans by 17% and lengthen life span by 5.1 years (Roger et al. 2012; Stone et al. 2016; Stone and Weaver 2018).

15.4.4 Potassium Supplementation Studies

Epidemiological studies have evaluated the effects of K from foods, while clinical intervention trials have primarily used K supplements. Several meta-analyses show a significant reduction in BP with increasing K supplementation (Beyer et al. 2006; Whelton et al. 1997; Cappuccio and MacGregor 1991; Geleijnse et al. 2003). In an early meta-analysis Cappuccio and MacGregor reviewed 19 clinical trials looking at the effect of K supplementation on BP in primarily hypertensive individuals (412 of

586 participants). With the average amount of K given at 86 mmol day^{-1} (~3300 mg day^{-1}; as primarily KCl) for an average duration of 39 days, researchers found that K supplementation significantly reduced SBP by 5.9 mm Hg and diastolic blood pressure (DBP) by 3.4 mm Hg. Greater reductions were found in individuals who were on supplementation for longer periods of time (Cappuccio and MacGregor 1991). Another regression analysis looked at the effect of K supplementation in both normotensive and hypertensive individuals. Researchers found an average K dose of 60–120 mmol day^{-1} (2500–5000 mg day^{-1}) reduced SBP and DBP by 4.4 and 2.5 mm Hg, respectively, in hypertensive patients, and by 1.8 and 1.0 mm Hg, respectively, in normotensive individuals (Whelton et al. 1997). As is evident, the effect of K supplementation on BP reduction is generally positive, but not consistent. According to a more recent meta-analysis conducted by Dickinson et al. (2006), K supplementation did not significantly reduce BP in those with hypertension, although this analysis was only based on five trials, and findings, while not statistically significant, did reveal reductions in both SBP and DBP (Beyer et al. 2006; Dickinson et al. 2006). In general, these outcomes show that the BP-lowering effects of K supplementation are greater in those with HTN and more pronounced in blacks compared to whites. Other noted factors that may influence the effects of K supplementation on BP include pre-treatment BP, age, gender, intake of Na and other ions (magnesium, calcium), weight, physical activity level, and concomitant medications. In addition, these analyses suggest the optimal K dose range as 1900–3700 mg day^{-1}, for lowering of approximately 2–6 mm Hg in SBP and 2–4 mm Hg in DBP (Houston 2011; Stone et al. 2016; Stone and Weaver 2018).

15.4.5 *Dietary Intake Clinical Trials*

Findings from the recent Agency for Healthcare Research and Quality (AHRQ) report on K intake and chronic disease concluded, with a moderate strength of evidence, that increasing K intake decreases BP, particularly in those with HTN (Newberry et al. 2018). Although, of the 18 randomized controlled trials assessed by the AHRQ, only 4 were dietary interventions, the rest involved K supplementation as described above.

Evidence from dietary interventions is extremely limited, with the majority of findings being extrapolated from The Dietary Approaches to Stop Hypertension (DASH) study (Appel et al. 1997). The DASH interventions determined that a diet higher in fruits and vegetables, fiber, and low-fat dairy products, and lower in saturated and total fat and Na could improve BP outcomes compared to the average American diet (Sacks and Campos 2010). Although the DASH diet does lead to a dramatic increase in K consumption (4100–4400 mg day^{-1}), due to its other dietary modifications, the beneficial effects on arterial pressure cannot be attributed to K alone. In an earlier study conducted by Chalmers and colleagues in an Australian cohort, researchers assessed the effects of both the reduction of Na and the increase of K in the diet on BP (Chalmers et al. 1986). Two-hundred-and-twelve hypertensive

(DBP between 90- and 100-mmHg) adults (age 52.3 ± 0.8 year; 181 males and 31 females) were recruited and placed in one of the 4 following diet groups: a normal diet group (control), a high K diet (>100 mmol K day^{-1} or > 3900 mg day^{-1}), a reduced Na diet (50–75 mmol Na^+ day^{-1} or 1150–1725 mg day^{-1}), or a high K/low Na diet. The duration of the diet intervention for this parallel design study was 12 weeks in which subjects were given nutrition coaching on how to adjust their diet choices based on their group (e.g., increasing fruit/vegetable intake, avoiding table salt and foods high in Na). Investigators found significant reductions in both SBP and DBP in each intervention group compared to controls, but no significant differences between diet manipulation groups, with reductions in the high K group being −7.7 ± 1.1 and −4.7 ± 0.7 mm Hg for SBP and DBP, respectively. Although high K intake did appear to reduce BP the lack of differences between groups points to the possibility of an overall diet effect. In a more recent study conducted on a UK cohort, Berry et al. assessed the effects of increased K intake from both dietary and supplement sources on BP in untreated pre-hypertensive individuals (DBP between 80 and 100 mm Hg) (Berry et al. 2010). In a cross-over design, subjects (n = 48, age 22–65 year) completed four, 6-week dietary interventions including a control diet, an additional 20 or 40 mmol K day^{-1} (780 or 1560 mg day^{-1}) from increased fruit and/or vegetable intake, and 40 mmol K citrate day^{-1} as capsules. Each treatment was followed by a washout of at least 5 weeks. Similar to the Chamlers study, subjects were counseled by nutrition professionals on how to regulate their food choices during each dietary intervention, primarily focused on increasing fruit and vegetable intake. Findings revealed no significant changes in the primary outcome measure of ambulatory BP between the control group and any of the interventions. The lack of control used to conduct these K dietary interventions is the primary limiting factor in their ability to adequately assess the true effect of increased dietary K intake on BP outcomes. A complete balance study with a controlled diet is necessary to accurately assess K retention, and its acute and prolonged effects on BP and related outcomes. Currently, the lack of evidence from clinical trials looking specifically at dietary K intake and its effect on BP points to a large gap in the K literature. More research is needed in this area to completely understand the effects of dietary K intake on the regulation of arterial pressure and the potential for health benefit (Stone et al. 2016; Stone and Weaver 2018; Weaver et al. 2018) (Table 15.1).

15.5 Potassium, Diabetes, and Glucose Control

Blood glucose levels are tightly regulated within a range of 70–100 mg dL^{-1}. After ingestion of a meal the rise in circulating glucose levels, along with other factors, stimulates the release of the hormone insulin from the pancreas. Insulin is secreted from the islets of Langerhans from the β-cells of the pancreas and has an action on target tissues (e.g., skeletal muscle, liver, adipose) to facilitate cellular glucose uptake. Antagonistically the hormone glucagon is secreted from the α-cells of the

Table 15.1 Published studies that showed an effect of additional dietary potassium (K) intake on blood pressure (BP) outcomes

Reference, duration	Participants	Design	K Form	K Dose	BP Outcome
Chalmers et al. (1986). 8-week run-in (control diet) followed by 12 weeks of intervention.	212, men (181) and women (31), mean age 52.3 ± 0.8 year). All hypertensives DBP: 90–100 mm Hg.	Clinical trial with 4 arm parallel design: Normal (control) diet, high K diet, low Na diet, and high K/low Na diet. (2 × 2 factorial).	K from increased intake of K rich dietary sources (free living diet, subjects counseled on what foods to eat depending on group).	K content of intervention diets: High K and high K/low Na >100 mmol day^{-1}, control/low Na not reported.	High K diet: SBP and DBP: −7.7 and − 4.7, respectively. Low Na diet: −8.9 and − 5.8. High K/low Na diet: −7.9 and − 4.2. All compared to control.
Appel et al. (1997) (DASH Trial). 3-week control diet run-in, followed by 8 weeks of intervention.	459, men and women (~50% each), mean age: 45 ± 10 year. SBP: <160 mm Hg, DBP: 80–95 mm Hg.	Clinical trial with 3 arm parallel design: Control diet, high fruit and vegetable diet, high fruit and vegetable diet + reduced fat (combination diet).	K from increased intake of fruits and vegetables (from controlled research diets).	K content of each diet: Control 1752, High fruit and vegetable 4101, Combination diet 4415 mg day^{-1}.	Combination diet: SBP and DBP: −5.5 and − 3.0 mmHg, respectively. High fruit and vegetable diet: SBP and DBP: −2.8 and − 1.1 mmHg, respectively; both compared to control.
Berry et al. (2010) 3-week run-in (control diet), four 6-week diet interventions with at least 5-week washout in between.	48, men (23) women (25), mean age: 45 ± 10 year. All hypertensives DBP: 80–100 mm Hg.	Clinical trial cross-over design with 4 interventions: Control diet, control diet + K supplement, additional intake (20 mmol day^{-1}) from fruits and vegetables, additional intake (40 mmol day^{-1}) from fruits and vegetables.	K supplement, K from increased intake of fruits and vegetables (free living diet, subjects counseled on what foods to eat depending on group).	K content of intervention diets: Control 15 mmol day^{-1}, control + K supplement: 40 mmol day^{-1}, control + additional intake from fruits and vegetables: 20 and 40 mmol day^{-1}.	No significant change in SBP or DBP within interventions (pre, post) or compared to control.

Macdonald-Clarke et al. (2016) 5-day periods of additional K, separated by at least 7 days of washout.	35, men and women, mean age 29.7 ± 11.2 year normotensives.	Single-blind, cross-over, randomized control trial with 9 interventions of additional K: 0 mmol/day, 20 mmol/day, 40 mmol/day from supplement or potato, and 40 mmol/day from French fries.	K supplement, K from increased intake of potatoes and French fries (from controlled research diets).	K content (additional) of intervention diets: 0 mmol/day (control = 60 mmol day^{-1} total; phases 1 and 5), 20, 40, 60 mmol day^{-1} from supplement or potato, 40 mmol day^{-1} from French fries.	No significant change in SBP or DBP within interventions or compared to control.

This table has been adapted and reproduced from the following: Stone, M.S., L. Martyn, and C.M. Weaver, *Potassium Intake, Bioavailability, Hypertension, and Glucose Control.* Nutrients, 2016. **8**(7). License: CC by 4.0

pancreas in response to low levels of blood glucose, stimulating the release of glucose from tissues and glucose production (gluconeogenesis) in the liver (Stipanuk 2006). Continually changing levels of insulin and glucagon are important signals in informing various physiological systems of the body's nutritional state (Stone et al. 2016).

Diabetes Mellitus (DM) is a degenerative disease associated with a lack of, or insufficient secretion, of insulin or an insensitivity to insulin stimulation in the cells of target tissues. DM comes in two forms: type 1, insulin-dependent, DM, or type 2, non-insulin-dependent DM. Type 1 DM is primarily characterized as an autoimmune disease in which the immune system attacks the cells of the pancreas leading to nearly complete β-cell destruction or extreme dysfunction. This results in essentially a complete inability to produce insulin, and the requirement of daily insulin injections to control blood glucose. Type 2 DM (T2DM) is more complex and is often the result of obesity coupled with poor dietary and lifestyle choices. In T2DM the pancreas may still produce insulin, often in increasing amounts in response to increases in glucose load, but this is often insufficient to maintain homeostatic glucose levels if intake becomes too high and frequent. Eventually, β-cell insulin granules become depleted and target tissues exhibit resistance to insulin stimulation, leaving blood glucose levels unchecked. Prolonged elevated blood glucose levels can be damaging to small vessels, especially in the brain, kidneys, eyes, and extremities, and can eventually lead to nerve damage and tissue death. While the use of drugs to increase insulin secretion and improve tissue insulin sensitivity can be effective, lifestyle changes including better dietary choices and increased physical activity will often lead to control of the disease (Delli and Lernmark 2016; Hupfeld and Olefsky 2016; Stone et al. 2016).

Potassium plays a role in blood glucose control by modulating the secretion of insulin from the pancreas. On the cellular level, K^+ efflux from ATP sensitive K^+ (K^+/ATP) channels influences β-cell excitability and holds membrane potential at low levels (~−60 mV) (Ekmekcioglu et al. 2016). Increases in blood glucose lead to increased β-cell glucose uptake and subsequent ATP generation, which in turn inhibit K^+/ATP channels. Decreased K^+ efflux leads to stimulation of voltage-gated Ca^{2+} (Ca^{2+}V) channels, cellular depolarization via Ca^{2+} influx, and increased insulin secretion. Potassium efflux through voltage-gated K^+ channels leads to repolarization and an inhibition of Ca^{2+}V channels, inhibiting insulin release. While experimentally supraphysiological concentrations of K^+ (≥10 mM) induce a depolarizing effect on β-cell membrane potential, the effects of extracellular K^+ at the upper end of the physiological range (5.5 mM) are unknown (Meissner et al. 1978; Stone et al. 2016).

15.5.1 Potassium and Glucose Control

Glucose intolerance can often be a result of severe hypokalemia due to a deficit in K balance that may occur in primary or secondary aldosteronism or prolonged treatment with diuretics (He and MacGregor 2008). The use of thiazide diuretics is widely considered the preferred initial pharmacological treatment for hypertension (Haddy et al. 2006). The tendency of thiazide diuretics to negatively influence glucose tolerance and increase the incidence of new-onset diabetes is well known. In a recent systematic quantitative review, researchers analyzed 59 clinical trials in which the relationship between the use of thiazide diuretics, hypokalemia, and glucose intolerance was strong (Zillich et al. 2006). Thiazide diuretics have a common side effect of lowering serum K and evidence shows that diuretic-induced hypokalemia may lead to impaired glucose tolerance via the reduction in insulin secretion in response to glucose loads (Chatterjee et al. 2012). In healthy individuals, there is also evidence to support the role of K in glucose control. Studies involving K depletion (e.g., low K diets) show that low levels of K can lead to glucose intolerance via impaired insulin secretion (Rowe et al. 1980; Sagild et al. 1961). In addition, when patients with thiazide-induced hypokalemia are given K supplements, the defects in insulin release in response to glucose loads are corrected, thus indicating that hypokalemia may be a significant contributing factor to the glucose abnormality (Helderman et al. 1983; Stone et al. 2016).

15.5.2 Potassium and Diabetes

The relationship between K intake and diabetes was examined in a prospective cohort study conducted by Colditz et al. (1992) looking at women ($n = 84, 360$; age 34–59 year) from the Nurse's Health Study. After a six-year follow-up, investigators found that high K^+ intake may be associated with a decreased risk for developing T2DM in women with a body mass index (BMI) of 29 kg m^{-2} or less (Colditz et al. 1992). When compared with women in the lowest quintile, women in the highest quintile for K^+ intake had a relative risk of 0.62 (p trend $= 0.008$) for T2DM. More recently, Chatterjee et al. assessed the association between K^+ intake and T2DM using data from the Coronary Artery Risk Development in Young Adults (CARDIA) study (Chatterjee et al. 2012). Researchers examined the relationship between urinary K^+ and diabetes risk for 1066 participants. Use of multivariate models adjusted for potential confounders including BMI, fruit and vegetable intake and other dietary factors revealed that those in the lowest quintile of K intake were more than twice as likely to develop diabetes compared to those in the highest quintile (HR 2.45; 95% CI 1.08, 5.59; p for trend $= 0.04$). Investigators also found that those in the lowest quintile of K intake were significantly more likely to develop diabetes than those in the highest quintile of K intake ($p = 0.008$). Of the 4754 participants, 373 (7.8%) developed T2DM during the follow-up period of 20 year, and, overall, the mean K intake of those who developed diabetes was significantly lower than those who did not (3393 vs. 3684 mg day^{-1}; $p = 0.002$). This same research group examined data from 12,209 individuals participating in the Atherosclerosis Risk in Communities (ARIC) cohort and found serum K^+ to be independently associated with diabetes risk. Using multivariate cross-sectional analyses, a significant inverse relationship between serum K^+ and fasting insulin levels was identified ($p < 0.01$) (Chatterjee et al. 2010). Dietary K^+ intake was significantly associated with diabetes risk in unadjusted models, with adults having serum K^+ levels lower than 4.0 mM at highest risk for DM incidence. This relationship continued to hold true after covariate adjustment (e.g., age, sex, race, BMI, serum magnesium, serum calcium, physical activity, hypertension, etc.) in multivariate models, with lower K^+ levels associated with higher BMI, larger waist circumference, lower serum magnesium levels, and higher fasting insulin levels as well (Chatterjee et al. 2010; Stone et al. 2016).

The relationship between K and T2DM also extends to the kalemic effects of insulin. Higher plasma insulin levels are associated with increased K^+ absorption into cells (DeFronzo et al. 1980), and without a threshold, as seen in glycemic response, these kalemic effects continue to increase as insulin levels rise. DeFronzo et al. examined this relationship using the insulin clamp technique and graded doses of insulin. Investigators found a dose-dependent decline in plasma K^+ concentration with increasing insulin dose, independent of glucose uptake. This effect is likely to be mediated by an increased sensitivity to intracellular Na, activation of Na^+-K^+ ATPase, and inhibition of K efflux (DeFronzo et al. 1980; Stone et al. 2016).

15.6 Potassium and Bone

Osteoporosis, or a severe reduction in bone mass leading to decreased bone health and increased fracture risk, is a global health problem with great financial impact. Over 200 million people worldwide suffer from osteoporosis, including 30% of postmenopausal women in both the US and Europe (Sözen et al. 2017). Peak bone mass is achieved by the third decade of life, after which bone loss begins, accelerating with aging in both men and women (Weaver and Fuchs 2014). The bone mass present at any given point during life is determined by factors that influence the acquisition, maintenance, or loss of bone throughout the lifespan, many of which are modifiable lifestyle factors (Weaver et al. 2018).

Adequate K intake may benefit overall bone health and has been proposed to do so through its effect on acid-base balance (Barzel 1995; Brandao-Burch et al. 2005). Support for the acid-base bone theory stems from the idea that the Western diet is high in meats and cereal grains and low in fruits and vegetables, creating an environment of low-grade metabolic acidosis (net acid excretion (NAE) = 75–100-mEq acid/day) (Barzel 1995). Buffering of this increased acid load via bone tissue-derived Ca salts, is proposed to lead to bone loss. Alkaline K salts produced from metabolizing fruits and vegetables or K supplements (potassium bicarbonate or citrate) are thought to provide bicarbonate precursors and help to maintain pH homeostasis (~7.35–7.45). The impact of excess systemic acid on bone is suggested to be mediated by two mechanisms: pH buffered through the dissolution of the bone matrix, and cell-based mechanisms (e.g., upregulation of bone-resorbing cell (osteoclast) activity) (Barzel 1995; Brandao-Burch et al. 2005). However, opposition to the acid-base balance theory exists. In a rat model, looking at the relationship between the inhibitory effect of vegetables on bone resorption and base excess, the addition of potassium citrate at levels that neutralized urinary acid excretion from an acidogenic diet had no effect on bone turnover (Muhlbauer et al. 2002). Researchers measured bone turnover via a urinary excretion tracer from prelabeled bone and concluded reductions in bone resorption via increased vegetable intake and subsequent base excess were not causally related. The authors suggested that bioactive compounds (e.g., flavonoids) in fruits and vegetables may be responsible for benefits to bone. Despite this, there is some consistency in the literature that increased K intake benefits bone, though the mechanisms behind this remain unclear (Weaver 2013; Stone et al. 2016; Weaver et al. 2018).

15.6.1 Potassium and Calcium Balance

Potassium intake has been associated with reduced urinary Ca^{2+} excretion. Clinical trials show persistently increased calciuria in both men and women given K^{+} supplements (bicarbonate or citrate) vs. similar Na supplements, suggesting K may have a role in bone benefit beyond acid balance (Lemann et al. 1989; Frassetto et al. 2005). In the kidney, Ca is reabsorbed via solvent drag in the PCT (60–70%) and the TAL (20%). Active reabsorption of Ca takes place in the DCT via specific transport proteins. Calcium is reabsorbed via the Ca^{2+} channel TRPV5 from the tubular fluid into the cell where it binds to the transfer protein calbindin 28 K and is shuttled across and out of the cell via the plasma membrane Ca^{2+} ATPase (PMCA) and Na^{+}/Ca^{2+} exchanger (NCX). High Na intakes have been shown to increase urinary Ca losses, with a loss of approximately 24–40 mg of Ca^{2+} for a Na^{+} intake of $\approx$2.3 g (Shils and Shike 2006). The mechanism for this is not well defined, but most likely involves Ca following Na excretion via solvent drag. Increased intracellular Na within the kidney tubular cells may also affect the dynamics of the NCX (which exchanges 3 Na for 1 Ca^{2+}), leading to its dysregulation and possible reversal. Increased intakes of K may have the opposite effect on Ca, in which paracellular reabsorption in the TAL is facilitated by movement of Na^{+}, K^{+}, and Cl^{-} across the $Na^{+}/K^{+}/2Cl^{-}$ cotransporter (NKCC) on the apical membrane. Potassium shuttled into the cell via NKCC is subsequently re-secreted into the lumen via ROMK, maintaining an electropositive lumen, facilitating the passive reabsorption of Ca, decreasing urinary loss, and improving Ca balance.

15.6.2 Potassium Bone Turnover and Bone Mineral Density

Beyond the effect of K on Ca balance, several studies have assessed the influence of K on biochemical markers of bone turnover. Studies have shown decreases in the bone resorption markers C- and N-telopeptide and procollagen type I N-terminal propeptide, with K supplementation (Dawson-Hughes et al. 2009; Marangella et al. 2004). In postmenopausal women, K bicarbonate at 60–120 mmol day^{-1} decreased urinary hydroxyproline excretion by 10%, while increasing serum osteocalcin, a marker of bone formation (Sebastian et al. 1994; Weaver 2013).

The relationship between increased K intake and bone mineral density shows conflicting results as well. Only three clinical trials have been reported all done in populations of postmenopausal women or the elderly (>60 year). One trial showed protection from bone mass density (BMD) loss in the spine, hip, and femoral neck, with a 30 mmol K day^{-1} dose as K citrate compared to KCl, but lacked a placebo control (Jehle et al. 2006). A second trial revealed no BMD benefit with increased intake from K citrate (55 or 18.5 mmol day^{-1}), fruits and vegetables (18.5 alkali mmol day^{-1}), or a placebo (Macdonald et al. 2008). And the third, and strongest, reported a 1.7% increase in spine BMD with K citrate supplementation (60 mmol K day^{-1}) compared to placebo (Jehle et al. 2006, 2013; Macdonald et al. 2008). While generally inconclusive, findings may reveal the significance of K form and dose in any potential benefit for BMD.

15.7 Opportunities for Future Interdisciplinary Efforts to Improve Potassium Recommendations of Agricultural Crops

There is still much to learn about the role of K in overall human health. The importance of K in normal physiology is clear, but how adequate to greater than adequate intakes can help facilitate benefit to these systems is not well understood. Increasing dietary K has potential benefit to lowering the risk of hypertension, and may provide benefit to normal kidney function, glucose control, and bone (Fig. 15.1). We need to understand more about bioavailability and retention of K from foods as well as other sources. Are there unidentified inhibitors to K absorption or food matrix effects? Do some anions that accompany K in foods have differential functional advantages? Organic salts of K appear to have more benefit to the bone, perhaps through effects on acid-base balance. The form seems less important for controlling blood pressure. Research on dietary K intake is likely to increase because it is an identified shortfall nutrient and increasing K consumption may have a marked influence on arterial pressure and hypertension, an important risk factor for all cardiovascular and related chronic diseases (Stone et al. 2016; Stone and Weaver 2018).

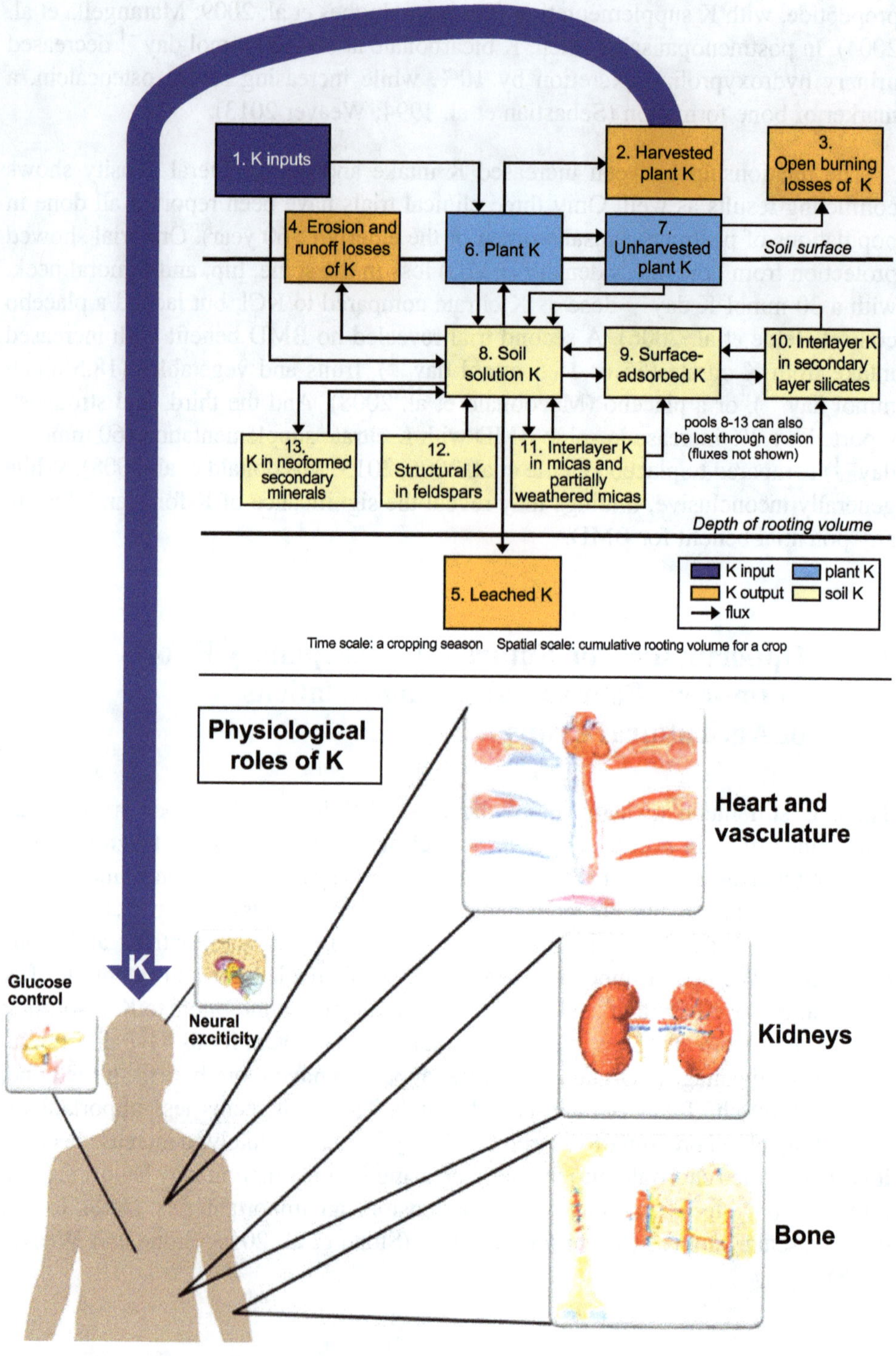

Fig. 15.1 A better understanding of how K consumed from the environment moves through the body including K storage and excretion, and how this may affect various physiological systems, which will in turn improve outcomes related to the health consequences of these relationships

References

Aburto NJ, Hanson S, Gutierrez H, Hooper L, Elliott P, Cappuccio FP (2013) Effect of increased potassium intake on cardiovascular risk factors and disease: systematic review and meta-analyses. BMJ 346:f1378. https://doi.org/10.1136/bmj.f1378

Appel LJ, Moore TJ, Obarzanek E, Vollmer WM, Svetkey LP, Sacks FM, Bray GA, Vogt TM, Cutler JA, Windhauser MM, Lin PH, Karanja N (1997) A clinical trial of the effects of dietary patterns on blood pressure. DASH collaborative research group. N Engl J Med 336:1117–1124. https://doi.org/10.1056/NEJM199704173361601

Aronson PS, Giebisch G (2011) Effects of pH on potassium: New explanations for old observations. J Am Soc Nephrol 22:1981–1989. https://doi.org/10.1681/ASN.2011040414

Barzel US (1995) The skeleton as an ion exchange system: implications for the role of acid-base imbalance in the genesis of osteoporosis. J Bone Miner Res 10:1431–1436. https://doi.org/10.1002/jbmr.5650101002

Bechgaard H, Shephard NW (1981) Bioavailability of potassium from controlled-release tablets with and without water loading. Eur J Clin Pharmacol 21:143–7. https://wwwncbinlmnihgov/pubmed/7341281. Accessed 17 May 2020

Benjamin EJ, Virani SS, Callaway CW, Chamberlain AM, Chang AR, Cheng S, Chiuve SE, Cushman M, Delling FN, Deo R, De Ferranti SD, Ferguson JF, Fornage M, Gillespie C, Isasi CR, Jiménez MC, Jordan LC, Judd SE, Lackland D, Lichtman JH, Lisabeth L, Liu S, Longenecker CT, Lutsey PL, Mackey JS, Matchar DB, Matsushita K, Mussolino ME, Nasir K, O'flaherty M, Palaniappan LP, Pandey A, Pandey DK, Reeves MJ, Ritchey MD, Rodriguez CJ, Roth GA, Rosamond WD, Sampson UKA, Satou GM, Shah SH, Spartano NL, Tirschwell DL, Tsao CW, Voeks JH, Willey JZ, Wilkins JT, Wu JH, Alger HM, Wong SS, Muntner P (2018) Heart disease and stroke statistics—2018 update: a report from the American Heart Association. Circulation 137:e67–e492. https://doi.org/10.1161/CIR.0000000000000558

Berry SE, Mulla UZ, Chowienczyk PJ, Sanders TA (2010) Increased potassium intake from fruit and vegetables or supplements does not lower blood pressure or improve vascular function in UK men and women with early hypertension: a randomised controlled trial. Br J Nutr 104:1839–1847. https://doi.org/10.1017/S0007114510002904

Betlach CJ, Arnold JD, Frost RW, Leese PT, Gonzalez MA (1987) Bioavailability and pharmacokinetics of a new sustained-release potassium chloride tablet. Pharm Res 4:409–11. https://wwwncbinlmnihgov/pubmed/3508550. Accessed 17 May 2020

Beyer FR, Dickinson HO, Nicolson DJ, Ford GA, Mason J (2006) Combined calcium, magnesium and potassium supplementation for the management of primary hypertension in adults. Cochrane Database Syst Rev 3:CD004805. https://doi.org/10.1002/14651858.CD004805.pub2

Bia MJ, Defronzo RA (1981) Extrarenal potassium homeostasis. Am J Phys 240:F257–F268. https://doi.org/10.1152/ajprenal.1981.240.4.F257

Brandao-Burch A, Utting JC, Orriss IR, Arnett TR (2005) Acidosis inhibits bone formation by osteoblasts in vitro by preventing mineralization. Calcif Tissue Int 77:167–174. https://doi.org/10.1007/s00223-004-0285-8

Cappuccio FP, Macgregor GA (1991) Does potassium supplementation lower blood pressure? A meta-analysis of published trials. J Hypertens 9:465–73. https://www.ncbi.nlm.nih.gov/pubmed/1649867. Accessed 17 May 2020

Chalmers J, Morgan T, Doyle A, Dickson B, Hopper J, Mathews J, Matthews G, Moulds R, Myers J, Nowson C, et al (1986) Australian National Health and Medical Research Council dietary salt study in mild hypertension. J Hypertens Suppl 4:S629–37. https://wwwncbinlmnihgov/pubmed/3475429. Accessed 17 May 2020

Chatterjee R, Yeh HC, Shafi T, Selvin E, Anderson C, Pankow JS, Miller E, Brancati F (2010) Serum and dietary potassium and risk of incident type 2 diabetes mellitus: the atherosclerosis risk in communities (ARIC) study. Arch Intern Med 170:1745–1751. https://doi.org/10.1001/archinternmed.2010.362

Chatterjee R, Colangelo LA, Yeh HC, Anderson CA, Daviglus ML, Liu K, Brancati FL (2012) Potassium intake and risk of incident type 2 diabetes mellitus: the coronary artery risk development in young adults study. Diabetologia 55:1295–1303. https://doi.org/10.1007/s00125-012-2487-3

Cogswell ME, Mugavero K, Bowman BA, Frieden TR (2016) Dietary sodium and cardiovascular disease risk--measurement matters. N Engl J Med 375:580–586. https://doi.org/10.1056/NEJMsb1607161

Colditz GA, Manson JE, Stampfer MJ, Rosner B, Willett WC, Speizer FE (1992) Diet and risk of clinical diabetes in women. Am J Clin Nutr 55:1018–1023. https://doi.org/10.1093/ajcn/55.5.1018

Dawson-Hughes B, Harris SS, Palermo NJ, Castaneda-Sceppa C, Rasmussen HM, Dallal GE (2009) Treatment with potassium bicarbonate lowers calcium excretion and bone resorption in older men and women. J Clin Endocrinol Metab 94:96–102. https://doi.org/10.1210/jc.2008-1662

Defronzo RA, Felig P, Ferrannini E, Wahren J (1980) Effect of graded doses of insulin on splanchnic and peripheral potassium metabolism in man. Am J Phys 238:E421–E427. https://doi.org/10.1152/ajpendo.1980.238.5.E421

Delli AJ, Lernmark Å (2016) Chapter 39 – type 1 (insulin-dependent) diabetes mellitus: etiology, pathogenesis, prediction, and prevention. In: Groot LJD, Kretser DMD, Giudice LC, Grossman AB, Melmed S, Potts JT, Weir GC (eds) Endocrinology: adult and pediatric, 7th edn). W.B. Saunders, Philadelphia. https://doi.org/10.1016/B978-0-323-18907-1.00039-1. Accessed 17 May 2020

Demigne C, Sabboh H, Remesy C, Meneton P (2004) Protective effects of high dietary potassium: nutritional and metabolic aspects. J Nutr 134:2903–2906. https://doi.org/10.1093/jn/134.11.2903

Desalvo KB, Olson R, Casavale KO (2016) Dietary guidelines for Americans. JAMA 315:457–458. https://doi.org/10.1001/jama.2015.18396

Dickinson HO, Nicolson DJ, Campbell F, Beyer FR, Mason J (2006) Potassium supplementation for the management of primary hypertension in adults. Cochrane Database Syst Rev 19: CD004641. https://doi.org/10.1002/14651858.CD004641.pub2

Edmonds CJ, Jasani B (1972) Total-body potassium in hypertensive patients during prolonged diuretic therapy. The Lancet 300:8–12. https://doi.org/10.1016/S0140-6736(72)91275-5. Accessed 17 May 2020

Ekmekcioglu C, Elmadfa I, Meyer AL, Moeslinger T (2016) The role of dietary potassium in hypertension and diabetes. J Physiol Biochem 72:93–106. https://doi.org/10.1007/s13105-015-0449-1

Elford J, Phillips A, Thomson AG, Shaper AG (1990) Migration and geographic variations in blood pressure in Britain. BMJ 300:291–295. https://doi.org/10.1136/bmj.300.6720.291

Frassetto L, Morris RC Jr, Sebastian A (2005) Long-term persistence of the urine calcium-lowering effect of potassium bicarbonate in postmenopausal women. J Clin Endocrinol Metab 90:831–834. https://doi.org/10.1210/jc.2004-1350

Fulgoni VL 3rd (2007) Limitations of data on fluid intake. J Am Coll Nutr 26, 588S–591S. https://wwwncbinlmnihgov/pubmed/17921470. Accessed 17 May 2020

Geleijnse JM, Kok FJ, Grobbee DE (2003) Blood pressure response to changes in sodium and potassium intake: a metaregression analysis of randomised trials. J Hum Hypertens 17:471–480. https://doi.org/10.1038/sj.jhh.1001575

Ginsburg JM (1962) Equilibration of potassium in blood and tissues. Am J Dig Dis 7:34–42. https://linkspringercom/article/101007%2FBF02231928. Accessed 17 May 2020

Ginsburg JM, Wilde WS (1954) Distribution kinetics of intravenous radiopotassium. Am J Phys 179:63–75. https://doi.org/10.1152/ajplegacy.1954.179.1.63

Greenlee M, Wingo CS, Mcdonough AA, Youn JH, Kone BC (2009) Narrative review: evolving concepts in potassium homeostasis and hypokalemia. Ann Intern Med 150:619–25. https://www.ncbi.nlm.nih.gov/pubmed/19414841. Accessed 17 May 2020

Gumz ML, Rabinowitz L, Wingo CS (2015) An integrated view of potassium homeostasis. N Engl J Med 373:60–72. https://doi.org/10.1056/NEJMra1313341

Haddy FJ, Vanhoutte PM, Feletou M (2006) Role of potassium in regulating blood flow and blood pressure. Am J Physiol Regul Integr Comp Physiol 290:R546–R552. https://doi.org/10.1152/ajpregu.00491.2005

He FJ, Macgregor GA (2008) Beneficial effects of potassium on human health. Physiol Plant 133:725–35. https://wwwncbinlmnihgov/pubmed/18724413. Accessed 17 May 2020

He FJ, Marciniak M, Carney C, Markandu ND, Anand V, Fraser WD, Dalton RN, Kaski JC, Macgregor GA (2010) Effects of potassium chloride and potassium bicarbonate on endothelial function, cardiovascular risk factors, and bone turnover in mild hypertensives. Hypertension 55:681–688. https://doi.org/10.1161/HYPERTENSIONAHA.109.147488

Helderman JH, Elahi D, Andersen DK, Raizes GS, Tobin JD, Shocken D, Andres R (1983) Prevention of the glucose intolerance of thiazide diuretics by maintenance of body potassium. Diabetes 32:106–11. https://wwwncbinlmnihgov/pubmed/6337892. Accessed 17 May 2020

Houston MC (2011) The importance of potassium in managing hypertension. Curr Hypertens Rep 13:309–317. https://doi.org/10.1007/s11906-011-0197-8

Hupfeld CJ, Olefsky JM (2016) Chapter 40 – type 2 diabetes mellitus: etiology, pathogenesis, and natural Histor. In: Groot LJD, Kretser DMD, Giudice LC, Grossman AB, Melmed S, Potts JT, Weir GC (eds) Endocrinology: adult and pediatric, 7th edn. W.B. Saunders, Philadelphia. https://doi.org/10.1016/B978-0-323-18907-1.00040-8

ICRP (1975) Physiological data for reference man. Chapter 3. In: Report of the task group on reference man. Pergamon Press, Oxford. https://doi.org/10.1016/S0074-2740(75)80023-7. ICRP publication 23:335–365

ICRP (2007) The 2007 recommendations of the international commission on radiological protection. ICRP publication 103. Ann ICRP 37:1–332. https://doi.org/10.1016/j.icrp.2007.10.003

Institute of Medicine (2005) Dietary reference intakes for water, Potassium, Sodium, Chloride, and Sulfate, The National Academies Press, Washington, DC. https://wwwnapedu/read/10925/chapter/1. Accessed 17 May 2020

INTERSALT (1988) Intersalt: an international study of electrolyte excretion and blood pressure. Results for 24 hour urinary sodium and potassium excretion. Intersalt cooperative research group. BMJ 297:319–328. https://doi.org/10.1136/bmj.297.6644.319

Jasani BM, Edmonds CJ (1971) Kinetics of potassium distribution in man using isotope dilution and whole-body counting. Metabolism 20:1099–1106. https://doi.org/10.1016/0026-0495(71)90034-5

Jehle S, Zanetti A, Muser J, Hulter HN, Krapf R (2006) Partial neutralization of the acidogenic Western diet with potassium citrate increases bone mass in postmenopausal women with osteopenia. J Am Soc Nephrol 17:3213–3222. https://doi.org/10.1681/ASN.2006030233

Jehle S, Hulter HN, Krapf R (2013) Effect of potassium citrate on bone density, microarchitecture, and fracture risk in healthy older adults without osteoporosis: a randomized controlled trial. J Clin Endocrinol Metab 98:207–217. https://doi.org/10.1210/jc.2012-3099

Khaw KT, Rose G (1982) Population study of blood pressure and associated factors in St Lucia, West Indies. Int J Epidemiol 11:372–377. https://doi.org/10.1093/ije/11.4.372

Leggett RW, Williams LR (1986) A model for the kinetics of potassium in healthy humans. Phys Med Biol 31:23–42. https://www.ncbi.nlm.nih.gov/pubmed/3952144. Accessed 17 May 2020

Lemann J Jr, Hornick LJ, Pleuss JA, Gray RW (1989) Oxalate is overestimated in alkaline urines collected during administration of bicarbonate with no specimen pH adjustment. Clin Chem 35:2107–10. https://www.ncbi.nlm.nih.gov/pubmed/2551541. Accessed 17 May 2020

Lopez AD, Mathers CD (2006) Measuring the global burden of disease and epidemiological transitions: 2002–2030. Ann Trop Med Parasitol 100:481–499. https://doi.org/10.1179/136485906X97417

Ludlow M (1993) Renal handling of potassium. ANNA J Am Nephrol Nurs Assoc 20:52–58

Lupton JR, Blumberg JB, L'abbe M, Ledoux M, Rice HB, Von Schacky C, Yaktine A, Griffiths JC (2016) Nutrient reference value: non-communicable disease endpoints--a conference report. Eur J Nutr 55(Suppl 1):S1–S10. https://doi.org/10.1007/s00394-016-1195-z

Macdonald HM, Black AJ, Aucott L, Duthie G, Duthie S, Sandison R, Hardcastle AC, Lanham New SA, Fraser WD, Reid DM (2008) Effect of potassium citrate supplementation or increased fruit and vegetable intake on bone metabolism in healthy postmenopausal women: a randomized controlled trial. Am J Clin Nutr 88:465–474. https://doi.org/10.1093/ajcn/88.2.465

Macdonald-Clarke CJ, Martin BR, Mccabe LD, Mccabe GP, Lachcik PJ, Wastney M, Weaver CM (2016) Bioavailability of potassium from potatoes and potassium gluconate: a randomized dose response trial. Am J Clin Nutr 104(2):346–353. https://doi.org/10.3945/ajcn.115.127225

Marangella M, Di Stefano M, Casalis S, Berutti S, D'amelio P, Isaia GC (2004) Effects of potassium citrate supplementation on bone metabolism. Calcif Tissue Int 74:330–335. https://doi.org/10.1007/s00223-003-0091-8

Meissner HP, Henquin JC, Preissler M (1978) Potassium dependence of the membrane potential of pancreatic B-cells. FEBS Lett 94:87–89. https://doi.org/10.1016/0014-5793(78)80912-0

Melikian AP, Cheng LK, Wright GJ, Cohen A, Bruce RE (1988) Bioavailability of potassium from three dosage forms: suspension, capsule, and solution. J Clin Pharmacol 28:1046–1050. https://doi.org/10.1002/j.1552-4604.1988.tb03128.x

Meneton P, Loffing J, Warnock DG (2004) Sodium and potassium handling by the aldosterone-sensitive distal nephron: the pivotal role of the distal and connecting tubule. Am J Physiol Renal Physiol 287:F593–F601. https://doi.org/10.1152/ajprenal.00454.2003

Millen BE, Abrams S, Adams-Campbell L, Anderson CA, Brenna JT, Campbell WW, Clinton S, Hu F, Nelson M, Neuhouser ML, Perez-Escamilla R, Siega-Riz AM, Story M, Lichtenstein AH (2016) The 2015 dietary guidelines advisory committee scientific report: development and major conclusions. Adv Nutr 7:438–444. https://doi.org/10.3945/an.116.012120

Mohrman DE, Heller LJ (2010) Cardiovascular physiology, 7th edn. McGraw-Hill Education, New York

Muhlbauer RC, Lozano A, Reinli A (2002) Onion and a mixture of vegetables, salads, and herbs affect bone resorption in the rat by a mechanism independent of their base excess. J Bone Miner Res 17:1230–1236. https://doi.org/10.1359/jbmr.2002.17.7.1230

National Academies of Sciences, E. & Medicine (2019) Dietary reference intakes for sodium and potassium. The National Academies Press, Washington, DC. https://doi.org/10.17226/25353

Newberry SJ, Chung M, Anderson CAM, Chen C, Fu Z, Tang A, Zhao N, Booth M, Marks J, Hollands S, Motala A, Larkin J, Shanman R, Hempel S (2018) Sodium and potassium intake: effects on chronic disease outcomes and risks. Agency for Healthcare Research and Quality, Rockville, MD. https://doi.org/10.23970/AHRQEPCCER206

Palmer BF (2015) Regulation of potassium homeostasis. Clin J Am Soc Nephrol 10:1050–1060. https://doi.org/10.2215/CJN.08580813

Penton D, Czogalla J, Loffing J (2015) Dietary potassium and the renal control of salt balance and blood pressure. Pflugers Arch 467:513–530. https://doi.org/10.1007/s00424-014-1673-1

Perez GO, Oster JR, Vaamonde CA (1981) Serum potassium concentration in acidemic states. Nephron 27:233–243. https://doi.org/10.1159/000182061

Roger VL, Go AS, Lloyd-Jones DM, Benjamin EJ, Berry JD, Borden WB, Bravata DM, Dai S, Ford ES, Fox CS, Fullerton HJ, Gillespie C, Hailpern SM, Heit JA, Howard VJ, Kissela BM, Kittner SJ, Lackland DT, Lichtman JH, Lisabeth LD, Makuc DM, Marcus GM, Marelli A, Matchar DB, Moy CS, Mozaffarian D, Mussolino ME, Nichol G, Paynter NP, Soliman EZ, Sorlie PD, Sotoodehnia N, Turan TN, Virani SS, Wong ND, Woo D, Turner MB (2012) Executive summary: heart disease and stroke statistics--2012 update: a report from the American heart association. Circulation 125:188–197. https://doi.org/10.1161/CIR.0b013e3182456d46

Rowe JW, Tobin JD, Rosa RM, Andres R (1980) Effect of experimental potassium deficiency on glucose and insulin metabolism. Metabolism 29:498–502. https://doi.org/10.1016/0026-0495(80)90074-8

Sacks FM, Campos H (2010) Dietary therapy in hypertension. N Engl J Med 362:2102–2112. https://doi.org/10.1056/NEJMct0911013

Sagild U, Andersen V, Andreasen PB (1961) Glucose tolerance and insulin responsiveness in experimental potassium depletion. Acta Med Scand 169:243–251. https://doi.org/10.1111/j.0954-6820.1961.tb07829.x

Sebastian A, Harris ST, Ottaway JH, Todd KM, Morris RC Jr (1994) Improved mineral balance and skeletal metabolism in postmenopausal women treated with potassium bicarbonate. N Engl J Med 330:1776–1781. https://doi.org/10.1056/NEJM199406233302502

Shils ME, Shike M (2006) Modern nutrition in health and disease, 10th edn. Lippincott Williams & Wilkins, Philadelphia. https://www.worldcat.org/title/modern-nutrition-in-health-and-disease/oclc/819172526?referer=di&ht=edition. Accessed 17 May 2020

Soleimani M, Bergman JA, Hosford MA, Mckinney TD (1990) Potassium depletion increases luminal Na+/H+ exchange and basolateral Na+:CO3=:HCO3- cotransport in rat renal cortex. J Clin Invest 86:1076–1083. https://doi.org/10.1172/JCI114810

Sözen T, Özışık L, Başaran NÇ (2017) An overview and management of osteoporosis. Eur J Rheumatol 4:46–56. https://doi.org/10.5152/eurjrheum.2016.048

Stamler R (1991) Implications of the INTERSALT study. Hypertension 17:I16–I20. https://doi.org/10.1161/01.HYP.17.1_Suppl.I16

Stipanuk MH (2006) Biochemical, physiological, and molecular aspects of human nutrition, 2nd edn. Saunders Elsevier, St Louis, MO. isbn 1 4160 0209 X

Stone M, Weaver C (2018) Potassium intake, metabolism, and hypertension SCAN's pulse. 37 (3):5–10. http://wwwscandpgorg/nutrition-info/pulse. Accessed 17 May 2020

Stone MS, Martyn L, Weaver CM (2016) Potassium intake, bioavailability, hypertension, and glucose control. Nutrients 8(7):444. https://doi.org/10.3390/nu8070444

Strohm D, Ellinger S, Leschik-Bonnet E, Maretzke F, Heseker H (2017) Revised reference values for potassium intake. Ann Nutr Metab 71:118–124. https://doi.org/10.1159/000479705

Surveyor I, Hughes D (1968) Discrepancies between whole-body potassium content and exchangeable potassium. J Lab Clin Med 71:464–472. https://www.translationalres.com/article/0022-2143(68)90100-5/pdf. Accessed 17 May 2020

Svetkey LP, Simons-Morton D, Vollmer WM, Appel LJ, Conlin PR, Ryan DH, Ard J, Kennedy BM (1999) Effects of dietary patterns on blood pressure: subgroup analysis of the Dietary Approaches to Stop Hypertension (DASH) randomized clinical trial. Arch Intern Med 159:285–293. https://www.ncbi.nlm.nih.gov/pubmed/9989541. Accessed 17 May 2020

Svetkey LP, Simons-Morton DG, Proschan MA, Sacks FM, Conlin PR, Harsha D, Moore TJ (2004) Effect of the dietary approaches to stop hypertension diet and reduced sodium intake on blood pressure control. J Clin Hypertens (Greenwich) 6:373–381. https://doi.org/10.1111/j.1524-6175.2004.03523.x

Unwin RJ, Luft FC, Shirley DG (2011) Pathophysiology and management of hypokalemia: a clinical perspective. Nat Rev Nephrol 7:75–84. https://doi.org/10.1038/nrneph.2010.175

Valentin J (2002) Basic anatomical and physiological data for use in radiological protection: reference values: ICRP publication 89. Ann ICRP 32:1–277. https://doi.org/10.1016/S0146-6453(03)00002-2

Veiras LC, Han J, Ralph DL, Mcdonough AA (2016) Potassium supplementation prevents sodium chloride cotransporter stimulation during angiotensin II hypertension. Hypertension 68:904–912. https://doi.org/10.1161/HYPERTENSIONAHA.116.07389

Weaver CM (2013) Potassium and health. Adv Nutr 4:368S–377S. https://doi.org/10.3945/an.112.003533

Weaver CM, Fuchs RK (2014) Chapter 12 – skeletal growth and development. In: Burr DB, Allen MR (eds) Basic and applied bone biology. Academic Press, San Diego, pp 245–260. ibsn: 978-0-12-416015-6

Weaver CM, Stone MS, Lobene AJ, Cladis DP, Hodges JK (2018) What is the evidence base for a potassium requirement? Nutr Today 53:184–195. https://doi.org/10.1097/nt.0000000000000298

Whelton PK, He J, Cutler JA, Brancati FL, Appel LJ, Follmann D, Klag MJ (1997) Effects of oral potassium on blood pressure. Meta-analysis of randomized controlled clinical trials. JAMA 277:1624–1632. https://doi.org/10.1001/jama.1997.03540440058033

World Health Organizatoin (2012) Guideline: potassium intake for adults and children. World Health Organizatoin, Geneva. isbn: 9789241504829

Youn JH (2013) Gut sensing of potassium intake and its role in potassium homeostasis. Semin Nephrol 33:248–256. https://doi.org/10.1016/j.semnephrol.2013.04.005

Young DB, Lin H, Mccabe RD (1995) Potassium's cardiovascular protective mechanisms. Am J Phys 268:R825–R837. https://doi.org/10.1152/ajpregu.1995.268.4.R825

Zillich AJ, Garg J, Basu S, Bakris GL, Carter BL (2006) Thiazide diuretics, potassium, and the development of diabetes: a quantitative review. Hypertension 48:219–224. https://doi.org/10.1161/01.HYP.0000231552.10054.aa

Thematic Glossary of Terms

K Pools

Harvested plant K K in plant material, such as grain or biomass, that has been removed from a given area of soil by crop harvest. It may be reported as kg K ha^{1} (Fig. 1.1, Box 2).

Interlayer K K that is bound by varying intensities between 2:1 layers of phyllosilicate minerals. These K ions may or may not hydrate like those in soil solution or those associated with the outer surfaces of mineral particles or colloids.

Interlayer K in micas and partially weathered micas K ions that are between 2:1 layers and that are not hydrated. When K occurs "deep" within mineral particles, it is largely isolated from chemical reactions that occur near the surface or edges of the mineral domain. However, interlayer K ions close to the edges of these minerals may be susceptible to release to the soil solution in response to crystal dissolution, ion exchange reactions, and diffusion gradients (Fig. 1.1, Box 11).

Interlayer K in secondary layer silicates Interlayer K in minerals such as illite, vermiculite, and smectite for which the layer charge is lower than that of primary micas. The lower layer charge means that the K ions are more likely to be hydrated and susceptible to cation exchange reactions. For a low-charge smectite like montmorillonite, this term overlaps with *surface-adsorbed K* (Fig. 1.1, Box 10).

K in neoformed K minerals K that is bound in crystals of newly formed minerals created by the precipitation of K^+ with other soil solution ions (Fig. 1.1, Box 13).

K pool K in one of several types of physical or chemical states in a soil or soil–crop system (Fig 1.1).

Leached K (plants) K removed from plants by the action of aqueous solutions, such as rain, dew, mist, and fog.

Leached K (soil) Soluble K^+ that is displaced below the rooting zone by water percolating through the soil (Fig. 1.1, Box 5).

Potassium cycle A schematic depicting pools of K in the soil–plant system and the directions of fluxes of K among those pools (Fig 1.1).

Residue K K in plant material such as crop residues, roots, leaf litter, or dead plants that have been returned to the soil surface, with or without incorporation by tillage. It is typically reported as kg K ha^{-1} (Fig. 1.1, Box 7).

Root-zone K K in the bulk volume of a soil where roots are located. It may be reported with units of kg K ha^{-1}, usually incorporating assumptions for a uniform rooting depth and bulk density of the soil horizons. Root-zone K includes K in the rhizosphere as well as K in the bulk soil that is not closely associated with root surfaces.

Soil solution K K that is dissolved in the aqueous liquid phase of a soil. This K is a monovalent cation that may be hydrated or part of an ion pair (Fig. 1.1, Box 8).

Structural K K in the structures of primary tectosilicate minerals like feldspars and K in the interlayer regions of primary layer silicates like biotite and muscovite (Fig. 1.1, Boxes 11 and 12).

Surface-adsorbed K K electrostatically associated with negatively charged planar surfaces of phyllosilicate minerals and surfaces of iron and aluminum oxides, as well as that associated with soil organic matter (Fig. 1.1, Box 9).

Key Minerals

Illite One kind of K-bearing mica that is typically dioctahedral and usually occurs in the clay size fraction of soils and sediments. In soils, it is generally a product of the physical and chemical weathering of primary micas. In marine sediments, illite may form when K derived from feldspar weathering enters deeply buried smectite-rich sediments.

Interstratified minerals The occurrence of multiple layer silicate minerals in the same mineral domain. For example, high-charge layers of mica and lower-charge layers of vermiculite or smectite may be contiguous with one another in the same particle (Fig. 7.2).

Micas Layer silicate minerals with a high layer charge per formula unit (typically about 1 mole of charge per 10-oxygen formula unit). The most common of these minerals are the primary minerals muscovite and biotite, but there are many others. Most micas host K ions between the aluminosilicate layers (Fig. 7.3a).

Smectite A class of layer silicate minerals composed of one alumina sheet between two silica sheets and has a low layer charge due to isomorphic substitution. Layer charge normally ranges from 0.2 to 0.6 moles per 10-oxygen formula unit. Smectites include montmorillonite and beidellite, among others. The lower the layer charge, the more the interlayer regions are likely to swell and allow entry of water molecules and hydrated ions (Fig 7.3b).

Vermiculite A layer silicate mineral that forms by the weathering of primary micas. It has a variable concentration of K in interlayer positions. Its layer charge is intermediate between that of micas and smectites. In soils, it normally occurs in the clay fraction (Fig 7.3b).

K Processes

Atmospheric deposition of K The quantity of K transferred from the atmosphere to a given area of land surface. Atmospheric deposition is the sum of wet deposition and dry deposition. *Wet deposition* is the quantity of K transferred from the atmosphere to a given area of land by rain, fog, or snow. *Dry deposition* is the quantity of K in atmospheric particles transferred to the soil surface under the influence of wind and gravity.

Diffusion of K Movement of dissolved K in water in response to a K concentration gradient in the water.

Erosion loss K lost from a soil when particles that it is associated with are removed by wind or water transport.

Flux of K The amount of K that moves from one state (pool) to another in a unit of time, often referenced to the masses or volumes of the respective states. Flux terms are related to the context and states for which they are used (e.g., the rate of transfer of K from the soil solution pool to the plant pool). Flux terms may also be used to describe movement of K across a tissue boundary, a cell membrane, a soil surface, or a crystal boundary. *Flux* may refer to movement out of or into a volume or mass of soil or plant tissue, with *efflux* used to describe movement *from* one state by referencing the initial state, while *influx* describes the movement *into* another state by referencing the receiving state. *Flux density* usually refers specifically to the movement across a boundary defined as an area, not a volume or mass.

Macropore flow The movement of water and solutes like K^+ through large, continuous pores in soil. This movement is largely controlled by gravity. It is included in the term *preferential flow*, one that also incorporates thin-film flows, where mobile water films at solid surfaces are only a few micrometers thick. Macropore flow typically involves limited interaction with the soil solid phase.

Mass flow of K Movement of dissolved K as water moves through soil in response to gravitational and capillary forces.

Matrix flow The movement of water and solutes like K^+ from soil volumes of higher total soil water potential to soil volumes of lower total soil water potential, driven both by gravity and by capillary forces on water molecules. In the context of plant nutrition, matrix flow is an example of mass flow in which potential interactions of water, solute, and the soil solid phase are maximized.

Mobility of K The ability of aqueous-phase K^+ ions to move in soil, either vertically or laterally, through both mass flow and diffusion processes.

Potassium fixation The process in which hydrated K^+ ions move from the solution phase to interlayer positions in phyllosilicate minerals, and then dehydrate as the mineral layers contract around them. The K^+ in this position is not readily soluble and is therefore considered nonbioavailable unless changes in the concentration of Al-complexing ions, pH, or redox potential of the soil solution promote its release during the growing season.

Runoff loss, subsurface The quantity of K in water that infiltrates the soil surface to shallow depths and then moves laterally in the direction of the topographic slope.

Runoff loss, surface The quantity of K in water moving laterally over the soil surface in the direction of the topographic slope.

Adsorption of K by Soil

Adsorption of K The retention of K^+ ions at or near the surface of solid-phase minerals or organic matter in a soil

Localities of K in layer silicate minerals Positions on the surface, edges, or interior of layer silicate domains where K^+ is thought to be differentially adsorbed.

e-Position An interlayer adsorption site, presumably near a particle edge, where hydrated K^+ may be adsorbed but can also move short distances out of the crystal and into the soil solution. K in this position may be partially extractable by exchange or displacement with NH_4^+, depending on the mineral, degree of prior drying, or other factors.

i-Position An interlayer adsorption site where dehydrated K^+ is retained and from which it may diffuse only slowly into the soil solution. K ions in this position may not be readily extractable with NH_4^+-based soil test solutions. K in these localities includes that in unweathered primary layer silicates like mica. In the context of soil testing, this K may be identified as *nonexchangeable K*, although that term also includes structural K in feldspars or other primary minerals.

p-Position Planar surfaces of phyllosilicate minerals where hydrated K^+ ions may be adsorbed by Coulombic forces in outer-sphere complexes. K associated with these surfaces is readily exchangeable or displaceable with NH_4^+ and other cations.

Wedge zone The interlayer volume of a 2:1 layer silicate near which two joined phyllosilicate mineral layers have begun to separate from one another due to changes in layer charge and adsorbed interlayer cations. Localities where K ions are retained in this volume of a mineral crystal are called **wedge sites**. This K is included in the concept of *nonexchangeable K*.

Soil Testing

Exchangeable K The mass of K that that can be extracted from a known mass of soil sample by cation exchange or displacement reactions, using a solution of a specified composition under a specific set of controlled conditions (e.g., temperature, shaking time, and solid:solution ratio). It is normally reported in units of mass or equivalents (mg or cmol K) per unit mass of soil (e.g., kg^{-1}). Because exchangeable K values in soil have been correlated with plant growth or crop yield, the term has been assumed to correlate with soil K that can be absorbed by a growing plant under unspecified soil conditions and periods. Its magnitude and reproducibility vary, depending on the specific chemical and physical characteristics of the extraction procedure as well as on the history of previous fertilization, method of sample collection, sample pretreatments, and soil mineralogy. One widely used method is the extraction of K by equilibrating a soil sample with 1 *M* NH_4 acetate buffered at pH 7.

Extractable K K that can be removed from a soil sample by a particular solution under standardized conditions.

Nonexchangeable K Soil K that is not extracted from a soil sample by soil tests that promote exchange or displacement of K^+ by another cation. In principle, this value is equal to *total soil K* minus *exchangeable K*. However, in some literature, nonexchangeable K has referred to the difference between K that can be dissolved from a soil sample by incomplete chemical dissolution extractions and *exchangeable K*. Incomplete extraction procedures include partial dissolution using boiling nitric acid or precipitation of solution K using sodium tetraphenylborate ($NaB(C_6H_5)_4$).

Soil K status An evaluation of a soil's capacity to provide sufficient K to meet demands for normal crop growth and development. Classes (e.g., very low, low, optimal, or high) are normally based on the soil-test values of soil samples collected from the rooting zone and may be specific for different crop genotypes and species. Other factors (e.g., soil mineralogy, parent material, drainage class, or regional climate) may also be incorporated. For example, soils described as K deficient could have low-medium K status, and K amendments would be recommended to improve crop growth and yield.

Soil-test K The quantity of K solubilized by the reagents in an extracting solution under standardized conditions. Commercially used soil testing extractions are designed to be rapid, inexpensive, and reproducible. The most common soil test for K is the exchangeable K test.

Total soil K The concentration of K determined by a procedure in which the mass of a soil sample is entirely dissolved, usually requiring both hydrofluoric acid and a strong acid like nitric acid.

Modeling

Labile K The quantity of K in the soil solution plus the quantity of K that is readily desorbed into solution from solid-phase surfaces.

Plant-available soil K The quantity of soil K that is capable of being absorbed by a plant during a growing season. In the context of soil testing and modeling, this term represents K that is *potentially* bioavailable to plants. See also *bioavailability*.

Plant K The total quantity of K that has accumulated in a plant.

Potassium accumulation The quantity of K in either the entire plant (roots, shoots, and seeds/storage organs) or in specified plant organs at a given time during the growing season.

Potassium harvest index The mass of harvested plant K divided by the mass of total K accumulation in the plant either at the point of maximum accumulation during the production season or at crop harvest. Typically, this measure does not consider the K contained in plant roots.

Potassium-holding capacity The maximum quantity of K that can be retained by a given volume of soil and not lost through runoff, leaching, or erosion.

Potassium input The quantity of K originating outside a given volume of soil that moves into that volume as a result of fertilization, leaching, irrigation, atmospheric deposition, runoff, erosion, evapotranspiration, or similar processes.

Potassium output or loss The quantity of K in a given volume of soil (or soil–plant system) that moves outside that volume in response to crop harvest, leaching, erosion, or similar processes. In some contexts, output may refer only to K removed by crop harvest.

Potassium release Movement of K from organic matter, mineral surfaces, or mineral interlayers to the soil solution, making it mobile or plant-available.

Remobilization The translocation of K from one plant tissue to another (e.g., from older to younger leaves, from vegetative material into grains or cotton lint, and from senescing leaves into other plant parts). In some literature, this concept is referred to as *resorption*.

Resorption See *remobilization*.

Throughfall Precipitation that passes through the plant canopy.

Unharvested plant K The quantity of plant K returned to the soil in the form of leaves, stems, root residues, and other plant organs or tissues (Fig. 1.1 Box 7).

Nutrient Use and Efficiency

Agronomic efficiency For a given nutrient, the difference in crop yield with and without nutrient inputs divided by the sum of nutrient sources. For example, the *agronomic efficiency of K fertilizer use* is defined as the difference in crop yield with and without K fertilizer inputs, divided by the amount of K fertilizer applied. A more general term, *agronomic K use efficiency*, represents crop production (biomass or harvested yield) per unit of all soil K (indigenous and applied in fertilizer) that is assumed to be bioavailable. The general term is numerically equal to the product of the *K uptake efficiency* and the *agronomic K utilization efficiency* (see below).

Apparent fertilizer recovery efficiency For a given nutrient, the increase in nutrient uptake attributable to fertilization divided by the sum of fertilizer inputs.

K uptake efficiency Mass of K in crop tissue per unit of soil K that is assumed to be bioavailable to the crop.

K utilization efficiency Crop yield per unit of K mass in crop tissue.

Nutrient balance An account of the total inputs, outputs, and transformations of a particular nutrient in an agroecosystem.

Partial factor productivity For a given nutrient, biomass yield divided by the sum of inputs of a selected factor (e.g., K).

Partial nutrient balance For a given nutrient, the sum of nutrient outputs divided by the sum of nutrient inputs.

Partial nutrient balance intensity For a given nutrient, the sum of nutrient inputs minus the sum of nutrient outputs.

Fertilizers and Fertilization

Banding Confining the placement of fertilizer or other agricultural inputs to defined zones in the soil profile, usually parallel to existing crop rows or where crop rows will be planted, and at varying depths below the soil surface.

Broadcasting Placement of fertilizer on the soil surface in an even but random distribution.

Fertigation Supplying nutrients to plants in water that is used to irrigate the plants.

Foliar nutrient application Application of nutrients directly to plant leaves in the form of aqueous solutions. To avoid tissue damage, foliar applications must be made in dilute nutrient concentrations. Therefore, for major nutrients like K, foliar application can only supplement and not substitute for uptake from the soil.

Potash fertilizer Forms of K that are used as soil or plant amendments, including potassium chloride (KCl), also called muriate of potash (MOP); potassium sulfate (K_2SO_4), also called sulfate of potash (SOP); potassium magnesium sulfate ($K_2SO_4 \cdot MgSO_4$), sometimes referred to as sulfate of potash magnesia (MgSOP or SOPM); potassium nitrate (KNO_3), also called nitrate of potash (NOP) or saltpeter; KOH, potassium hydroxide, diluted in aqueous form; and mixed sodium-potassium nitrate ($NaNO_3 + KNO_3$), also called Chilean saltpeter. By historical convention in some countries, the K concentration in commercially available potash fertilizers is usually reported in K_2O equivalents (as % K_2O), although there is no potassium oxide in the fertilizer.

Reserve That part of the *reserve base* which could be economically extracted or produced at the time of determination.

Reserve base That part of an identified *resource* that meets specified minimum physical and chemical criteria related to current mining and production practices, including those for grade, quality, thickness, and depth.

Resource A concentration of naturally occurring solid, liquid, or gaseous material in or on the Earth's crust in such a form and amount that economic extraction of a commodity from the concentration is currently or potentially feasible.

Tissue Testing

Critical concentration The tissue concentration of a nutrient that differentiates between suboptimal and supra-optimal (luxury consumption) nutrient uptake, also called *critical percentage* (Fig. 9.11).

Luxury consumption Supra-optimal nutrient tissue concentrations where there is no relationship between increasing tissue concentration and biomass or crop yield.

Plant K deficiency Tissue K concentrations at which plant growth is suboptimal and where increasing tissue K will result in increases in biomass, grain yield, crop quality, and/or plant health.

Sufficiency range (plants) A continuous range of tissue concentrations of a nutrient that are related to adequacy for optimal growth and development of that plant species.

Tissue test K Determination of the K concentration in plant tissue to evaluate the nutrient status of the plant.

Plants and Plant Processes

Bioavailable K The flux of K into a living organism over a specified period defines what is bioavailable from the organism's perspective. The term may be used to refer to K flux into soil flora and fauna, not just plants. From the soil's perspective, *bioavailability* is most often used as a *potential* term, i.e., the amount of a nutrient in soil that is capable of being taken up by organisms (e.g., plants, soil biota, or microbial communities) over the period of a growing season. Soil tests are intended to index bioavailable K from the soil's perspective. Bioavailable K in either sense varies with the forms of K in soil as well as with plant species or other organisms, soil properties, and edaphic conditions.

Chlorosis The condition when plant cells are unable to produce sufficient chlorophyll.

Chlorotic lesion A localized region of plant tissue consisting of cells in which chlorophyll is not produced.

Demand stage Division of the growing season of a plant into periods when demand for a particular nutrient waxes or wanes, depending on the need for the nutrient during growth; examples are germination, vegetative, and reproductive stages.

Flux equilibrium in nutrient uptake When the rate of nutrient uptake from the soil solution by the plant equals the rate of nutrient replenishment from the soil solid phase to the soil solution.

Genotype All or part of the collection of biological information in genes that are inherited from one organism by its descendants.

Guttation An exudation of xylem sap from leaves when leaf stomata are closed, due to osmotic pressure from roots.

Necrosis Premature cell death.

Necrotic lesion A localized region of plant tissue that shows physiological or disease-related cell changes that lead to autolysis, in which cell components are decomposed by the plant's own enzymes.

Phenology The expression of plant traits relating to different development stages. The rate of phenological development will be influenced by climatic factors as well as the availability of water and essential nutrients.

Phenotype The expression of genetic information in an individual organism.

Red edge A rapid increase in reflectance at the end of the visible and the beginning of the infra-red spectrum.

Synchrony The degree to which bioavailability and uptake of a nutrient in soil matches a plant's demand for that nutrient during the course of the growing season.

Transporter proteins Proteins that are embedded in cell membranes and that facilitate the movement of ions and small molecules into and out of the cell.

Tillering The development of multiple stems on a seedling sprouted from a single seed.

Uniculm species A monocot species that has a single stem with no tillers.

Rhizosphere and Roots

Exudates A wide range of organic compounds secreted into the soil by roots, including organic acid anions, amino acids and peptides, monosaccharides and oligosaccharides, hormones and growth factors, lipids, phenols, and sterols. Roots also release hydronium ions as well as nutrient cations and anions.

Rhizosphere The volume of soil immediately adjacent to and influenced by plant roots. For most nutrients, the extent of the soil affected by root processes ranges from ~0.5 to ~4 mm from the root surface.

Rhizosphere pH The pH of water in the rhizosphere immediately adjacent to roots. Root processes may *acidify* that solution when the net uptake of charges associated with cations (including K^+) is greater than the net uptake of charges of anions taken up by the root. Conversely, net uptake of anionic charges (e.g., those of nutrient anions such as nitrate and phosphate) that exceeds the net uptake of cationic charges may lead to net *alkalinization*. Changes in pH as a result of root activity depend on a number of factors, including concentrations of nutrient cations and anions that are absorbed by the root, the buffer capacity of the soil, the initial pH of the solution, respiration rates of CO_2, and the rates of diffusion of molecules and ions released to the solution.

Root hairs Elongated, tubular extensions of single cells from a root's epidermis. Because of their high surface area and lack of a cuticle, root hairs are primarily responsible for water and nutrient absorption from the soil. They are also responsible for the release of exudates into the soil solution.

Root system architecture (RSA) The typical angles and degree of branching of the roots of a plant.

Root system plasticity The ability of a growing plant to adapt its root system architecture and root length to access nutrients and water in different parts of the soil profile as it grows.

Permissions

All chapters in this book were first published in IPRAC, by Springer; hereby published with permission under the Creative Commons Attribution License or equivalent. Every chapter published in this book has been scrutinized by our experts. Their significance has been extensively debated. The topics covered herein carry significant information for a comprehensive understanding. They may even be implemented as practical applications or may be referred to as a beginning point for further studies.

The contributors of this book come from diverse backgrounds, making this book a truly international effort. We would like to thank all the contributing authors for lending their expertise to make the book truly unique. They have played a crucial role in the development of this book. Without their invaluable contributions this book wouldn't have been possible. They have made vital efforts to compile up to date information on the varied aspects of this subject to make this book a valuable addition to the collection of many professionals and students.

This book was conceptualized with the vision of imparting up-to-date and integrated information in this field. To ensure the same, a matchless editorial board was set up. Every individual on the board went through rigorous rounds of assessment to prove their worth. After which they invested a large part of their time researching and compiling the most relevant data for our readers.

The editorial board has been involved in producing this book since its inception. They have spent rigorous hours researching and exploring the diverse topics which have resulted in the successful publishing of this book. They have passed on their knowledge of decades through this book. To expedite this challenging task, the publisher supported the team at every step. A small team of assistant editors was also appointed to further simplify the editing procedure and attain best results for the readers.

Apart from the editorial board, the designing team has also invested a significant amount of their time in understanding the subject and creating the most relevant covers. They scrutinized every image to scout for the most suitable representation of the subject and create an appropriate cover for the book.

The publishing team has been an ardent support to the editorial, designing and production team. Their endless efforts to recruit the best for this project, has resulted in the accomplishment of this book. They are a veteran in the field of academics and their pool of knowledge is as vast as their experience in printing. Their expertise and guidance has proved useful at every step. Their uncompromising quality standards have made this book an exceptional effort. Their encouragement from time to time has been an inspiration for everyone.

The publisher and the editorial board hope that this book will prove to be a valuable piece of knowledge for students, practitioners and scholars across the globe.

Index

Printed in the USA
CPSIA information can be obtained
at www.ICGtesting.com
LVHW082154201023
761596LV00009B/1459